Hilbert Space Methods in Signal Processing

This lively and accessible book describes the theory and applications of Hilbert spaces, and also presents the history of the subject to reveal the ideas behind theorems and the human struggle that led to them.

The authors begin by establishing the concept of "countably infinite," which is central to the proper understanding of separable Hilbert spaces. Fundamental ideas such as convergence, completeness, and dense sets are first demonstrated through simple familiar examples and then formalized. Having addressed fundamental topics in Hilbert spaces, the authors then go on to cover the theory of bounded, compact, and integral operators at an advanced but accessible level. Finally, the theory is put into action, considering signal processing on the unit sphere, as well as reproducing kernel Hilbert spaces. The text is interspersed with historical comments about central figures in the development of the theory, which helps to bring the subject to life.

Rodney A. Kennedy is a Professor in the Research School of Engineering and the Head of the Applied Signal Processing research group at the Australian National University, Canberra. He has won a number of prizes in engineering and mathematics, including UNSW University and ATERB Medals. He has supervised more than 40 Ph.D. students and co-authored approximately 300 research papers. He is a Fellow of the IEEE.

Parastoo Sadeghi is a Fellow in the Research School of Engineering, at the Australian National University, Canberra. She has published around 90 refereed journal and conference papers, and received two IEEE Region 10 paper awards. She is a Senior Member of the IEEE.

Hilbert Space Methods in Signal Processing

RODNEY A. KENNEDY

Australian National University, Canberra

PARASTOO SADEGHI

Australian National University, Canberra

CAMBRIDGE
UNIVERSITY PRESS

CAMBRIDGE
UNIVERSITY PRESS

University Printing House, Cambridge CB2 8BS, United Kingdom

Published in the United States of America by Cambridge University Press, New York

Cambridge University Press is part of the University of Cambridge.

It furthers the University's mission by disseminating knowledge in the pursuit of education, learning and research at the highest international levels of excellence.

www.cambridge.org
Information on this title: www.cambridge.org/9781107010031

© Cambridge University Press 2013

First published 2013

A catalogue record for this publication is available from the British Library

ISBN 978-1-107-01003-1 Hardback

For our parents
Joan and William
Akram and Mostafa, and for our children
Lachlan, Kelan, and Aiden

Contents

II Operators

3 Introduction to operators

Preface

This is our book on the theory of Hilbert spaces, its methods and usefulness in signal processing research. It is pitched at a graduate student level, but relies only on undergraduate background material. There are many fine books on Hilbert spaces and our intention is not to generate another book to stick on the pile or to be used to level a desk. So from the onset, we have sought to synthesize the book with special goals in mind.

The needs and concerns of researchers in engineering differ from those of the pure sciences. It is difficult to put the finger on what distinguishes the engineering approach that we take. In the end, if a potential use emerges from any result, however abstract, then an engineer would tend to attach greater value to that result. This may serve to distinguish the emphasis given by a mathematician who may be interested in the proof of a foundational concept that links deeply with other areas of mathematics or is part of a long-standing human intellectual endeavor — not that engineering, in comparison, concerns less intellectual pursuits. As an example, Carleson in 1966 proved a conjecture by Luzin in 1915 concerning the almost-everywhere convergence of Fourier series of continuous functions. Carleson's theorem, as it is called, has its roots in the questions Fourier asked himself, in French presumably, about the nature of convergence of the series named after him. As a result it is important for mathematics, but less clear for engineers.

However, there is an important observation to be made here in that from the time of Fourier's first results in 1807 to Carleson's results in 1966, it was more than 150 years and from Fourier to today it is more than 200 years. In these long intervals of time, a lot of very bright people have been thinking about and refining ideas. So to learn any new topic, such as Hilbert spaces, is rather unnatural because it hides the human struggle to understand. Most mathematical treatments have the technical material laid out in a very concise and logical form. As a result the material comes across as a bit dry, at least to many people wanting or needing to learn. It does need livening up. So this book makes an attempt to inject a bit of life into the learning process. There are some mildly risqué characterizations of the founders of the theory and the style of writing is intentionally light and more reflective of the lecturing style than a written discourse. Of these founders we have immense admiration.

As said above, learning is a human endeavor and the material we are learning is the culmination of centuries of efforts of significant people who will be remembered and revered in future centuries, much more than celebrities and world leaders of today. Therefore it is of considerable interest to know why the ideas took so long and what were the stumbling blocks on the way. So one of the special goals of the book is to become comfortable with ideas and get a feeling for how to think about Hilbert spaces in the right way. Armed with the right elementary notions, self-study and taking on more advanced material and extensions are possible. An example is the theory on sets and infinity that Cantor laid down starting in 1873. This led to a proper understanding of the important constructions in Fourier series and generalizations to Hilbert spaces. So in this book we do spend some time on the nature of infinity, which might seem a very non-engineering concern. We defend this because countable infinity is a pillar of the theory, at least to better understand in what sense a Fourier series can represent a function, to understand separability of a space, to understand what breaks when we combine an infinite number of very smooth functions and end up with something unexpected. Cantor's work is important because of the influence it had on Hilbert and subsequent developments. In addition, we have been intrigued by Cantor's personal struggles and confrontations with other mathematicians, which contributed to him spending some time in an asylum. And this tends to be true of a number of the central figures that we meet on our journey, such as Cauchy, Bessel, Schmidt and Hilbert. They have an otherworldliness to their personalities and tend to have amusing anecdotes supporting their eccentricities.

Another goal in the book is to not let rigor dominate the material and look to reveal the nub of each result. Once an idea or concept is formulated, then the more technical results can follow, but often we direct the reader to two beautiful books more suited to mathematicians, the elegant (Helmberg, 1969) and the sublime classic (Riesz and Sz.-Nagy, 1990). Mathematician Weyl Hermann (1885–1955) was quoted as saying:

"My work always tried to unite the truth with the beautiful, but
when I had to choose one or the other, I usually chose the beautiful."

This reveals a lot about Weyl — a student of Hilbert, his successor at Göttingen, and a leading mathematician of the twentieth century – and what really motivated his work and his fallibility. We are not in his league, nor aspire to be, but we have a similar weakness for revealing intuitive and direct demonstrations for results even when the approach might lack absolute rigor or skip some technicalities.

The book comes in three parts as revealed in the table of contents. As an alternative to the contents, in engineering terms, Part I is mostly about signals, Part II is about systems and Part III puts the theory into action. Part III reflects material closer to our research interests, but distinguishes itself from the standard time domain signals and systems which are a feature of many engineering texts. We provide quite a lengthy treatment of signals and systems where the domain is the 2-sphere where the power of Hilbert spaces lays bare strong analogies with time domain signal processing. The final chapter of the book, Chapter 10, provides an accessible, somewhat original, treatment on reproducing kernel Hilbert spaces (RKHS) which draws on many theoretical aspects developed in all other chapters.

Some of the research results presented in Part III, especially in Chapter 8 and Chapter 9, have stemmed from joint work with our former and current PhD students. This research was supported under Australian Research Council's Discovery Projects funding scheme (project number DP1094350). We are also grateful to Zubair Khalid for proofreading many parts of the book and providing useful feedback and for some simulation results in Chapter 8. We acknowledge the support and encouragement of our colleagues and students at the Australian National University. Last, but not least, special thanks go to Andrew Hore, our cartoonist/illustrator, `http://funnyworksoz.com/`, who did a wonderful job with the illustration of characters that you see throughout the book. We do hope that, like us, you enjoy these lively and detailed illustrations.

PART I
Hilbert Spaces

1 Introduction

1.1 Introduction to Hilbert spaces

1.1.1 The basic idea

Hilbert spaces are the means by which the "ordinary experience of Euclidean concepts can be extended meaningfully into the idealized constructions of more complex abstract mathematics" (Bernkopf, 2008).

If our global plan is to abstract Euclidean concepts to more general mathematical constructions, then we better think of what it is in Euclidean space that is so desirable in the first place. An answer is geometry — in geometry one talks about points, lines, distances and angles, and these are familiar objects that our brains are well-adept to recognize and easily manipulate. Through imagery we use pictures to visualize solutions to problems posed in geometry. We may still follow Descartes and use algebra to furnish a proof, but typically through spatial reasoning we either make the breakthrough or see the solution to a problem as being plausible. Contrary to any preconception you may have, Hilbert spaces are about making obtuse problems have obvious answers when viewed using geometrical concepts.

The elements of Euclidean geometry such as points, distance and angle between points are abstracted in Hilbert spaces so that we can treat sets of objects such as functions in the same manner as we do points (and vectors) in 3D space. Hilbert spaces encapsulate the powerful idea that in many regards abstract objects such as functions can be treated just like vectors.

To others, less fond of mathematics, Hilbert spaces also encapsulate the logical extension of real and complex analysis to a wider sphere of suffering.

The theory of Hilbert spaces is one that succeeds in drawing together apparently different theories under a common framework via abstraction. The value of studying Hilbert spaces is not in providing new tools, but in showing how simple and familiar tools can be employed to tackle broad classes of problems. As a theory it highlights that there is a huge amount of redundancy in the literature. As a theory it requires relatively few ideas, but those ideas are deep.

1.1.2 Application domains

The usual application domains for Hilbert spaces are integral and differential equations, generalized functions and partial differential equations, quantum mechanics, orthogonal polynomials and functions, optimization and approximation theory. In signal processing and engineering: wavelets, optimization problems,

Euclid of Alexandria, (c. 300 BC) — "Euclid" is the anglicized name of Eukleides, who lived around 300 BC and is the "Father of Geometry," a title bestowed because of his very influential "Elements," which has served as a textbook on geometry and mathematical reasoning for more than two millennia. The style of Elements is in the form of definitions, axioms, theorems, and proofs. To the mathematically challenged this must come as the most torturous manuscript imaginable.

Not much is known about Euclid and his life, but along with Archimedes he is regarded as one of the greatest ancient mathematicians. His approach to geometry, which was seen as capturing directly statements about physical reality, has led to the terminology "Euclidean space." Over 23 centuries Euclidean space has been re-visited, recast, refined and generalized with notable extensions being "analytic geometry" (Cartesian coordinates) of René Descartes (1596–1650), and an axiomatic system for geometry, the 1899 "Grundlagen der Geometrie," of David Hilbert. The generalization of Euclidean space in the field of functional analysis is associated with a different abstraction by Hilbert and other researchers in the 1900s and called Hilbert space.

optimal control, filtering and equalization, signal processing on 2-sphere, Shannon information theory, communication theory, linear and nonlinear stability theory, and many more.

1.1.3   Broadbrush structure

Notion (cocktail party definition). *A Hilbert space is a complete inner product space. This is fine, except we are yet to define precisely what we mean by space, inner product and the adjective complete. But at a cocktail party where the objective is to impress strangers, particularly of the opposite sex, then it doesn't matter.*   □

Notion (broad definition). *The term "vector" is ingrained in early mathematical education as an ordered finite list of scalars, but in Hilbert space it is a more general notion. We will alternatively use the term point in lieu of vector when the situation is not ambiguous. So when working in Hilbert spaces the word vector might represent a conventional vector, a sequence or a function (and even more general objects).*   □

There are four key parts to a Hilbert space: vector space, norm, inner product and completeness. We can hear the minimalists screaming already.[1] To have a degree of comfort with Hilbert space is to have a clear notion of what these four things really mean and we will shortly move in the direction to address any deficiency. For the moment we are only interested in knowing what these mean in a general, possibly vague, way.

Vector space

Vector spaces should be familiar and align with the notions developed when dealing with the arithmetic of conventional vectors. Given two N-dimensional complex-valued vectors $a = (\alpha_1, \alpha_2, \ldots, \alpha_N)'$ and $b = (\beta_1, \beta_2, \ldots, \beta_N)'$, where $'$ denotes transpose, vector spaces encapsulate the banal aspects like

$$\gamma a + \delta b = \gamma(\alpha_1, \alpha_2, \ldots, \alpha_N)' + \delta(\beta_1, \beta_2, \ldots, \beta_N)'$$
$$= (\gamma\alpha_1 + \delta\beta_1, \gamma\alpha_2 + \delta\beta_2, \ldots, \gamma\alpha_N + \delta\beta_N)', \quad \gamma, \delta \in \mathbb{C}, \tag{1.1}$$

where $\mathbb{C}$ denotes the set of complex numbers. (Also, $\mathbb{R}$ denotes the set of real numbers.)

Norm

The norm is a means to measure the size of vectors and define "convergence" when we have sequences of vectors. The norm generalizes the notion of Euclidean distance in $\mathbb{R}^N$

$$\|a\| = \left(\alpha_1^2 + \alpha_2^2 + \cdots + \alpha_N^2\right)^{1/2}. \tag{1.2}$$

---

[1]   One of the drivers in mathematics is to provide a minimalist list of notions in preference to something that might be clearer and less clever. Engineers tend to think in terms of robustness and redundancy and this is our preferred approach in this book.

When armed with a norm we have a means to determine the distance between two vectors. The norm is responsible for a substantial part of the action in Hilbert spaces. It provides a measure of closeness, defines convergence and is necessary to make sense of "completeness."

Inner product

Inner product is a means to abstractly define orthogonality of vectors, projections and angles. We point out now that an inner product will be defined in such a way that it naturally induces a norm. That is, in the above list we are not implying that it is necessary to specify a norm in addition to specifying the inner product.[2] In Euclidean space $\mathbb{R}^N$, an inner product can be defined as

$$\langle a, b \rangle = \alpha_1 \beta_1 + \alpha_2 \beta_2 + \cdots + \alpha_N \beta_N, \tag{1.3}$$

which induces (1.2) through $\|a\| = \langle a, a \rangle^{1/2}$. When armed with an inner product we can do everything we can with a norm. Finally, we mention that the inner product is also called a scalar product or the dot product.

Completeness

Completeness is a subtle concept associated with the norm to guarantee the vector space is big enough by including the natural limits of converging vector sequences.

> **Remark 1.1 (Banach space).** In the above list were we to discard the inner product we still end up with something quite powerful and useful, known as a *Banach space*, named after Stefan Banach (1892–1945). A Banach space is a *complete normed space*. This means a vector space equipped with a norm and we have completeness. All Hilbert spaces are Banach spaces, but not all Banach spaces are Hilbert spaces. In a Banach space the norm needs to satisfy an additional condition known as the parallelogram equality to be a Hilbert space. That is, there is a condition on the norm that has a geometric interpretation. Loosely, we would say the norm needs to be a Euclidean-like norm and emulate Euclidean distance, but generally in a more abstract setting. When any result in Hilbert space does not strictly require the existence of an inner product, then that result will naturally belong to Banach space theory.   □

> **Remark 1.2 (Finite-dimensional spaces).** There exists a rich set of texts and works which make the theory of finite-dimensional vector spaces painfully bland. Hilbert spaces subsume such finite-dimensional vector spaces.
>
> Be wary of salesmen trying to sell you completeness in a finite-dimensional space — completeness is automatic if the Hilbert space is finite-dimensional. More generally any finite-dimensional Banach space is automatically complete.

---

[2]   So in the Hilbert space context the norm is like a free set of steak knives, it comes at no cost.

This is actually a non-trivial result. In real analysis, the property that a sequence of real numbers which is Cauchy necessarily converges is the nub of completeness. In summary, real numbers can be shown to be complete and finite-dimensional spaces inherit that completeness.   □

Function spaces

The arena where the Hilbert space concept brings new insights is in the treatment of functions, that is, where our "points" are now whole functions in a *function space*. This means we are generally (but not always) considering *infinite-dimensional vector spaces*. Hilbert spaces do not need to be infinite-dimensional, but represent a degree of overkill in abstraction when we want to consider finite-dimensional vector spaces.

The preferred way to think about functions in Hilbert space is as points or vectors in space rather than as a mapping. So in functional analysis, when one says a "point or vector in Hilbert space" one means a "function." The mental image is a mathematical point in space (albeit infinite-dimensional space) — akin to a conventional vector — not a squiggly line. Using the conventional vector as an analog to guide thinking is very effective. That is, when dealing with Hilbert space, it is very profitable to repeatedly ask the question: what does this correspond to in the case of a finite-dimensional vector space?

> **Remark 1.3.** A finite N-vector can be regarded as a function defined on $\{1, 2, \ldots, N\}$. Therefore, the terminology function space and functions in Hilbert space can be used to cover both cases of finite-dimensional and infinite-dimensional spaces.   □

It is tempting to think that function spaces must be infinite-dimensional or that infinite-dimensional Hilbert spaces must be function spaces. Both associations are wrong as we now illustrate.

First, one example of a Hilbert space has elements that are sequences of real numbers which satisfy certain conditions

$$\{\alpha_1, \alpha_2, \alpha_3, \ldots\},$$

which can be added and scaled in obvious ways. When performing algebraic manipulations on sequences, they should be treated as column vectors of infinite size. This Hilbert space is infinite-dimensional, but does not involve what is generally understood to be functions.

The second space we shall consider, to sharpen our thinking, is the space of linear combinations of two functions, $1/2\pi$ (constant function) and $(1/\pi) \cos \theta$. That is, all elements of the space look like

$$f(\theta) = \frac{\alpha}{2\pi} + \frac{\beta}{\pi} \cos \theta, \quad \alpha, \beta \in \mathbb{R}.$$

In the end, computations regarding the norms and inner products of such elements reduce to linear algebra of 2-vectors (α, β) and as such are not different from $\mathbb{R}^2$. So even though this is a function space, it is only two-dimensional.

Problem

1.1. What besides a conventional vector, a sequence or a function might be further examples for an abstract vector?

1.1.4   Historical comments

The theory of Hilbert spaces was initiated by David Hilbert (1862–1943), in the early twentieth century in the context of the study of *integral equations*.[3] Of course, he did not decide to write a few papers and name the theoretical construct after him. Nor did he solely develop the theory. It is generally regarded that a number of people developed the theory of Hilbert spaces, especially Erik Ivar Fredholm (1866–1927), whose work directly influenced Hilbert. Hilbert's work was simplified, generalized and abstracted further by Hilbert's student Erhard Schmidt (1876–1959). Other researchers at the same time developed key results including Frigyes Riesz (1880–1956) in his work (Riesz, 1907) and Ernst Sigismund Fischer (1875–1954) in his contribution (Fischer, 1907). The term "Hilbert space," or at least the German equivalent, is generally attributed to John von Neumann (1903–1957) in 1929.

It is tempting to regard Hilbert space theory as a generalization of familiar Euclidean space, which is true. Yet the theory developed quite late in mathematics. The delays in the development of the theory were manyfold and we will step through these as they align with conceptual barriers that need to be hurdled to fully understand the theory at a sufficiently advanced level.

The theory came together due to a number of factors. The first factor related to how to deal with the *infinite* (which will be explored more fully below). The second factor concerned the ongoing dispute about the meaning of *Fourier series*, developed by Jean Baptiste Joseph Fourier (1768–1830). That is, what class of functions does the Fourier series expansion converge to and how does this relate to the original function. The third factor concerned *integration*. For function spaces the inner product is defined in terms of integrals. However, a sufficiently general notion of an integral was late in arriving and the preferred notion was due to Henri Lebesgue (1875–1941) in the early twentieth century.[4] The Lebesgue integral (and associated measure theory) is needed for a rigorous development of the most useful classes of Hilbert spaces. However, the more subtle aspects of the Lebesgue integral are not essential to come to grips with Hilbert space from an application viewpoint. We will, however, highlight the nature of the subtleties later. Finally, we mention that the development of Hilbert spaces received a significant boost from the co-development of quantum mechanics in the 1920s, largely through the work on operators by von Neumann.[5]

---

[3] Integral equations are a natural complement to differential equations and arise, for example, in the study of existence and uniqueness of functions which are solutions of partial differential equations such as the wave equation. Convolution and Fourier transform equations also belong to this class.

[4] Lebesgue is pronounced in the French style, that is, with pursed lips, with every second letter silent and the remainder mumbled.

[5] Again the most useful type of operator, called a compact operator, emerged much earlier and, in fact, inspired Hilbert to develop his first results in Hilbert spaces.

Hilbert, David (1862–1943) — Hilbert was a famous German mathematician who originally hailed from Königsberg (no longer part of Germany and now renamed) and spent the majority of his life in Göttingen. His biography records that he had a large 6m blackboard constructed in his back-yard along with a covered walkway, which enabled him to do his work outdoors in any weather. At the age of 45, he learned to ride a bike and combined riding with weeding and pruning trees as part of a ritualistic style of behavior when deep in thought on some mathematical problem in his back-yard (Reid, 1986). He had a propensity to get elementary work, such as calculus, garbled and confused in lectures sometimes leading to fiascos. He was described as "slow to understand" (Reid, 1986, p. 172). His thinking was more strongly directed towards existence-style arguments versus constructive ones. He elucidated the difference in lectures by saying "Among those who are in this lecture hall, there is one who has the least number of hairs."

Hilbert was deeply influential in the development of many fields of mathematics and mathematical physics. His influence came through the problems he worked on, his major breakthroughs, and making Göttingen a major center for mathematical research. The disproportionate strength of German mathematics in the world scene at the time is a tribute to Hilbert and Felix Klein (1849–1925) who shared the vision of establishing Göttingen as the world's leading mathematics research center. The importance of such individuals was driven home by the virtual destruction of Göttingen with political and social changes, which caused many key people, who were to carry on Hilbert's legacy, to leave Germany. It was a tragedy for Hilbert to see in 1934 (after his retirement) what he, Klein and Hermann Minkowski (1864–1909) had built was destroyed in a short period when he lamented (possibly in anger) "Mathematics in Göttingen? There is really none any more."

1.2   Infinite dimensions

It has been eluded to that the more relevant and interesting Hilbert spaces are infinite-dimensional. In Hilbert space theory it is critical to have the correct notion of infinity and it is not sufficient to regard the symbol ∞ as being something obvious. In the following, we are going to explore the meaning behind the various types of infinity. Although seemingly a large digression, it is only conceptually challenging and not technically challenging. It is highly relevant to understanding infinite-dimensional Hilbert spaces.

Sanity is optional

Mathematicians had a lot of trouble dealing with the infinite sets and finding the most sensible approach to the topic was left to Georg Cantor (1845–1918) in the late nineteenth century, who explored the boundary between sanity and insanity (Dauben, 1990; Aczel, 2001). To understand his *transfinite cardinals*[6] does not cause insanity, but it probably helps to be well down that track. Cantor's ideas met initially a lot of resistance, but now are seen to be profound (or at least profoundly crazy). Hilbert was greatly influenced by the ideas of Cantor, as might be gleaned from the problem of the "Continuum Hypothesis," which was first amongst Hilbert's list of 23 unsolved problems in the Paris conference of the International Congress of Mathematicians in 1900.[7]

1.2.1   Why understand and study infinity?

We now consider some attributes about infinite sets which underpin the structure of infinite-dimensional Hilbert spaces. In case you have met the theory of transfinite cardinals before and you want to skip this material then the coverage is: countable or denumerably infinite (transfinite) sets, $\aleph_0$; cardinality of the continuum; integers are countable but not dense in the reals; rationals are countable *and* dense in the reals; the continuum has a cardinality which is equivalent to the set of all subsets of a countable set, which may be written $\mathfrak{c}$; the existence of transfinite cardinals beyond $\mathfrak{c}$; etc. That is, to skip this material is only recommended if you have a familiarity with the general conceptual and arithmetical properties of transfinite cardinals.

Of critical importance in what is known as a *separable Hilbert space* is the existence of a *countable dense set*. Having a *countable* set of vectors and having

---

[6]   Transfinite means beyond finite, i.e., infinite. The expression transfinite is preferred over infinite since it is less well recognized and, therefore, more likely to impress strangers. In short, the theory says there are different sizes of infinity with the smallest corresponding to the cardinality of the integers.

[7]   Initially Hilbert had ten problems which were later expanded to 23. The Continuum Hypothesis is the hypothesis that there is no infinite set whose cardinality or size is strictly between that of the integers and that of the real numbers. The meaning behind the Continuum Hypothesis can be easily understood in the context of these notes. Kurt Gödel (1906–1978) showed in 1940 that the Continuum Hypothesis cannot be disproved from the standard set theory axiom system. Paul Joseph Cohen (1934–2007) showed in 1963 that the Continuum Hypothesis cannot be proven from those same axioms either. Therefore, mathematicians regard the problem of the Continuum Hypothesis resolved. Gödel starved himself to death believing people were trying to poison him.

Cantor, Georg (1845–1918) — Cantor was a German mathematician and born in Saint Petersburg, Russia. He initiated the theory of sets and the theory of transfinite numbers/cardinals. Famously, Cantor was institutionalized to an asylum a number of times most likely suffering depression brought on by being unremittingly assailed by his contemporaries in mathematics. His correspondence with colleagues reflects that he and his work were under constant criticism, particularly from Leopold Kronecker (1823–1891) in Berlin.

Cantor was involved in generating innovative and philosophically deep work, which challenged conventional thinking. Hilbert was one of his supporters, recognizing the significance of the work, and remarked, albeit somewhat too late in 1925 when giving a talk on "On the infinite" for a celebration in honor of Karl Weierstrass (1815–1897), "No one shall drive us out of this paradise that Cantor created for us" (Reid, 1986, pp. 176–177). It is now generally regarded that his work was a building block of modern mathematics and revolutionized many mathematical fields. Had he known what his impact was, then he might have enjoyed a better fate than dying in an institution in 1918. A comprehensive and scholarly biography of Cantor and his work is (Dauben, 1990) and a more popular account can be found in (Aczel, 2001).

that set *dense* yields a type of spanning property giving a natural generalization of what happens in finite-dimensional spaces. Knowing what *countable* and *dense* mean is important and will be explained later. These concepts derive from and are mimicked in the simpler analogous structure of rational numbers within the real numbers. The analogy is so faithful and can guide our intuition in function spaces and hence it justifies a digression to hone the concepts.

1.2.2   Primer in transfinite cardinals

Primary or elementary school children know infinity is pretty big. Some know that infinity plus one, two times infinity and infinity squared are at least as big or even bigger.[8] But some grown-ups have doubts. What is the nature of any formal assertion involving infinity? Ultimately how do you measure infinite sets or compare infinite sets? Satisfactory resolution of these questions had to wait for the Theory of Transfinite Cardinals developed by Cantor in 1874.

With finite sets there are two natural ways to check if they have the same number of members:

- count the members, call this the cardinality, and see if the two sets have the same cardinality; or

- pair off the members from each set without leftovers (requiring no need to compute the total number of elements in each set).

It is the latter technique that can be used on infinite sets to determine if one set is larger or equal to another in cardinality. The former way, the one we tend to prefer to use, turns out to be only sensible for finite sets.

The basic tools of working with transfinite sets is to either find a clever mapping taking one set to the other (and hence assert that one is "equal" to the other) or establish a contradiction that one set cannot be put in one-to-one correspondence with another (and hence assert that one is "bigger" than the other).

Natural numbers and integers

Consider the three infinite sets

$$\mathbb{N} = \{1, 2, 3, \ldots\}, \tag{1.4a}$$

$$\mathbb{Z}^{\star} = \{0, 1, 2, 3, \ldots\}, \tag{1.4b}$$

$$\mathbb{Z} = \{\ldots, -3, -2, -1, 0, 1, 2, 3, \ldots\}, \tag{1.4c}$$

corresponding to the set of all natural numbers, non-negative integers and integers, respectively. Cantor calls the cardinality of the set of all natural numbers $\aleph_0$ (aleph null); see Figure 1.1.[9] It is also called the countably infinite, or denumerable, or denumerably infinite, or equipollent to the ordinal numbers —

---

[8]   This is essential for transfinite taunting.

[9]   That is, we could imagine replacing the symbol ∞ with $\aleph_0$ although, fortunately no-one follows this practice.

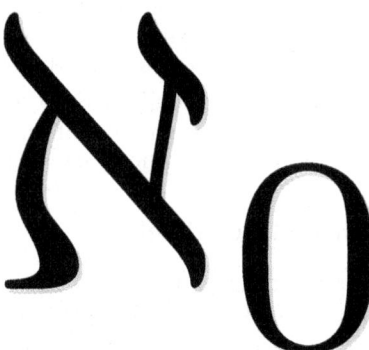

Figure 1.1 The smallest infinity, but still plenty big. The cardinality of the set of all natural numbers $\mathbb{N}$, $\aleph_0 = |\mathbb{N}|$.

so you can use the various terms at different cocktail parties without sounding repetitive. We prefer the term countable and use the symbol $\aleph_0$ (that finite sets in a sense are countable should not lead to confusion).

In terms of pairing, it is obvious we can pair off natural numbers (1.4a) with the non-negative integers (1.4b), that is, $1 \in \mathbb{N}$ is paired with $0 \in \mathbb{Z}^\star$, $2 \in \mathbb{N}$ is paired with $1 \in \mathbb{Z}^\star$, and so on. For the integers (1.4c) we rearrange them by interleaving the negative integers with the natural numbers yielding

$$\mathbb{Z} = \{0, -1, +1, -2, +2, -3, +3, \ldots\} \tag{1.5}$$

and so too is countable. So all these three infinite sets have cardinality $\aleph_0$ and, therefore, should be regarded as being of equal size (or more correctly having the same cardinality).

To summarize, a set S is countable if and only if its elements can be written as an unending sequence $\{s_1, s_2, s_3, \ldots\}$ — that the index starts at 1 or 0 is unimportant. Since any infinite set contains a countable subset (by ordered selection from an inexhaustible supply), it follows that $\aleph_0$ is the "smallest" transfinite number. So two questions arise:

Q1: Are there infinite sets that are non-trivially different from the integers with cardinality $\aleph_0$?

Q2: Are there infinite sets with cardinality greater than $\aleph_0$?

Both answers are yes, provided you accept the arguments that Cantor developed.

Rational and algebraic numbers

Cantor showed that the rational numbers, $\mathbb{Q}$, and some other infinite sets are countable. To show the rationals are countable, it is sufficient to find a way of ordering them such that each rational number appears once. It is sufficient to show we can do this for the non-negative rationals with any well-defined ordering,

Table 1.1 An ordering of the (proper) nonnegative rationals, $\mathbb{Q}^\star$, which places them in one-to-one correspondence with the natural numbers $\mathbb{N}$. The height is defined as $p + q$ where $p \in \mathbb{Z}^\star$ and $q \in \mathbb{N}$, (1.6). Combinations of p and q which have common factor are not proper and excluded.

p/q	0/1	1/1	1/2	2/1	1/3	3/1	1/4	2/3	3/2	4/1	1/5	5/1	$\cdots$
Height	1	2	3	3	4	4	5	5	5	5	6	6	$\cdots$
$\mathbb{Q}^\star$	0	1	$\frac{1}{2}$	2	$\frac{1}{3}$	3	$\frac{1}{4}$	$\frac{2}{3}$	$\frac{3}{2}$	4	$\frac{1}{5}$	5	$\cdots$
$\mathbb{N}$	1	2	3	4	5	6	7	8	9	10	11	12	$\cdots$

since we can adopt the same trick that worked for the integers that took (1.4c) to (1.5).

Take a non-negative proper rational number[10] written in the form $p/q \in \mathbb{Q}^\star$ (non-negative rationals), where $p \in \mathbb{Z}^\star$ and $q \in \mathbb{N}$, such that p and $q > 0$ are relatively prime. Define the "height" as

$$h \triangleq p + q \in \mathbb{N}. \tag{1.6}$$

For any height, h, there are only a finite number of rationals (at most h) and clearly these can be ordered, for example, from the lowest p increasing. Then it is a matter only to list the proper non-negative rationals according to height. For example, for height 5 we have $\{1/4, 2/3, 3/2, 4/1\}$ and for height 6 we have $\{1/5, 5/1\}$ where we have omitted the remaining terms of height 6, that is, $\{2/4, 3/3, 4/2\}$, since they are improper. Then we have the ordering which exhausts the non-negative proper rationals, the first part of which is shown in Table 1.1 and in Figure 1.2.

By interleaving the negative rationals with the non-negative rationals, $\mathbb{Q}^\star$, it is clear that all rationals, $\mathbb{Q}$, are countable.

This may seem fine as a cure for insomnia, but this is somewhat counter-intuitive. The set of rationals is non-trivially different from the set of integers. Rationals are dense, unlike the integers, meaning that between any two rationals there exist other rationals.[11] Because of this property we well might have expected, but now we know better, that the transfinite cardinality of the set of rational numbers exceeds $\aleph_0$. Where else might our dodgy intuition lead us astray? We suspect that if we now add the irrationals to the rationals then the cardinality may well be larger. Here we need to be careful about what we mean by irrationals.

Integer lattice, $\mathbb{Z}^2$

Before moving onto more significant infinite sets, we can quickly consider the cardinality of the set of points on a two-dimensional integer lattice, that is,

[10] For example, 2/4 is improper as it can be reduced to 1/2. For the construction, against convention, we take 7/1 as the proper form for 7. Furthermore, by convention, we treat 0 as a special case and deem 0/1 as the only proper representation of zero (discarding 0/2, 0/3, etc.).

[11] In fact, there is an infinite number of other rational numbers between any two rationals.

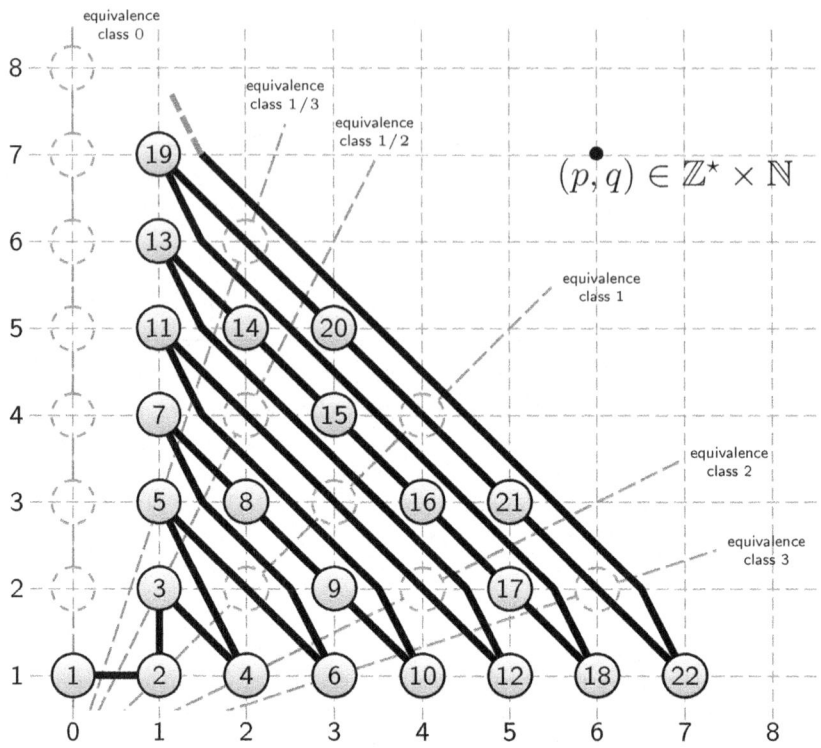

Figure 1.2 An ordering of the (proper) nonnegative rationals which places them in one-to-one correspondence with the natural numbers $\mathbb{N}$. Proper rationals corresponding to combinations of $p \in \mathbb{Z}^{\star}$ and $q \in \mathbb{N}$ with no common factor are depicted on an integer grid at point (p, q) with a solid circle with the natural number index inside. Improper points are depicted with dashed circles and un-numbered. Numbers corresponding to equal height lie on diagonals at $135°$. Improper rationals can be grouped with a proper rational into equivalence classes representing the same rational number lying on the indicated dashed lines of constant slope.

ordered pairs of integers which we write $\mathbb{Z}^2 \triangleq \mathbb{Z} \times \mathbb{Z}$. This case is very similar to the rational case considered above. In Figure 1.2, the rationals were identified with a subset of points of the lattice $\mathbb{Z}^{\star} \times \mathbb{N}$. It is not difficult to see that two-dimensional lattices formed by the Cartesian product of any of $\mathbb{N}$, $\mathbb{Z}^{\star}$ or $\mathbb{Z}$ can be put into one-to-one correspondence with each other. For this reason, to consider the cardinality of the lattice $\mathbb{Z}^2$, it is sufficient to consider the lattice $\mathbb{N}^2 \triangleq \mathbb{N} \times \mathbb{N}$. In Figure 1.3 we have shown two mappings to demonstrate that the set $\mathbb{N}^2$, and hence $\mathbb{Z}^2$, is countable.

Algebraic and transcendental numbers

Numbers like $\sqrt{2}$ are irrational. Such numbers can be represented as the roots of a polynomial with integer coefficients (or equivalently with rational coefficients). This leads to the following definition.

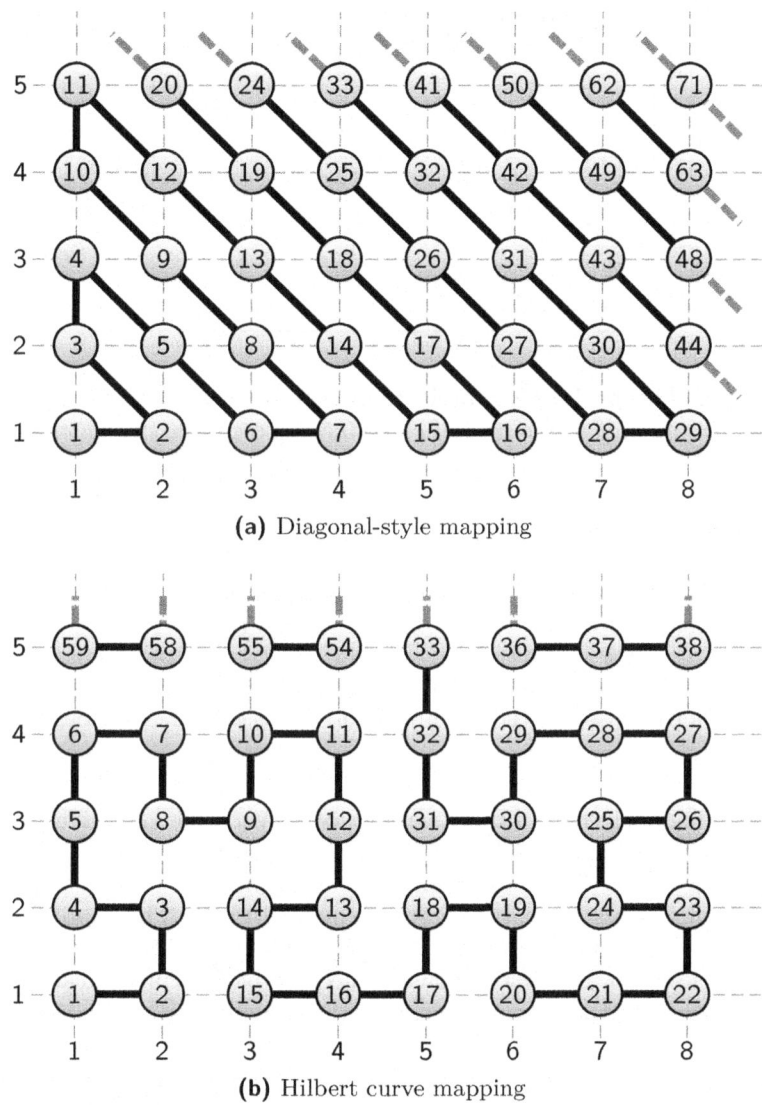

(a) Diagonal-style mapping

(b) Hilbert curve mapping

Figure 1.3 Two examples of orderings of the natural number lattice, that is, one-to-one mappings of $\mathbb{N}^2 \triangleq \mathbb{N} \times \mathbb{N}$ to $\mathbb{N}$. **(a)** Common diagonal-style mapping of $\mathbb{N}^2 \triangleq \mathbb{N} \times \mathbb{N}$ to $\mathbb{N}$. **(b)** Hilbert curve mapping of $\mathbb{N}^2 \triangleq \mathbb{N} \times \mathbb{N}$ to $\mathbb{N}$ based on Hilbert's 1891 "space filling" fractal curve. It is defined recursively. The first order curve is the U-shaped set of four connected nodes 1 through 4. The second order curve is U-shaped set of four connected first order curves (1 through 4, 5 through 8, 9 through 12, and 13 through 16), etc.

Table 1.2 Algebraic properties of $\aleph_0$ with examples, where the set of natural numbers $\mathbb{N}$ is taken as our "reference" countable set.

Operation	Symbolic notation	Example
Addition	$\aleph_0 + 1 = \aleph_0$	$\mathbb{Z}^\star = \mathbb{N} \cup \{0\}$
	$\aleph_0 + r = \aleph_0$	$\mathcal{A} = \mathbb{N} \cup \{-r+1, \ldots, 0\}$
	$\aleph_0 + \aleph_0 = \aleph_0$	$\mathbb{Z} = \mathbb{Z}^\star \cup \{-1, -2, -3, \ldots\}$
Multiplication	$\aleph_0 \cdot 2 = \aleph_0$	$\mathbb{Z} = \mathbb{Z}^\star \cup \{-1, -2, -3, \ldots\}$
	$\aleph_0 \cdot r = \aleph_0$	$\mathcal{A} = \{(p,q) \colon p, q, r \in \mathbb{N}, \ p \leq r\}$
	$\aleph_0 \cdot \aleph_0 = \aleph_0$	$\mathcal{A} = \{(p,q) \colon p, q \in \mathbb{N}\}$
Exponentiation	$r^{\aleph_0} = \mathfrak{c}$	$\mathbb{R}$

Definition 1.1 (Algebraic numbers). *A complex number is said to be an algebraic number if it is a root of some integer-coefficient polynomial equation*

$$f(x) = p_0 x^n + p_1 x^{n-1} + \cdots + p_{n-1} x + p_n = 0,$$

where $p_0 \in \mathbb{N}$ and $p_1, p_2, \ldots, p_n \in \mathbb{Z}$. □

Definition 1.2 (Transcendental numbers). *A complex number which is not algebraic is said to be a transcendental number.* □

That algebraic numbers are countable follows by extending the concept of height, used in the argument for the case of the non-negative rationals, to

$$h = n + p_0 + |p_1| + |p_2| + \cdots + |p_n| \in \mathbb{N}.$$

There are only a finite number of polynomials of a given height and, therefore, only a finite number of algebraic numbers (the roots, possibly repeated) of a given height. This means that enlarging the rationals to include all algebraic numbers still leads to a set of cardinality $\aleph_0$. This is somewhat surprising.

Before moving on beyond the countably infinite we note the algebraic properties of $\aleph_0$ in Table 1.2 (which need to be interpreted carefully). The last row of the table dealing with exponentiation will be discussed later.

Problems _____

1.2 (Hilbert's hotel). Hilbert presented the following paradox about infinity. In a hotel with a finite number of rooms, once it is full, no more guests can be accommodated. Now imagine a hotel — Hilbert's hotel — with a countably infinite number of rooms. Given the rooms are numbered $1, 2, 3, \ldots$, and fully occupied devise a strategy to accommodate a countable number of additional guests (without sharing).

1.3. It is holiday season. A bus arrives at Hilbert's hotel which is full with an infinite number of camera-toting tourists. But it is only the first in an infinite number of such buses. Can Hilbert Hotel accommodate all the guests assuming no sharing of rooms? How?

1.4. Prove every transcendental number is irrational.

1.2.3   Uncountably infinite sets

Fortunately, in the interest of variety, the situation changes when we augment the real transcendental numbers. This augmentation leads to the complete set of real numbers, also called the reals or the continuum.[12] Joseph Louiville (1809–1882) constructed the first transcendental number in 1844; see (Stewart, 1973, p. 68):

$$\xi = \sum_{n=1}^{\infty} 10^{-n!}$$

$$= 0.11000100000000000000000001 \underbrace{000 \cdots 000}_{95 \text{ zeros}} 1 \underbrace{000 \cdots 000}_{599 \text{ zeros}} 1000 \cdots .$$

The more interesting numbers e and π have been shown to be transcendental. Furthermore, transcendentals, such as: e^{π}, $2^{\sqrt{2}}$, $\sin 1$, $\log q$ with $q \in \mathbb{Q}$, $\Gamma(1/3)$ and $\Gamma(1/4)$, were hard to track down. Cantor established the following key result:

Theorem 1.1 (The real numbers are not countable). *The cardinality of the set of all real numbers, the reals, is not countable.* □

Proof (sketch). The proof follows by contradiction. Suppose all real numbers can be placed in a sequence $\{\alpha_1, \alpha_2, \alpha_3, \ldots\}$. These numbers can be written uniquely (essentially) in a non-terminating decimal. Then one can construct a real number that cannot possibly belong to the list. The construction is called the Cantor diagonal process. Without loss of generality,[13] we can look at the real interval $[0, 1)$. Suppose the list looks like

$$\alpha_1 = 0.\mathbf{2}70\,986\,113\,250 \cdots ,$$
$$\alpha_2 = 0.4\mathbf{0}2\,139\,270\,649 \cdots ,$$
$$\alpha_3 = 0.08\mathbf{1}\,139\,616\,783 \cdots ,$$
$$\alpha_4 = 0.537\,\mathbf{5}94\,426\,181 \cdots , \tag{1.7}$$
$$\alpha_5 = 0.821\,2\mathbf{7}5\,476\,892 \cdots ,$$
$$\alpha_6 = 0.172\,18\mathbf{9}\,432\,914 \cdots ,$$
$$\vdots$$

Now, construct a number $\alpha_0 \in [0, 1)$ such that its first digit after the decimal point *differs* from the first digit after decimal point in α_1 (bold digit 2 in (1.7)), its second digit after the decimal point *differs* from the second digit after decimal point in α_2 (bold digit 0 in (1.7)), and so on. An example is

$$\alpha_0 = 0.\mathbf{415698} \cdots .$$

---

[12] Not to be confused with the "Continuum" from Star Trek.
[13] Simple geometrical transformations can show that the real line has the same cardinality as the real numbers in any line segment by a mapping based on tan and arctan.

We can assert that α_0 cannot be in the list. Because if it were equal to α_n for some n, then it meant that the n-th digit after the decimal point in α_0 had to be identical to that in α_n, which is impossible according to the specific construction of α_0. Hence, according to the definition α_0 should be on the list, because it is just a number between zero and one and it cannot be on the list at the same time according to its construction. No list avoids this problem. Therefore, the set of real numbers cannot be countable.   □

In short, the reals constitute a set with larger cardinality than $\aleph_0$. So the infinity associated with the reals is mind-bogglingly huge relative to the integers or rationals or even the algebraic numbers. Even a countable number of countable sets does not come close (still is $\aleph_0$).

When juxtaposed with the difficulty of establishing even relatively few transcendental numbers, the Cantor result is remarkable. He basically showed that the real line is almost completely composed of transcendental numbers and the countable subsets make up an almost irrelevantly insignificant proportion. It seemed like all the real numbers are transcendental except the real numbers we know about. This shows the value of having existence proofs in one's armory, in addition to constructive methods.

However, there is something quite significant in the property that we have an entity, in this case the real line, for which we have poor understanding of the typical members (transcendentals), and much better understanding of the atypical elements (algebraic numbers including the rationals and integers). The deep idea is that we can use the atypical well-defined elements, such as rational numbers, to characterize and arbitrarily well approximate the obscure but typical elements (transcendental reals).

> **Remark 1.4.** This is the idea we need to come to grips with later when we more carefully consider functions in function spaces. The unpleasant reality is that the typical function is something we do not well understand and may not be even able to draw or visualize. The atypical functions contain all the functions we can imagine and generally delude ourselves as being representative.   □

Cardinality of $\mathbb{R}^N$

As an aside, it can be shown that the cardinality of points in a 3D unit cube is the same as the cardinality of points in a line segment $[0, 1)$, again by looking at the strings of digits that represent them. For example, a random point in the cube say at Cartesian coordinates

$$(0.\mathbf{231}\,\mathbf{496}\,02\cdots, 0.780\,291\,52\cdots, 0.531\,440\,26\cdots)$$

can be interleaved to form a single real number on the line segment $[0, 1)$, such as,

$$0.\underline{2}75\,\underline{3}83\,\underline{1}01\,\underline{4}24\,\underline{9}94\,\underline{6}10\,\underline{0}52\,\underline{2}26\cdots,$$

and conversely any point in $[0, 1)$ has a decimal representation, which can be deinterleaved to generate three coordinates of a point in the cube. This is shown conceptually in Figure 1.4. All points in the cube can be represented in this way

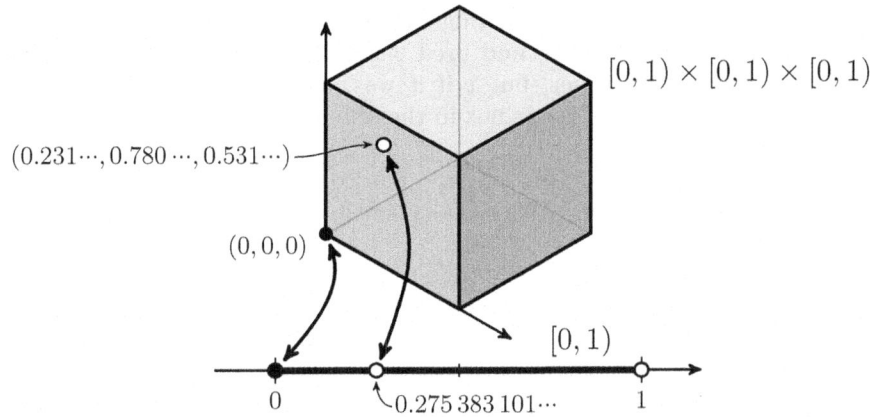

$[0,1) \times [0,1) \times [0,1)$

$(0.231\cdots, 0.780\cdots, 0.531\cdots)$

$(0,0,0)$

$[0,1)$

0 $0.275\,383\,101\cdots$ 1

Figure 1.4 Cardinality of points in a 3D cube, $[0,1) \times [0,1) \times [0,1)$, is the same as the cardinality of points in a line segment $[0,1)$. The decimal expansions of the 3D coordinates of a point in the cube are interleaved to form a (almost unique) point on the line segment $[0,1)$.

as can all points on $[0,1)$. Hence, the transfinite cardinality of boxes in $\mathbb{R}^N$, or $\mathbb{C}^N$, does not increase with dimension N.

Remark 1.5. There is an annoying technicality of working with decimal expansions, which is deliberately glossed over in the previous developments. There can be more than one decimal expansion for a given real number. For example, we can write either 0.4 or $0.399999\cdots$, and they represent the same real number. This would screw up the cube mapping in ways that would excite a mathematician, but not others. In fact this type of oversight was made by Cantor, which only increases his endearment to us. $\square$

1.2.4   Continuum as a power set

Recall, from Section 1.2.2, that the basic tools of working with transfinite sets are to either find a clever mapping taking one set to the other, or establishing by contradiction that one set cannot be put in one-to-one correspondence with another (one bigger than the other).

Here we want to fill out our knowledge a bit further. Three things will be explored, albeit briefly:

- that the cardinality of the reals, $\mathfrak{c}$, is equivalent to the set of all possible subsets (including countable subsets) of a countable set;

- that there are sets with cardinality greater than $\mathfrak{c}$;

- that there exists the concept of measure and the bizarre property that sets can in one sense be everywhere and at the same time nowhere significant.

If you consider the set of all subsets of the natural numbers, including all infinite subsets, this set has the same cardinality as the real numbers. This was proven by Cantor by noting that if you wrote a real number between zero and one in binary form, and used 0 for the first digit after the decimal point if 0 was not in the set, but 1 if it was, and used 0 for the second digit after the decimal point if 1 was not in the set, but 1 if it was, and so on, then you would have exactly one string of digits, that is, one unique binary representation of a real number between zero and one corresponding to each possible subset of the natural numbers. Hence, it is sensible to write, particularly if you are on the fringe of sanity, $\mathfrak{c} = 2^{\aleph_0}$.

Exponentiation of the transfinite sets leads to transfinite sets of higher cardinality whereas addition and multiplication do not.

Problems

1.5. Give an argument to justify the notation $\mathfrak{c} = 2^{\aleph_0}$. (Hint: think about a finite set first.)

1.6. Show $\mathfrak{c} = 10^{\aleph_0}$. Generalize the result.

1.7. Similar to Table 1.2, make a table of the algebraic properties of $\mathfrak{c}$ with examples given for addition, multiplication and exponentiation operations.

1.8. Prove that set of all continuous functions $f(x)$ defined over the interval $0 \leq x \leq 1$ has a cardinal number equal to $\mathfrak{c}$.

1.9. Prove that the set of all single-valued functions $f(x)$ defined over the interval $0 \leq x \leq 1$ has a cardinal number greater than $\mathfrak{c}$. This class of functions necessarily includes discontinuous functions.

1.2.5 Countable sets and integration

If we take the interval $0 \leq x \leq 1$ then this constitutes an interval of length 1. This length is associated with the real numbers in that interval. What length might we associate with the rational numbers in the interval $0 \leq x \leq 1$, particularly given they are dense in the reals, that is, they are everywhere at every scale? This question cannot be answered because we are being too vague. What is lacking is a suitable notion of length which can cater for non-trivial sets, such as the rationals or algebraic numbers or more esoteric constructions.

Consider three functions defined on the real line interval $0 \leq x \leq 1$, which we wish to integrate. Here the notion of integration is to determine the area under the "curve." Figure 1.5 shows representations of the three functions which are defined as follows:

$f_1(x)$ is zero everywhere except at the point $x = 0.4$ where it takes value 1,

$f_2(x)$ is zero everywhere except on the interval $x \in [0.4 - \epsilon/2, 0.4 + \epsilon/2]$ where it takes value 1 for some $\epsilon > 0$,

$f_3(x)$ is zero everywhere except at points corresponding to the rationals where it takes value 1, that is, $\mathbf{1}_{\mathbb{Q}}(x)$, which gives 1 if $x \in \mathbb{Q}$ and 0 otherwise.

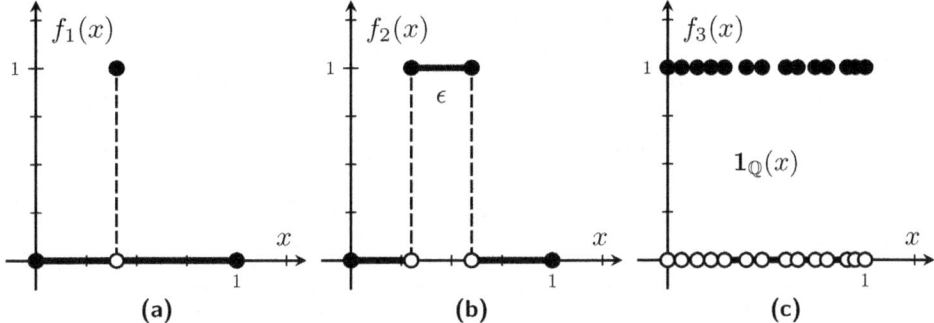

Figure 1.5 Three functions defined on the real line interval $0 \leq x \leq 1$ which are to be integrated: **(a)** $f_1(x)$ — zero except it takes value 1 at one point; **(b)** $f_2(x)$ — zero except it takes value 1 on ϵ-interval; and **(c)** $f_3(x)$ — zero except it takes value 1 on set $\mathbb{Q}$.

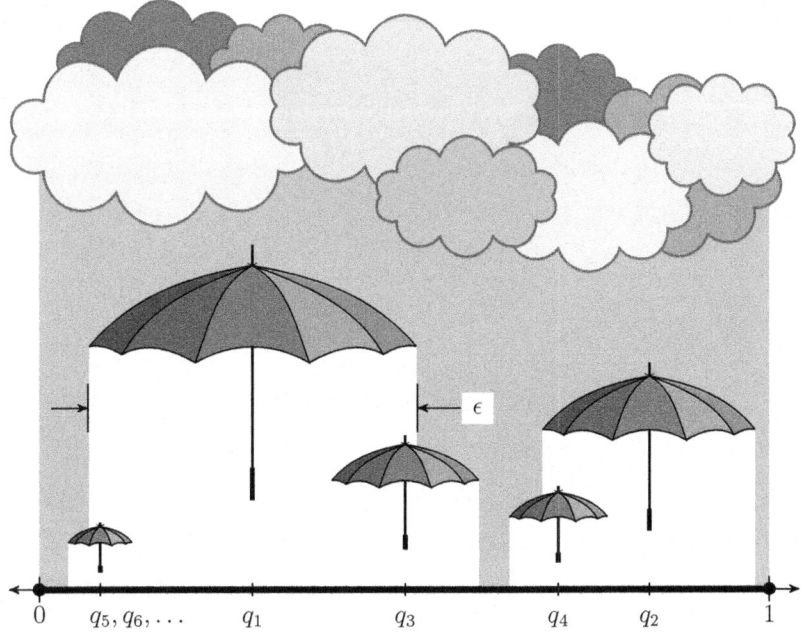

Figure 1.6 Schematic for Problem 1.10. All rationals within the interval $[0, 1]$, a countable number of them, are given umbrellas of varying widths, with the ϵ the widest, $\epsilon/2$ the next, etc. (diagram is not to scale and shows only the first 5 rationals at nominal locations). There is continuum rain on $[0, 1]$ and each rational can stay dry. How much of $[0, 1]$ gets wet as the widths of the umbrellas goes to zero (letting $\epsilon \to 0$)?

For the first case, $f_1(\cdot)$, we can believe that the integral should be zero. This follows since the second function, $f_2(\cdot)$, dominates it and has integral/area ϵ. Then letting $\epsilon \to 0$ shows the first integral should be zero. If we try the same trick on the third case, $f_3(\cdot)$, then have "$\infty \times \epsilon$" and Cantor waves his finger at us.[14] The following (worked) problem resolves the $f_3(\cdot)$ case.

Problems _____

1.10 (Rain and rational umbrellas). The rationals are out on the real line and want to stay dry — it's raining on the continuum interval $[0, 1]$. You give each rational point a finite-width umbrella to stay dry, but of different widths; see Figure 1.6. You give the first rational, your favorite rational, an ϵ-umbrella (width of ϵ). The second rational gets an $\epsilon/2$-umbrella, the third gets an $\epsilon/4$-umbrella, and so on. Since each umbrella covers a finite interval then each rational is happy and dry. The questions is: how much of the continuum gets wet as $\epsilon \to 0$?

1.11. Determine the integral of the function that is zero everywhere except at points corresponding to *algebraic* numbers where it takes value 1. (See the text following Definition 1.1, p. 16, for clarification of the definition and cardinality of algebraic numbers.)

_____

---

[14] Cantor still doesn't calm down even when we write "$\aleph_0 \times \epsilon$."

Take-home messages – infinite dimensions

M1: The basic tools of working with transfinite sets is to either find a clever mapping taking one set to the other or establishing by contradiction that one set cannot be put in one-to-one correspondence with another (one is bigger than the other). There are different types of infinity. $\aleph_0$, the cardinality of set of integers $\mathbb{N}$, is the "smallest" transfinite number.

M2: A set is countable if we are able to write its elements as an unending sequence with a one-to-one pairing or correspondence with the elements in $\mathbb{N}$. Examples: $\mathbb{Z}$, $\mathbb{Q}$, algebraic numbers.

M3: It can be shown by contradiction that the set of real numbers $\mathbb{R}$ cannot be put in one-to-one correspondence with $\mathbb{N}$ and is larger in cardinality. Hence $|\mathbb{R}| = \mathfrak{c} > |\mathbb{N}| = \aleph_0$.

M4: Cantor showed that the real line is almost completely composed of transcendental numbers and the countable subsets make up an almost irrelevantly insignificant proportion. It seems like all the real numbers are transcendental except the real numbers we know about (that is integer, rational and algebraic numbers). Although seemingly insignificant in size, rationals can arbitrarily well approximate real numbers. Hence, they are functionally important.

M5: Addition and multiplication of transfinite sets lead to transfinite sets of the same cardinality. Examples: $\mathbb{N}^r$ has the same cardinality as $\mathbb{N}$, $\mathbb{R}^r$ has the same cardinality as $\mathbb{R}$.

M6: Exponentiation of the transfinite sets leads to transfinite sets of higher cardinality. For every transfinite set $\mathbb{S}$, the power set of $\mathbb{S}$, i.e., the set of all subsets of $\mathbb{S}$, is larger than $\mathbb{S}$ itself. Examples: $\mathfrak{c} = 2^{\aleph_0} > \aleph_0$. The set of all single-valued functions $f(x)$ defined over the interval $0 \leq x \leq 1$ has a cardinality larger than $\mathfrak{c}$.

2 Spaces

2.1 Space hierarchy: algebraic, metric, geometric

There is a hierarchy: vector spaces, normed spaces and inner product spaces. The first defines algebraic properties, the second augments pure metric properties (distance) and the third further augments geometric properties (angles).

By the term *vector spaces* we do not mean conventional vectors in $\mathbb{R}^N$ or $\mathbb{C}^N$ (vectors of length N). Rather the term really means "spaces of vector-like objects," that is, we are looking to define the most general sets of objects which mimic the most basic properties of vectors. The scalar field associated with the vector space, just as for the conventional vectors, can either be the reals $\mathbb{R}$ or the complex numbers $\mathbb{C}$ (the latter understood if not explicitly stated).

So what does "space" actually refer to? The term "space" has a definite conceptual meaning and it is best not to use the term too loosely. Space typically means a pairing of:

- a set of like-objects (vectors or abstract points); and

- a bunch of basic rules (mappings) for combining these like-objects (vectors) to create other like-objects (vectors).

For example, we would like to add vectors and expect to get another vector.

Hilbert spaces are a special type of vector spaces which inherit the algebraic structure of vector spaces and augment this with a metric and geometric superstructure. The last key attribute of Hilbert spaces is the elusive notion of *completeness*. As we will point out later, completeness is tied to the norm because the norm determines convergence.

2.2 Complex vector space

We can readily jump into the all too familiar definition.

Definition 2.1 (Complex vector space). *A complex vector space $\mathcal{H}$ is a set of elements called vectors together with two operations:*

1: *addition, which associates a vector $f + g \in \mathcal{H}$ with any two vectors $f, g \in \mathcal{H}$ (closed under addition);*

2: *scalar multiplication, which associates a vector $\alpha f \in \mathcal{H}$ with a vector $f \in \mathcal{H}$ and any complex scalar $\alpha \in \mathbb{C}$ (closed under multiplication by a scalar).*

For any $f, g, h \in \mathcal{H}$ and any $\alpha, \beta \in \mathbb{C}$ the following also hold:

$$f + g = g + f,$$
$$(f + g) + h = f + (g + h),$$
$$\exists o \in \mathcal{H} \quad such \ that \quad f + o = f,$$
$$\alpha(f + g) = \alpha f + \alpha g, \tag{2.1}$$
$$(\alpha + \beta)f = \alpha f + \beta f,$$
$$(\alpha\beta)f = \alpha(\beta f),$$
$$0f = o, \quad 1f = f,$$

where o is the special unique "zero" vector.   □

Definition 2.2 (Linear combination). *Let $\mathcal{H}$ be a vector space over a scalar field and let $f_1, f_2, \ldots, f_n$ be n elements of $\mathcal{H}$. The finite sum*

$$\sum_{k=1}^{n} \alpha_k f_k$$

for some $\alpha_1, \alpha_2, \ldots, \alpha_n$ in the scalar field is called a linear combination of vectors $f_1, f_2, \ldots, f_n$.   □

Definition 2.3 (Linear independence and dependence). *Vectors $f_1, f_2, \ldots, f_n$ are called linearly independent if from*

$$\sum_{k=1}^{n} \alpha_k f_k = o,$$

one can conclude that

$$\alpha_1 = \alpha_2 = \cdots = \alpha_n = 0.$$

In other words, $f_1, f_2, \ldots, f_n$ are linearly independent if no vector f_k is a linear combination of other vectors in the group. We call an infinite number of vectors linearly independent if every finite subset of those vectors is linearly independent.

If, on the other hand, there exist $\alpha_1, \alpha_2, \ldots, \alpha_n$, not all equal to zero, such that the linear combination

$$\sum_{k=1}^{n} \alpha_k f_k = o,$$

we say that $f_1, f_2, \ldots, f_n$ are linearly dependent.   □

Remark 2.1. Every group of vectors that includes the zero element o in $\mathcal{H}$ is necessarily dependent.   □

A deep observation regarding Definition 2.1 is that the rules talk about combining two or three vectors only. It is clear that by re-applying the rules, (2.1), then one can infer that combining, say, one thousand vectors the result is well defined and still a vector. However, the rules are silent on what happens if we want to do an unlimited number of operations such as additions. This is the nub of what we need to study. The vector space in isolation is not cut out to serve our needs. If we add an infinite number of vectors then sometimes we get a vector and sometimes it is unbounded (and, therefore, technically not a vector). In what ultimately appears in real applications, we need to have a means to deal with infinite vector addition and we need some tools to make sense of this.

Problem

2.1. Provide an example where two vectors are linearly independent in one scalar field and linearly dependent in another.

2.3   Normed spaces and Banach spaces

Many of the topological properties of Hilbert spaces hinge on the concept of norm. The norm is a powerful concept that allows us to define space neighborhoods of a vector, converging sequences, Cauchy sequences and ultimately completeness. Therefore, we will separately treat normed vector spaces and dwell in such spaces for a while. We can then apply our knowledge of normed spaces to inner product spaces and in particular to Hilbert spaces.

2.3.1   Norm and normed space

Normed spaces are built upon vector spaces by incorporating the following definition of the norm.

Definition 2.4 (Norm). *Let $\mathcal{H}$ be a complex vector space. A functional mapping $f \in \mathcal{H}$ into $\mathbb{R}$ is called a norm and is denoted by $\|f\|$ if it satisfies the following conditions:*

$$\|f\| = 0 \iff f = o, \tag{2.2}$$

$$\|\alpha f\| = |\alpha|\,\|f\|, \quad \forall f \in \mathcal{H} \text{ and } \forall \alpha \in \mathbb{C}, \tag{2.3}$$

$$f + g \leq \|f\| + \|g\|, \quad \forall f, g \in \mathcal{H}. \tag{2.4}$$

□

The last condition is called the *triangle inequality* and shown in Figure 2.1. Airlines rely on it. Flat-Earth airlines expect that the direct route between cities is shorter (or no longer) than every route via an intermediate city.

Remark 2.2. In our concept of a norm, we think of it as being a distance measure generating either a positive real number or zero (that is, a nonnegative number). This is a property that we can infer from the conditions in the above Definition 2.4; see Problem 2.2 below.   □

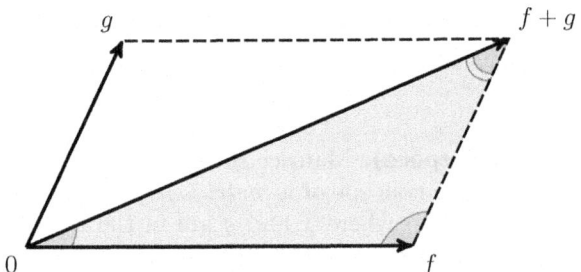

(a) Abstract form: $\|f+g\| \le \|f\| + \|g\|$ where f and g belong to a normed space.

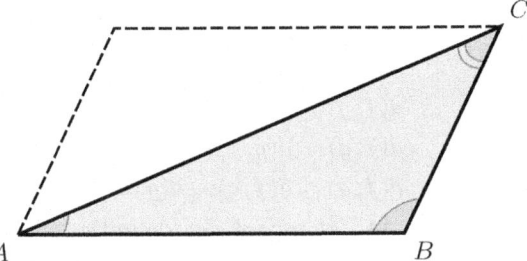

(b) Euclidean form: $AC \le AB + BC$ where AB, BC and AC are the lengths in a conventional Euclidean sense.

Figure 2.1 Two versions of the triangle inequality. **(a)** Triangle inequality in abstract form $\|f + g\| \le \|f\| + \|g\|$. The norm of the sum of two vectors (vector addition) is at most as large as the sum of the norms. **(b)** Triangle inequality corresponding to the Euclidean plane statement from geometry "the sum of the lengths of any two sides of a triangle must be greater than the length of the remaining side."

Example 2.1. Consider the vector space of complex-valued continuous functions, $C[0,1]$, defined on the closed interval $[0,1]$. The following mapping

$$\|f\| = \max_{x \in [0,1]} |f(x)|$$

satisfies all the properties for being a norm in $C[0,1]$. Specifically,

$$\|f\| = \max_{x \in [0,1]} |f(x)| = 0 \iff f(x) = 0 \quad \forall x \in [0,1],$$

$$\|\alpha f\| = \max_{x \in [0,1]} |\alpha f(x)| = |\alpha| \max_{x \in [0,1]} |f(x)| = |\alpha| \, \|f\|,$$

$$\|f + g\| = \max_{x \in [0,1]} |f(x) + g(x)| \le \max_{x \in [0,1]} |f(x)| + \max_{x \in [0,1]} |g(x)|.$$

This type of norm is in fact a special type of norm called the *sup-norm*. We will revisit this and the normed space of continuous functions when we discuss different types of convergence and explore completeness further. □

Definition 2.5 (Normed space). *A vector space $\mathcal{H}$ with a norm is called a normed space.* □

A normed space is sometimes written $\{\mathcal{H}, \|\cdot\|\}$, which emphasizes well the two-part nature of a normed space.

2.2. By considering $g = -f$ in Definition 2.4 or otherwise, show

$$\|f\| > 0, \quad \forall f \neq o.$$

2.3 (Metric space). Maurice René Fréchet (1878–1973) had marvellous hair and developed the concept of a *metric space* in 1906 which combines a *set*, $\mathcal{M}$, and a *metric*, $d(f, g)$ where f and g are in the set. The metric, which is analogous to the norm in a normed space, measures the distance between two elements, $f, g \in \mathcal{M}$, and is defined by

$$d(f, g) \colon \mathcal{M} \times \mathcal{M} \longmapsto \mathbb{R},$$

satisfying

$$d(f, g) \geq 0,$$
$$d(f, g) = 0 \iff f = g,$$
$$d(f, g) = d(g, f),$$
$$d(f, g) \leq d(f, g) + d(g, h) \quad \text{(triangle inequality)},$$

where $f, g, h \in \mathcal{M}$. Compare the definitions of a normed space and a metric space to determine which concept is more general.

2.3.2   Convergence concepts in normed spaces

Convergence of a sequence is familiar from standard analysis whereby the elements of a sequence of numbers get closer and closer to some fixed number. The closeness measure is just a distance. The norm in a normed space plays the same role as measuring distance and this motivates the following definition:

Definition 2.6 (Convergence in normed spaces). *A sequence $\{f_n\}_{n=1}^{\infty}$ in a normed space $\mathcal{H}$ converges to some $f \in \mathcal{H}$, called the limit, if $\forall \epsilon > 0, \exists N$ such that*

$$\|f_n - f\| < \epsilon, \quad \forall n > N.$$

$\square$

Remark 2.3. When $\{f_n\}_{n=1}^{\infty}$ converges to the limit $f \in \mathcal{H}$ we write

$$\lim_{n \to \infty} f_n = f \quad \text{or} \quad f_n \to f,$$

where it is implicit (but necessary) that $f_n \in \mathcal{H}$ and $f \in \mathcal{H}$.

$\square$

In reviewing the Definition 2.6, we may ask what happens if the limit $f \notin \mathcal{H}$. That is, suppose we have a sequence $\{f_n\}_{n=1}^{\infty}$ in $\mathcal{H}$ (meaning every $f_n \in \mathcal{H}$) satisfying the condition $\forall \epsilon > 0, \exists N$ such that $\|f_n - f\| < \epsilon, \forall n > N$, but $f \in \mathcal{H}$ is violated. The answer to this question is that in this case the sequence *does not converge*. So there are really two important requirements to convergence:

R1: the difference between the f_n and the limit f needs to get closer and closer to zero as n increases;

R2: equally importantly, the limit f itself needs to be in the space $\mathcal{H}$.

Apparently, it is possible for every f_n to be within the space and "at the last moment" $f \equiv f_\infty$ pops out of the space, but is in some sense is arbitrarily close to the space. It is useful to have an example and study it in detail, and that is furnished in Section 2.3.4.

So there is a beauty pageant for women where every year the winner tends to get more and more beautiful and we are saying in the limit the winner is ugly or a male or both (the usual combination).[1] Shallow people around the world would plummet into despair.

Nonetheless, violation of the $f \notin \mathcal{H}$ condition appears somewhat pathological. The gut feeling is that this should not occur in properly formulated spaces or beauty pageants. What we do expect is if sequences are settling down and getting closer and closer together, then this is what we should like to mean by convergence. Unfortunately no such luck because convergence is defined differently (and for sound reasons). Such a sequence is not guaranteed to converge (in the absence of further conditions), but such sequences have a name.

Definition 2.7 (Cauchy/fundamental sequence).   *A sequence $\{f_n\}_{n=1}^\infty$ in a normed space $\mathcal{H}$ is called a Cauchy sequence if $\forall \epsilon > 0$, $\exists N$ such that*

$$\|f_m - f_n\| < \epsilon, \quad \forall m, n > N.$$

$\square$

Remark 2.4.   Cauchy sequences are also often referred to as *fundamental sequences*. $\square$

Remark 2.5.   In the definition of a Cauchy sequence there is no reference to any limit. This is significant because we may not know what the limit is or even if it exists (that is, is an element in the normed space $\mathcal{H}$). $\square$

Problems

2.4. Show that every convergent sequence is a Cauchy sequence.

2.5. Show that not every Cauchy sequence is convergent.

2.3.3   Denseness and separability

By drawing on our rationals-reals analogy, we introduce two central concepts in normed spaces: namely denseness and separability. We will see in Section 2.3.6, p. 35, how denseness is tied up to the notion of completeness. Just like everywhere you go you run across dense people, so too elements of a dense set are everywhere you look.

---

[1]   For example, a slim waist may be deemed to be an essential element of beauty and over time the waist lines of the sequence of winners get slimmer and slimmer. Historical records from the future reveal that in 2084 a point is reached where trolleys of life support become a feature of beauty pageants given the dangerously narrowing waists of contestants and their pet Chihuahuas. In the limit, there is no waist and pageants becomes a parade of bisected corpses.

Cauchy, Augustin-Louis (1789–1857) — Cauchy was a French mathematician with a prolific output of 789 or so original works which have been collected into 27 volumes that took others almost a century to compile between 1882 and 1970. During periods of self-induced overwork he abandoned evening work so as to recover. His early education was as an engineer, but Lagrange and others convinced him that his calling was in mathematics.

As a person he tended to drive everyone nuts. The volume of his submitted works put such a strain on the French Academy of Science's financial resources, in printing costs, that it forced them to place restrictions on the length and number of submitted articles. Undeterred Cauchy submitted papers up to 300 pages to other journals. The rough treatment he gave others led the Norwegian mathematician Abel to proclaim "Cauchy is mad and there is nothing that can be done about him, although, right now, he is the only one who knows how mathematics should be done." For further reading see (Bell, 1986).

Definition 2.8 (Dense set). *A set $\mathcal{D}$ is said to be dense in a normed space $\mathcal{H}$ if for each point $f \in \mathcal{H}$ and each $\epsilon > 0$ there exists $d \in \mathcal{D}$ with $\|f - d\| < \epsilon$.*   □

Rational numbers can arbitrarily well approximate real numbers. Whatever tolerance $\epsilon > 0$ is specified there is always a rational number within ϵ in absolute value of any given real number (simply truncate the decimal expansion). We say that the set of rational numbers is dense in the space of real numbers. Rational numbers are also countable, that is, they can be put in one-to-one correspondence with the integers — a non-obvious property that we established in Section 1.2.2, p. 11.

Intuitively, dense means there is at least one member of the set in the neighborhood of any point in the normed space. You cannot avoid them, and this is good news. This means members of the dense set can be used as effective approximations to every member of the normed space.[2]

The dense set can be a strict subset of the normed space. In fact, it is precisely when it is a strict subset that it is of greatest value. This leads to a fundamental definition.

Definition 2.9 (Separable normed space). *A normed space is separable if it contains a countable dense set.*   □

The set of real numbers, $\mathbb{R}$, considered as a normed space (where the absolute value is the norm), is separable since the rationals are dense in $\mathbb{R}$, and the rationals are countable.

Do we care if a normed space contains a dense set or not? The answer is yes and it really is quite crucial. A dense set guarantees that for any point in the space (not necessarily within the dense set) we can construct a sequence (wholly within the dense set) that converges to that point. Conversely, a converging sequence wholly within the dense has its limit within the normed space (its *closure*). Therefore, a dense set is capable of capturing the essence of a normed space. A countable dense set is icing on the cake because it acts like a backbone. As was discussed in Section 1.2.3, p. 17, this allows us to approximate arbitrarily well any typical, but poorly understood, real number with an atypical, but much better understood, rational number. A similar thing applies to typical and atypical functions in function spaces. We will see later that denseness has a geometric significance and is the reason Fourier series representations can work. The fact that you can get away with a sum (albeit an infinite sum) of "basis functions" to represent almost any function is ultimately guaranteed by separability.

Remark 2.6. Countable sets can be placed in a sequence. With sequences we can formally talk about convergence.   □

---

[2]   These concepts are also tied up with the notion of *accumulation points* or limit points of dense set $\mathcal{D}$. Accumulation points fill in the gaps between the dense points and also build the boundary points of open sets in space. The set consisting of the points of the dense set $\mathcal{D}$ and the accumulation points of $\mathcal{D}$ is called the *closure* of $\mathcal{D}$. It is the smallest closed set containing $\mathcal{D}$. Then an alternative definition of $\mathcal{D}$ being dense in a space $\mathcal{H}$ is if the closure of $\mathcal{D}$ equals $\mathcal{H}$.

2.3.4   Completeness of the real numbers

As was said before, one central attribute of Hilbert spaces is the elusive notion of *completeness*. Having defined what convergence means in normed spaces, it is worth spending some time carefully developing the notion as it will play a key role in the more abstract Hilbert spaces such as ones involving functions. Completeness is an abstraction of something very simple, but rarely made clear in texts. We will again draw on rationals-reals analogy that will be insightful and sharpen our understanding:

1: Consider the space of rational numbers with the absolute value as the norm used to determine if two numbers are close. Then the sequence of rational numbers

$$\{1, 1.4, 1.41, 1.414, 1.4142, 1.41421, 1.414213, 1.4142135, \ldots\} \qquad (2.5)$$

has the property that successive numbers[3] are getting closer and closer with the limiting value being $\sqrt{2}$. Now in a well-defined sense this sequence does not converge! The argument is as follows. Consider the rational numbers. Even though every point in the sequence (2.5) is rational, the limit is $\sqrt{2}$, which is irrational. In this case we can say that the normed space of rational numbers is not complete. The critical issue is whether the limiting value lies within the original space. If all sequences which have a limit converge to a point in the space, then that space is complete.

2: There are two profound points we can assert illustrated by the above:

- when we consider or study an *infinite* number of well-defined objects (rationals) we cannot expect to end up with something equally well-defined (reals, which generally are not rational)

- in the opposite direction, it might be that we are interested in more complicated objects (reals) and the insight is that we can study them from the perspective of a more robust understanding of simpler objects (rationals)

3: Consider another example where we have a sequence of numbers which is growing without bound, such as $\alpha_n = e^n$. We would say one of the following: it diverges or it diverges to infinity or it does not converge. Why doesn't it converge? Such a question seems stupid. What you could say is the sequence does not converge in the reals because the reals, by definition, exclude $\pm$ infinity. The problem is not infinity per se, but that infinity is not actually in the space, that is, infinity is not in the reals.

What do we do if a normed space is not complete? We expand it to include more points so that the enlarged space is complete and there are no missing points. In the first example, all we need to do is throw in the irrational numbers, that is, the real numbers are the completion of the rational numbers. Ultimately completeness is tied to the distance because it determines convergence. It is often much easier proving lack of completeness than proving completeness.

---

[3]   In fact, the distance between any pair of numbers, not necessarily adjacent ones, further down the sequence becomes smaller and smaller the further out we go.

In short, the space of reals is complete because every Cauchy sequence in the space converges to a point within reals.

The fact that rational numbers are dense in the space of reals is no coincidence. We will see in Section 2.3.6 that we can turn any normed space, which is not necessarily complete, into an at least equally big space which is complete by carefully adding the missing points. Then the original space will become a dense set within its completion.

2.3.5   Completeness in normed spaces

Guided by the analogy to the above study of rationals and reals, we begin with a well-defined normed space of functions. Such a function space with norm can be thought of as playing the role of the rationals with absolute value as the norm. The space will be shown to be incomplete. As a consequence more functions will be added, retaining the norm, to create a complete space. This complete space with norm can be thought of as playing the role of the reals with absolute value as the norm.

Define the real interval $\Omega \triangleq [\alpha, \beta]$ and consider the space of complex-valued continuous functions $C(\Omega)$, (analogous to the rationals) on Ω, with norm (analogous to the absolute value)

$$\|f\| \triangleq \left(\int_\Omega |f(x)|^2 dx \right)^{1/2}. \tag{2.6}$$

Then it is easy to contrive a sequence of continuous functions on Ω, say

$$f_n(x) \triangleq \arctan\big((n(x - (\alpha + \beta)/2)\big), \quad n \in \mathbb{N}, \tag{2.7}$$

as shown in Figure 2.2, whose limit is not continuous because it is given by the step function

$$\lim_{n \to \infty} f_n(x) = \begin{cases} -\pi/2 & \alpha \leq x < (\alpha + \beta)/2, \\ 0 & x = (\alpha + \beta)/2, \\ +\pi/2 & (\alpha + \beta)/2 < x \leq \beta, \end{cases} \tag{2.8}$$

which is discontinuous at the midpoint. Since such a step function is not a continuous function, the sequence does not converge.[4] To finish the analogy we have to devise a broader class of functions (analogous to the reals) than the continuous functions (analogous to the rationals). The answer is the space of square integrable functions on Ω, denoted $L^2(\Omega)$, that is, ones satisfying a finite mean square value property

$$\int_\Omega |f(x)|^2 dx < \infty,$$

where the integral is interpreted in the sense of Lebesgue. The function space $L^2(\Omega)$ is the completion of the space of continuous functions on Ω, with respect to $L^2(\Omega)$-norm $\|f\|$ (where the integral in (2.6) is in the sense of Lebesgue). That

---

[4]   Even though the limit may have a well-defined form we still say that the sequence does not converge, which really means: does not converge to a point in the space. More specifically, a sequence of continuous functions does not necessarily converge to a continuous function.

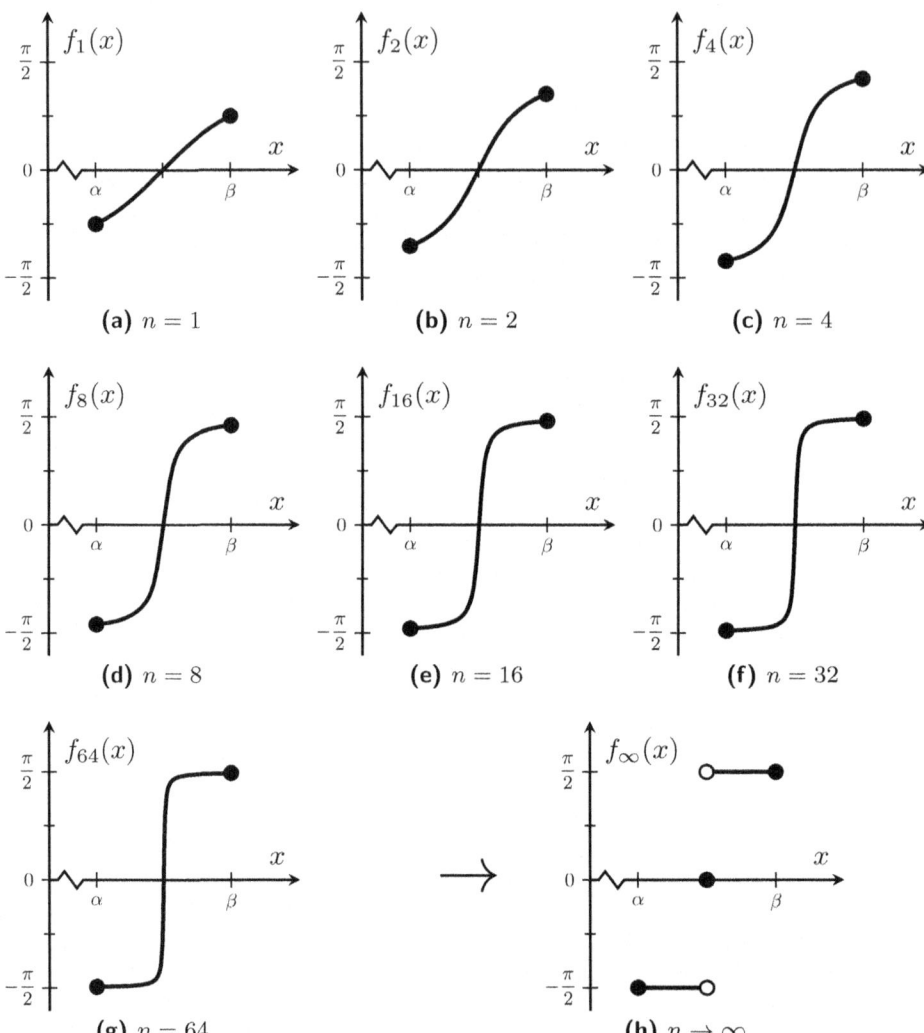

Figure 2.2 A sequence of continuous functions on the interval $\Omega \triangleq [\alpha, \beta]$ whose "limit" is a discontinuous step function. Plotted here are $f_n \in C(\Omega)$ given in (2.7) for $n = 1, 2, 4, \ldots, 64$, and the discontinuous step function $f_\infty \equiv \lim_{n \to \infty} f_n \notin C(\Omega)$ in (2.8).

is, the class of functions which have finite mean square value is a superset of the class of continuous functions.

We now come back to the notion of denseness. Functions in $L^2(\Omega)$ can always be arbitrarily well approximated by continuous functions to any desired accuracy when measured in the $L^2(\Omega)$-norm. In this sense we say that the set of continuous functions on Ω is dense in $L^2(\Omega)$ and again this is not a coincidence. When we carefully expand the space of continuous functions on Ω to its completion $L^2(\Omega)$, then continuous functions on Ω constitute a dense set in $L^2(\Omega)$.

Less clear is the potential separability of $L^2(\Omega)$. We give a quick argument to make it plausible. We can ask if the space of continuous functions on Ω, which are dense in $L^2(\Omega)$, with the $L^2(\Omega)$-norm (here we can use a Riemann integral) is itself countable. This can be indicated in two hops. First, one can find a polynomial function arbitrarily close to any given continuous function. Second one can find a polynomial with rational coefficients arbitrarily close to the polynomial. The polynomials on Ω with rational coefficients, considered as a set, are countable. Working back up the chain, we can assert that, indeed, $L^2(\Omega)$ is separable. (The preceding argument does not constitute a proof.)

2.3.6   Completion of spaces

It can be proven that any normed space $\mathcal{H}$ (which is possibly not complete) can be turned into a complete normed space $\mathcal{H}_c$ (with norm $\|\cdot\|_c$) with the following properties (Helmberg, 1969, pp. 21–22):

P1: $\mathcal{H} \subset \mathcal{H}_c$;

P2: $\|f\|_c = \|f\|$ for all $f \in \mathcal{H}$;

P3: $\mathcal{H}$ is everywhere dense in $\mathcal{H}_c$.

$\mathcal{H}_c$ is called "the" completion of $\mathcal{H}$. This is because one can prove that if there are two spaces $\mathcal{H}_{c_1}$ and $\mathcal{H}_{c_2}$ that complete $\mathcal{H}$, then one can find a one-to-one mapping between the two spaces which preserves the norm.

Roughly speaking, the process of completion is as follows. For every Cauchy sequence in the original space $\mathcal{H}$ that does not converge to a point in $\mathcal{H}$, one augments a new point (its limit point) that completes it with the norm determined uniquely by the second (continuity) property above. The only tricky part is to associate the same completion point to all Cauchy sequences in $\mathcal{H}$ that come arbitrarily close to each other (Helmberg, 1969).[5]

2.3.7   Complete normed spaces — Banach spaces

With all the preceding discussion, we are now in a position to formally define completeness of a normed space.

---

[5]   The beauty pageant analogy is as follows. We just need to broaden our definition of beautiful and nice body features so that it now includes their limit points (what was originally deemed to be ugly bisected corpses!). We need to be careful to associate the same corpse to those whose body features have come arbitrarily close to each other over the years as a result of tough competition. The original set of "human" contestants is dense in the new space, because any ugly bisected corpse can now be arbitrarily well approximated by a member of the set!

Banach, Stefan (1892–1945) — Banach was Polish and one of the most influential mathematicians of the twentieth century. Banach treated his body like a temple. His outstanding intellect was in spiritual harmony with his body. It was cruel fate that he succumbed to lung cancer at a tender age of 53. Possibly, just possibly, his habit of smoking four or five packs of cigarettes a day may have contributed. Interestingly there are a number of photos showing him smoking — clearly it would have been distressing for him to refrain from smoking in the few seconds for the photo capture. Perhaps recognizing the hazard of smoking so much, Banach compensated with a comparable intake of alcohol.

Banach liked to do his mathematics with his mates in the cafés of Lwów in what was then Poland. It is recorded that he was very difficult to out-drink and out-smoke, and that sessions in the cafés could run for 17 hours and into the night. He led discussions, preferred to sit by the band and was clearly a fine role model. In the cafés, research results were scribbled on the marble table-tops and erased by the waiters after each session. At the Scottish café mathematical problems were set and prizes offered, including: bottles of wines, beers, fondue in Geneva, a live goose, bottles of whisky, a kilogram of bacon, and champagne.

Definition 2.10 (Banach space). *A normed space $\mathcal{H}$ is called complete if every Cauchy sequence of vectors in $\mathcal{H}$ converges to a vector in $\mathcal{H}$. A complete normed space is called a Banach space.* □

2.4   Inner product spaces and Hilbert spaces

2.4.1   Inner product

Definition 2.11 (Inner product). *Let $\mathcal{H}$ be a complex vector space. A complex bilinear function/mapping $\langle\cdot,\cdot\rangle\colon \mathcal{H}\times\mathcal{H}\longmapsto \mathbb{C}$ is called an* inner product *if for any $f, f_1, f_2, g, g_1, g_2 \in \mathcal{H}$ and $\alpha_1, \alpha_2 \in \mathbb{C}$ the following conditions are satisfied:*

$$\langle f, g\rangle = \overline{\langle g, f\rangle},$$

$$\langle \alpha_1 f_1 + \alpha_2 f_2, g\rangle = \alpha_1\langle f_1, g\rangle + \alpha_2\langle f_2, g\rangle,$$

$$\langle f, f\rangle \geq 0, \quad and \quad \langle f, f\rangle = 0 \quad implies \quad f = o,$$

where $\overline{(\cdot)}$ denotes complex conjugate. □

Remark 2.7. From the above conditions we can deduce

$$\langle f, \alpha_1 g_1 + \alpha_2 g_2\rangle = \overline{\alpha_1}\langle f, g_1\rangle + \overline{\alpha_2}\langle f, g_2\rangle,$$

which is sometimes given as an alternative condition to the second one in the definition above. □

Once we have an inner product then a valid norm can be constructed or induced.

Definition 2.12 (Induced norm). *Given an inner product $\langle\cdot,\cdot\rangle$ in a vector space $\mathcal{H}$, the norm can be defined by*

$$\boxed{\|f\| = \langle f, f\rangle^{1/2},} \tag{2.9}$$

or more succinctly $\|\cdot\| = \langle\cdot,\cdot\rangle^{1/2}$. □

We have a small job to do to prove this definition actually defines a norm. That is, it satisfies the conditions in Definition 2.4, p. 26, as asserted. The gap is to establish the triangle inequality holds and this can be demonstrated from the famous and invaluable *Cauchy-Schwarz inequality*.[6]

Theorem 2.1 (Cauchy-Schwarz inequality). *For any two elements of an inner product space $f, g \in \mathcal{H}$, we have*

$$\boxed{|\langle f, g\rangle| \leq \|f\|\,\|g\|.} \tag{2.10}$$

□

---

[6] The integral form of the inequality was actually first obtained by Bunyakovsky (Bouniakowsky, 1859), 29 years before Schwarz, which is an inequity not an inequality. Then again, if people have unpronounceable names then they can hardly expect to get naming rights on key results.

Proof. We know that

$$\|f - \alpha g\|^2 \geq 0, \qquad \forall \alpha \in \mathbb{C}.$$

Expanding $\|f - \alpha g\|^2$ into

$$\|f - \alpha g\|^2 = \|f\|^2 + |\alpha|^2 \|g\|^2 - \overline{\alpha}\langle f, g \rangle - \alpha \langle g, f \rangle$$

and choosing $\alpha = \langle f, g \rangle \|g\|^{-2}$ results in

$$\|f\|^2 - |\langle f, g \rangle|^2 \|g\|^{-2} \geq 0,$$

from which we obtain Cauchy-Schwarz inequality.   □

Cauchy-Schwarz inequality has a nice geometric meaning: in $\langle f, g \rangle$ is the information about the "angle" between two vectors. For real-valued vectors

$$-1 \leq \cos \theta(f, g) \triangleq \frac{\langle f, g \rangle}{\|f\| \|g\|} \leq 1.$$

For complex-valued vectors and noting that $\mathfrak{Re}(\langle f, g \rangle) \leq |\langle f, g \rangle|$ (where $\mathfrak{Re}(\cdot)$ takes the real part of its argument), we have

$$-1 \leq \cos \theta(f, g) \triangleq \frac{\mathfrak{Re}(\langle f, g \rangle)}{\|f\| \|g\|} \leq 1. \tag{2.11}$$

The equalities above are achieved when vectors f and g are aligned in the same or opposite direction. This interpretation can guide us in inner product spaces.

Problem
_____

2.6. Using Theorem 2.1 prove triangular inequality for the norm induced by the inner product.

_____

2.4.2   Inner product spaces

Recall a normed space is a vector space with a norm. Following the same pattern we have:

Definition 2.13 (Inner product space). *A vector space $\mathcal{H}$ with an inner product is called an inner product space or a pre-Hilbert space, and may be denoted by $\{\mathcal{H}, \langle \cdot, \cdot \rangle\}$.*   □

Example 2.2. The space of complex numbers $\mathbb{C}$ is an inner product space, with $\langle \alpha, \beta \rangle \triangleq \alpha \overline{\beta}$.   □

Example 2.3. The space $\mathbb{C}^N$ of ordered N-tuples $a = (\alpha_1, \alpha_2, \ldots, \alpha_N)'$ and $b = (\beta_1, \beta_2, \ldots, \beta_N)'$ of complex numbers, with inner product

$$\langle a, b \rangle = \sum_{n=1}^{N} \alpha_n \overline{\beta_n},$$

is an inner product space.   □

2.4.3   When is a normed space an inner product space?

Now we learnt from Theorem 2.1 (Cauchy-Schwarz inequality) that an inner product and the norm (induced from that inner product) satisfy the Cauchy-Schwarz inequality (2.10). Another way of stating the same result is to say that an inner product space can be regarded as a special type of normed space. Which raises the question: can a normed space be an inner product space? That is, given a norm can we construct a suitable inner product? Now that does not mean we can cook up any plausible-looking inner product and use it with the norm. What it means is, if we come up with a candidate inner product then:

1: it needs to be a valid inner product according to Definition 2.11;

2: it needs to induce the same/given norm according to Definition 2.12, that is, satisfy (2.9).

Towards resolving this issue we give a critical property of the norm that appears necessarily for inner product spaces:

Theorem 2.2 (Parallelogram law). *For any two elements of an inner product space $f, g \in \mathcal{H}$, we have*

$$\|f + g\|^2 + \|f - g\|^2 = 2(\|f\|^2 + \|g\|^2),\qquad(2.12)$$

where $\|f\| = \langle f, f \rangle^{1/2}$. □

Proof. We expand the left and right-hand sides of (2.12) and compare. The two components of the left-hand side are:

$$\begin{aligned}
\|f \pm g\|^2 &= \langle f \pm g, f \pm g \rangle \\
&= \langle f, f \rangle \pm \langle f, g \rangle \pm \langle g, f \rangle + \langle g, g \rangle \\
&= \|f\|^2 \pm \langle f, g \rangle \pm \langle g, f \rangle + \|g\|^2.
\end{aligned}$$

Summing these two terms yields the right-hand side of (2.12). □

Parallelogram law is shown in Figure 2.3.

Now let us examine a normed space that cannot be reformed into an inner product space.[7]

Example 2.4 ($C[0,1]$ is not an inner product space). Consider the normed space of complex-valued continuous functions, $C[0,1]$, defined on the closed interval $[0,1]$ with norm given by

$$\|f\| = \max_{x \in [0,1]} |f(x)|.$$

Define two functions on $x \in [0,1]$:

$$f(x) = 1, \quad \text{and} \quad g(x) = x,$$

---

[7] No surprise there. Otherwise we would not need to distinguish between normed and inner product spaces.

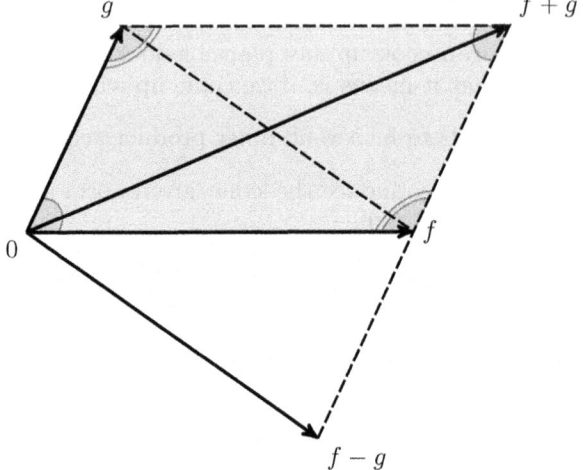

(a) Abstract form: $\|f + g\|^2 + \|f - g\|^2 = 2(\|f\|^2 + \|g\|^2)$ where f and g belong to a normed space.

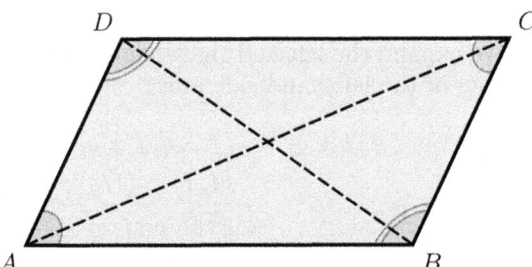

(b) Euclidean form: $(AC)^2 + (BD)^2 = 2((AB)^2 + (BC)^2)$ where AB, BC, AC and BD are the lengths in a conventional Euclidean sense.

Figure 2.3 Two versions of the parallelogram law. **(a)** Parallelogram law in abstract form $\|f+g\|^2 + \|f-g\|^2 = 2(\|f\|^2 + \|g\|^2)$. **(b)** Parallelogram law corresponding to the Euclidean plane statement from geometry "sum of the squares of the lengths of the four sides of a parallelogram equals the sum of the squares of the lengths of the two diagonals."

where evidently $f, g \in C[0, 1]$ (they are continuous functions) and $\|f\| = \|g\| = 1$. Now

$$\|f + g\|^2 + \|f - g\|^2 = 2^2 + 1^2 = 5$$

and

$$2(\|f\|^2 + \|g\|^2) = 2(1^2 + 1^2) = 4.$$

Thus the parallelogram law (2.12) is violated. We conclude that $C[0, 1]$ is not an inner product space because it cannot be furnished with a consistent definition of an inner product.   □

So it appears that the parallelogram law can be used to show the existence of normed spaces that cannot be augmented to become inner product spaces. But, somewhat surprisingly, the parallelogram law characterizes the situation completely as the following result asserts:

Theorem 2.3 (Normed to inner product space). *Given a complex normed space $\mathcal{H}$ where the norm satisfies the parallelogram law*

$$\|f + g\|^2 + \|f - g\|^2 = 2(\|f\|^2 + \|g\|^2),$$

for all $f, g \in \mathcal{H}$, then $\mathcal{H}$ can be made a complex inner product space with the inner product given by

$$\langle f, g \rangle = \frac{1}{4}\left(\|f + g\|^2 - \|f - g\|^2 + i\|f + ig\|^2 - i\|f - ig\|^2\right). \qquad (2.13)$$

This condition is known as the polarization identity.   □

Our terminology, which is not standard, is that in an inner product space the natural induced norm is a "geometric norm." With such a special norm we can make sense of orthogonality. Following this line of reasoning through to its logical conclusion, we should be able to manufacture an inner product and hence an inner product space from a norm and normed space whenever we have the additional parallelogram law (2.12).

Remark 2.8. In the case where the scalars are the reals, that is, we have a real normed space $\mathcal{H}$, then

$$\langle f, g \rangle = \frac{1}{4}\left(\|f + g\|^2 - \|f - g\|^2\right). \qquad (2.14)$$

That is, the inner product can be expressed in terms of the norm and $\mathcal{H}$ can be made a real inner product space.   □

2.7. Consider the space of real N-vectors $\mathbb{R}^N$ furnished with norm

$$\|a\|^2 \triangleq \sum_{n=1}^{N} \alpha_n^2.$$

Verify that (2.14) yields the standard definition of the inner product, that is,

$$\langle a, b \rangle \triangleq \frac{1}{4}\left(\|a + b\|^2 - \|a - b\|^2\right) = \sum_{n=1}^{N} \alpha_n \beta_n,$$

where $a = (\alpha_1, \alpha_2, \ldots, \alpha_N)' \in \mathbb{R}^N$ and $b = (\beta_1, \beta_2, \ldots, \beta_N)' \in \mathbb{R}^N$.

2.8. Consider the space of complex N-vectors, $\mathbb{C}^N$, with norm

$$\|a\|^2 \triangleq \sum_{n=1}^{N} |\alpha_n|^2.$$

Verify that (2.13) yields the standard definition of the inner product, that is,

$$\langle a, b \rangle \triangleq \frac{1}{4}\left(\|a + b\|^2 - \|a - b\|^2 + i\|a + ib\|^2 - i\|a - ib\|^2\right) = \sum_{n=1}^{N} \alpha_n \overline{\beta_n}.$$

where $a = (\alpha_1, \alpha_2, \ldots, \alpha_N)' \in \mathbb{C}^N$ and $b = (\beta_1, \beta_2, \ldots, \beta_N)' \in \mathbb{C}^N$.

2.9. Consider a normed space of functions defined on the closed interval $[0, 1]$ with norm given by

$$\|f\| \triangleq \int_0^1 |f(x)| \, dx, \tag{2.15}$$

such that $\|f\| < \infty$. This space includes the two functions

$$f(x) = 1, \quad \text{which satisfies} \quad \|f\| = \int_0^1 1 \, dx = 1,$$

$$g(x) = x, \quad \text{which satisfies} \quad \|g\| = \int_0^1 |x| \, dx = 1/2,$$

and similarly, $\|f + g\| = 3/2$ and $\|f - g\| = 1/2$. Verify that the parallelogram law (2.12) is satisfied. Does there exist an inner product for this space whose induced norm is given by (2.15)? Discuss.

2.10. Consider the continuous functions $h(x) = 1 - x$ and $g(x) = x$ defined on the closed interval $[0, 1]$ and the following three norms:

$$\|f\|_1 \triangleq \int_0^1 |f(x)| \, dx,$$

$$\|f\|_2 \triangleq \left(\int_0^1 |f(x)|^2 dx\right)^{1/2},$$

$$\|f\|_{\max} \triangleq \max_{x \in [0,1]} |f(x)|.$$

Confirm the parallelogram law (2.12) for these continuous functions with $\|\cdot\|_2$, but not the other two norms.

2.4.4   Orthonormal sets and sequences

The following definitions have familiar counterparts in Euclidean geometry and will be used later when we discuss orthogonal subspaces and complete orthonormal sequences or bases. In the following definitions $\mathcal{H}$ denotes an inner product space.

Definition 2.14 (Orthogonal vectors). *Two vectors $f, g \in \mathcal{H}$ are called orthogonal if*

$$\langle f, g \rangle = 0.$$

Orthogonal vectors are denoted by $f \perp g$.   □

Definition 2.15 (Orthonormal vectors). *Two orthogonal vectors $f, g \in \mathcal{H}$ that have the additional property*

$$\|f\| = \|g\| = 1$$

are called orthonormal vectors.   □

Definition 2.16 (Orthonormal set). *A subset $\mathcal{O}$ of non-zero vectors in $\mathcal{H}$ is called an orthonormal set if*

$$\langle f, g \rangle = \begin{cases} 1 & \text{if } f = g, \\ 0 & \text{otherwise}, \end{cases}$$

for all $f, g \in \mathcal{O}$.   □

In this definition, $\mathcal{O}$ can be finite, countably infinite or uncountably infinite. When it is countably infinite, then $\mathcal{O}$ can be arranged in a sequence and we refine the definition as follows using the Kronecker delta function $\delta_{n,m}$ (defined as $\delta_{n,m} = 1$ when $m = n$ and zero otherwise).

Definition 2.17 (Orthonormal sequence). *A sequence of non-zero vectors*

$$\{\varphi_n\}_{n=1}^{\infty}$$

in $\mathcal{H}$ is called an orthonormal sequence if

$$\langle \varphi_m, \varphi_n \rangle = \delta_{n,m},$$

for all $m, n \in \mathbb{N}$.   □

We attempt to avoid unnecessary duplication in having statements for orthogonal vectors in addition to orthonormal vectors, because we can make any two orthogonal vectors orthonormal by scaling them with the reciprocal of their norm.

Although orthogonality is a stronger form of independence and hence might sound too restricting (see Problem 2.11), it greatly simplifies computation. This property is common between finite and infinite-dimensional spaces. In principle,

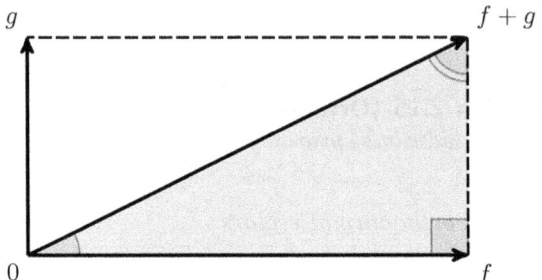

(a) Abstract form: $\|f + g\|^2 = \|f\|^2 + \|g\|^2$ where f and g are orthogonal and belong to a normed space.

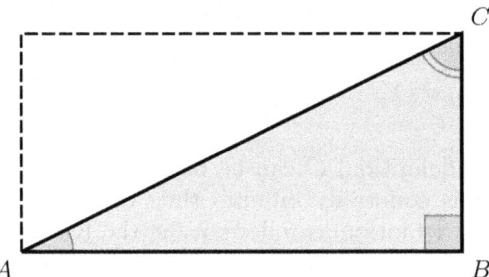

(b) Euclidean form: $(AC)^2 = (AB)^2 + (BC)^2$ where AB, BC and AC are the lengths in a conventional Euclidean sense.

Figure 2.4 Two versions of the Pythagorean theorem. **(a)** Pythagorean theorem in abstract form $\|f + g\|^2 = \|f\|^2 + \|g\|^2$. **(b)** Pythagorean theorem corresponding to the Euclidean plane statement from geometry "in any right triangle, the square of the hypotenuse is equal to the sum of the squares of the remaining two sides."

one could work with independent vectors, but the simplicity of the resulting theory makes it compelling to develop a theory in the stronger form of independence using orthogonality (when an inner product is available).

Pythagorean theorem is a familiar concept in Euclidean geometry, which also applies to general inner product spaces and will play a key role in the Fourier series expansion later on.

Theorem 2.4 (Pythagorean theorem). *The following property holds for the norm of any two orthogonal vectors f and g, in an inner product space $\mathcal{H}$:*

$$\|f + g\|^2 = \|f\|^2 + \|g\|^2, \quad \textit{whenever} \quad f \perp g.$$

$\square$

Proof. When $f \perp g$ then $\|f + g\| = \|f - g\|$ and the result follows from the parallelogram law, Theorem 2.2, p. 39. $\square$

> **Remark 2.9.** The Pythagorean theorem can be extended to more than two orthogonal vectors. $\square$

In Figure 2.4 we have illustrated the abstract form of the Pythagorean theorem involving vectors in an inner product space and compare it to the standard Euclidean form.

Problems ―――

2.11. This exercise shows that orthogonality is a stronger condition than linear independence, which was defined in Definition 2.3, p. 25. Prove that a family $\mathcal{O}$ of non-zero orthogonal vectors is linearly independent.

2.12. Show that any two orthonormal functions are distance $\sqrt{2}$ apart. That is, whenever

$$f \perp g \quad \text{and} \quad \|f\| = \|g\| = 1,$$

one has

$$\|f - g\|^2 = 2.$$

Take-home messages – spaces

M1: There is a hierarchy: vector spaces, normed spaces and inner product spaces. The first defines algebraic properties, the second augments metric or distance properties, and the third further augments geometric properties, such as orthogonality.

M2: Completeness is tied to the norm. The critical issue is whether a sequence of vectors in a space converges (in norm) to a vector within the original space.

M3: Dense set means there is at least one member of the set in the neighborhood of any point in the normed space. This means members of the dense set can be used as effective approximations to the every member of the normed space. A dense subset acts like a backbone for the normed space. Examples:

- The set of rational numbers is dense in the set of real numbers.
- Functions in $L^2(\Omega)$ can always be arbitrarily well approximated by continuous functions to any desired accuracy when measured in the $L^2(\Omega)$-norm.

M4: A normed space is separable if it contains a countable dense set.

M5: It is necessary and sufficient that a norm satisfy an additional property called the parallelogram law to be an inner product space. This means, in this case, it is possible to define an inner product using the norm alone and thereby define an inner product space from the normed space.

2.4.5   The space ℓ^2

The space ℓ^2 of all sequences $a = \{\alpha_1, \alpha_2, \alpha_3, \ldots\}$ of complex numbers, also denoted by $a = \{\alpha_k\}_{k=1}^{\infty}$, such that they are square summable,

$$\sum_{k=1}^{\infty} |\alpha_k|^2 < \infty,$$

with the inner product between $a = \{\alpha_k\}_{k=1}^{\infty}$ and $b = \{\beta_k\}_{k=1}^{\infty}$ defined by

$$\langle a, b \rangle = \sum_{k=1}^{\infty} \alpha_k \overline{\beta_k}$$

is an inner product space.

The space ℓ^2 is of central importance when dealing with separable Hilbert spaces. This will be revealed later in Section 2.12 and after we have introduced complete orthonormal sequences. But for the moment, it suffices to introduce a countably infinite orthonormal sequence within ℓ^2 with elements $e_1 = \{1, 0, 0, \ldots\}$, $e_2 = \{0, 1, 0, \ldots\}$ and so on, or more succinctly

$$e_k \triangleq \{\delta_{k,n}\}_{n=1}^{\infty},$$

where $\delta_{k,n}$ is the Kronecker delta function. Each element a in ℓ^2 written as $a = \{\alpha_1, \alpha_2, \alpha_3, \ldots\}$ has countably many components and can be expressed as a weighted sum of orthonormal sequences

$$a = \sum_{k=1}^{\infty} \alpha_k e_k.$$

We say that ℓ^2 is an *infinite-dimensional* inner product space. We will come back to this point at a later time.

2.4.6   The space $L^2(\Omega)$

Consider the set of square integrable functions defined on Ω. Unless otherwise stated, Ω refers to the closed interval on the real interval $[\alpha, \beta]$. The limits α and β can be finite, but $\alpha = -\infty$ or $\beta = +\infty$ or both are also allowed, in which case the appropriate open interval is meant. A functions $f(x) \in L^2(\Omega)$ satisfies

$$\|f\| = \left(\int_{\Omega} |f(x)|^2 dx \right)^{1/2} < \infty, \tag{2.16}$$

where this and any other integral is understood in the sense of Lebesgue. This equation serves to define both the definition of the norm and the condition for functions to be part of the space. Then equipping this normed space with the inner product

$$\boxed{\langle f, g \rangle \triangleq \int_{\Omega} f(x) \overline{g(x)} \, dx} \tag{2.17}$$

induces the norm (2.16) through $\|f\| = \langle f, f \rangle^{1/2}$. This inner product space is called $L^2(\Omega)$.

When are functions equivalent?

We saw before in the definition of the norm, Definition 2.4, p. 26, that

$$\|f\| = 0 \iff f = o.$$

The role of this condition in Definition 2.4, p. 26, can be viewed as a property that *separates* the vectors in our space, so that

$$\|f - g\| = 0 \implies f - g = o \quad \text{or} \quad f = g \qquad (2.18)$$

(the last condition follows from Definition 2.1, p. 24). In words, if the norm of the difference of two vectors is zero then the vectors are the same. This is plainly most sensible and well informed by our intuition. But for $L^2(\Omega)$ this causes some grief. As we explored in Section 1.2.5, p. 20, if two functions differed, for example, "only" at a countable number of points then this did not affect integration. Then the question is: what does all this mean given the norm in (2.16) is defined in terms of integration? There is a problem looming here.

Consider two functions on domain $\Omega = [0, 2\pi]$, (that is, $\alpha = 0$ and $\beta = 2\pi$), as follows

$$f(x) = \sin x, \quad x \in [0, 2\pi] \qquad (2.19)$$

and

$$g(x) = \begin{cases} 2 & \text{for } x \in \mathbb{Q} \cap [0, 2\pi], \\ \sin x & \text{otherwise}, \end{cases} \qquad (2.20)$$

where $\mathbb{Q} \cap [0, 2\pi]$ is the countable set of rational numbers restricted to the interval $[0, 2\pi]$ (of course we could have written $\mathbb{Q} \cap [0, 2\pi)$ given 2π is not rational). Using the norm in (2.16) one finds $\|f\| = \sqrt{\pi}$ and, by the arguments in Section 1.2.5, p. 20, $\|g\| = \sqrt{\pi}$ and $\|f - g\| = 0$. Based on (2.18), we can infer

$$\text{``} f = g, \text{''} \qquad (2.21)$$

which is evidently not the case comparing (2.19) with (2.20). So what is going wrong? One solution is to do nothing and live comfortably in ignorance[8] and might be preferable to the revelations that come. In comparison with (2.21) we can infer

$$\|f - g\| = 0 \not\implies f(x) = g(x), \quad \forall x \in [0, 2\pi]. \qquad (2.22)$$

We can say that the norm in (2.16) cannot resolve down to a pointwise comparison of functions.[9] Of course, pointwise equality is a very strong type of equality, so that indeed in the opposite direction we are confident that, for some $h(x)$ and $d(x)$,

$$h(x) = d(x), \forall x \in [0, 2\pi] \implies \|h - d\| = 0.$$

At a semantic level, we might interpret $f - g = o$ and $f = g$ as *not* meaning they are the "same" or "identical" or "pointwise equal," but "indistinguishable with respect to the norm." We can live with that, particularly since in this

---

[8] This forms a great life strategy in general.
[9] If you do insist on pointwise equality of functions, then that is another space altogether and not $L^2(\Omega)$.

case f looks sane and familiar and g is somewhat crazy with a "rational beard." Vincent van Gogh (1853–1890) had a beard and we know what he was like. If g had a shave, it would look like f (mind you, handing Vincent a razor was not a smart move). So the pattern is that hairy and/or crazy functions are essentially equivalent to a shaved smoother function.

Almost everywhere equivalence

Going a bit deeper, pretty clearly there are an unlimited number of functions like g that are indistinguishable from f (with respect to the norm). So we could define an *equivalence relation* based on the norm of the difference of functions and this is explored in Problem 2.14 below. In this way we have an equivalence class of functions and f might be regarded as the sanest or best-groomed representative of this class (which is just a set). So f and g can be bundled together with all other functions h that satisfy $\|f - h\| = 0$ and these resulting equivalence classes are distinct and partition $L^2(\Omega)$. They can be regarded as the true elements of $L^2(\Omega)$ and not functions per se. So we can claim that f above is not strictly an element, that is vector, of the space $L^2(\Omega)$. We do not need to get hung-up on this point. All is well, we live in a society with many crazy people. So too can we live with $L^2(\Omega)$, which has some (if not a majority) of crazy functions, albeit housed within the asylum of equivalence classes of functions.

In summary, equality of functions, shown in (2.21), really means

$$f = g \iff \|f - g\| = 0. \tag{2.23}$$

That is, it is equality (equivalence) under the norm and not pointwise equivalence.[10] So we need to be aware that the equality symbol, $=$, when used on the left-hand side of (2.23) is not the conventional naive, that is, pointwise equality.

Clearly it is annoying to know how to refer to the elements of $L^2(\Omega)$ without getting too pedantic and emulating a mathematician. For f and g satisfying (2.23) it is possible to say that f and g "differ only on a set of measure zero." The rationals are an example of a set which is countable and their area under integration can be safely ignored (or shaved off). A better terminology is to say f equals g "almost everywhere" with the abbreviation "a.e." So a clearer version of (2.21) is

$$\boxed{f = g \text{ a.e.} \iff \|f - g\| = 0.} \tag{2.24}$$

We can also say

$$g(x) = \sin x \text{ a.e.}$$

on the domain $x \in [0, 2\pi]$. In this way the equality symbol, "$=$," used on the left-hand side of (2.24) is the conventional naive (pointwise) equality.

---

[10] Unless the norm is the sup-norm, in which case we will be dealing with a different space and not $L^2(\Omega)$.

Problems

2.13 (Zero function). Discuss the accuracy of the statement: In $L^2(0,1)$ the zero vector o is the zero function. By the zero function we mean $o(x) = 0$ for all $x \in [0,1]$. [Hint: look at Figure 1.5, p. 21.]

2.14 (Equivalence relation). Let f, g and h be functions and the norm be given by (2.16), then define a relation $\sim$ as follows

$$f \sim g \iff \|f - g\| = 0.$$

Show that $\sim$ is an equivalence relation, that is, it satisfies:

$$f \sim f \qquad \qquad \text{(Reflexivity)},$$

$$\text{if } f \sim g \quad \text{then} \quad g \sim f \qquad \qquad \text{(Symmetry)},$$

$$\text{if } f \sim g \text{ and } g \sim h \quad \text{then} \quad f \sim h \qquad \text{(Transitivity)}.$$

2.15. Consider the domain $[0,1]$. We have seen that the rationals constitute a countable set and have used it to define points in the domain of a function, call these "exception points," where the function can take arbitrary finite values and not affect the norm and integration.

Are the rationals the only such choice? Is it possible to have a countable set of exception points in the domain $[0,1]$, none of which are rational?

2.4.7　Inner product and orthogonality with weighting in $L^2(\Omega)$

Sometimes it is useful to work with a slightly different inner product in $L^2(\Omega)$, than that given in (2.17), by incorporating a weight function into the definition of the inner product. The definition below is a generalization of the inner product in $L^2(\Omega)$[11]

$$\langle f, g \rangle_w = \int_\Omega w(x) f(x) \overline{g(x)} \, dx, \tag{2.25}$$

where $w(x) \geq 0$ is a real, non-negative weighting function and reduces to the normal (unweighted) inner product when $w(x) = 1$. In addition, $w(x)$ should be such that the inner product implicit in (2.25) satisfies all the conditions of an inner product.

Two functions φ_m and φ_n in $L^2(\Omega)$ are said to be orthonormal with weight $w(x)$ if

$$\langle \varphi_m, \varphi_n \rangle_w = \int_\Omega w(x) \varphi_m(x) \overline{\varphi_n(x)} \, dx = \delta_{n,m}, \quad \forall m, n \in \mathbb{N}. \tag{2.26}$$

---

[11] Any constants, whose value can even depend on f or g, can be absorbed into the function definition. We will soon see examples of this where index-dependent normalization is applied to a sequence of orthogonal functions to make them orthonormal. So we do not explicitly need a weighted inner product in this case. In the definition, we assume that the weight $w(x)$ is a given fixed function of x, independent of the inputs to the inner product, $f(x)$ and $g(x)$.

Riesz, Frigyes (1880–1956) — Riesz was a Hungarian mathematician who made a number of foundational contributions to the development of functional analysis somewhat picking up the disparate pieces others had developed (mostly Lebesgue, Fréchet, Hilbert and Schmidt) and bringing the work together in a compelling way. Much of what we now regard as a standard rigorous treatment of Hilbert space is due to Riesz. Further he developed the extensions to normed spaces and sharpened the ideas on weak convergence, somewhat anticipating the work of Banach. He wrote with his student the influential book (Riesz and Sz.-Nagy, 1990), which reflects his mastery of the subject, and many key results presented therein are drawn from his own research.

There is an account of Riesz's unusual (but largely sensible) means of giving his lectures, where a junior academic read from his handbook and another assistant wrote equations on the board whilst Riesz stood by nodding occasionally.

2.4.8   Complete inner product spaces — Hilbert spaces

Definition 2.18 (Hilbert space). *A complete inner product space is called a Hilbert space.*   □

Example 2.5. Since $\mathbb{C}$ is complete, it is a Hilbert space.   □

Example 2.6. A finite-dimensional inner product space is a Hilbert space. This is why we have said that in the finite case the Hilbert space concept is overkill as completeness is automatic.   □

Example 2.7. $\mathbb{C}^N$ is a Hilbert space.   □

Example 2.8. Spaces ℓ^2 and $L^2(\Omega)$ defined in Section 2.4.5 and Section 2.4.6 are both Hilbert spaces. Completeness is not obvious, but can be established (Helmberg, 1969, pp. 23–24, 29–30). It is much easier proving lack of completeness than proving completeness.   □

Example 2.9. Let $\mathcal{H}_F$ be the space of sequences $a = \{\alpha_1, \alpha_2, \alpha_3, \ldots\}$ of complex numbers such that only a finite number of terms is non-zero. We will call this the space of finite support sequences, with inner product

$$\langle a, b \rangle = \sum_{k=1}^{\infty} \alpha_k \overline{\beta_k}.$$

$\mathcal{H}_F$ is an inner product space, but is not a Hilbert space since it is not complete. Lack of completeness is established by construction of a sequence of points in the space which is Cauchy, but whose limit lies outside the space. By "point in the space" we mean a finite support sequence, so a sequence of points in the space is really a sequence of (finite support) sequences. Towards establishing a contradiction, define the n-th finite support sequence as

$$a_n \triangleq \left\{ 1, \frac{1}{2}, \frac{1}{3}, \ldots, \frac{1}{n}, 0, 0, \ldots \right\}. \tag{2.27}$$

Then $\{a_1, a_2, a_3, \ldots\}$ is a Cauchy sequence since

$$\lim_{n,m \to \infty} \|a_n - a_m\| = \lim_{n,m \to \infty} \left(\sum_{k=m+1}^{n} \frac{1}{k^2} \right)^{1/2} = 0, \quad \text{for} \quad m < n. \tag{2.28}$$

However, the sequence does not converge in $\mathcal{H}_F$, because its limit is not of finite support and, therefore, not in $\mathcal{H}_F$. Note that the above sequence converges in ℓ^2 since the limit a_n as $n \to \infty$ is square summable. The Hilbert space ℓ^2 can be regarded as the *completion* of $\mathcal{H}_F$.   □

2.5   Orthonormal polynomials and functions

Orthogonal (orthonormal) polynomials are polynomials, that is, functions with terminating or finite series expansions (Lebedev, 1972) that are orthogonal (orthonormal) to each other. The term orthogonal function is used for orthogonal (orthonormal) but non-polynomial functions.

Table 2.1 Legendre polynomials $P_\ell(x)$ for $\ell \in \{0, 1, \dots, 11\}$

Degree ℓ	Legendre polynomial $P_\ell(x)$
0	1
1	x
2	$\frac{1}{2}(3x^2 - 1)$
3	$\frac{1}{2}(5x^3 - 3x)$
4	$\frac{1}{8}(35x^4 - 30x^2 + 3)$
5	$\frac{1}{8}(63x^5 - 70x^3 + 15x)$
6	$\frac{1}{16}(231x^6 - 315x^4 + 105x^2 - 5)$
7	$\frac{1}{16}(429x^7 - 693x^5 + 315x^3 - 35x)$
8	$\frac{1}{128}(6435x^8 - 12012x^6 + 6930x^4 - 1260x^2 + 35)$
9	$\frac{1}{128}(12155x^9 - 25740x^7 + 18018x^5 - 4620x^3 + 315x)$
10	$\frac{1}{256}(46189x^{10} - 109395x^8 + 90090x^6 - 30030x^4 + 3465x^2 - 63)$
11	$\frac{1}{256}(88179x^{11} - 230945x^9 + 218790x^7 - 90090x^5 + 15015x^3 - 693x)$

Here we are trying to give an overview of some major classes of countably infinite sets of orthogonal polynomials. The intention is to highlight the common structure in elements and motivate, by raising the ennui, a better formulation. This better formulation will provide a compelling representation for vectors in a Hilbert space, analogous to an orthogonal vector basis in a finite-dimensional space. We will see later that the explicit functions that constitute orthonormal sets are not actually responsible for the various properties normally or classically attributed to them such as the Parseval relation and various convergence results. Rather, as we shall see later, these properties are a direct consequence of the axiomatic foundations that can be provided to orthonormal sets using Hilbert space theory via an economical number of concepts.

In any case, it is useful to know these orthogonal polynomials as they crop up in a plethora of applications.

2.5.1   Legendre polynomials

Legendre polynomials are denoted $P_\ell(x)$ for non-negative integer degree $\ell \in \{0, 1, 2, \dots\}$, or more succinctly $\ell \in \mathbb{Z}^\star$, and the first twelve are explicitly given in Table 2.1 (Lebedev, 1972; Riley et al., 2006) and further coefficients have been tabulated in (Sloane, 2002).

These polynomials are shown in Figure 2.5 for $\ell \in \{0, 1, \dots, 11\}$. The non-negative index ℓ is called the degree and corresponds to the degree of the polynomial, e.g., $P_3(x)$ is a cubic polynomial in x. Although well defined for all x, in applications their behavior on the finite interval $-1 \le x \le 1$ is of greatest interest.

There are a number of ways to define the Legendre polynomials, but the most popular way is through the Rodrigues formula (Lebedev, 1972; Riley et al., 2006),

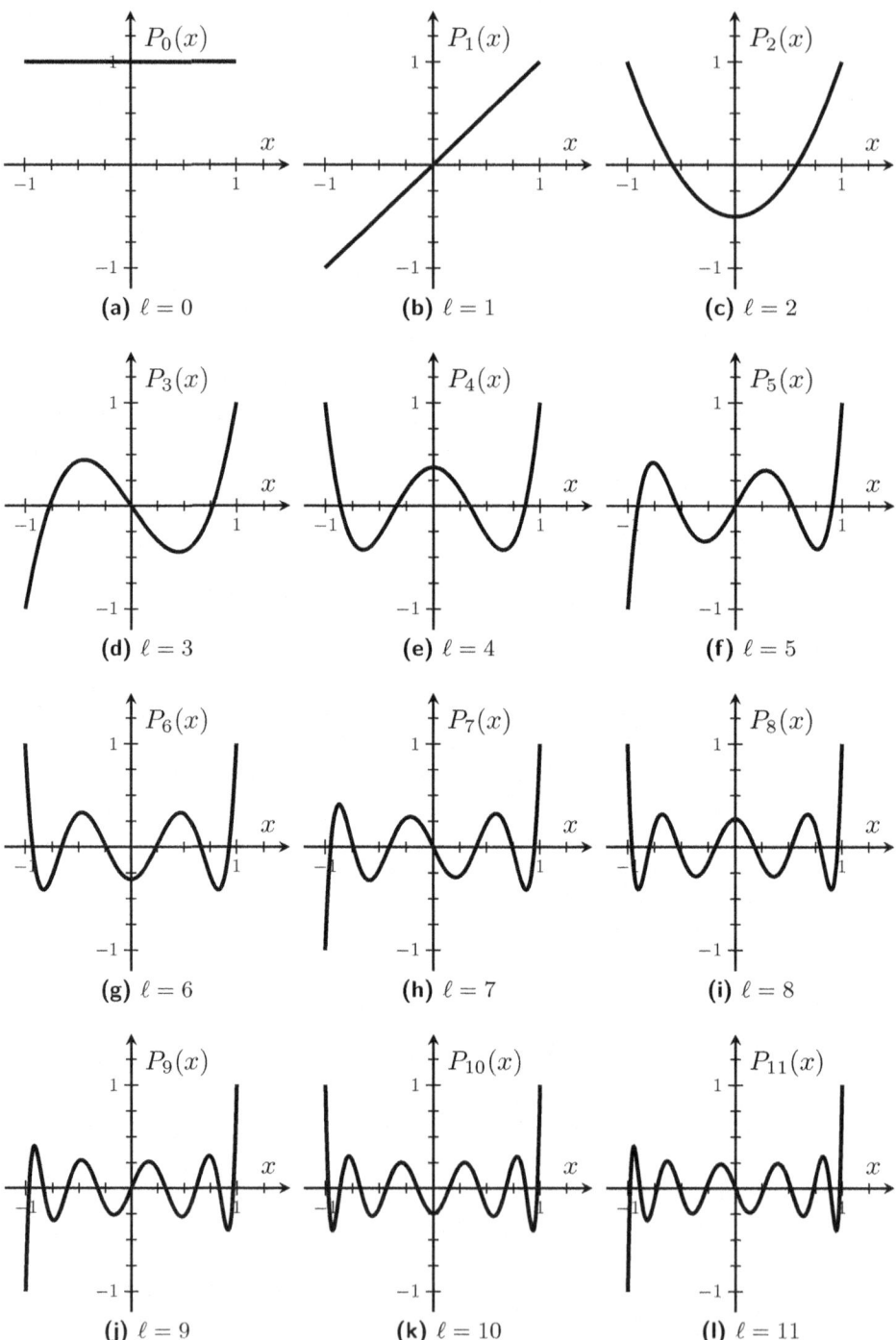

Figure 2.5 Legendre polynomials $P_\ell(x)$ for degree $\ell = 0, 1, 2, \ldots, 11$ as given in (2.29).

given by

$$P_\ell(x) = \frac{1}{2^\ell \ell!} \frac{d^\ell}{dx^\ell} (x^2 - 1)^\ell, \quad \ell \in \{0, 1, 2, \ldots\}, \tag{2.29}$$

but they can also be defined through the solutions of a certain differential equation.

In $L^2(-1, 1)$, these polynomials satisfy the orthogonality relation

$$\langle P_\ell, P_p \rangle = \int_{-1}^{+1} P_\ell(x) P_p(x) \, dx = 0, \quad \ell \neq p. \tag{2.30}$$

That is, they are orthogonal on the interval $[-1, +1]$ with unity weight $w(x) = 1$.

When $\ell = m$ we have

$$\int_{-1}^{+1} P_\ell^2(x) \, dx = \frac{2}{2\ell + 1}, \quad \ell \in \{0, 1, 2, \ldots\}, \tag{2.31}$$

which implies they are, by convention, unnormalized. The normalized form is

$$\zeta_\ell(x) \triangleq \sqrt{\ell + \tfrac{1}{2}} P_\ell(x), \quad \ell \in \{0, 1, 2, \ldots\}, \tag{2.32}$$

which are orthonormal on $[-1, +1]$ (with unity weight), that is,

$$\langle \zeta_\ell, \zeta_p \rangle = \int_{-1}^{+1} \zeta_\ell(x) \zeta_p(x) \, dx = \delta_{\ell,p}, \quad \ell, p \in \{0, 1, 2, \ldots\},$$

where $\delta_{\ell,p}$ is the Kronecker delta.

We can represent a sufficiently regular real function $f(x)$, defined on the interval $[-1, +1]$, in a series of Legendre polynomials (Lebedev, 1972)

$$f(x) = \sum_{\ell=0}^{\infty} \alpha_\ell P_\ell(x), \quad -1 \leq x \leq 1. \tag{2.33}$$

Directly from orthogonality (2.30) and (2.31) we see that the coefficients are real and given by

$$\alpha_\ell = (\ell + \tfrac{1}{2}) \int_{-1}^{+1} f(x) P_\ell(x) \, dx, \quad \ell \in \{0, 1, 2, \ldots\}. \tag{2.34}$$

To make (2.33) and (2.34) more precise, one notion of sufficiently regular is captured by the following theorem from (Lebedev, 1972, p. 55):

Theorem 2.5. *If the real function $f(x)$ is piecewise smooth in $[-1, +1]$ and if the integral*

$$\int_{-1}^{+1} f^2(x) \, dx$$

is finite, then the series (2.33), *with coefficients α_ℓ* (2.34), *converges to $f(x)$ at every continuity point of $f(x)$. At points of discontinuity the series converges to*

$$\tfrac{1}{2}\big(f(x^+) + f(x^-)\big), \tag{2.35}$$

where $f(x^+)$ and $f(x^-)$ mean the limit from the right and left, respectively. $\square$

Table 2.2 Legendre polynomials $P_\ell(x)$ coefficients for $\ell \in \{0, 1, \ldots, 4\}$. The second row shows the leading fraction of the polynomial.

Degree ℓ	0	1	2	3	4
$\times$	1	1	$\frac{1}{2}$	$\frac{1}{2}$	$\frac{1}{8}$
1	1	0	-1	0	3
x	0	1	0	-3	0
x^2	0	0	3	0	-30
x^3	0	0	0	5	0
x^4	0	0	0	0	35

Proof. See (Lebedev, 1972, pp. 53–58). □

Remark 2.10. Indeed not only does (2.35) hold at points of discontinuity, but also at points of continuity since $f(x^+) = f(x^-) = f(x)$. □

Problems

2.16. In Table 2.1 apart from the leading fraction, the Legendre polynomial coefficients are integer-valued and show a particular pattern which can be generated through the recurrence relation

$$(\ell + 1)P_{\ell+1}(x) = (2\ell + 1)xP_\ell(x) - P_{\ell-1}(x),$$

with $P_0(x) = 1$ and $P_1(x) = x$ to start the recursion.

Tabulate the coefficients up to order 15 against the power in the polynomial in the format shown in Table 2.2.

2.17. Prove the orthogonality of the Legendre polynomials (2.30) and (2.31).

2.18. Verify the Legendre polynomial coefficient expression (2.34) using (2.30) and (2.31) or otherwise.

2.19. Consider the sequence of orthogonal polynomials given by

$$\{P_0(x), P_2(x), P_4(x), \ldots\}.$$

All of these polynomials satisfy (2.30) and (2.31). The coefficients are given by (2.34) for even indices, in other words $\alpha_{2\ell}$. Why cannot a general piecewise smooth function be represented in terms of a series using coefficients $\alpha_{2\ell}$? What type of piecewise smooth functions can be represented in this way?

Table 2.3 Hermite polynomials $H_n(x)$ for $n \in \{0, 1, 2, \ldots, 11\}$

Degree n	Hermite polynomial $H_n(x)$
0	1
1	$2x$
2	$4x^2 - 2$
3	$8x^3 - 12x$
4	$16x^4 - 48x^2 + 12$
5	$32x^5 - 160x^3 + 120x$
6	$64x^6 - 480x^4 + 720x^2 - 120$
7	$128x^7 - 1344x^5 + 3360x^3 - 1680x$
8	$256x^8 - 3584x^6 + 13440x^4 - 13440x^2 + 1680$
9	$512x^9 - 9216x^7 + 48384x^5 - 80640x^3 + 30240x$
10	$1024x^{10} - 23040x^8 + 161280x^6 - 403200x^4 + 302400x^2 - 30240$
11	$2048x^{11} - 56320x^9 + 506880x^7 - 1774080x^5 + 2217600x^3 - 665280x$

2.5.2   Hermite polynomials

Fortunately, the first name of Charles Hermite (1822–1901) was not Veg, lest he be confused with a popular Australian dark brown food paste made from yeast extract. The *Hermite polynomials* are defined on the real line $\mathbb{R} \equiv (-\infty, \infty)$ and are denoted $H_n(x)$ for $n \in \{0, 1, 2, \ldots\}$ (Courant and Hilbert, 1966; Lebedev, 1972; Riley et al., 2006).[12] The first twelve are explicitly given in Table 2.3 and further coefficients have been tabulated in (Sloane, 2001).

It can be shown that $H_n(x)$ is a polynomial of degree n containing terms of only even or only odd powers depending on whether n is even or odd. There are a number of ways to formally define them, such as

$$H_n(x) \triangleq (-1)^n e^{x^2} \frac{d^n}{dx^n} \left(e^{-x^2} \right), \quad n \in \{0, 1, 2, \ldots\},$$

or through the solutions of a certain physically significant differential equation (Lebedev, 1972).

These polynomials satisfy the orthogonality relation

$$\langle H_m, H_n \rangle_w = \int_{-\infty}^{\infty} e^{-x^2} H_m(x) H_n(x) \, dx = 0, \quad m \neq n. \qquad (2.36)$$

That is, they are orthogonal on the real line $\mathbb{R}$ with weight function

$$w(x) = e^{-x^2}.$$

Remark 2.11. From (2.36), one can deduce that the actual functions that belong to $L^2(\mathbb{R}) \triangleq L^2(-\infty, \infty)$ *without* weighting are of the (unnormalized)

---

[12] Given in the French pronunciation the "H" in Hermite is silent, then the proper pronunciation of $H_n(x)$ is in dispute.

form
$$e^{-x^2/2}H_n(x),$$

and hence not polynomials. Nevertheless, $e^{-x^2/2}$ does not depend on n, which warrants explicitly dealing with the simpler Hermite polynomials, following historical convention. $\quad\square$

Remark 2.12. The effect of the weighting $e^{-x^2/2}$ is to strongly attenuate the integrand in (2.36) for large $|x|$. This explains why the Hermite polynomials can grow like $O(x^n)$ as $|x| \to \infty$ and the integral stay finite and physically meaningful. As a consequence, one does not see the Hermite polynomials appear in isolation of the weighting in most practical and physical problems. For this reason we do not plot the Hermite polynomials, seen in Table 2.3, because they are relatively uninteresting. We do, however, plot the related *Hermite functions* which incorporate the weighting later in Problem 2.21. $\quad\square$

When $m = n$ we have
$$\|H_n\|_w^2 = \int_{-\infty}^{\infty} e^{-x^2} H_n^2(x)\,dx = 2^n n! \sqrt{\pi}, \quad n \in \{0,1,2,\ldots\}, \tag{2.37}$$

which implies they are, by convention, not normalized. The normalized form is
$$\zeta_n(x) \triangleq (2^n n! \sqrt{\pi})^{-1/2} H_n(x), \quad n \in \{0,1,2,\ldots\}, \tag{2.38}$$

which are orthonormal on $\mathbb{R}$ with weight function $w(x) = e^{-x^2}$, that is,
$$\langle \zeta_m, \zeta_n \rangle_w = \int_{-\infty}^{\infty} e^{-x^2} \zeta_m(x)\zeta_n(x)\,dx = \delta_{n,m}, \quad m,n \in \{0,1,2,\ldots\}. \tag{2.39}$$

We can expand a sufficiently regular real function $f(x)$, defined on the real line $\mathbb{R}$, in a series of Hermite polynomials
$$f(x) = \sum_{n=0}^{\infty} \alpha_n H_n(x), \quad x \in \mathbb{R}. \tag{2.40}$$

Directly from orthogonality (2.36) and (2.37) (see Problem 2.20) we have
$$\alpha_n = \frac{1}{2^n n! \sqrt{\pi}} \int_{-\infty}^{\infty} e^{-x^2} f(x) H_n(x)\,dx, \quad n \in \{0,1,2,\ldots\}. \tag{2.41}$$

One notion of sufficiently regular is captured by the following theorem from (Lebedev, 1972, p. 71):

Theorem 2.6. *If the real function $f(x)$ defined in the interval $(-\infty, \infty)$ is piecewise smooth in every finite interval $[-\beta, \beta]$, and if the integral*
$$\int_{-\infty}^{\infty} e^{-x^2} f^2(x)\,dx$$

is finite, then the series (2.40), with coefficients α_n (2.41), converges to $f(x)$ at every continuity point of $f(x)$. At points of discontinuity the series converges to
$$\frac{1}{2}\left(f(x^+) + f(x^-)\right),$$

where $f(x^+)$ and $f(x^-)$ mean the limit from the right and left, respectively. $\quad\square$

Proof. See (Lebedev, 1972, pp. 68–73).   □

> **Remark 2.13.** There are further classical classes of orthogonal polynomials such as Jacobi, Chebyshev and, more recently, Gegenbauer and Koshlyakov. The development for the Legendre and Hermite polynomials, given above, share similarities which can be extended in a repetitive way to the other classes of orthogonal polynomials. Later we will show how to unify this development and abstract away the details. Next we shall consider some important classes of orthogonal functions which are not polynomials.   □

Problems

2.20. Verify the Hermite polynomial coefficient expression (2.41) using (2.36) and (2.37) or otherwise.

2.21 (Hermite functions). Use the expressions for the Hermite polynomials to show that the set of functions, called the *Hermite functions*,

$$\varrho_n(x) \triangleq (2^n n! \sqrt{\pi})^{-1/2} e^{-x^2/2} H_n(x), \quad n \in \{0, 1, 2, \ldots\},$$

are orthonormal on $L^2(\mathbb{R})$ (that is, with weighting $w(x) = 1$). These Hermite functions are shown in Figure 2.6.

2.5.3   Complex exponential functions

The simplest examples of complex-valued orthonormal functions are the complex exponential functions (Courant and Hilbert, 1966, p. 50). In this case, we are interested in finite domain, symmetric about 0 and the functions are indexed by the integers $\mathbb{Z}$. We take them as normalized and given by

$$\varphi_n(x) \triangleq \frac{e^{inx}}{\sqrt{2\pi}}, \quad n \in \mathbb{Z}$$

and are defined on the interval $-\pi \leq x \leq \pi$. In $L^2(-\pi, \pi)$, these complex-valued functions satisfy the orthogonality relation

$$\langle \varphi_n, \varphi_m \rangle = \int_{-\pi}^{\pi} \frac{e^{imx}}{\sqrt{2\pi}} \frac{e^{-inx}}{\sqrt{2\pi}} \, dx = 0, \quad m \neq n. \tag{2.42}$$

That is, they are orthogonal on the interval $[-\pi, +\pi]$ with unity weight $w(x) = 1$. When $m = n$ we have

$$\|\varphi_n\|^2 = \frac{1}{2\pi} \int_{-\pi}^{\pi} e^{i(n-n)x} \, dx = 1, \quad n \in \mathbb{Z}, \tag{2.43}$$

which, with (2.42), shows they are orthonormal.

Note that the interval is symmetric about the origin. Hence, by normalizing the interval to $[-1, +1]$ we have an alternative set of orthonormal functions to

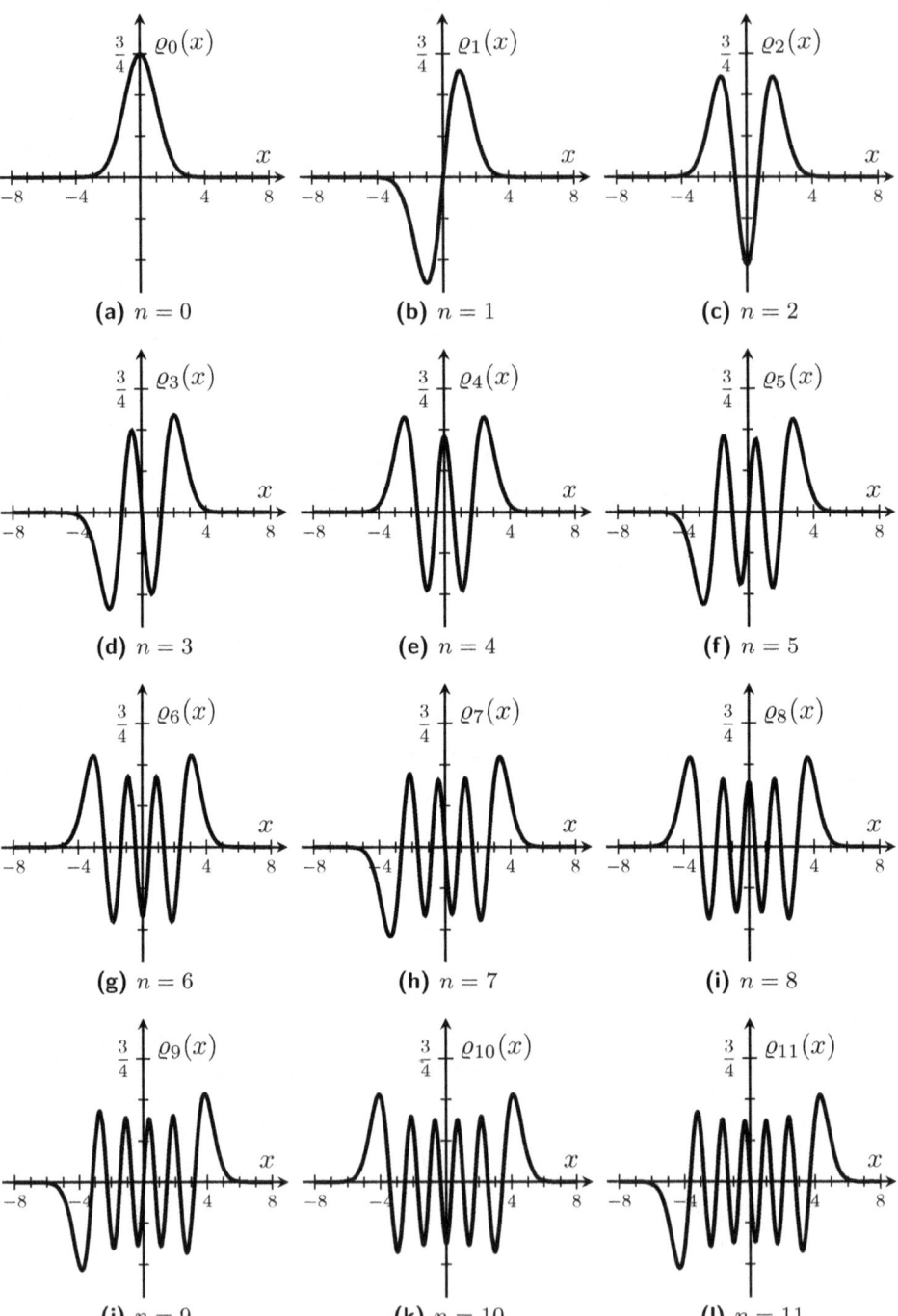

Figure 2.6 Hermite functions $\varrho_n(x) \triangleq (2^n n! \sqrt{\pi})^{-1/2} e^{-x^2/2} H_n(x)$, plotted for degree $n = 0, 1, 2, \ldots, 11$ on interval $-9 \leq x \leq 9$, which are normalized square root Gaussian weighted Hermite polynomials, $H_n(x)$.

Table 2.4 Associated Legendre functions $P_\ell^m(x)$ for low degrees $\ell \in \{0, 1, 2, 3, 4\}$ and orders $m \in \{-\ell, -\ell+1, \ldots, \ell\}$. For $m = 0$, $P_\ell^0 = P_\ell$ and Table 2.2 can also be used. For $m \in \{-\ell, -\ell+1 \ldots, -1\}$, (2.45) can be used to relate $P_\ell^{-m}(x)$ to $P_\ell^m(x)$.

Degree ℓ	Order m	$P_\ell^m(x)$	$P_\ell^{-m}(x)$
0	0	1	1
1	0	x	x
1	1	$-(1-x^2)^{1/2}$	$\frac{1}{2}(1-x^2)^{1/2}$
2	0	$\frac{1}{2}(3x^2-1)$	$\frac{1}{2}(3x^2-1)$
2	1	$-3x(1-x^2)^{1/2}$	$\frac{1}{2}x(1-x^2)^{1/2}$
2	2	$3(1-x^2)$	$\frac{1}{8}(1-x^2)$
3	0	$\frac{1}{2}(5x^3-3x)$	$\frac{1}{2}(5x^3-3x)$
3	1	$-\frac{3}{2}(5x^2-1)(1-x^2)^{1/2}$	$\frac{1}{8}(5x^2-1)(1-x^2)^{1/2}$
3	2	$15x(1-x^2)$	$\frac{1}{8}x(1-x^2)$
3	3	$-15(1-x^2)^{3/2}$	$\frac{1}{48}(1-x^2)^{3/2}$
4	0	$\frac{1}{8}(35x^4-30x^2+3)$	$\frac{1}{8}(35x^4-30x^2+3)$
4	1	$-\frac{5}{2}(7x^3-3x)(1-x^2)^{1/2}$	$\frac{1}{8}(7x^3-3x)(1-x^2)^{1/2}$
4	2	$\frac{15}{2}(7x^2-1)(1-x^2)$	$\frac{1}{48}(7x^2-1)(1-x^2)$
4	3	$-105x(1-x^2)^{3/2}$	$\frac{1}{48}x(1-x^2)^{3/2}$
4	4	$105(1-x^2)^2$	$\frac{1}{384}(1-x^2)^2$

that provided by the Legendre polynomials in, what we will see to be, essentially the same space. Furthermore, given the periodic nature of the complex exponential functions, there is orthogonality over any length 2π interval of the real line, so the interval $[-\pi, +\pi]$ is somewhat arbitrary and any contiguous 2π range can be used.

2.5.4 Associated Legendre functions

By generalizing the differential equation satisfied by the *Legendre polynomials*, $P_\ell(x)$ where ℓ is the degree, we arrive at a more general class of orthogonal functions called the *associated Legendre functions* defined on the interval $[-1, +1]$. As the name implies these need not be polynomials (but approximately half of them are, as we shall see below). The associated Legendre functions are denoted $P_\ell^m(x)$, where m is known as the order and ℓ is the degree. For $m = 0$ they reduce to the Legendre polynomials given in Section 2.5.1. The low-order associated Legendre functions, when $m > 0$, are shown in Table 2.4. For negative orders, shown in the rightmost column, they can be inferred from the positive orders shown in the second rightmost column (see below).

There are a number of ways to define the associated Legendre functions, through the solutions of a certain differential equation, or through a formula such as

$$P_\ell^m(x) = (-1)^m \left(1 - x^2\right)^{m/2} \frac{d^m}{dx^m} P_\ell(x)$$

$$= \frac{(-1)^m}{2^\ell \ell!} \left(x^2 - 1\right)^{m/2} \frac{d^{\ell+m}}{dx^{\ell+m}} \left(x^2 - 1\right)^\ell, \quad m \in \{0, 1, \dots, \ell\}. \tag{2.44}$$

And for negative order

$$P_\ell^{-m}(x) = (-1)^m \frac{(\ell - m)!}{(\ell + m)!} P_\ell^m(x), \quad m \in \{0, 1, \dots, \ell\}. \tag{2.45}$$

In $L^2(-1, 1)$, for any $m \in \{-\ell, -\ell + 1, \dots, \ell\}$, these functions satisfy

$$\langle P_\ell^m, P_n^m \rangle = \int_{-1}^{+1} P_\ell^m(x) P_n^m(x) \, dx = 0, \quad n \neq \ell. \tag{2.46}$$

That is, they are orthogonal on the interval $[-1, +1]$ with unity weight $w(x) = 1$. When $\ell = n$ we have

$$\|P_\ell^m\|^2 = \int_{-1}^{+1} \left(P_\ell^m(x)\right)^2 dx = \frac{2}{2\ell + 1} \frac{(\ell + m)!}{(\ell - m)!},$$

$$\ell \in \{m, m + 1, m + 2, \dots\}, \tag{2.47}$$

which implies they are, by convention, not normalized. The normalized form is

$$\varphi_\ell^m(x) \triangleq \sqrt{\frac{2\ell + 1}{2} \frac{(\ell - m)!}{(\ell + m)!}} P_\ell^m(x), \tag{2.48}$$

whereupon the functions $\{\varphi_\ell^m(x)\}_{\ell=m}^\infty$, for any fixed m, are orthonormal on $[-1, +1]$ with unity weight $w(x) = 1$. So what we have is not one set of orthogonal functions on $[-1, +1]$, but an infinite family parameterized by m, as suggested by our notation in (2.48). The orthonormal function index is ℓ, which satisfies $\ell = m, m + 1, m + 2, \dots$. Hence, we have a countable set of orthonormal function sets to choose from:[13]

$$\{\varphi_\ell^0(x)\}_{\ell=0}^\infty, \quad \{\varphi_\ell^1(x)\}_{\ell=1}^\infty, \quad \{\varphi_\ell^2(x)\}_{\ell=2}^\infty, \quad \{\varphi_\ell^3(x)\}_{\ell=3}^\infty, \quad \cdots.$$

Furthermore, whenever we have m as an even integer then these are sets of orthonormal *polynomials*, as can be inferred from (2.44) and corroborated by Table 2.4.[14]

Finally, we flag that the associated Legendre functions play a critical role in the definition of spherical harmonics given later.

---

[13] The fact that the starting index varies is inconsequential to the construction.

[14] That is why the use of "associated Legendre polynomials" instead of "associated Legendre functions" is not uncommon in the literature, but is not strictly accurate.

2.6   Subspaces

2.6.1   Preamble

Having created enough boredom in the last few pages, it is high time that we asked ourselves what is common between different orthonormal sets in $L^2(\Omega)$ and how they can be better formulated. The key is *complete orthonormal sequences*, also called *maximal orthonormal sequences* or *basis*. But at this stage, these sound like cocktail party definitions. Deeper understanding of the geometrical properties of Hilbert spaces is required, which will come to your rescue if there is someone in the party who starts asking you smart questions. Here is a flavor of the main topics that will be covered.

Our starting point is to study and understand "smaller" non-trivial Hilbert subspaces that are contained within an original Hilbert space. Non-trivial Hilbert subspaces combine topological and algebraic aspects of Hilbert spaces. There is a rather subtle difference between subspaces in familiar N-dimensional Euclidean spaces $\mathbb{C}^N$ and those in Hilbert spaces, which arises from the extra requirement of closedness. So in Hilbert spaces there can be a distinction between manifolds and closed manifolds or simply subspaces.

Among all non-trivial subspaces of a Hilbert space, orthogonal subspaces are particularly useful. Orthogonal subspaces provide us with a "vector sum spanning" property, meaning that any vector in their span (to be defined) can be written as a sum of unique vectors in the orthogonal subspaces. This is not true in general for non-orthogonal subspaces.

A special class of orthogonal subspaces are closely linked with separable Hilbert spaces, our favorite type of Hilbert space. Countable sets of orthogonal subspaces that can span the whole Hilbert space as a vector sum are special. In fact, there is an equivalence between the existence of such countable set of orthogonal subspaces and the separability of space. Each of these orthogonal subspaces is created by a single unit-norm "seed vector" that is orthogonal to the seed of all other subspaces. But this is not enough to make the class of orthogonal subspaces special. One should not be able to come up with even one extra non-zero seed vector to create a non-trivial orthogonal subspace to the original subspaces. Moreover, removing even a single seed and its corresponding subspace from the set would result in violating the ability to span the whole space. The collection of orthonormal seed vectors is called complete orthonormal sequences, maximal orthonormal sequences, or basis.[15] This special class of orthogonal subspaces and their corresponding complete orthonormal sequences equip us with the powerful Fourier series representations and Parseval relation.

We hope to have now created enough curiosity to motivate further reading.

2.6.2   Subsets, manifolds and subspaces

A subspace of a Hilbert space is an extension of the concept of subspace in Euclidean spaces $\mathbb{R}^N$, $\mathbb{C}^N$ or other familiar finite-dimensional vector spaces.

---

[15] Our preferred terminology is complete orthonormal sequences and uses the same adjective "complete" as in Hilbert spaces, which some authors believe leads to confusion. At an abstract level the use is consistent and according to the above discussion marries well with the common English usage of the word.

The natural expectation for a subset $\mathfrak{M} \subset \mathcal{H}$ to be a subspace is that every linear combination $\alpha f + \beta g$ of two elements f and g in $\mathfrak{M}$ stays in the same subset. But this is not enough. A subset $\mathfrak{M}$ of $\mathcal{H}$ satisfying only this condition is called a manifold. In $\mathbb{C}^N$, manifolds are automatically subspaces. In Hilbert spaces, every subspace is a manifold, but not vice versa.

For a manifold $\mathfrak{M}$ of a Hilbert space to be a subspace closedness of the subset is also required. A closed set contains all its *accumulation or boundary points*, the definition of which becomes more clear in Example 2.11. Closedness of a manifold guarantees that the subspace is actually a Hilbert space (is complete). This is because every Cauchy sequence in $\mathfrak{M} \subset \mathcal{H}$ converges to a vector in $\mathcal{H}$ which must also belong to $\mathfrak{M}$ (if $\mathfrak{M}$ is closed then it has to contain all its limit points).

> **Example 2.10.** Referring back to Example 2.9, we can see that the subset of finite-support sequences forms a manifold (the sum of two finite-support sequences is also a finite-support sequence). But as was shown before, this subset is not closed and hence cannot be a subspace of ℓ^2. □

> **Example 2.11.** An example of a manifold in ℓ^2 which is also a subspace is the set $\mathfrak{M}$ of all sequences such as $a = \{\alpha_k\}_{k=1}^{\infty}$ for which $\alpha_{2k} = 0$. Closedness is established as follows. Suppose that $b = \{\beta_k\}_{k=1}^{\infty}$ is an accumulation point. Then according to the definition of accumulation point (Helmberg, 1969, p. 16), one can find an element of the subset such as a arbitrarily close to b (for any $\epsilon > 0$, there exists $a \in \mathfrak{M}$ for which $\|a-b\| = (\sum_{k=1}^{\infty}|\alpha_k - \beta_k|^2)^{1/2} < \epsilon$). From which we conclude that $|\beta_{2k}| = |\beta_{2k}-0| = |\beta_{2k}-\alpha_{2k}| < \|a-b\| < \epsilon$. Hence, $\beta_{2k} = 0$ and all accumulation points of $\mathfrak{M}$ also belong to $\mathfrak{M}$. □

> **Example 2.12.** An example of a subspace of $L^2(\Omega)$ is the subset of all functions $f(x)$ in $L^2(\Omega)$ for which $f(x) = 0$ a.e. (almost everywhere) on a subinterval of Ω. □

It is possible to generate a manifold from an arbitrary countable subset $\mathfrak{N}$ of a Hilbert space. This is done by generating all linear combinations of the elements in the given subset. For example, for a subset $\mathfrak{N}$ of $\mathcal{H}$ with elements $f_1, f_2, \ldots, f_n$, the set of all linear combinations

$$\mathfrak{M} = \Big\{ \sum_{k=1}^{n} \alpha_k f_k : f_k \in \mathfrak{N},\ \alpha_k \in \mathbb{C},\ \text{for } 1 \leq k \leq n,\ n \geq 1 \Big\} \qquad (2.49)$$

forms a manifold.

What should we do if a manifold $\mathfrak{M}$ in a Hilbert space is not a subspace? Similar to the discussion on completeness, we just need to expand the manifold to include all its accumulation points so that the enlarged manifold is closed and hence complete. This is the "smallest" subspace which contains the manifold.

From the above discussion, we have now the tools to turn any subset $\mathfrak{N} \subset \mathcal{H}$ to a subspace of $\mathcal{H}$. We first form a manifold $\mathfrak{M}$ from $\mathfrak{N}$ by generating all linear combinations of its elements and then turn it into a subspace by including all the accumulation points of $\mathfrak{M}$. This is called the subspace spanned by the subset $\mathfrak{N}$ denoted by $\bigvee \mathfrak{N}$ and is, in fact, the smallest subspace containing $\mathfrak{N}$.

Problem

2.22. Prove that in ℓ^2, the set
$$\mathfrak{M}_k = \left\{ \alpha e_k \colon \alpha \in \mathbb{C}, \ e_k = \{\delta_{k,n}\}_{n=1}^{\infty} \right\}$$
is a subspace, where $\delta_{k,n}$ is the Kronecker delta function. That is, if we start with the original subset containing a single unit-norm "seed vector" e_k ($\mathfrak{N}_k = \{e_k\}$), we arrive at $\mathfrak{M}_k$ which is closed and hence is equal to its span $\bigvee e_k$. e_k and $\bigvee e_k$ have a particular significance in the discussions in Section 2.7.

2.6.3   Vector sums, orthogonal subspaces and projections

Definition 2.19 (Vector sum of subspaces). *The vector sum of two subspaces* $\mathfrak{M}_1$ *and* $\mathfrak{M}_2$ *is defined as*

$$\mathfrak{M}_1 + \mathfrak{M}_2 = \{f + g \colon f \in \mathfrak{M}_1, \ g \in \mathfrak{M}_2\},$$

which can be extended to a countable vector sum of subspaces, denoted

$$\sum_{k=1}^{\infty} \mathfrak{M}_k,$$

in a similar fashion.   □

The vector sum of subspaces is a manifold, but as much as we would have liked it to be, is not necessarily a subspace by itself. One needs to make the vector sum closed by including all its accumulation points. Orthogonal subspaces, to be defined below, are a special class of subspaces for which their vector sum is a subspace itself.

Definition 2.20 (Orthogonal subspaces). *Two subspaces* $\mathfrak{M}_1$ *and* $\mathfrak{M}_2$ *are orthogonal if every element* $f \in \mathfrak{M}_1$ *is orthogonal to every element* $g \in \mathfrak{M}_2$ *or* $f \perp g$. *This is denoted in short by*
$$\mathfrak{M}_1 \perp \mathfrak{M}_2.$$

The result is extendable to a countable set of mutually orthogonal subspaces.   □

Example 2.13. In ℓ^2, the two sets
$$\mathfrak{M}_e = \left\{ a = \{\alpha_k\}_{k=1}^{\infty} \colon \alpha_{2k} = 0 \right\}$$
and
$$\mathfrak{M}_o = \left\{ b = \{\beta_k\}_{k=1}^{\infty} \colon \beta_{2k-1} = 0 \right\}$$
are two orthogonal subspaces.   □

We are now ready to prove the following theorem linking orthogonal subspaces to vector sums.

Theorem 2.7 (Vector sum of orthogonal subspaces). *If* $\mathfrak{M}_1$ *and* $\mathfrak{M}_2$ *are two orthogonal subspaces, then the manifold generated by their vector sum* $\mathfrak{M}_1 + \mathfrak{M}_2$ *is also a subspace.*   □

Proof. Consider a Cauchy sequence $\{f_k\}_{k=1}^{\infty}$ where f_k's belong to the vector sum $\mathfrak{M}_1 + \mathfrak{M}_2$. From the definition of Cauchy sequences we can say $\lim_{m,n \to \infty} \|f_m - f_n\| = 0$. Since f_m and f_n belong to the vector sum, we can write them as $f_m = g_m + h_m$ and $f_n = g_n + h_n$ where $g_m, g_n \in \mathfrak{M}_1$ and $h_m, h_n \in \mathfrak{M}_2$. We now invoke the Pythagorean theorem, Theorem 2.4, p. 45, for orthogonal vectors to say that

$$\lim_{m,n \to \infty} \|f_m - f_n\|^2 = \lim_{m,n \to \infty} \|g_m - g_n\|^2 + \lim_{m,n \to \infty} \|h_m - h_n\|^2 = 0.$$

Therefore, sequences $\{g_k\}_{k=1}^{\infty}$ and $\{h_k\}_{k=1}^{\infty}$ are also Cauchy sequences in their corresponding subspaces and since the subspaces are closed by definition, they converge to points $g \in \mathfrak{M}_1$ and $h \in \mathfrak{M}_2$. Therefore, $\{f_k\}_{k=1}^{\infty}$ also converges to $g + h$, which, according to the definition of vector sum, must belong to $\mathfrak{M}_1 + \mathfrak{M}_2$ and hence the manifold is closed. □

> **Example 2.14.** The vector sum of orthogonal subspaces $\mathfrak{M}_k$ and $\mathfrak{M}_j$ defined in Problem 2.22, which is denoted by
>
> $$\mathfrak{M}_k + \mathfrak{M}_j = \{\alpha e_k + \beta e_j : \alpha, \beta \in \mathbb{C}\},$$
>
> is another subspace in ℓ^2. □

Definition 2.21 (Orthogonal complement of a subspace). *The orthogonal complement of a subspace $\mathfrak{M}$ in $\mathcal{H}$ is denoted by $\mathfrak{M}^{\perp}$ and is the set of all vectors in the original Hilbert space $\mathcal{H}$ which are orthogonal to $\mathfrak{M}$.* □

Orthogonal complement of a subspace is a subspace in its own right. The original subspace $\mathfrak{M}$ may be so big that $\mathfrak{M}^{\perp} = \{o\}$. That is, the only vector orthogonal to it is the trivial zero vector.

Problem

2.23. Prove that the subspaces $\mathfrak{M}_k = \{\alpha e_k : \alpha \in \mathbb{C}, \ e_k = \{\delta_{k,n}\}_{n=0}^{\infty}\}$ defined in Problem 2.22 are mutually orthogonal to each other. That is,

$$\mathfrak{M}_k \perp \mathfrak{M}_j \quad \text{for all} \quad k \neq j.$$

2.6.4   Projection

It turns out that for any given subspace $\mathfrak{M}$ and for any given vector $f \in \mathcal{H}$, one can break f into two unique components. The first component belongs to $\mathfrak{M}$, which is also called the projection of f into $\mathfrak{M}$ and denoted by $f_1 = \mathcal{P}_{\mathfrak{M}} f$. The remaining component $f - f_1$ is orthogonal to $\mathfrak{M}$ and hence belongs to $\mathfrak{M}^{\perp}$ according to its definition. The unique property of the projection component f_1 is that it minimizes the norm of the remaining component $\|f - f_1\|$ in its orthogonal subspace. That is to say, other than f_1, there is no other part of f which could have been possibly represented by the elements in $\mathfrak{M}$, at least as much as norm is concerned. We summarize this in the following theorem.

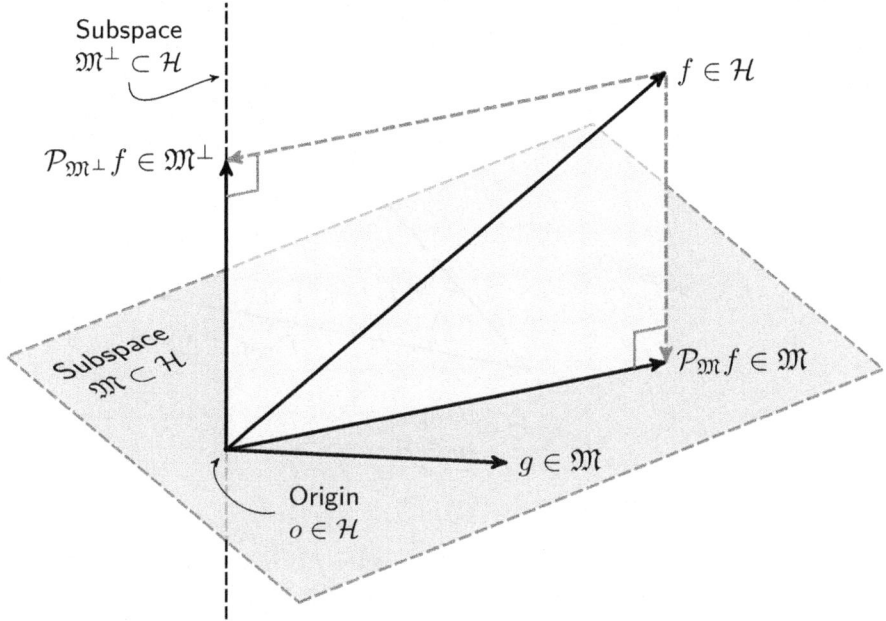

Figure 2.7 The projection theorem, $f \in \mathcal{H}$ can be written as the sum of $\mathcal{P}_{\mathfrak{M}} f \in \mathfrak{M}$ and $\mathcal{P}_{\mathfrak{M}^{\perp}} f \in \mathfrak{M}^{\perp}$. The vector $\mathcal{P}_{\mathfrak{M}} f$ is the closest vector $g \in \mathfrak{M}$ in subspace $\mathfrak{M}$ in norm to $f \in \mathcal{H}$.

Theorem 2.8 (Projection theorem). *If $\mathfrak{M}$ is a subspace of $\mathcal{H}$, then every element of $\mathcal{H}$ such as f can be uniquely represented as follows*

$$f = f_1 + f_2,$$

where $f_1 = \mathcal{P}_{\mathfrak{M}} f \in \mathfrak{M}$ and $f_2 = \mathcal{P}_{\mathfrak{M}^{\perp}} f \in \mathfrak{M}^{\perp}$. Moreover, f_1 has the property that

$$\|f - f_1\| = \inf_{g \in \mathfrak{M}} \|f - g\|.$$

$\square$

The projection theorem is depicted abstractly in Figure 2.7, where $\mathcal{H}$, $\mathfrak{M}$ and $\mathfrak{M}^{\perp}$ are in general infinite-dimensional.

Remark 2.14. Another interpretation of the projection theorem is that the original Hilbert space is a vector sum of $\mathfrak{M}$ and $\mathfrak{M}^{\perp}$:

$$\mathcal{H} = \mathfrak{M} + \mathfrak{M}^{\perp}.$$

More generally, if we have a countable set of mutually orthogonal subspaces $\{\mathfrak{M}_k\}_{k=1}^{\infty}$, any vector f in their vector sum $\sum_{k=1}^{\infty} \mathfrak{M}_k$ subspace is represented as $f = \sum_{k=1}^{\infty} f_k$ where f_k is a unique vector in $\mathfrak{M}_k$ and is the projection of f upon $\mathfrak{M}_k$.

$\square$

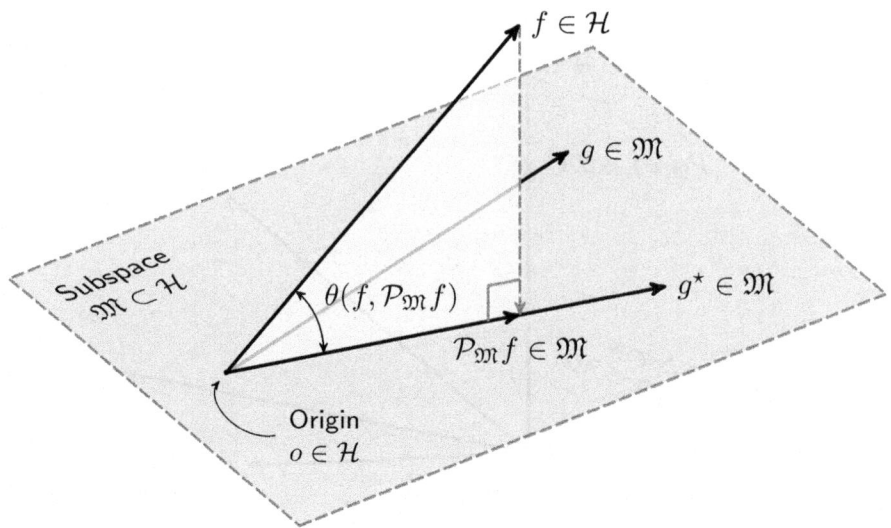

Figure 2.8 Infimum angle between vector $f \in \mathcal{H}$ and subspace $\mathfrak{M}$ is given by $\theta(f, \mathcal{P}_{\mathfrak{M}}f)$, which is attained by any $g^\star = \alpha \mathcal{P}_{\mathfrak{M}}f$ for any real $\alpha > 0$. In this illustration $\alpha = 1.5$ and $g \in \mathfrak{M}$ represents any vector in the subspace $\mathfrak{M}$.

Infimum angle between f and $\mathfrak{M}$

The projection theorem has also another nice geometrical implication, which we discuss now. It can help us answer the following question:

Fix vector $f \in \mathcal{H}$ and subspace $\mathfrak{M}$. Among all vectors such as $g \in \mathfrak{M}$, which one or ones attain the infimum angle defined in (2.11), p. 38, between f and g? That is, find $g^\star$ such that

$$g^\star = \inf_{g \in \mathfrak{M}} \cos \theta(f, g) = \inf_{g \in \mathfrak{M}} \frac{\mathfrak{Re}(\langle f, g \rangle)}{\|f\| \|g\|}.$$

Remarkably, the answer to this question is the projection of f into $\mathfrak{M}$ or a multiple of it such as $g^\star = \alpha \mathcal{P}_{\mathfrak{M}}f$ for real $\alpha > 0$. The geometrical interpretation of this result is that, among all vectors in $\mathfrak{M}$, vectors of the form $\alpha \mathcal{P}_{\mathfrak{M}}f$ are most closely aligned with f. No other type of vectors in $\mathfrak{M}$ can be as similar to or aligned with f, as far as the inner product is concerned. In other words, its projection $\mathcal{P}_{\mathfrak{M}}f$ is everything from f that can possibly live in $\mathfrak{M}$, which is the nub of projection theorem. This is abstractly shown in Figure 2.8.

The proof of this result is not too difficult and uses the projection theorem and Cauchy-Schwarz inequality. First, from the projection theorem we note that

$$\langle f, g \rangle = \langle f - \mathcal{P}_{\mathfrak{M}}f + \mathcal{P}_{\mathfrak{M}}f, g \rangle = \langle f - \mathcal{P}_{\mathfrak{M}}f, g \rangle + \langle \mathcal{P}_{\mathfrak{M}}f, g \rangle = \langle \mathcal{P}_{\mathfrak{M}}f, g \rangle,$$

since $f - \mathcal{P}_{\mathfrak{M}}f$ is orthogonal to all $g \in \mathfrak{M}$. Therefore, we can write

$$\mathfrak{Re}(\langle f, g \rangle) \leq |\langle f, g \rangle| = |\langle \mathcal{P}_{\mathfrak{M}}f, g \rangle|. \tag{2.50}$$

Also by the same token

$$\langle f, \mathcal{P}_{\mathfrak{M}}f \rangle = \langle f - \mathcal{P}_{\mathfrak{M}}f + \mathcal{P}_{\mathfrak{M}}f, \mathcal{P}_{\mathfrak{M}}f \rangle = \langle \mathcal{P}_{\mathfrak{M}}f, \mathcal{P}_{\mathfrak{M}}f \rangle = \|\mathcal{P}_{\mathfrak{M}}f\|^2. \tag{2.51}$$

Now from Cauchy-Schwarz inequality

$$|\langle \mathcal{P}_{\mathfrak{M}} f, g \rangle| \leq \|\mathcal{P}_{\mathfrak{M}} f\| \|g\|. \tag{2.52}$$

Therefore, combining the three results in (2.50)-(2.52) we have

$$\frac{\mathfrak{Re}(\langle f, g \rangle)}{\|f\| \|g\|} \leq \frac{\|\mathcal{P}_{\mathfrak{M}} f\|}{\|f\|} = \frac{\mathfrak{Re}(\langle f, \mathcal{P}_{\mathfrak{M}} f \rangle)}{\|f\| \|\mathcal{P}_{\mathfrak{M}} f\|}.$$

Since $\cos \theta$ is a decreasing function in $\Omega = [0, \pi]$, we can assert that

$$\theta(f, g) \geq \theta(f, \mathcal{P}_{\mathfrak{M}} f),$$

where the equality is achieved whenever $g = \alpha \mathcal{P}_{\mathfrak{M}} f$.

2.6.5 Completeness of subspace sequences

Now imagine that we are given a special countable set of mutually orthogonal non-zero subspaces $\{\mathfrak{M}_k\}_{k=1}^{\infty}$ of $\mathcal{H}$ such that

$$\mathcal{H} = \sum_{k=1}^{\infty} \mathfrak{M}_k + \mathfrak{M}^{\perp}$$

and $\mathfrak{M}^{\perp} = \{o\}$. That is, this special class of orthogonal subspaces spans the whole space $\mathcal{H}$. These subspaces are complete in the sense that we cannot add any more non-trivial orthogonal subspace to them, because simply there is none. We cannot remove any subspace from the set either without violating the above spanning property. If we removed $\mathfrak{M}_1$ from our set, let's say, then

$$\mathcal{H} = \sum_{k=2}^{\infty} \mathfrak{M}_k + \mathfrak{M}_1.$$

Therefore, a non-zero vector in $\mathfrak{M}_1 \subset \mathcal{H}$ (whose existence is guaranteed by the assumption that $\mathfrak{M}_k \neq \{o\}$) cannot be represented by projection into subspaces $\{\mathfrak{M}_k\}_{k=2}^{\infty}$.

In the context of projection theorem, if we keep projecting a vector $f \in \mathcal{H}$ onto $\mathfrak{M}_1, \mathfrak{M}_2, \mathfrak{M}_3, \ldots$, we will be able to fully capture f in terms of sum of unique projected vectors in $\mathfrak{M}_1, \mathfrak{M}_2, \mathfrak{M}_3, \ldots$, and there is no part in f that is left out (again as long as norm is concerned) and this is a good thing.

Through examples and worked problems, we saw that the set of subspaces of the form $\mathfrak{M}_k = \{\alpha e_k : \alpha \in \mathbb{C}, \ e_k = \{\delta_{k,n}\}_{n=0}^{\infty}\}$ in ℓ^2 are mutually orthogonal to each other. This set of subspaces is complete. Because if there exists a subspace $\mathfrak{M}$ orthogonal to all $\mathfrak{M}_k$, any vector a in $\mathfrak{M}$ is orthogonal to all vectors in all $\mathfrak{M}_k$. Therefore, $\langle a, e_k \rangle = 0$ for all k, which means that $a = o$. We cannot help but note that there is a single representative element e_k which is responsible for generating the entire subspace $\mathfrak{M}_k$. So instead of directly working with bulky complete orthogonal subspaces, we can work with elegant complete orthonormal sequences $\{e_k\}_{k=1}^{\infty}$ in ℓ^2 and invoke (2.49) to generate the (closed) manifold whenever needed.

2.7 Complete orthonormal sequences

In this section, we will formalize the preceding discussion to a good extent. The master plan is to find a compelling representation for vectors in a separable Hilbert space, analogous to an orthogonal vector basis in a finite-dimensional space.

If the Hilbert space $\mathcal{H}$ were finite-dimensional, then there would be only a finite number of mutually orthogonal vectors and we could not have a orthonormal sequence with more than a finite number of terms equal to the dimension. Hence, the terminology orthonormal sequence is strictly reserved for infinite-dimensional Hilbert spaces.

Ultimately, it is of interest to know to what extent an orthonormal sequence can indeed represent the whole of a Hilbert space $\mathcal{H}$ in the sense of acting as a type of basis. Orthonormal sequences can be complete only in separable Hilbert spaces and we will give a flavor of why this is so later in Section 2.12. If the original Hilbert space is non-separable (there is no countable dense set in the space), then there is no hope of accurately representing all elements of the space by an orthonormal sequence. But beware that separability is not a guarantee for completeness of a given orthonormal sequences. This is later clarified through an example.

2.7.1 Definitions

Definition 2.22 (Complete orthonormal sequence). *An orthonormal sequence* $\{\varphi_n\}_{n=1}^{\infty}$ *is complete in a (separable) Hilbert space* $\mathcal{H}$ *if any of the following equivalent statements are true:*

P1: *$\nexists$ vector $\psi \neq o$ in $\mathcal{H}$ for which $\{\varphi_n\}_{n=1}^{\infty} \cup \psi$ forms an orthogonal set.*

P2: *Span of $\{\varphi_n\}_{n=1}^{\infty}$ is the whole space. That is, $\mathcal{H} = \bigvee \{\varphi_n\}_{n=1}^{\infty}$.*

P3: $f = \displaystyle\sum_{n=1}^{\infty} \langle f, \varphi_n \rangle \varphi_n, \forall f \in \mathcal{H}$ *(Fourier series).*

P4: *$\forall \epsilon > 0, \ \exists N_0 < \infty$ such that $\left\| f - \displaystyle\sum_{n=1}^{N} \langle f, \varphi_n \rangle \varphi_n \right\| < \epsilon, \forall N > N_0, \forall f \in \mathcal{H}$.*

P5: $\|f\|^2 = \displaystyle\sum_{n=1}^{\infty} |\langle f, \varphi_n \rangle|^2, \ \forall f \in \mathcal{H}$ *(Parseval relation).*

P6: *If $\langle f, \varphi_n \rangle = 0, \ \forall n \in \mathbb{N}$, then $f = o$.*

P7: $\langle f, g \rangle = \displaystyle\sum_{n=1}^{\infty} \langle f, \varphi_n \rangle \overline{\langle g, \varphi_n \rangle}, \ \forall f, g \in \mathcal{H}$ *(Generalized Parseval relation).*

$\square$

In short, these expressions capture the notion that you cannot augment further orthonormal vectors to a *complete orthonormal sequence*. The key equation is the expansion, which should be committed to memory

$$\boxed{f = \sum_{n=1}^{\infty} \langle f, \varphi_n \rangle \varphi_n,}$$
                                                                                (2.53)

which holds for all $f \in \mathcal{H}$ whenever $\{\varphi_n\}_{n=1}^{\infty}$ is complete. In (2.53), the $\langle f, \varphi_n \rangle$ are scalars and we have the following definition, which needs little explanation.

Definition 2.23 (Fourier series). *In a separable Hilbert space, the expansion*

$$f = \sum_{n=1}^{\infty} \langle f, \varphi_n \rangle \varphi_n$$
                                                                                (2.54)

of any $f \in \mathcal{H}$ in terms of a complete orthonormal sequence $\{\varphi_n\}_{n=1}^{\infty}$ is called a Fourier series expansion and the coefficients

$$\langle f, \varphi_n \rangle \in \mathbb{C}$$

are called the Fourier series coefficients.   □

The geometric interpretation of the Fourier series coefficients is that they represent the projection of f along φ_n in direct analogy with the Euclidean picture.

All orthogonal sequences in an uncountably infinite-dimensional non-separable Hilbert space are necessarily incomplete. For a separable Hilbert space we illustrate with a meaningful (as distinct from mathematically contrived) example that not all countably infinite orthogonal sequences are guaranteed to be complete.

Example 2.15 (Incomplete orthogonal sequences). Consider the real Hilbert space $\mathcal{H}$ of functions in $L^2(-\pi, \pi)$. Define the orthonormal and countably infinite Fourier cosine sequence

$$\{\varphi_n^{(\text{even})}\}_{n=0}^{\infty}, \quad \text{where} \quad \varphi_n^{(\text{even})} \triangleq \frac{\cos(nx)}{\sqrt{2}}.$$
                                                                                (2.55)

It is incomplete since every $\sin(mx) \in \mathcal{H}$ for $m \in \mathbb{N}$, or any linear combination of the same, is orthogonal to (2.55). Since all constituents of (2.55) are even functions the sequence will be orthogonal to any almost-everywhere odd function.

We can construct a complete orthonormal sequence by augmenting to (2.55) the Fourier sine sequence

$$\{\varphi_m^{(\text{odd})}\}_{m=1}^{\infty}, \quad \text{where} \quad \varphi_m^{(\text{odd})} \triangleq \frac{\sin(mx)}{\sqrt{2}}.$$
                                                                                (2.56)

As a digression, consider redefining the Hilbert space $\mathcal{H}$ such that (2.55) is a complete orthonormal sequence in $\mathcal{H}$. One possibility is to retain the inner product, but consider the (strict) linear subspace of almost everywhere even functions in $L^2(-\pi, \pi)$.   □

2.24 (Generalized Parseval relation). Prove the generalized Parseval relation

$$\langle f, g \rangle = \sum_{n=1}^{\infty} \langle f, \varphi_n \rangle \overline{\langle g, \varphi_n \rangle}, \quad \forall f, g \in \mathcal{H}. \tag{2.57}$$

2.7.2   Fourier coefficients and Bessel's inequality

Now we dig a little deeper into the definitions and concepts introduced in the previous section. Suppose we have a finite orthonormal set $\{\varphi_n\}_{n=1}^{N}$ in a Hilbert space $\mathcal{H}$ and an arbitrary $f \in \mathcal{H}$. We want to see how well we can approximate f using linear combinations of our finite orthonormal set where we measure the error using the norm. Then we have the following:

$$\begin{aligned}
\left\| f - \sum_{n=1}^{N} \alpha_n \varphi_n \right\|^2 &= \left\langle f - \sum_{n=1}^{N} \alpha_n \varphi_n, f - \sum_{n=1}^{N} \alpha_n \varphi_n \right\rangle \\
&= \langle f, f \rangle + \sum_{n=1}^{N} \alpha_n \overline{\alpha_n} - \sum_{n=1}^{N} \alpha_n \langle \varphi_n, f \rangle - \sum_{n=1}^{N} \overline{\alpha_n} \langle f, \varphi_n \rangle.
\end{aligned} \tag{2.58}$$

The minimum of this is attained when

$$\alpha_n = \langle f, \varphi_n \rangle, \quad n \in \{1, 2, \dots, N\}.$$

Hence the optimal coefficients in the linear combination of our finite orthonormal set are the Fourier series coefficients.

Putting the Fourier series coefficients into (2.58) yields

$$\left\| f - \sum_{n=1}^{N} \langle f, \varphi_n \rangle \varphi_n \right\|^2 = \|f\|^2 - \sum_{n=1}^{N} |\langle f, \varphi_n \rangle|^2,$$

from which we glean Bessel's inequality

$$\sum_{n=1}^{N} |\langle f, \varphi_n \rangle|^2 \leq \|f\|^2,$$

which is a close cousin of the Parseval relation given earlier. The minimization can be expressed also through

$$\left\| f - \sum_{n=1}^{N} \langle f, \varphi_n \rangle \varphi_n \right\|^2 \leq \left\| f - \sum_{n=1}^{N} \alpha_n \varphi_n \right\|^2, \quad \forall \{\alpha_n\}_{n=1}^{N}, \ \forall f \in \mathcal{H}.$$

Clearly in no sense can the finite orthonormal set $\{\varphi_n\}_{n=1}^{N}$ determine the whole Hilbert space $\mathcal{H}$. Letting $N \to \infty$ we obtain

$$\sum_{n=1}^{\infty} |\langle f, \varphi_n \rangle|^2 \leq \|f\|^2, \quad \forall f \in \mathcal{H}, \tag{2.59}$$

which means Bessel's inequality remains valid. It would be a mistake to think (2.59) should have an equality sign. This is precisely the point about completeness. Having a countably infinite number of orthonormal vectors is not a guarantee of completeness. When the Bessel's inequality becomes the Parseval relation then we have completeness.

Problem

2.25. Prove that every orthonormal sequence in an uncountably infinite non-separable Hilbert space is necessarily incomplete.

Take-home messages – orthonormal sequences

M1: Orthogonality is a stronger form of independence between vectors, which greatly simplifies computation, both in finite and infinite dimensions.

M2: The master plan is to find a compelling representation for vectors in a Hilbert space, analogous to an orthogonal vector basis in a finite-dimensional space.

M3: For this purpose, we need to know to what extent an orthonormal sequence can indeed represent the whole of a Hilbert space $\mathcal{H}$ in the sense of acting as a type of basis.

M4: This question can only make sense if the Hilbert space is separable in the first place. All orthogonal sequences in an uncountably infinite-dimensional non-separable Hilbert space are necessarily incomplete.

M5: A separable Hilbert space is a necessary, but not sufficient, condition for an orthonormal sequence to be complete. It is possible to find incomplete orthonormal sequences in a separable Hilbert space.

M6: Completeness of an orthonormal sequence can be verified by checking a number of equivalent conditions, most notably the Fourier series and Parseval relation.

M7: The geometric interpretation of the Fourier series coefficients is that they represent the projection of f along φ_n in direct analogy with the Euclidean picture.

M8: When approximating a vector in a Hilbert space $\mathcal{H}$ based on linear combinations of a finite orthonormal set, the optimal coefficients (that minimize the norm of error) are the Fourier series coefficients.

2.8   On convergence

2.8.1   Strong convergence

When dealing with infinite series, equalities such as those appearing in Definition 2.22 need to be interpreted correctly. For example, item 3 in Definition 2.22 strictly means

$$f = \sum_{n=1}^{\infty} \langle f, \varphi_n \rangle \varphi_n \iff \lim_{N \to \infty} \left\| f - \sum_{n=1}^{N} \langle f, \varphi_n \rangle \varphi_n \right\| = 0, \qquad (2.60)$$

where $\|\cdot\|$ is the norm in $\mathcal{H}$.

For example, if $\mathcal{H} = L^2(-\pi, \pi)$ and $\{\varphi_n\}_{n=1}^{\infty}$ is a complete orthonormal sequence in $\mathcal{H}$ then we mean

$$\lim_{N \to \infty} \int_{-\pi}^{\pi} \left| f(t) - \sum_{n=1}^{N} \alpha_n \varphi_n(t) \right|^2 dt = 0,$$

where

$$\alpha_n \triangleq \langle f, \varphi_n \rangle = \int_{-\pi}^{\pi} f(\tau) \overline{\varphi_n(\tau)} \, d\tau.$$

Convergence such as in (2.60) is referred to as *strong convergence*, also called *convergence in the mean*, and is formally defined as follows.

Definition 2.24 (Strong convergence). *A sequence of vectors $\{f_n\}_{n=1}^{\infty}$ in a Hilbert space $\mathcal{H}$ is called strongly convergent to a vector $f \in \mathcal{H}$ if*

$$\|f_n - f\| \to 0$$

as $n \to \infty$. In notation we write $f_n \to f$.   □

Example 2.16. For a complete orthonormal sequence $\{\varphi_n\}_{n=1}^{\infty}$ in a Hilbert space $\mathcal{H}$, the series

$$f_n = \sum_{m=1}^{n} \langle f, \varphi_m \rangle \varphi_m$$

converges strongly to f, for all $f \in \mathcal{H}$.   □

The sense in which the limit of $f_n \in \mathcal{H}$, say given by $f^\star$, actually equals $f \in \mathcal{H}$ is less than one might initially hope, but one comes resigned to the fact that the manner in which they do differ is not of essential concern. Why is it called strong? Because it is not *weak* as in weak convergence, which is the concept to follow.

It is helpful to have a sense of why strong convergence is also called "convergence in the mean" (Fischer, 1907)[16] because this is an intuitively sound terminology, at least for an important class of problems.

Historically, one of the main drivers for the development of functional analysis was the study of Fourier series representations of functions defined on a finite

---

[16] "Sur la convergence en moyenne."

interval or, equivalently, periodic functions on the real line. As a concrete example, the vectors are the exponential functions (orthonormal functions, as in Section 2.5.3), the sequence is composed of the partial sums of the Fourier series, and the limit is the infinite Fourier series. The Fourier series representation of a sufficiently well-behaved function[17] is known to differ from that function only at points of discontinuity. In comparison with the value $f(x)$ taken by a function at x, the limit of the infinite Fourier series converges to

$$(f(x^+) + f(x^-))/2, \tag{2.61}$$

where $f(x^+)$ and $f(x^-)$ mean the limiting value of the function as we approach $f(x)$ from the right and left, respectively. Such convergence was seen in Theorem 2.5 and Theorem 2.6. By (2.61) we are not asserting that the value x corresponds to a point of discontinuity. To the contrary, this formula prescribes the value for *every* x. So in (2.61) we see that the value of the limit can be characterized as the arithmetic mean of the limiting values from the right and left. Intuitively one can think of the Fourier series as being a smoother version of the original function.

2.8.2   Weak convergence

Given that we are resigned to, or not aversely disturbed by, the lack of strict pointwise equality of the converging sequence $\{f_n\}_{n=1}^{\infty}$ in a Hilbert space $\mathcal{H}$ to another point $f \in \mathcal{H}$ (strong convergence), then it is reasonable to explore dropping our standards even further.

Definition 2.25 (Weak convergence). *A sequence of vectors $\{f_n\}_{n=1}^{\infty}$ in a Hilbert space $\mathcal{H}$ is called weakly convergent to a vector $f \in \mathcal{H}$ if*

$$\langle f_n, g \rangle \to \langle f, g \rangle$$

as $n \to \infty$, for every $g \in \mathcal{H}$ and is denoted by $f_n \rightharpoonup f$. ☐

One rationale for this definition is that in a Hilbert space the primary operations we carry out involve inner products. Now if there is no difference between using the limit of $\{f_n\}$ as $n \to \infty$ or using $f \in \mathcal{H}$ when taking the inner product with any other member $g \in \mathcal{H}$ then why discriminate between them? There need not be any value in resolving the difference. In short, weak convergence, in the context of Hilbert spaces, is a statement about being able to resolve points in a Hilbert space using an inner product microscope and strong convergence is a more powerful microscope that uses a norm microscope. Obviously, if the resolving power of the norm is better than the inner product then it comes as no surprise that strong convergence implies weak convergence as captured in the following theorem.

Theorem 2.9. *A strongly convergent sequence $\{f_n\}_{n=1}^{\infty}$ in a Hilbert space $\mathcal{H}$ is weakly convergent to the same limit. That is, if $\|f_n - f\| \to 0$ as $n \to \infty$ then $\langle f_n, g \rangle \to \langle f, g \rangle$ as $n \to \infty$, for every $g \in \mathcal{H}$.* ☐

---

[17] By sufficiently well-behaved function we mean it is sufficiently smooth such that various limits exist. An example is a piecewise continuous function. Refer also to Theorem 2.5, p. 55, and Theorem 2.6, p. 58.

Proof. For the inner product, we have

$$|\langle f_n, g\rangle - \langle f, g\rangle| = |\langle f_n - f, g\rangle|$$
$$\leq \|f_n - f\| \, \|g\| \tag{2.62}$$

by the Cauchy-Schwarz inequality given in Theorem 2.1, p. 37. By strong convergence in (2.62) $\|f_n - f\| \to 0$ as $n \to \infty$. Therefore, we have weak convergence. In the above, the limit f is the same for both weak and strong convergence. $\square$

Remark 2.15. In shorthand this theorem says $f_n \to f \implies f_n \rightharpoonup f$. $\square$

In general, the converse is not true (a dodgy weak inner product microscope is no match for a strong norm microscope). Example 2.17 gives an example.

Example 2.17. Riesz and Sz.-Nagy show that a series with elements

$$f_n \equiv f_n(x) \triangleq \sin(nx) \in L^2(0, 2\pi) \tag{2.63}$$

exhibits weak convergence to the zero function as $n \to \infty$, but not strong convergence (Riesz and Sz.-Nagy, 1990). This is illustrated in Figure 2.9. Weak convergence follows because $f_n(x)$ arbitrarily quickly "modulates" (multiplies, as in the integrand of the inner product) any given function $g(x)$ the integral of which tends to zero. That is, $f_n(x)$ weakly converges to the zero function — to the fuzzy eyes of the inner product $\sin(nx)$ looks like the zero function as $n \to \infty$. However, $\|f_n(x) - f_m(x)\| = \pi$ for all m, n and so the $\{f_n(x)\}_{n=1}^{\infty}$ sequence is not Cauchy and, hence, there is no limit in the mean. It does not converge strongly. $\square$

The sequence (2.63) is an instance of a more general simple-to-prove result on orthonormal sequences, which is worth expanding on for two reasons:

1: it reveals an initially counterintuitive result, which reflects that weak convergence does not meet one's expectations of convergence;

2: this result has an important role to play when we look later at operators (compact operators in particular).

Theorem 2.10. *An orthonormal sequence, $\{\varphi_n\}_{n=1}^{\infty}$, in a Hilbert space $\mathcal{H}$ converges weakly to zero. That is,*

$$\langle \varphi_n, f\rangle \to \langle o, f\rangle \equiv 0 \quad as \quad n \to \infty, \quad \forall f \in \mathcal{H}. \tag{2.64}$$

This can be written $\varphi_n \rightharpoonup o$. $\square$

Before we prove this result, we should note that this is all rather bizarre. Normally when we think of orthonormal sequences, they spread out to cover the complete space. They are not converging to anything, they are not becoming more and more similar to any element in the space, if anything, they are maximally avoiding each other and being persistently dissimilar. The theorem does not match our expectation about "convergence." In the end convergence means strong convergence (convergence in the mean). Weak convergence is a more esoteric convergence, but it is important in a number of theoretical constructions.

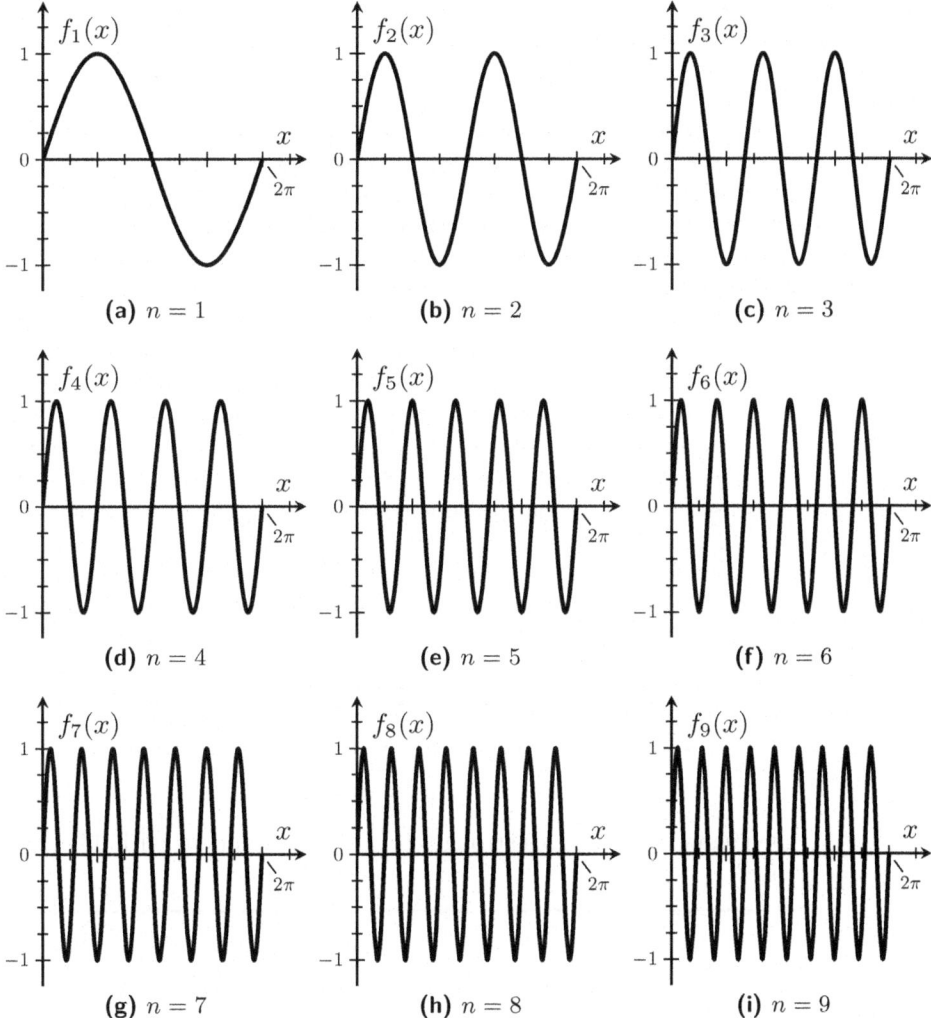

Figure 2.9 Sequence $f_n(x) = \sin(nx)$ that converges weakly to the zero function, but does not converge strongly (to anything, that is, it does not converge).

Proof. By Bessel's inequality, (2.59), we have that the sequence $\{\langle f, \varphi_n \rangle\}_{n=1}^{\infty}$ converges for all $f \in \mathcal{H}$ (whereby $\|f\| < \infty$), which implies

$$\sum_{n=1}^{\infty} |\langle f, \varphi_n \rangle|^2 < \infty, \quad \forall f \in \mathcal{H}.$$

Therefore, $|\langle f, \varphi_n \rangle|^2 \to 0$ as $n \to \infty$, which implies $\langle f, \varphi_n \rangle \to 0$ as $n \to \infty$, or in a form directly amenable to the weak convergence definition: $\langle \varphi_n, f \rangle \to \langle o, f \rangle$ as $n \to \infty$, for every $f \in \mathcal{H}$. $\qquad \square$

Note that the sequence does not need to be complete, but if it were then (2.59) would then hold with equality and the conclusion is unchanged. For the same sequence there is no sense of strong convergence since $\|\varphi_n - \varphi_m\|^2 = 2$, whenever $n \neq m$, so the sequence is not Cauchy (which, by completeness, would have implied strong convergence). In Problem 2.26 it is shown that a specific additional condition is sufficient to permit one to infer strong convergence from weak convergence.

Remark 2.16 (Riesz representation theorem). Defining weak convergence in terms of an inner product in Hilbert space is an entry point to a more general notion of weak convergence. The more general concept of weak convergence involves the concept of a bounded linear functional, which in the case of Hilbert space is non-trivially equivalent to the inner product expression in Definition 2.25 by the Riesz representation theorem (Debnath and Mikusiński, 1999, p. 126). This important theorem was shown independently by Riesz and Fischer in 1907; see (Riesz, 1907; Fischer, 1907). $\qquad \square$

Problem _____

2.26. If f_n converges weakly to f and we have the additional assumption that

$$\lim_{n \to \infty} \|f_n\| = \|f\|,$$

then show f_n converges strongly to f. What happens with Theorem 2.10 in this case?

_____

2.8.3   Pointwise convergence

There is another notion of convergence that crops up primarily when dealing with $L^2(\Omega)$, the space of square integrable functions defined on the finite interval $\Omega = [\alpha, \beta]$. It is called pointwise convergence.

Strong convergence does not imply pointwise convergence. Pointwise convergence is a much stronger notion of convergence again (than strong convergence) and corresponds to the naive view of convergence in the form that held up understanding Fourier series for over multiple decades.

Definition 2.26 (Pointwise convergence). *A sequence of vectors* $\{f_n\}_{n=1}^\infty$ *in the normed space* $L^2(\Omega)$ *for* $\Omega = [\alpha, \beta]$ *is pointwise convergent to a vector* $f \in \mathcal{H}$ *if for every* $\epsilon > 0$ *and any given* $x \in \Omega$, *there exists an* $N(x)$ *for which*

$$\left| f_n(x) - f(x) \right| < \epsilon \quad \text{for all} \quad n > N(x).$$

□

The sequence of continuous functions, given in Section 2.3.5, p. 33, and repeated here

$$f_n(x) \triangleq \arctan\big((n(x - (\alpha + \beta)/2))\big), \quad n \in \mathbb{N},$$

is a good example of pointwise convergence to

$$\lim_{n \to \infty} f_n(x) = \begin{cases} -\pi/2 & \alpha \le x < (\alpha + \beta)/2, \\ 0 & x = (\alpha + \beta)/2, \\ +\pi/2 & (\alpha + \beta)/2 < x \le \beta, \end{cases}$$

which is no longer continuous, but still resides in $L^2(\Omega)$.

The strong convergence implicit within the definition of complete orthogonal sequences, Definition 2.22, p. 70, does not imply pointwise convergence. This can be corroborated by the following. At points of discontinuity of a sufficiently regular function, the conventional Fourier series converges to the average of the left and right limits. In contrast, the actual value of the function at the discontinuity can be any finite value (in the same spirit as Theorem 2.5 and Theorem 2.6).

Establishing pointwise convergence or more importantly the degree of pointwise convergence is a hard problem. That Hilbert spaces can brush aside this nasty issue, by working with strong and weak convergence, is one of the great advantages of the theory. Nonetheless, a major result by Carleson in 1966 established that Fourier series of functions in $L^2(-\pi, \pi)$ converge pointwise almost everywhere (which is as much as can be hoped). But that is largely an intellectual challenge and somewhat misses the point of working with strong convergence and having a large set of relatively clean and aesthetically nice results.

Pointwise convergence has very little to do with Hilbert space and below we explore whether there is a way to capture pointwise convergence in a more general well-behaved space such as a Banach space. That is, can we define a non-Hilbert space norm that is equivalent to pointwise convergence? In the end, there is a norm which is quite close, but does fall short. It is better to abandon pointwise convergence for uniform convergence using the sup-norm.

But before defining what uniform convergence and sup-norm actually are, we have the following case study, which is going to seem being pedantic, but the point is actually very deep. In fact, to understand what follows should mean that vast swaths of difficulty evaporate. It is a beautiful illustration of the concepts.

Exploring completeness

How should we evaluate the statement: "the space of continuous functions on Ω is not complete"?

Firstly, this statement is somewhat meaningless because no reference is made to a norm. A "space" is actually a pair — continuous functions are the linear space portion but we also need to specify the norm. Given a norm, then we saw in Section 2.3.6, p. 35, how one can turn an incomplete normed space into a complete one, while keeping the essence of the normed space intact. However, the completion in general will include non-continuous functions for many choices of norm, such as the $L^2(\Omega)$ norm, as we have seen from examples.

A second consideration with this statement is whether we mean *all* continuous functions or a space of a subset of the continuous functions. Both avenues can be explored but here we are interested in the former case.

So the question is can we find a norm for which all the continuous functions on Ω are complete, that is, all Cauchy sequences of continuous functions converge to continuous functions. The answer to this question is yes and the norm is called the sup-norm, which induces a uniform and not pointwise convergence.

Problem

2.27. The functions $f_n(x) = \sin(nx)$ are in $L^2(0, 2\pi)$ for all n, but the series
$$\left\{\sin(nx)\right\}_{n=0}^{\infty}$$
tends to a point which is not in $L^2(0, 2\pi)$. Hence, $L^2(0, 2\pi)$ is incomplete. Please explain.

2.8.4   Uniform convergence

Here we define uniform convergence and again we limit our attention to function spaces and not the general abstract normed space context.

Definition 2.27 (Uniform convergence). *A sequence of vectors $\{f_n\}_{n=1}^{\infty}$ is uniformly convergent to a vector f if for every $\epsilon > 0$ there exists an N, independent of x, for which*
$$\left|f_n(x) - f(x)\right| < \epsilon,$$
for all $n > N$.   □

Note that in the definition of uniform convergence, the threshold N is not a function of x. This is what fundamentally differentiates uniform convergence from pointwise convergence where $N(x)$ could depend on x. Uniform convergence implies pointwise convergence ($N(x) = N$ for all x in Definition 2.26, p. 79), but the opposite is not true in general.

There is an equivalence between uniform convergence and sup-norm convergence, where sup-norm is defined below
$$\left\|f(x)\right\|_{\infty} \triangleq \sup_{x \in \Omega}\left|f(x)\right|,$$
where x belongs to some domain Ω, say $[\alpha, \beta]$ and the space on which sup-norm is defined is denoted by $\mathcal{H}_{\infty}$. That is, if a sequence of vectors $\{f_n\}_{n=1}^{\infty}$ is uniformly convergent to a vector $f \in \mathcal{H}_{\infty}$, then they are also convergent in the sup-norm:
$$f_n \underset{\text{unif}}{\to} f \iff \lim_{n \to \infty}\left\|f_n(x) - f(x)\right\|_{\infty} = 0.$$

Hence, in a complete normed space $\mathcal{H}_\infty$ defined through the sup-norm, every Cauchy sequence uniformly converges to a vector in $\mathcal{H}_\infty$ and hence is also pointwise convergent to a vector in the space.

Finally, it can be proven that if a sequence of continuous functions uniformly converges to a function f, then f is also continuous. That is why continuous functions can be complete or not complete depending on what norm is considered. They are certainly not complete in the $L^2(\Omega)$ space using the $L^2(\Omega)$-norm, but they are complete in a normed space induced by the sup-norm.

Problem

2.28. Prove the equivalence between the sup-norm and uniform convergence.

Take-home messages – convergence

M1: A complete orthonormal sequence guarantees strong convergence of a point f in the Hilbert space to its Fourier series expansion. That is, the norm of error between them tends to zero.

M2: In general, the Fourier series converges to the arithmetic mean of the limiting values of f from left and right.

M3: Hence, at points of discontinuity of f, the Fourier series expansion is known to differ from f. Intuitively one can think of the Fourier series expansion as being a smoother version of the original function. If the original f is sufficiently smooth, this difference does not become an issue.

M4: Weak convergence in the context of Hilbert spaces is a statement about being able to resolve points in a Hilbert space using an inner product microscope and strong convergence is a more powerful microscope that uses a norm microscope.

M5: An orthonormal sequence in a Hilbert space converges weakly to zero.

M6: One of the great advantages of the Hilbert space theory is that it can brush aside pointwise convergence, by working with strong and weak convergence.

M7: Continuous functions are not complete in the $L^2(\Omega)$ space using the $L^2(\Omega)$-norm, but they are complete in a normed space induced by the sup-norm.

M8: Relation between various convergence results: uniform convergence $\implies$ pointwise convergence $\implies$ strong convergence $\implies$ weak convergence. Weak convergence $\implies$ strong convergence if $\lim_{n\to\infty}\|f_n\| = \|f\|$.

2.9   Examples of complete orthonormal sequences

2.9.1   Legendre polynomials

Consider the space $L^2(-1, +1)$ of square integrable real functions defined on the real interval $[-1, +1]$, that is, real functions $f(x)$ satisfying

$$\|f\|^2 = \int_{-1}^{+1} f^2(x) \, dx < \infty, \tag{2.65}$$

where this and any other integral is understood in the sense of Lebesgue. Then equipping $L^2(-1, +1)$ with the inner product

$$\langle f, g \rangle \triangleq \int_{-1}^{+1} f(x)g(x) \, dx \tag{2.66}$$

induces the norm (2.65) through $\|f\|^2 = \langle f, f \rangle$. This defines a real separable Hilbert space $\mathcal{H}$. Recall that separable means there exists a countable set dense in $\mathcal{H}$.

The *normalized Legendre polynomials*, met in Section 2.5.1, p. 53,

$$\zeta_\ell(x) \triangleq \sqrt{\ell + \tfrac{1}{2}} P_\ell(x), \quad \ell \in \{0, 1, 2, \dots\},$$

is an example of a complete orthonormal sequence (denumerably infinite) for $\mathcal{H}$ — the vector space of functions in $L^2(-1, +1)$ with inner product (2.66). That is, $\{\zeta_\ell(x)\}_{\ell=0}^{\infty}$ defines a complete orthonormal sequence and any member of $f \in \mathcal{H}$ can be arbitrarily well approximated by a series of the form

$$\sum_{\ell=0}^{\infty} \langle f, \zeta_\ell \rangle \zeta_\ell(x),$$

which is understood in the sense of strong convergence or convergence in the mean. Note, by convention, we are indexing our orthonormal sequence from 0 instead of 1.

The normalized Legendre polynomials are by no means the only such complete orthonormal set. In a finite-dimensional vector space, such as $\mathbb{R}^N$, there are an infinite variety of orthonormal bases and these can be transformed to one another through unitary transformations (rotations and flips). For our separable infinite-dimensional Hilbert space this is also the case.

2.9.2   Bessel functions

Bessel functions of the first kind, denoted by $J_\nu(x)$, are the solutions to the following differential equation

$$x^2 \frac{d^2y}{dx^2} + x\frac{dy}{dx} + (x^2 - \nu^2)y = 0,$$

where $x \in [0, \infty)$ and ν is the function parameter. The most important physical cases occur when ν is an integer (cylindrical problems such as heat conduction or electromagnetic propagation in cylinders) or a half-integer (spherical problems) (Watson, 1995).

The representation as solutions to the above differential equation is by no means the only way Bessel functions can be expressed and there are various (recursive or otherwise) relations and representations, which may be useful for certain applications. Their treatment can be found in the definitive reference (Watson, 1995). It suffices to mention their Taylor series expansion around $x = 0$

$$J_\nu(x) = \sum_{n=0}^{\infty} \frac{(-1)^n}{n!\,\Gamma(\nu + n + 1)} \left(\frac{x}{2}\right)^{\nu+2n},$$

where $\Gamma(\cdot)$ is the Gamma function. Bessel functions for some non-negative integers from $\nu = 0$ to $\nu = 64$ are plotted in Figure 2.10. Apart from the zero-order Bessel function, $J_0(x)$, which takes the value of 1 at $x = 0$, $J_\nu(0) = 0$ for integer $\nu \neq 0$. For a particular ν, there are countably many zeros of Bessel functions, which occur at irregular intervals and should be computed numerically. We denote them by $z_{\nu,m}$ for $m = 1, 2, 3, \ldots$. That is, $J_\nu(z_{\nu,m}) = 0$. We can observe from Figure 2.10 that there is some kind of "bandpass" property associated with Bessel functions. They kick off later as the order ν increases and their peak magnitude will eventually die down as $1/\sqrt{x}$. The late kick-off property for high-order Bessel functions is crucial in establishing the dimensionality of multipath fields (Kennedy et al., 2007).

Orthogonality

The significance of zeros of Bessel functions is in defining the following weighted orthogonality relationship

$$\int_0^1 x J_\nu(z_{\nu,m}x) J_\nu(z_{\nu,n}x)\,dx = \frac{1}{2}\delta_{n,m}\left(J_{\nu+1}(z_{\nu,m})\right)^2$$

with the inner product weight of $w(x) = x$. In fact, the orthogonality stems from a more general relation (Riley et al., 2006)

$$\int_a^b x J_\nu(\alpha x) J_\nu(\beta x)\,dx = \frac{1}{\alpha^2 - \beta^2}\left[\beta x J_\nu(\alpha x) J_\nu'(\beta x) - \alpha x J_\nu(\beta x) J_\nu'(\alpha x)\right]_a^b,$$

where $J_\nu'(\cdot)$ denotes the derivative of $J_\nu(\cdot)$. Whenever $J_\nu(\alpha x)$ and $J_\nu(\beta x)$ vanish at $x = a$ and $x = b$, we obtain an orthogonality relation.

Fourier-Bessel expansion

For a given ν, $\{J_\nu(z_{\nu,m}x)\}_{m=0}^{\infty}$ forms a complete orthonormal sequence over $(0, 1)$ and hence any real square integrable function on the interval $\Omega = (0, 1)$ can be expanded into

$$f(x) = \sum_{n=0}^{\infty} \alpha_n J_\nu(z_{\nu,n}x),$$

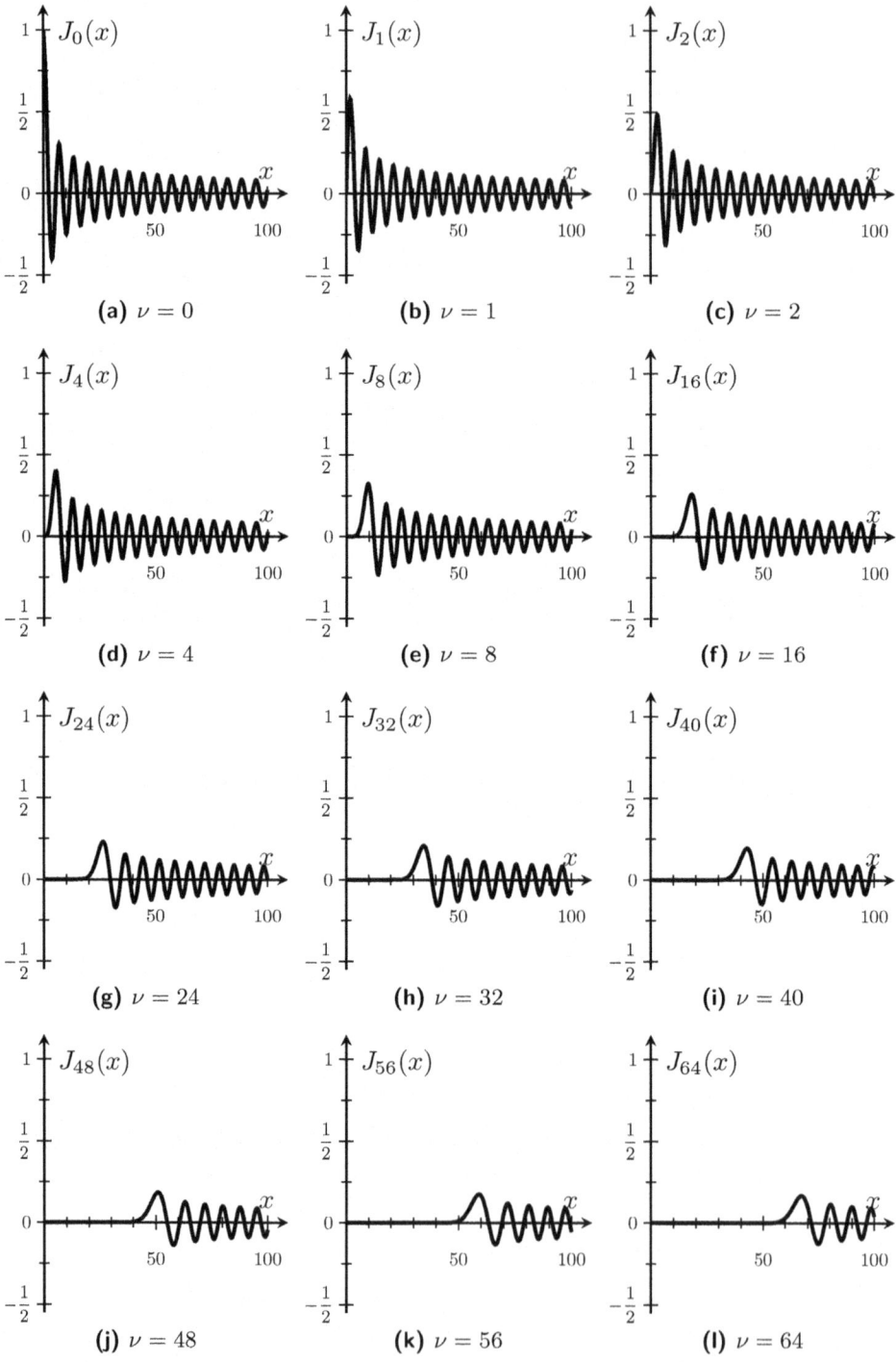

Figure 2.10 Bessel functions $J_\nu(x)$ for a set of non-negative integer ν values from $\nu = 0$ to $\nu = 64$, defined for $x \geq 0$ but only shown on the interval $x \in [0, 100]$.

where α is the "Fourier-Bessel" coefficient given by

$$\alpha_n \triangleq \frac{2}{(J_{\nu+1}(z_{\nu,n}))^2} \int_0^1 x f(x) J_\nu(z_{\nu,n}x)\, dx, \quad n \in \{0,1,2,\ldots\},$$

with weighting $w(x) = x$.

> **Remark 2.17.** There is no claim that Bessel functions for different integers
> $\nu \in \{0,1,2,\ldots\}$ are orthogonal over any interval. For example, the functions
> shown in Figure 2.10 are almost certainly *not* orthogonal. However, as shown
> above, the way to get an orthogonal sequence is to take a single "mother"
> Bessel function, for example, $\nu = 4$, and dilate it (scale its domain) so that
> the appropriate zero lines up with $x = 1$. We do this scaling of the given
> mother Bessel function for each of its countable non-uniformly spaced zeros
> to generate our orthogonal sequence of functions on the interval $(0,1)$ with
> weighting $w(x) = x$. Although the zeros are non-uniformly spaced, it is
> known that their spacing is asymptotically equal to π.   □

2.9.3   Complex exponential functions

Classical complex Fourier series functions can be utilized as representations:

- for functions on the interval $[0, 2\pi)$ on the real line;

- for periodic functions with period 2π on the real line; and

- for functions on the unit circle.

It is the last utilization that we wish to explore and subsequently generalize.

The unit circle embedded within a two-dimensional real space and parameter-
ized by $0 \le \phi < 2\pi$ is called the 1-sphere (actually a circle) and denoted by $\mathbb{S}^1$.
We have the following result.

Theorem 2.11. *Let $\mathcal{H}$ denote the vector space of square integrable functions on
unit circle $\mathbb{S}^1$, $L^2(\mathbb{S}^1)$, with inner product*

$$\langle f, g \rangle = \int_0^{2\pi} f(\phi)\overline{g(\phi)}\, d\phi, \quad f, g \in L^2(\mathbb{S}^1).$$

Then the complex exponential functions, the circular harmonics, in $\mathcal{H}$

$$\left\{ \frac{e^{in\phi}}{\sqrt{2}} \right\}_{n \in \mathbb{Z}} \tag{2.67}$$

form a complete orthonormal sequence in $\mathcal{H}$ — a separable Hilbert space.   □

> **Remark 2.18.** Clearly with the reordering of the integers defined in (1.5), p. 12,
> we can rewrite (2.67) in the form compatible with our general notation and
> indexing of orthonormal sequences:
>
> $$\{\varphi_n\}_{n=1}^\infty, \quad \text{where} \quad \varphi_n \triangleq \frac{e^{i\xi(n)\phi}}{\sqrt{2}},$$

Bessel, Friedrich Wilhelm (1784–1846) — Bessel was a German mathematician and astronomer. Carl Friedrich Gauss (1777–1855) was his thesis advisor and in the future direction his mathematical genealogy links with Cantor. One of his claims to fame was being the first person in 1838 to measure the distance to a star (other than the Sun) — the star 61 Cygni was measured to be 10.4 light-years away and was in error by less than 10%.

Bessel's name is associated with a plethora of special functions named after him which he introduced in 1824. Mathematicians, including the great Leonhard Euler (1707–1783), had earlier discovered these functions, but lacked Bessel's good looks and outstanding crop of hair. Bessel functions appear naturally as the radial term in a plethora of important physical problems when expressed in cylindrical or spherical coordinates. Bessel's name also appears in the inequality named after him which he derived in 1828 for the trigonometric Fourier series. In life, Bessel "turned his back on the prospect of affluence, chose poverty and the stars," and took up work running his observatory (Cajori, 1985).

with

$$\xi(n)\colon \mathbb{N} \longmapsto \mathbb{Z}$$

$$\xi(n) \triangleq (-1)^{n-1} \times \left\lfloor \frac{n}{2} \right\rfloor,$$

where $\lfloor \cdot \rfloor$ is the floor function. Of course, this is only of theoretical interest and no one in their right mind would use such a mapping.   □

2.9.4   Spherical harmonic functions

Spherical harmonic functions or simply spherical harmonics appear as the generalization of the classical complex Fourier series functions given in the previous subsection. In fact, Hilbert obtained his doctorate in 1885, with a dissertation titled "Über invariante Eigenschaften spezieller binärer Formen, insbesondere der Kugelfunktionen."[18] So we are in good company to study these functions.[19]

Consider the complex vector space of square integrable functions defined on the unit sphere or 2-sphere $\mathbb{S}^2$. That is, the 2D spherical surface embedded in 3D, parameterized by $0 \leq \theta \leq \pi$, the co-latitude, and $0 \leq \phi < 2\pi$, the azimuth, with Cartesian representation

$$(\sin\theta\cos\phi,\ \sin\theta\sin\phi,\ \cos\theta)'.$$

Any point on $\mathbb{S}^2$ is a unit vector and, hence, can be denoted $\widehat{\boldsymbol{u}}$ and satisfies $\|\widehat{\boldsymbol{u}}\| = 1$. The surface element on $\mathbb{S}^2$ is given by

$$ds(\widehat{\boldsymbol{u}}) \triangleq \sin\theta\, d\theta\, d\phi.$$

Then we have the following complete orthonormal sequence expressed in terms of the associated Legendre functions $P_\ell^m(\cos\theta)$; see Section 2.5.4, p. 61.

Theorem 2.12 (Completeness of spherical harmonics). *Let $L^2(\mathbb{S}^2)$ denote the vector space of square integrable functions on the 2-sphere $\mathbb{S}^2$, with inner product*

$$\langle f, g \rangle \triangleq \int_0^{2\pi}\!\!\int_0^\pi f(\theta,\phi)\overline{g(\theta,\phi)}\ \sin\theta\, d\theta\, d\phi, \quad f, g \in L^2(\mathbb{S}^2). \qquad (2.68)$$

Then the spherical harmonics in $L^2(\mathbb{S}^2)$

$$Y_\ell^m(\theta,\phi) \triangleq \sqrt{\frac{2\ell+1}{4\pi}\frac{(\ell-m)!}{(\ell+m)!}}\, P_\ell^m(\cos\theta)e^{im\phi},$$

$$\ell \in \{0,1,2,\ldots\},\ m \in \{-\ell, -\ell+1, \ldots, \ell\}, \qquad (2.69)$$

form a complete orthonormal sequence in $L^2(\mathbb{S}^2)$ with $\langle Y_\ell^m, Y_p^q \rangle = \delta_{\ell,p}\delta_{m,q}$. $L^2(\mathbb{S}^2)$ is a separable Hilbert space.   □

---

[18] Which in any language, other than German, would come across as an insult directed at your mother. Actually it means "On the invariant properties of special binary forms, in particular the spherical harmonic functions." Hilbert's advisor was Ferdinand von Lindemann (1852–1939), who is famous for proving that π was transcendental. An accessible proof for the transcendence of π is given in (Stewart, 1973, pp. 74–77).

[19] We will investigate spherical harmonics at some length in Part III and the treatment here is at an introductory level.

Proof. See (Colton and Kress, 1998, Theorem 2.7, pp. 25–26).   □

> **Remark 2.19.** The spherical harmonics in (Colton and Kress, 1998) differ in a non-critical way from that defined above, but this does not change the result; see Section 7.4.1, p. 198.   □

> **Remark 2.20.** Index $\ell \in \{0, 1, 2, \ldots\}$ is called the *degree* and index $m \in \{-\ell, -\ell+1, \ldots, \ell\}$ is called the *order* of the spherical harmonics.   □

Any $f \in L^2(\mathbb{S}^2)$ can be expressed

$$
\begin{aligned}
f(\theta, \phi) &= \sum_{\ell=0}^{\infty} \sum_{m=-\ell}^{\ell} \alpha_\ell^m Y_\ell^m(\theta, \phi) \\
&= \sum_{\ell,m} \alpha_\ell^m Y_\ell^m(\theta, \phi),
\end{aligned}
\tag{2.70}
$$

where we introduce the shorthand

$$
\sum_{\ell,m} \triangleq \sum_{\ell=0}^{\infty} \sum_{m=-\ell}^{\ell} \quad \text{and} \quad \sum_{\ell,m}^{L} \triangleq \sum_{\ell=0}^{L} \sum_{m=-\ell}^{\ell},
\tag{2.71}
$$

where the latter represents a truncation of degree ℓ to a maximum of L. The Fourier coefficients are given by

$$
\begin{aligned}
\alpha_\ell^m &= \int_0^{2\pi} \int_0^{\pi} f(\theta, \phi) \overline{Y_\ell^m(\theta, \phi)} \sin\theta \, d\theta \, d\phi \\
&= \langle f, Y_\ell^m \rangle
\end{aligned}
\tag{2.72}
$$

and the integral (2.72) is called the *spherical harmonic transform* of f. As usual (2.70) represents an equality in the sense of convergence in the mean or strong convergence, Definition 2.24, p. 74,

$$
\lim_{L \to \infty} \left\| f - \sum_{\ell,m}^{L} \alpha_\ell^m Y_\ell^m \right\| = 0,
$$

where the norm is the one induced from (2.68), that is,

$$
\|f\|^2 = \int_0^{2\pi} \int_0^{\pi} |f(\theta, \phi)|^2 \sin\theta \, d\theta \, d\phi,
$$

noting $\sin\theta \geq 0$ for $0 \leq \theta \leq \pi$.

We now recast these equations in the form compatible with our general notation and indexing of orthonormal sequences. To begin, we note that the double indexing over ℓ and m forms a triangular array with ℓ horizontal and m vertical. This is shown in Figure 2.11, which indicates a natural ordering that puts the indices, ℓ and m, in a one-to-one correspondence with an index n ranging over $\mathbb{Z}^\star$. Algebraically this bijection is defined by

$$
n = \ell(\ell+1) + m,
\tag{2.73}
$$

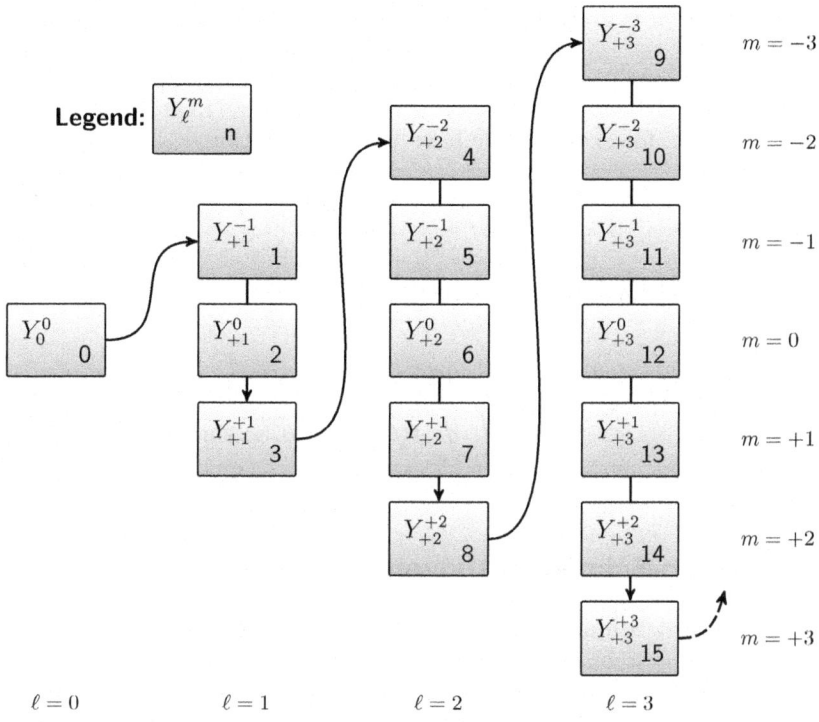

Figure 2.11 Linear ordering in the triangular array pairs. This shows for the spherical harmonics that the double summation $\sum_{\ell,m} \equiv \sum_{\ell=0}^{\infty} \sum_{m=-\ell}^{\ell}$ can be reordered into the form $\sum_{n=0}^{\infty}$.

with inverse

$$\ell = \lfloor \sqrt{n} \rfloor,$$
$$m = n - \lfloor \sqrt{n} \rfloor (\lfloor \sqrt{n} \rfloor + 1),$$

where $\lfloor \cdot \rfloor$ is the floor function. For example, $n = 10$ yields $\ell = \lfloor \sqrt{10} \rfloor = 3$ and $m = 10 - 3 \times 4 = -2$, that is,

$$\varphi_{10}(\theta, \phi) = Y_3^{-2}(\theta, \phi) = \sqrt{\frac{7}{480\pi}} P_3^2(\cos\theta) e^{-i2\phi}$$

$$= \sqrt{\frac{105}{8\pi}} \cos\theta \sin^2\theta e^{-i2\phi}$$

using (2.69).

Hence, we can define a complete orthonormal set through the obtuse formula

$$\{\varphi_n\}_{n=0}^{\infty}, \quad \text{where} \quad \varphi_n(\theta, \phi) \triangleq Y_{\lfloor \sqrt{n} \rfloor}^{n - \lfloor \sqrt{n} \rfloor (\lfloor \sqrt{n} \rfloor + 1)}(\theta, \phi). \tag{2.74}$$

This means that (2.70) is in the standard form

$$f = \sum_{n=0}^{\infty} \langle f, \varphi_n \rangle \varphi_n,$$

where n is given by (2.73) and φ_n by (2.74). The Fourier coefficients (2.72) are

$$\alpha_{\lfloor\sqrt{n}\rfloor}^{n-\lfloor\sqrt{n}\rfloor(\lfloor\sqrt{n}\rfloor+1)} = \langle f, \varphi_n\rangle.$$

Hence, for example, we have Parseval relation

$$\|f\|^2 = \sum_{n=0}^{\infty} |\langle f, \varphi_n\rangle|^2, \quad \forall f \in L^2(\mathbb{S}^2),$$

which is preferable to writing something like

$$\int_0^{2\pi}\int_0^{\pi} |f(\theta,\phi)|^2 \sin\theta\, d\theta\, d\phi =$$
$$\sum_{\ell,m} \left| \int_0^{2\pi}\int_0^{\pi} f(\theta,\phi)\overline{Y_\ell^m(\theta,\phi)} \sin\theta\, d\theta\, d\phi \right|^2, \quad \forall f \in L^2(\mathbb{S}^2). \quad (2.75)$$

2.10   Gram-Schmidt orthogonalization

Erhard Schmidt (1876–1959), Hilbert's student, as well as laying down in a more systematic way what we know refer to as Hilbert space (Pietsch, 2010), developed an orthogonalization procedure as a means to establish the existence of a countably infinite complete orthonormal sequence for the separable Hilbert space $L^2(\Omega)$ with the usual inner product. It is a straightforward procedure that tends to be a little tedious to describe. Before the formal description, we present an explicit construction which partially illuminates the procedure.

2.10.1   Legendre polynomial construction

Consider the interval $[-1, +1]$. We construct a set of basis functions starting from the non-orthogonal, but independent, power polynomials $\{1, x, x^2, x^3, \ldots\}$.[20] We can simplify our work by observing that we can partition the power polynomials into the even functions and odd functions since these two sets are already orthogonal. Beginning with constant function 1 we need to normalize it so

$$\varphi_0(x) \triangleq \sqrt{\frac{1}{2}}.$$

Next is x, which is already orthogonal to $\varphi_0(x)$. So we just normalize it

$$\varphi_1(x) \triangleq \sqrt{\frac{3}{2}}x.$$

To make x^2 orthogonal to $\varphi_0(x)$ we need to subtract off its projection onto $\varphi_0(x)$

$$\int_{-1}^{+1} \sqrt{\frac{1}{2}}x^2 dx = 1/3,$$

---

[20] That they are independent follows from the property that no power can be expressed as a linear combination of any other power.

Schmidt, Erhard (1876–1959) — Schmidt's advisor was Hilbert. In fact, the geometric development of Hilbert space closer to the modern form began with Schmidt, who introduced the norm notation, convergence concepts and made the first references to function spaces. Hilbert had the breakthrough ideas in his study of integral equations with a symmetric kernel in 1904, which in turn were inspired by Fredholm's work on integral equations. It should also be noted that Riesz and Fischer, along with Schmidt, progressed the notion of Hilbert space towards its modern formulation in the years up to 1910.

Walter Ledermann, who came in contact with Schmidt later in his career in Berlin, said he was a man of impressive appearance, charismatic and had a sense of humor. He was also regarded as eccentric, but that trait is not unusual in mathematicians. Schmidt organized his lectures late in the day and sometimes wore a dress suit so as to enjoy evening engagements in the city outside the university. He was very highly regarded as a person by all who knew him: "Schmidt belonged to the aristocrats among the mathematicians." For further reading and a good summary of what Schmidt did that built on the work of Hilbert; see (Pietsch, 2010).

then normalize leading to

$$\varphi_2(x) \triangleq \sqrt{\frac{5}{8}} (3x^2 - 1).$$

By projecting off the lower-order normalized polynomials this procedure leads to the normalized Legendre polynomials. Note that there is a lot of flexibility in the choice of the initial independent set, the inner product and the order of construction. The above computation happens to be the one that yields the normalized Legendre polynomials. In principle, any of the orthonormal sets of functions can be constructed this way. Despite being a construction, the *Gram-Schmidt orthogonalization* procedure has greater theoretical value.

2.10.2   Orthogonalization procedure

Let there be an indexed set of independent vectors $\{f_n\}_{n=1}^{\infty}$ in our Hilbert space $\mathcal{H}$. This set can be finite or countably infinite and does not need to be complete. Without loss of generality we will take the set to be a sequence, that is, countably infinite $\{f_n\}_{n=1}^{\infty}$. The goal is to replace this set with a second set of orthonormal vectors $\{\varphi_n\}$ which span the same subspace in the sense that each of the f_n can be expressed as a linear combination of the $\{\varphi_n\}$ and vice versa. In fact the procedure, because of its recursive nature, leads to a much stronger spanning property that $\{f_n\}_{n=1}^{N}$ spans the same space as $\{\varphi_n\}_{n=1}^{N}$ for all $N \in \mathbb{N}$.

We start by defining

$$\varphi_1 \triangleq \frac{f_1}{\|f_1\|}.$$

The single element sets $\{f_1\}$ and $\{\varphi_1\}$ span the same subspace.

Now we remove from the next element $\{f_2\}$ its projection in the direction of $\{\varphi_1\}$

$$\nu_2 \triangleq f_2 - \langle f_2, \varphi_1 \rangle$$

and normalize this

$$\varphi_2 \triangleq \frac{\nu_2}{\|\nu_2\|}.$$

This ensures that $\{\varphi_2\}$ is orthogonal to $\{\varphi_1\}$ and normalized. The two-element sets $\{f_1, f_2\}$ and $\{\varphi_1, \varphi_2\}$ span the same subspace.

Now we remove from the next element $\{f_3\}$ its projection in the direction of $\{\varphi_1\}$ and $\{\varphi_2\}$

$$\nu_3 \triangleq f_3 - \langle f_3, \varphi_1 \rangle - \langle f_3, \varphi_2 \rangle$$

and normalize this

$$\varphi_3 \triangleq \frac{\nu_3}{\|\nu_3\|}.$$

This ensures that $\{\varphi_3\}$ is orthogonal to $\{\varphi_1\}$, and $\{\varphi_2\}$ is normalized. The three-element sets $\{f_1, f_2, f_3\}$ and $\{\varphi_1, \varphi_2, \varphi_3\}$ span the same subspace.

This process continues such that at the $(n+1)$-th step

$$\nu_{n+1} \triangleq f_{n+1} - \sum_{m=1}^{n} \langle f_{n+1}, \varphi_m \rangle,$$

which can be normalized

$$\varphi_{n+1} \triangleq \frac{\nu_{n+1}}{\|\nu_{n+1}\|}.$$

This ensures that $\{\varphi_{n+1}\}$ is orthogonal to $\{\varphi_1, \varphi_2, \ldots, \varphi_n\}$ and $\{\varphi_{n+1}\}$ is normalized. The $(n+1)$-element sets $\{f_1, f_2, \ldots, f_{n+1}\}$ and $\{\varphi_1, \varphi_2, \ldots, \varphi_{n+1}\}$ span the same subspace.

> **Remark 2.21.** Assume we are interested in functions defined on an interval of the real line. Just as for the example with the Legendre polynomials, if the initial choice of the independent set corresponds to the power polynomials this implies that the resulting orthonormal set will also be polynomials. ☐

> **Remark 2.22.** The above procedure can be continued indefinitely. However, it is restricted to constructing countable orthonormal sequences. ☐

2.11 Completeness relation

With any orthonormal sequence of functions there is an identity among the functions in the sequence called the "completeness relation," which arises when the sequence is complete. An example of a completeness relation is the one for Fourier series on the interval $[-\pi, +\pi]$

$$\frac{1}{2\pi} \sum_{n=-\infty}^{\infty} e^{inx} e^{-iny} = \delta(x - y), \tag{2.76}$$

where $\delta(\cdot)$ is the Dirac delta function . In this case, we can further simplify (2.76) (for example, by letting $z = x - y$), but for other orthonormal sequences this degree of simplification is not expected.[21]

That completeness can be captured/defined in a different, but equivalent way, is one aspect of the result we show next. The other aspect is that it gives a very useful identity that complements the orthonormality identity $\langle \varphi_n, \varphi_m \rangle = \delta_{n,m}$.

Theorem 2.13 (Completeness relation). *Completeness of any orthonormal expansion* $\{\varphi_n\}$ *on* $L^2(\Omega)$, *(2.53), is equivalent to:*

$$\boxed{\sum_{n=1}^{\infty} \varphi_n(x)\overline{\varphi_n(y)} = \delta(x - y).} \tag{2.77}$$

☐

Proof (sketch). For comparison, we recall the "sifting property" for the Dirac delta function (Oppenheim et al., 1996)

$$f(x) = \int_\Omega f(y)\delta(x - y)\, dy, \quad \forall f \in L^2(\Omega). \tag{2.78}$$

---

[21] In particular, for each $\varphi_n(x)$, there is no expectation in general that $\varphi_n(x)\overline{\varphi_n(y)}$ is a function of the difference $x - y$.

Starting with (2.53), we can expand the Fourier coefficient and exchange the order of the integral and summation to yield

$$f(x) = \sum_{n=1}^{\infty} \int_{\Omega} f(y)\overline{\varphi_n(y)}\,dy\,\varphi_n(x)$$

$$= \int_{\Omega} \Big(\sum_{n=1}^{\infty} \varphi_n(x)\overline{\varphi_n(y)} \Big) f(y)\,dy, \quad \forall f \in L^2(\Omega),$$

which can be compared with (2.78) to infer (2.77). To rigorously prove this requires the use of theory of distributions, noting that the Dirac delta function is not in $L^2(\Omega)$. $\qquad\square$

Remark 2.23. Completeness in (2.77) is like saying there are enough functions to be able to synthesize a Dirac delta function. In fact all functions in the expansion are required, no more, no less. $\qquad\square$

If the domain is infinite, that is, $\Omega \equiv (\alpha, \beta)$ is infinite, then the completeness relation is expressed in terms of an integral and continuous Dirac delta function instead of a summation. For example,

$$\frac{1}{2\pi} \int_{-\infty}^{\infty} e^{ikx} e^{-iky}\,dk = \delta(x - y), \quad x, y \in \mathbb{R}.$$

Remark 2.24. Equation (2.77) is in the form (2.53) so we can infer the Fourier coefficients of a shifted Dirac delta function, that is,

$$f_\eta(x) = \delta(x - \eta)$$

has Fourier coefficients

$$\langle f_\eta, \varphi_n \rangle = \overline{\varphi_n(\eta)}.$$

However, the Dirac delta function is not in $L^2(\Omega)$ so this view needs more careful justification than we have considered here. $\qquad\square$

2.11.1   Completeness relation with weighting

To obtain the weighted completeness relation we follow the computation as in the un-weighted case

$$f(x) = \sum_{n=1}^{\infty} \int_{\Omega} w(y)f(y)\overline{\varphi_n(y)}\,dy\,\varphi_n(x)$$

$$= \int_{\Omega} \Big(\sum_{n=1}^{\infty} w(y)\varphi_n(x)\overline{\varphi_n(y)} \Big) f(y)\,dy, \quad \forall f \in L^2(\Omega),$$

where $w(x) \geq 0$ is a real, non-negative weighting function. From this we can directly infer

$$w(y) \sum_{n=1}^{\infty} \varphi_n(x)\overline{\varphi_n(y)} = \delta(x - y). \qquad (2.79)$$

However, given $w(\cdot)$ is real and $\delta(\cdot)$ is real and even, there are a number of equivalent expressions to (2.79). By such a transformation we can readily obtain

$$
w(x) \sum_{n=1}^{\infty} \varphi_n(x)\overline{\varphi_n(y)} = \delta(x - y). \tag{2.80}
$$

In fact, a slightly more general (and directly useful) completeness relation with index n dependent weighting is possible; see Problem 2.30.

Example 2.18. The spherical harmonics, $\{Y_\ell^m\}$, with the orthonormality given by

$$
\int_0^{2\pi}\int_0^\pi Y_\ell^m(\theta, \phi)\overline{Y_p^q(\theta, \phi)} \sin\theta \, d\theta \, d\phi = \delta_{\ell,p}\delta_{m,q}, \tag{2.81}
$$

have an associated weighting function given by $w(\theta, \phi) = \sin\theta$. In this case the *spherical harmonics completeness relation* is given by

$$
\sin\theta \sum_{\ell=0}^{\infty} \sum_{m=-\ell}^{\ell} Y_\ell^m(\theta, \phi)\overline{Y_\ell^m(\vartheta, \varphi)} = \delta(\theta - \vartheta)\delta(\phi - \varphi). \tag{2.82}
$$

Some authors transfer the $\sin\theta$ term into $\delta(\cos\theta - \cos\vartheta)$ term; see (2.86) in Problem 2.34.   □

Remark 2.25. In (2.82) there appears to be an issue with $\theta = 0$ or $\theta = \pi$, as $\sin\theta = 0$ in those cases. However, when used under the integral sign it is cancelled by the leading portion of the differential term $\sin\theta \, d\theta \, d\phi$ in (2.81). This is also somewhat addressed in Problem 2.34.   □

Problems

2.29. Show that

$$
\langle f, g \rangle = \sum_{n=1}^{\infty} \sum_{m=1}^{\infty} \langle f, \varphi_n \rangle \langle \varphi_n, \zeta_m \rangle \langle \zeta_m, g \rangle \tag{2.83}
$$

for complete orthonormal sequences $\{\varphi_n\}_{n=1}^{\infty}$ and $\{\zeta_m\}_{m=1}^{\infty}$. What is the generalization?

2.30. Prove the more general form of the completeness relation with index-dependent weighting, (2.80), given by

$$
\sum_{n=1}^{\infty} w_n(x)\varphi_n(x)\overline{\varphi_n(y)} = \delta(x - y). \tag{2.84}
$$

2.31 (Legendre polynomials). The Legendre polynomials, $\{P_\ell(x)\}_{\ell=0}^{\infty}$, satisfy the orthogonality (2.32), p. 55,

$$
(\ell + \tfrac{1}{2}) \int_{-1}^{+1} P_\ell(x)P_m(x) \, dx = \delta_{\ell,m}.
$$

Show that the *Legendre polynomials completeness relation* takes the form

$$\sum_{\ell=0}^{\infty}(\ell + \tfrac{1}{2})P_\ell(x)P_\ell(y) = \delta(x-y).$$

2.32 (Hermite polynomials). Show the *Hermite polynomials completeness relation* is given by

$$e^{-x^2}\sum_{n=0}^{\infty}\frac{1}{2^n n!\sqrt{\pi}}H_n(x)H_n(y) = \delta(x-y),$$

where the Hermite polynomials are defined in Section 2.5.2, p. 57.

2.33 (Generalized Laguerre polynomials). The *generalized Laguerre polynomials* are a set of orthogonal functions on the non-negative real line $\mathbb{R}^\star \triangleq [0,\infty)$ parameterized by $\alpha \geq 0$, written $L_n^{(\alpha)}(x)$ for $n \in \{0,1,2,\ldots\}$. They satisfy the orthogonality equation

$$\int_0^{\infty} x^\alpha e^{-x}L_n^{(\alpha)}(x)L_m^{(\alpha)}(x)\,dx = \frac{\Gamma(n+\alpha+1)}{n!}\delta_{n,m}, \qquad (2.85)$$

where $\Gamma(\cdot)$ is the Gamma function. Show that the *generalized Laguerre polynomials completeness relation* is given by

$$x^\alpha e^{-x}\sum_{n=0}^{\infty}\frac{n!}{\Gamma(n+\alpha+1)}L_n^{(\alpha)}(x)L_n^{(\alpha)}(y) = \delta(x-y).$$

Note that the regular *Laguerre polynomials*, written $L_n(x)$ for $n \in \{0,1,2,\ldots\}$, are recovered in the case $\alpha = 0$ and the *Laguerre polynomials completeness relation* is

$$e^{-x}\sum_{n=0}^{\infty}L_n(x)L_n(y) = \delta(x-y).$$

2.34 (Spherical harmonics). In the spherical coordinate system prove the identity

$$\frac{1}{\sin\theta}\delta(\theta - \vartheta) = \delta(\cos\theta - \cos\vartheta), \qquad (2.86)$$

and thereby show the *spherical harmonics completeness relation*

$$\sum_{\ell=0}^{\infty}\sum_{m=-\ell}^{\ell}Y_\ell^m(\theta,\phi)\overline{Y_\ell^m(\vartheta,\varphi)} = \delta(\cos\theta - \cos\vartheta)\delta(\phi - \varphi), \qquad (2.87)$$

which is an alternative to (2.82).

2.12 Taxonomy of Hilbert spaces

A Hilbert space comes in one of two general flavors: it is either a *non-separable Hilbert space* or a *separable Hilbert space*. Broadly speaking this classification partitions Hilbert spaces into the esoteric in the former case, and more concrete and useful in the latter case.

2.12.1   Non-separable Hilbert spaces

Non-separable Hilbert spaces are less commonly useful and we will content ourselves with a contrived example given in (Debnath and Mikusiński, 1999)[p. 127].
The construction at its heart relies essentially on the property

$$\aleph_0 \cdot \aleph_0 = \aleph_0,$$

given in Table 1.2, p. 16, which gives the algebraic properties of $\aleph_0$.

Consider the space of all complex-valued functions defined to be zero everywhere on the real line $\mathbb{R}$ except at a countable number of points such that

$$\sum_{x \in S_f} |f(x)|^2 < \infty, \tag{2.88}$$

where

$$S_f \triangleq \{x \in \mathbb{R} \colon f(x) \neq 0\} \subset \mathbb{R}$$

denotes the countable subset of $\mathbb{R}$ (that depends on f). Such functions have
countably infinite support and are zero a.e. Then

$$\langle f, g \rangle \triangleq \sum_{x \in S_f \cap S_g} f(x)\overline{g(x)}$$

defines an inner product whose induced norm reduces to (2.88). If the two functions f and g share no common points in their support then they are orthogonal.
They can also be orthogonal if they share at least two common points and the
inner product is zero (in an ℓ^2 type of way).

Assume this space is separable and consider a candidate complete orthonormal
sequence of functions in the space given by $\{\varphi_n\}_{n=1}^\infty$. Since there are a countable
($\aleph_0$) number of φ_n (by separability) and each is non-zero only at a countable
($\aleph_0$) number of points then such an orthonormal sequence can at most cover
a countable ($\aleph_0 \cdot \aleph_0 = \aleph_0$) set of points on the real line. Call this set S_φ.
Now construct a function which is non-zero at a countable number of points
in the complement of set S_φ (this is always possible since $\mathbb{R}$ is uncountable).
This function is orthogonal to $\{\varphi_n\}_{n=1}^\infty$, which contradicts the sequence being
complete. This holds for any candidate $\{\varphi_n\}_{n=1}^\infty$ and so by contradiction the
space cannot be separable.

2.12.2   Separable Hilbert spaces

Separable Hilbert spaces are subsumed within the notion of separable normed
spaces; see Section 2.3.3, p. 29. This is because the inner product has no role
to play in determining separability — having a norm is sufficient. But there is
a surprise. As much as we would like to think there are a plethora of different
manifestations of countably infinite-dimensional Hilbert spaces there are only
two (one real one corresponding to the choice of a real scalar field and one
complex one corresponding to the choice of a complex scalar field). What this
actually means will be clarified. We build up towards this interesting result with
a few preliminary results and definitions.

Theorem 2.14. *Every infinite orthogonal set in a separable Hilbert space $\mathcal{H}$ is countable.*   □

Proof. Let $\mathcal{O}$ be an orthogonal set, not necessarily countable, in a separable Hilbert space. Without loss of generality we can normalize the elements of $\mathcal{O}$, that is, we take $\mathcal{O}$ be an orthonormal set.[22] By Problem 2.12, p. 45, we know that for every $f, g \in \mathcal{O}$ we have $\|f - g\| = 2$ whenever $f \neq g$.

Construct spherical neighborhoods of radius less than $1/\sqrt{2}$ centered on every point in $\mathcal{O}$. With such a choice of radius, all spherical neighborhoods are disjoint, that is, have no points in common. Call these neighborhoods the "$\mathcal{O}$-balls," which are understood to be disjoint. The cardinality of $\mathcal{O}$ clearly equals the cardinality of the set of all $\mathcal{O}$-balls. Hence if we can show the $\mathcal{O}$-balls are countable then we are done.

By definition any dense subset of $\mathcal{H}$ has at least one point in every neighborhood. Therefore, every dense subset has at least one point in each of the $\mathcal{O}$-balls. All such points are necessarily distinct since none of the $\mathcal{O}$-balls intersect.

Since $\mathcal{H}$ is a separable Hilbert space, by definition there exists at least one dense subset $\mathcal{D}$ which is countable. We must have at least one point from this countable dense subset in each of the $\mathcal{O}$-balls. Hence we have a one-to-one correspondence between members of a countable set, $\mathcal{D}$, and every $\mathcal{O}$-Ball and, hence, with every point in $\mathcal{O}$. Therefore, the $\mathcal{O}$-balls are countable.   □

> **Remark 2.26.** If the orthogonal set had finite cardinality then it is a not an issue. Hence, it is common to simply state: every orthogonal set in a separable Hilbert space $\mathcal{H}$ is countable.   □

> **Remark 2.27.** Suppose we start with a complete orthonormal sequence and discard members such that we still have a countable number left. We conclude that a countable orthonormal set need not be complete.   □

Theorem 2.15 (Separable Hilbert space). *If a Hilbert space $\mathcal{H}$ contains a complete orthonormal sequence then $\mathcal{H}$ is separable.*   □

Proof. A proof can be found in (Kreyszig, 1978).   □

The above theorem is sometimes used in defining a separable Hilbert space. Given our development to this point, this theorem should now be regarded as intuitively reasonable.

Definition 2.28 (Hilbert dimension). *The Hilbert dimension of a Hilbert space $\mathcal{H}$ is the cardinality of any complete orthonormal set.*   □

All complete orthonormal sets can be shown to have the same cardinality. Hence Definition 2.28 can also be phrased in terms of *all* complete orthonormal sets. In the context of our work here it is clear that the Hilbert dimension for $\mathbb{R}^N$ or $\mathbb{C}^N$ is N, and the Hilbert dimension for a separable real or complex Hilbert

---

[22] That is, there is an obvious bijection between the set of points in an orthogonal set and the normalized versions of the same points. Therefore, these sets have the same cardinality.

space is $\aleph_0$. An infinite-dimensional Hilbert space is separable if and only if its dimension is $\aleph_0$.

We now dive down into the structure and form of Hilbert spaces. A Hilbert space which corresponds to a relabeling of the points of another Hilbert space is not strictly a different Hilbert space. In the interest of filtering out such clones and avoiding senseless repetition we make the following definition.

Definition 2.29 (Hilbert space isomorphism). *A Hilbert space $\mathcal{H}_1$ is said to be isomorphic to a Hilbert space $\mathcal{H}_2$ if there exists a bijective linear mapping $\mathcal{L}$ from $\mathcal{H}_1$ to $\mathcal{H}_2$ such that*

$$\langle \mathcal{L}f, \mathcal{L}g \rangle = \langle f, g \rangle,$$

for every $f, g \in \mathcal{H}_1$. Such a mapping $\mathcal{L}$ is called a Hilbert space isomorphism between $\mathcal{H}_1$ and $\mathcal{H}_2$. □

Determining whether two Hilbert spaces are isomorphic reduces, surprisingly, to checking only the dimension of the Hilbert space. The following results are for the field of complex scalars. An analogous result holds for real scalars.

Theorem 2.16 (Hilbert space classification). *Two complex Hilbert spaces $\mathcal{H}_1$ and $\mathcal{H}_2$ are isomorphic if and only if they have the same Hilbert dimension.* □

That is, Hilbert spaces can be completely classified — there is a unique Hilbert space up to isomorphism for every cardinality of the basis. More specifically in the case of separable Hilbert spaces we have an embellishment.

Theorem 2.17 (Separable Hilbert space classification). *Let $\mathcal{H}$ be a separable Hilbert space. Then we have the following: (a) If $\mathcal{H}$ is infinite-dimensional, then it is isomorphic to ℓ^2. (b) If $\mathcal{H}$ has finite dimension N, then it is isomorphic to $\mathbb{C}^N$.* □

Proof. The proof is by explicit construction of an isomorphism from $\mathcal{H}$ to ℓ^2. (One can guess that it is the generalized Parseval relation (2.57), p. 72, which does the trick.) Let $\{\varphi_n\}_{n=1}^{\infty}$ be a complete orthonormal sequence in $\mathcal{H}$. For $f \in \mathcal{H}$ define the linear mapping

$$\mathcal{L}f \triangleq \left\{ \langle f, \varphi_n \rangle \right\}_{n=1}^{\infty},$$

from which we infer[23]

$$\langle \mathcal{L}f, \mathcal{L}g \rangle \triangleq \left\langle \left\{ \langle f, \varphi_n \rangle \right\}_{n=1}^{\infty}, \left\{ \langle g, \varphi_n \rangle \right\}_{n=1}^{\infty} \right\rangle$$

$$= \sum_{n=1}^{\infty} \langle f, \varphi_n \rangle \overline{\langle g, \varphi_n \rangle} = \sum_{n=1}^{\infty} \langle f, \langle g, \varphi_n \rangle \varphi_n \rangle$$

$$= \left\langle f, \sum_{n=1}^{\infty} \langle g, \varphi_n \rangle \varphi_n \right\rangle = \langle f, g \rangle.$$

Thus, $\mathcal{L}$ is an isomorphism from $\mathcal{H}$ onto ℓ^2. □

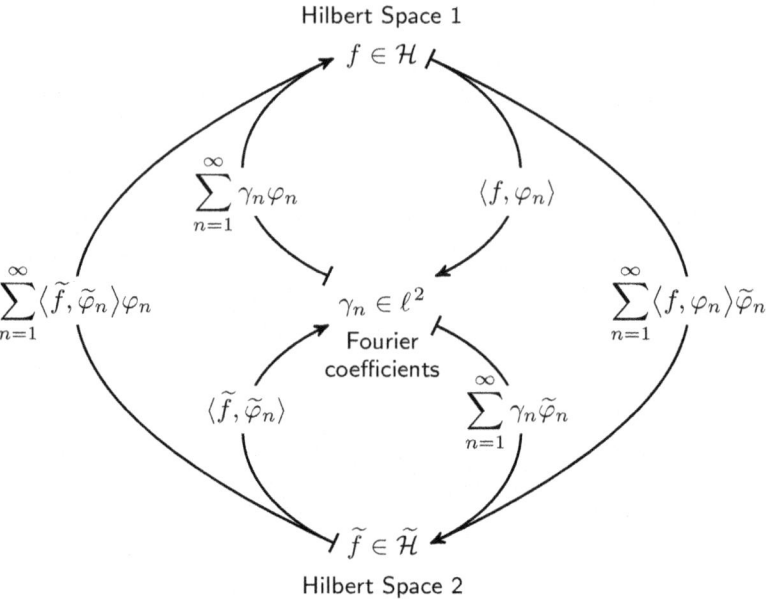

Figure 2.12 Isomorphic separable Hilbert spaces $\mathcal{H}$, ℓ^2 and $\widetilde{\mathcal{H}}$. In establishing the isomorphism we have shown a set of Fourier coefficients γ_n which are shared between the two Hilbert spaces. In essence, $\gamma_n = \langle f, \varphi_n \rangle = \langle \widetilde{f}, \widetilde{\varphi}_n \rangle$.

Remark 2.28. Since all infinite-dimensional separable Hilbert spaces are isomorphic to ℓ^2 it is clear that all infinite-dimensional Hilbert spaces are isomorphic to each other — this isomorphism is captured in Figure 2.12. Let $\mathcal{H}$ and $\widetilde{\mathcal{H}}$ be any two infinite-dimensional separable Hilbert spaces. Let $\{\varphi_n\}_{n=1}^{\infty}$ and $\{\widetilde{\varphi}_n\}_{n=1}^{\infty}$ be complete orthonormal sequences in $\mathcal{H}$ and $\widetilde{\mathcal{H}}$, respectively. Following the links in Figure 2.12 we infer that

$$\mathcal{L}f \triangleq \sum_{n=1}^{\infty} \langle f, \varphi_n \rangle \widetilde{\varphi}_n \longmapsto \widetilde{f} \in \widetilde{\mathcal{H}} \tag{2.89}$$

$$\widetilde{\mathcal{L}}\widetilde{f} \triangleq \sum_{n=1}^{\infty} \langle \widetilde{f}, \widetilde{\varphi}_n \rangle \varphi_n \longmapsto f \in \mathcal{H} \tag{2.90}$$

define bijections, and hence isomorphisms, between $\mathcal{H}$ and $\widetilde{\mathcal{H}}$, distilled in Figure 2.13. Note that the inner product in (2.89) is the one in $\mathcal{H}$, and the inner product in (2.90) is the one in $\widetilde{\mathcal{H}}$. There is no ambiguity because it is implicit by examining the arguments.   □

Remark 2.29. In Figure 2.13, $\{\varphi_n\}_{n=1}^{\infty}$ and $\{\widetilde{\varphi}_n\}_{n=1}^{\infty}$ are *any* complete orthonormal sequences in their respective spaces. Hence, there is an unlimited number of ways to establish the isomorphism.   □

---

[23] In the inner product expressions, the particular space, whether it is an abstract Hilbert space or ℓ^2, can be inferred from the arguments.

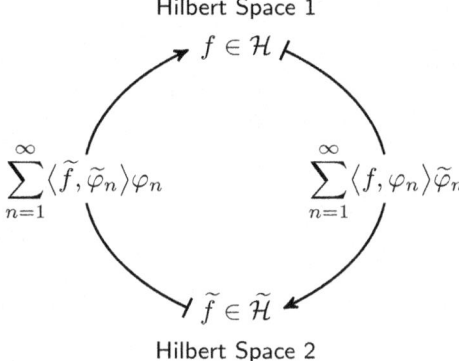

Figure 2.13 Bijection between separable Hilbert spaces $\mathcal{H}$ and $\widetilde{\mathcal{H}}$ (this figure is a distillation of Figure 2.12) by omitting the intermediate Fourier coefficients and the associated isomorphic ℓ^2 space. The isomorphism can be established by mapping a given $f \in \mathcal{H}$ to $\widetilde{f} \in \widetilde{\mathcal{H}}$ (top to bottom) by using the Fourier coefficients relative to the $\{\varphi_n\}_{n=1}^{\infty}$ orthonormal sequence in $\mathcal{H}$ to weight the $\{\widetilde{\varphi}_n\}_{n=1}^{\infty}$ orthonormal sequence in $\widetilde{\mathcal{H}}$, and vice versa (bottom to top).

Clone interpretation of isomorphic Hilbert spaces

How do we interpret these results? Table 2.5 may help. Two isomorphic Hilbert spaces are algebraically (vector space-wise), metrically (norm-wise) and geometrically (inner-product-wise) indistinguishable. This means, apart from the labeling and interpretation we give to the points or elements of the space, we are deceiving ourselves if we think that two isomorphic Hilbert spaces are (structurally) different. This is because there is actually only one complex infinite-dimensional separable Hilbert space. They are all clones. Similarly there is only one real infinite-dimensional separable Hilbert space.

In finite dimensions, for every dimension there is one complex Hilbert space and one real Hilbert space. Furthermore, there is no sense in which we need to consider separability for finite-dimensional Hilbert spaces, but it is proper to regard these spaces as separable (since they are finitely countable).

There is a second, necessarily equivalent, means to interpret separable Hilbert spaces. One means to motivate Hilbert spaces is as the infinite-dimensional generalization of N-dimensional (finite-dimensional) vector spaces. The most naive way we could conceive of answering this call is to replace vectors with sequences, that is, go from objects with the N-tuple structure

$$(\alpha_1, \alpha_2, \ldots, \alpha_N)'$$

to objects with the sequence structure as $N \to \infty$

$$\{\alpha_1, \alpha_2, \alpha_3, \ldots\}, \tag{2.91}$$

which is not that imaginative. After all there are wildernesses of exotic functions to hunt and explore — given that families of functions also seem to require an

Table 2.5 Hilbert spaces ℓ^2 and $L^2(\Omega)$ have equivalent and isometric structures. Here for $L^2(\Omega)$, $\alpha_n = \langle f, \varphi_n \rangle$ and $\beta_n = \langle g, \varphi_n \rangle$.

Hilbert spaces		
Vector space condition	**Fourier series**	**Inner product**
$a \in \ell^2 \iff \displaystyle\sum_{n=1}^{\infty} |\alpha_n|^2 < \infty$	$a = \displaystyle\sum_{n=1}^{\infty} \alpha_n e_n$	$\langle a, b \rangle = \displaystyle\sum_{n=1}^{\infty} \alpha_n \overline{\beta_n}$
$b \in \ell^2 \iff \displaystyle\sum_{n=1}^{\infty} |\beta_n|^2 < \infty$	$b = \displaystyle\sum_{n=1}^{\infty} \beta_n e_n$	
$f \in L^2(\Omega) \iff \displaystyle\int_\Omega |f(x)|^2 dx < \infty$	$f(x) = \displaystyle\sum_{n=1}^{\infty} \alpha_n \varphi_n(x)$	$\langle f, g \rangle = \displaystyle\int_\Omega f(x)\overline{g(x)}\,dx$
$g \in L^2(\Omega) \iff \displaystyle\int_\Omega |g(x)|^2 dx < \infty$	$g(x) = \displaystyle\sum_{n=1}^{\infty} \beta_n \varphi_n(x)$	

infinite number of dimensions for their characterization. The reality of separable Hilbert spaces is that they essentially *are just* the boring sequences (2.91), despite any esoteric clothing we don on them. That may seem like a bit of a let-down.

Functions are just labels

This leads to an interesting interpretation of what it really means when we have a function space as our archetype model for a separable Hilbert space. The functions are not essential elements to the properties of the space and are really just labels for the elements of our abstract Hilbert space. Of course, these functions have a real use aligned with the application we have in mind. Functions are also useful because they are more interesting labels than sequences. Furthermore, were we to stray into non-separable spaces then the sequences, because of their countable nature, are not expected to be adequate.

In a separable Hilbert space, each choice of complete orthonormal sequence $\{\varphi_n\}_{n=1}^{\infty}$ induces a different, but equivalent, mapping to ℓ^2. What do these differences mean? In the same way that unitary transformations in $\mathbb{C}^N$ correspond to different choices of basis vectors obtained by rotating and flipping axes, the generalization to the sequence case means that transforming between different choices of complete orthonormal sequences play the same role as the finite-dimensional unitary transformations.

Problem

2.35. Use the relationships shown in Figure 2.12, or otherwise, to establish algebraically that

$$\langle f, g \rangle = \langle \widetilde{f}, \widetilde{g} \rangle,$$

where $\widetilde{f} = \mathcal{L}f$, given in (2.89) and $g = \widetilde{\mathcal{L}}\widetilde{g}$, given in (2.90).

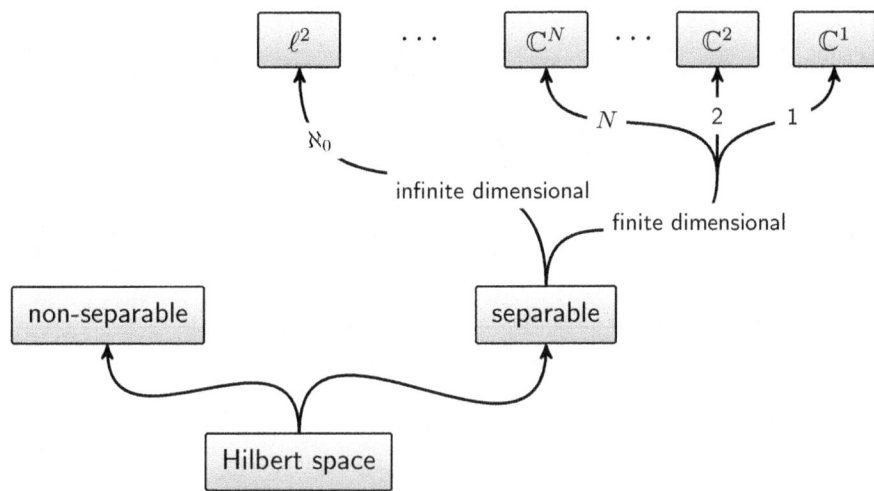

Figure 2.14 Partial taxonomy of complex Hilbert spaces. Hilbert spaces can be either separable or non-separable and we do not care much for the latter. In the separable case we have the canonical spaces $\mathbb{C}^N$ and ℓ^2 which are of Hilbert dimension N and $\aleph_0$, respectively. Any Hilbert space is isomorphic to either $\mathbb{C}^N$ and ℓ^2 only depending on the Hilbert dimension, N and $\aleph_0$, respectively.

2.12.3   The big (enough) picture

Bringing the results of the previous sections together we have Figure 2.14. It shows a partial taxonomy of complex Hilbert spaces. The picture is identical for real Hilbert spaces. We do not have much to say about non-separable Hilbert spaces and if the construction in Section 2.12.1 is anything to go by then these spaces do not appear useful or relevant. For the separable Hilbert spaces we have one for each Hilbert dimension, that is, for all finite Hilbert dimensions and one for $\aleph_0$ Hilbert dimension. That there is only one (countably) infinite-dimensional Hilbert space and a countable number of finite-dimensional Hilbert spaces is somewhat weird. No wonder Cantor died in a psychiatric clinic.

Take-home messages – story so far

M1: For engineers, almost all the exciting action takes place in separable Hilbert spaces.

M2: A Hilbert space is separable if and only if the cardinality of its complete orthonormal set, also referred to as its dimension, is countable.

M3: The dimension of Hilbert spaces is the only factor that determines whether there exists a bijective and linear mapping, or isomorphism, from one space onto another.

M4: Since the dimension of all denumerably countable separable Hilbert spaces is $\aleph_0$, all such separable Hilbert spaces are structurally the same and equivalent to a representative space, such as the ℓ^2 space.

M5: In a separable Hilbert space all objects can be faithfully represented by a sequence. The elements of the sequence for an object are the projections (or inner products) of that object along a complete orthonormal sequence (or basis) of the space.

PART II
Operators

3 Introduction to operators

3.1 Preamble

In the first part of the book we were mainly concerned with spaces, most notably separable Hilbert spaces, and the elements within them. It could be argued that physical processes always lead to separable spaces — if it is good enough for quantum physicists, it is good enough for us. It does not matter whether these elements are functions in an *archetype* Hilbert space $L^2(\Omega)$, sequences in the *canonical* Hilbert space ℓ^2, or esoteric elements in some *abstract* Hilbert space $\mathcal{H}$ with the same dimension, as these spaces are clones of each other.

In this part we are interested in *operations* on elements of separable Hilbert spaces. In fact, we went through all the pain to be able to work with and manipulate these elements. Due to its underlying engineering application, $L^2(\Omega)$ will be our concrete reference model. Functions in $L^2(\Omega)$ are real-world finite-energy *signals* which can depend on time (Ω being the real interval $[\alpha, \beta]$), 2-D space ($\Omega \times \Omega \subset \mathbb{R}^2$), 2-sphere $\mathbb{S}^2$, etc. The airline check-in agent definition of an operator is then *a mapping which takes a function or a signal as an input and generates a function or a signal as an output.* Talking in engineering language again, an operator is nothing but a *system* that transforms input signals into output signals. For example, the Fourier transform is an operator. A low-pass filter, which maps an input time-domain signal to a smoother version, is an operator. In differential calculus, derivative of a function is an operator.

Most systems that engineers deal with (or hope to deal with) are *linear* or can be approximated as (piecewise) linear. Therefore, operators in this book should almost always be understood as linear, though we will make an effort at the beginning to explicitly state if the result is general or is specific to linear operators. Also, engineers (or at least civilian engineers) are employed to design systems that are stable and do not blow up the input. In the hope to save as many engineering jobs as possible, we will study *bounded* linear operators.

Our goal in this part is to study main classes of linear operators and their properties and understand why they are important and how they can be used. We know from the previous part that using an appropriate set of complete orthonormal or basis functions, we can comfortably ignore that the signal originally lived in $L^2(\Omega)$ and deal with its corresponding Fourier series in ℓ^2. Due to such isomorphic relationship between $L^2(\Omega)$ and ℓ^2, and in general any Hilbert space $\mathcal{H}$ with the same dimension, our intuition says that there must a close relationship between how linear operators are defined for these spaces. We will formalize such relationship for bounded linear operators — a deep and important relationship which allows us to freely work with whichever representation of the space and operator that is most convenient and understandable.

Oftentimes, working with an abstract form is useful for some general derivations and provides some notational economy. Bounded linear operators in ℓ^2 are generalizations of a matrix with infinite dimensionality,[1] since we only need to understand how each of the orthonormal functions in the input space gets mapped to each of the orthonormal functions in the output space to fully define the action of the operator. This matrix viewpoint can undoubtedly guide us in our thinking and solving engineering problems. When we meet integral operators, working with functions in an archetype Hilbert space $L^2(\Omega)$ is needed.

3.1.1 A note on notation

So far in the book, we used the following notation:

N1: Normal Latin mathematics symbols, such as f, g, a, b, etc., to refer to elements of a Hilbert space $\mathcal{H}$, $L^2(\Omega)$ or ℓ^2.

N2: Normal Greek mathematics symbols α, β to refer to scalars (usually) in complex field $\mathbb{C}$.

N3: Curly brackets to represent sequences. In particular, an element of ℓ^2 such as a was written as a sequence $a = \{\alpha_n\}_{n=1}^{\infty}$ without appealing to their vector nature of infinite size.

From now on, we often need to deal with matrix representation of operators and multiplications of matrices and vectors is needed. Furthermore, we often need to alternately work back and forth with a function $f \in \mathcal{H}$ and its Fourier series coefficients and hence require some concise and clear form of relating the function to the vector containing its Fourier coefficients. Therefore, we make some amendments to our notations as follows.

N4: As much as possible and when the context is clear, we will continue to use normal Latin mathematics symbols, such as f, g, a, b, etc., to refer to elements of a Hilbert space $\mathcal{H}$, $L^2(\Omega)$ or ℓ^2. We will also continue to use sequences such as $a = \{\alpha_n\}_{n=1}^{\infty}$ for elements of ℓ^2. Specifically, this will be our preference when they are not involved in any matrix multiplication or related algebraic manipulations.

N5: We will use bold-face capital Latin letters to refer to matrices, such as $\mathbf{B}$, with element at row n and column m denoted by $b_{n,m}$. We use bold-face small Latin letters to refer to column vectors, such as $\mathbf{a}$. For example, if $a \in \ell^2$, its column vector representation would be $\mathbf{a} = (\alpha_1, \alpha_2, \ldots)'$, where $'$ denotes transpose. The inner product between two elements $a, b \in \ell^2$ in vector form is written as

$$\langle a, b \rangle = \sum_{n=1}^{\infty} \alpha_n \overline{\beta}_n = \mathbf{a}'\overline{\mathbf{b}}.$$

---

[1] The point of operators is to deal with the infinite-dimensional cases and not to revisit and to obscure standard linear algebra. In fact, the interesting features of operators occur when the infinite-dimensional varieties exhibit features not present or degenerate in the finite-dimensional case.

N6: We will use $(f)_n$ to refer to the n-th Fourier series coefficient of f. That is, $(f)_n = \langle f, \varphi_n \rangle$. The parenthesis is used to distinguish the n-th Fourier series coefficient of f from an indexed element such as $f_n \in \mathcal{H}$. The column vector containing all Fourier coefficients of f is denoted by

$$\mathbf{f} = \big((f)_1, (f)_2, (f)_3, \dots \big)'.$$

3.2   Basic presentation and properties of operators

Engineers like to represent systems as "boxes" on paper. We will now reward them with boxed representation of operators in abstract and archetype Hilbert spaces in Figure 3.1. Here, we have taken input from the entire Hilbert space and assumed input and output Hilbert spaces to be identical. This is not strictly necessary, but not much is lost in doing so.[2]

For a general operator, denoted by $\mathcal{L}$, mapping an input f to an output g, we write

$$g = \mathcal{L}f$$

and for the archetype case we may emphasize the functions variable, x, as in the notation

$$g(x) = \big(\mathcal{L}f\big)(x).$$

When composing general operators, applying $\mathcal{L}_1$ first, then $\mathcal{L}_2$ second, etc., in notation we order right to left, just like for matrices,

$$\mathcal{L}_N \circ \cdots \circ \mathcal{L}_2 \circ \mathcal{L}_1 \, f \equiv \Big(\mathcal{L}_N \cdots \big(\mathcal{L}_2(\mathcal{L}_1 \, f)\big)\Big).$$

Two operators $\mathcal{L}_1$ defined on $\mathcal{H}_1$ and $\mathcal{L}_2$ defined on $\mathcal{H}_2$ are said to be *equal*, denoted by $\mathcal{L}_1 = \mathcal{L}_2$, if their domain spaces are identical, $\mathcal{H}_1 = \mathcal{H}_2$, and for every $f \in \mathcal{H}_1 = \mathcal{H}_2$ the output of the operators are the same, $\mathcal{L}_1 f = \mathcal{L}_2 f$. This agrees with our intuition.

Linearity of an operator means that the superposition principle applies:

$$\mathcal{L}\Big(\sum_{n=1}^{N} \alpha_n f_n\Big) = \sum_{n=1}^{N} \alpha_n \big((\mathcal{L}f_n\big), \tag{3.1}$$

for scalars α_n, which we take as complex numbers. In (3.1) one cannot take $N \to \infty$ without care.[3] However, this does hold when the (linear) operator $\mathcal{L}$ is well behaved in some sense (continuity or boundedness, to be considered shortly).

Two trivial linear operators are the identity operator $\mathcal{I}$ for which

$$\mathcal{I}f = f$$

---

[2] A "mapping" is a more general notion than operator, where the domain $\mathcal{D}$ on which it is defined can be a strict subset of the Hilbert space $\mathcal{H}$ and the output space $\mathcal{H}'$ need not be the same as the input space. For most of this part and unless otherwise stated, we will be interested in the case where $\mathcal{D} = \mathcal{H}$ and $\mathcal{H} = \mathcal{H}'$. Such a mapping is then called an operator *on* $\mathcal{H}$.

[3] Not just because $\{\alpha_n\}_{n=1}^{\infty}$ may not, as a series, converge.

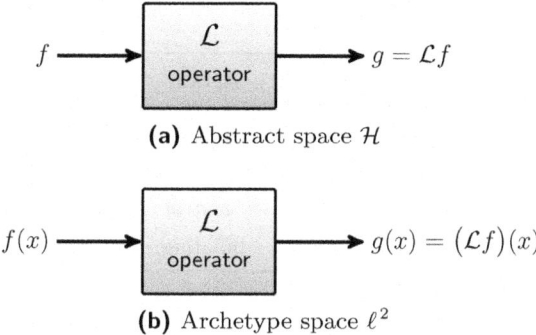

(a) Abstract space $\mathcal{H}$

(b) Archetype space ℓ^2

Figure 3.1 Action of a general operator. **(a)** Abstract Hilbert space $\mathcal{H}$ case, which maps input element $f \in \mathcal{H}$ to output element $g \in \mathcal{H}$. **(b)** Archetype Hilbert space $L^2(\Omega)$ case, which maps an input function $f(x) \in L^2(\Omega)$ to an output function $g(x) \in L^2(\Omega)$.

and the zero operator $\mathcal{O}$ for which

$$\mathcal{O}f = o,$$

where o denotes the zero element/vector, for all $f \in \mathcal{H}$.

Linearity of two or more operators on the same Hilbert space $\mathcal{H}$ means that we can add them or scale them in familiar ways for all $f \in \mathcal{H}$:

$$(\mathcal{L}_1 + \mathcal{L}_2)f \triangleq \mathcal{L}_1 f + \mathcal{L}_2 f, \tag{3.2}$$

$$(\alpha \mathcal{L}_1)f \triangleq \alpha \mathcal{L}_1 f. \tag{3.3}$$

Using the above relations, one can show that the set of all linear operators forms a linear space. For example,

$$(\mathcal{L}_1 + \mathcal{L}_2) \circ \mathcal{L}_3 = \mathcal{L}_1 \circ \mathcal{L}_3 + \mathcal{L}_2 \circ \mathcal{L}_3,$$
$$\mathcal{I} \circ \mathcal{L}_1 = \mathcal{L}_1 \circ \mathcal{I} = \mathcal{L}_1,$$

and

$$\mathcal{O} \circ \mathcal{L}_1 = \mathcal{L}_1 \circ \mathcal{O} = \mathcal{O}.$$

3.3   Classification of linear operators

As a roadmap, we make a broad classification of linear operators into four categories: finite rank, compact [completely continuous], bounded [continuous], and unbounded. This ordering has some sort of meaning; for example, a compact operator (whatever that is) is bounded, a finite rank operator is compact, and so on. We think of how an operator squeezes a range of inputs when generating the output when classifying linear infinite-dimensional operators:

Notion (Finite rank operator). *A finite rank operator is the Western consumer of operators, it discards all but a finite-dimensional subspace of functions as*

Table 3.1 Infinite-dimensional operator classification

Operator	Action	Example
Finite rank	Severe squeezing	Finite truncation operator
Compact	Squeezing	Integral operator
Bounded	Neutral	Identity operator
Unbounded	Inflating	Derivative operator

useful output. The input can be an entire infinite-dimensional space, but the output can live only on a finite-dimensional subspace. □

Notion (Compact [completely continuous] operator). *A compact operator is very similar to a finite rank operator, but some of the output can leak into infinite dimensions in a very restricted way — it is essentially a finite rank operator or can be well approximated by a finite rank operator. Both finite rank and compact operators are like smoothing operators and in this sense squeeze the infinite dynamic range of the input to form the restricted dynamic range of the output. These operators were first called* completely continuous transformations *and older books refer to them as such. As the name implies not only are they continuous but completely continuous.*[4] □

Notion (Bounded [continuous] operator). *A bounded operator means that the output is never more than some maximum gain of the input, measured in the norm (to be made precise shortly). Bounded operators need not do any squeezing and are best thought of as relatively neutral in the squeezing stakes. The identity operator is clearly bounded (gain is unity we hear you think), but in infinite dimensions it is not compact (whereas in finite dimensions the identity or any bounded operator is compact). It is a little odd to think that some operators are more primal than the identity operator. Bounded operators are also called* continuous operators. *Linearity plus boundedness implies continuity, as we shall show, and linearity plus continuity implies boundedness. They are equivalent notions for linear operators.* □

Notion (Unbounded operator). *An unbounded operator has the propensity to inflate the input. The standard example is the derivative operator where taking inputs with steeper and steeper slopes induces an output of greater and greater norm, whereupon there should be no gain to constrain the output in terms of the input. We are not going to delve into unbounded operators, but provide an adequate springboard to that topic.* □

The discussion above is summarized in Table 3.1. We more carefully visit these notions in the next chapters.

---

[4] This probably implies that finite rank operators should be called "completely and utterly continuous/"

Take-home messages – operators

M1: In engineering terms, an operator is a system that takes input signals and maps them into output signals.

M2: Unexpected properties of operators pop up when the domain of signals is an infinite-dimensional separable Hilbert space, as opposed to a finite-dimensional space. An example is the non-equivalence between bounded and compact operators.

M3: Operators can be broadly categorized into linear and non-linear, and into bounded and unbounded operators. We are only interested in linear and bounded operators.

M4: Two special subclasses of linear and bounded operators are compact operators and finite-rank operators, which both have squeezing or smoothing characteristics. The input signal cannot leak out to all infinite dimensions in the operator output in an unrestricted way.

4 Bounded operators

4.1 Definitions

We begin with some definitions.

Definition 4.1 (Operator norm). *The norm of linear operator $\mathcal{L}$ on Hilbert space $\mathcal{H}$ is defined as*

$$\|\mathcal{L}\|_{\text{op}} \triangleq \sup_{f \in \mathcal{H},\, f \neq o} \frac{\|\mathcal{L}f\|}{\|f\|} = \sup_{f \in \mathcal{H},\, f \neq o} \left\| \mathcal{L} \frac{f}{\|f\|} \right\| = \sup_{f \in \mathcal{H},\, \|f\|=1} \|\mathcal{L}f\|. \tag{4.1}$$

□

Definition 4.2 (Bounded operator). *A linear operator $\mathcal{B}$ on Hilbert space $\mathcal{H}$ is bounded if there exists a constant $B \geq 0$ such that*

$$\|\mathcal{B}f\| \leq B\|f\|, \quad \forall f \in \mathcal{H}. \tag{4.2}$$

□

Note that in this book linearity is always implied when we talk about bounded operators. The operator norm is the smallest constant B (least upper bound) such that (4.2) holds.

Clearly the operator norm is only finite for bounded operators. Note that it is linearity that allows us to invoke the second and third equalities in (4.1) and makes it sufficient to search over $\|f\| = 1$ rather than all $f \in \mathcal{H}$. Also in (4.1) the left-hand side norm is the operator norm and the right-hand side is the Hilbert space norm. The following equation is a direct reflection of the definitions:

$$\|\mathcal{L}f\| \leq \|\mathcal{L}\|_{\text{op}} \|f\|, \quad \forall f \in \mathcal{H}.$$

In words, the operator norm is the maximum gain, or "gain," of the operator, the maximum increase in the norm of the output relative to the norm of the input. In the extreme one-dimensional case the operator is just a complex number α — so the output is a fixed scaling of the input and the operator norm is $|\alpha|$.

Example 4.1 (Right shift operator). Let the operator $\mathcal{S}_R$ on ℓ^2 for all $a = \{\alpha_n\}_{n=1}^{\infty} \in \ell^2$ be defined as $\mathcal{S}_R a = \{\alpha_{n-1}\}_{n=1}^{\infty}$ and $\alpha_0 = 0$. In other words,

$\mathcal{S}_R$ is the *right shift operator* turning $\{\alpha_1, \alpha_2, \alpha_3, \ldots\}$ into $\{0, \alpha_1, \alpha_2, \ldots\}$. It is easy to verify that $\mathcal{S}_R$ is linear. Noting

$$\|\mathcal{S}_R a\|^2 = \langle \mathcal{S}_R a, \mathcal{S}_R a \rangle = \sum_{n=1}^{\infty} |\alpha_{n-1}|^2 = \langle a, a \rangle = \|a\|^2, \quad \forall a \in \ell^2,$$

we conclude that the right shift operator is bounded with norm one. That is, $\|\mathcal{S}_R\|_{\text{op}} = 1$. □

Example 4.2 (Truncation operator). Let the *truncation* operator $\mathcal{T}_N$ on ℓ^2 for all $a = \{\alpha_n\}_{n=1}^{\infty} \in \ell^2$ be defined as $\mathcal{T}_N a = \{\alpha_n\}_{n=1}^{N}$ for $N \in \mathbb{N}$. It is easy to verify that $\mathcal{T}_N$ is linear. Noting

$$\|\mathcal{T}_N a\|^2 = \sum_{n=1}^{N} |\alpha_n|^2 \leq \|a\|^2, \quad \forall a \in \ell^2,$$

we conclude that $\|\mathcal{T}_N\|_{\text{op}} \leq 1$. Now taking the unit norm element $a = \{1, 0, 0, \ldots\}$, we see that $\mathcal{T}_N a = a$ and therefore $\|\mathcal{T}_N\|_{\text{op}} = 1$. □

Problem

4.1. Turn the right shift operator into an unbounded linear operator (hint: apply a proper index-dependent gain to each shifted element).

4.2 Invertibility

Definition 4.3 (Invertible bounded operator). *A bounded operator $\mathcal{B}$ on Hilbert space $\mathcal{H}$ is invertible if there exists an operator denoted by $\mathcal{B}^{-1}$ on $\mathcal{H}$ such that*

$$\mathcal{B} \circ \mathcal{B}^{-1} = \mathcal{B}^{-1} \circ \mathcal{B} = \mathcal{I}. \tag{4.3}$$

$\mathcal{B}^{-1}$ *is called the inverse of $\mathcal{B}$.* □

Remark 4.1. Note that the inverse operator of a bounded operator need not be bounded. In fact compact operators, which we will meet in the next chapter, are bounded and can have an inverse. Such invertible compact operators necessarily have unbounded inverses. □

From the two examples above, we can immediately tell that the truncation operator is not invertible. Once components $\alpha_{N+1}, \alpha_{N+2}, \ldots$ are lost in the output due to truncation, they are lost forever. We cannot get them back.

But what about the right shift operator? One is tempted to nominate the *left shift operator* $\mathcal{S}_L$ with the operation on $a = \{\alpha_n\}_{n=1}^{\infty} \in \ell^2$ as $\mathcal{S}_L a = \{\alpha_{n+1}\}_{n=1}^{\infty}$ as the candidate for the inverse. However from the definition, we note that it is important to have the inverse operation applicable from both left and right. While we have $\mathcal{S}_L \circ \mathcal{S}_R = \mathcal{I}$, the opposite is not true: we cannot reverse the left shift operator because the first element α_1 is lost.

4.3   Boundedness and continuity

As we briefly mentioned before, for linear operators there is an equivalence between boundedness and continuity of the operator. Before, elaborating on this, we first formalize what is meant by a continuous operator. Continuity is the first step towards understanding topological aspects of operators. Roughly speaking, a continuous operator maps input elements that are close to each other to output elements that are also close to each other. This definition does not require the operator to be linear.

Definition 4.4 (Continuous operator). *An operator, $\mathcal{B}$, on Hilbert space $\mathcal{H}$ is continuous if for every $\epsilon > 0$ and for all $f_0 \in \mathcal{H}$ there exists a $\delta > 0$ such that*

$$\|f - f_0\| < \delta \implies \|\mathcal{B}f - \mathcal{B}f_0\| < \epsilon, \quad \forall f \in \mathcal{H}. \tag{4.4}$$

$\square$

An important consequence of continuity for a *linear* operator $\mathcal{B}$ is that one can switch the order of operator and limit of a converging sequence. That is, if the sequence of elements $\{f_n\}_{n=1}^{\infty}$ in $\mathcal{H}$ converges to $f_0 \in \mathcal{H}$, there is no difference between operating on the limit f_0 or taking the limit of operations on the sequence. The result is the same:

$$\text{Continuity implies} \quad \mathcal{B}f_0 = \mathcal{B}\big(\lim_{n\to\infty} f_n\big) = \lim_{n\to\infty} \mathcal{B}f_n.$$

Now we are in a position to talk about the relationship between continuity and boundedness for linear operators. A bounded linear operator $\mathcal{B}$ on $\mathcal{H}$ is also continuous,

$$\|\mathcal{B}f_n - \mathcal{B}f_0\| = \big\|\mathcal{B}(f_n - f_0)\big\| \le \|\mathcal{B}\|_{\text{op}}\|f_n - f_0\|,$$

where the equality is invoked due to linearity of the operator. If f_n converges in the mean to f (viz., $\|f_n - f\| \to 0$), then $\mathcal{B}f_n$ converges in the mean to $\mathcal{B}f$ (viz., $\|\mathcal{B}f_n - \mathcal{B}f\| \to 0$).[1] Topologically, the norm of a bounded linear operator quantifies an upper bound on how far input elements that are close can get away from each other in the output. Conversely, it can be proven that a continuous and linear operator $\mathcal{B}$ on $\mathcal{H}$ is bounded.

The technical issue that we alluded to in letting $N \to \infty$ in (3.1) is resolved for bounded linear operators. That is,

$$\lim_{N\to\infty} \big(\mathcal{L}\sum_{n=1}^{N} \alpha_n f_n\big) = \mathcal{L}\big(\lim_{N\to\infty} \sum_{n=1}^{N} \alpha_n f_n\big) = \lim_{N\to\infty} \sum_{n=1}^{N} \alpha_n\big(\mathcal{L}f_n\big) = \sum_{n=1}^{\infty} \alpha_n\big(\mathcal{L}f_n\big),$$

where the first equality is invoked because the operator is continuous and we can swap the order of limit and operator and the second equality follows from linearity.

---

[1] More generally, one should start with continuity of an operator about a single point $f_0 \in \mathcal{H}$, but it is linearity of the operator that makes the result global as captured in Problem 4.2.

Problems

4.2. Show that if a linear operator is continuous at a point (usually taken as the origin, that is, $f_0 = o$) then this implies it is continuous everywhere.

4.3. Show that a nowhere continuous operator is unbounded.

4.4   Convergence of a sequence of bounded operators

Often it will be the case that to understand the action of a bounded operator, $\mathcal{B}$, it is best to consider the actions of a sequence of more elementary bounded operators, $\{\mathcal{B}_n\}_{n=1}^{\infty}$, that converge in some sense to $\mathcal{B}$ as n goes to infinity. Understanding how the operator converges helps understanding its topological behavior. One very useful sequence of operators is a sequence of *finite rank* operators that we will come across later. In this case the output of the operator has a finite dimension and the general idea is that we might be able to understand the behavior of a general operator, $\mathcal{B}$, if it is well approximated by a sequence of finite rank operators of increasing dimension which converges to $\mathcal{B}$.

Operators can converge in a number of ways:

Definition 4.5 (Weak convergence of operators). *The sequence $\mathcal{B}_n f$ converges weakly to $\mathcal{B}f$ if*

$$\langle \mathcal{B}_n f, g \rangle \to \langle \mathcal{B}f, g \rangle \ \text{as } n \to \infty, \quad \forall f, g \in \mathcal{H}. \tag{4.5}$$

Or, equivalently,

$$\lim_{n \to \infty} \langle \mathcal{B}_n f, g \rangle \to \langle \mathcal{B}f, g \rangle, \quad \forall f, g \in \mathcal{H}. \tag{4.6}$$

In notation, we write this as

$$\boxed{\mathcal{B}_n f \rightharpoonup \mathcal{B}f,}$$

where the limit $n \to \infty$ is understood.   □

Definition 4.6 (Strong convergence of operators). *The sequence $\mathcal{B}_n f$ converges strongly (in the mean) to $\mathcal{B}f$ if*

$$\| \mathcal{B}_n f - \mathcal{B}f \| \to 0 \ \text{as } n \to \infty, \quad \forall f \in \mathcal{H}.$$

Or, equivalently,

$$\lim_{n \to \infty} \| \mathcal{B}_n f - \mathcal{B}f \| \to 0, \quad \forall f \in \mathcal{H}.$$

In notation, we write this as

$$\boxed{\mathcal{B}_n f \to \mathcal{B}f,}$$

where the limit $n \to \infty$ is understood.   □

Definition 4.7 (Convergence in (operator) norm). *The sequence* $\|\mathcal{B}_n - \mathcal{B}\|_{\text{op}} \to$ 0 *or, equivalently,*

$$\lim_{n \to \infty} \sup_{\|f\|=1} \|\mathcal{B}_n f - \mathcal{B} f\| \to 0.$$

☐

The first two definitions relate to the inner product and norm of elements of the Hilbert space, $\mathcal{H}$, as acted on by the operator rather than explicitly on the operator, and so naturally carry on from the definitions of strong and weak convergence given in Section 2.8.1, p. 74, and Section 2.8.2, p. 75. There, in Theorem 2.9, it was shown that strong convergence implies weak convergence:

$$\mathcal{B}_n f \to \mathcal{B} f \implies \mathcal{B}_n f \rightharpoonup \mathcal{B} f.$$

The third and new definition is that of convergence in norm. For bounded operators, we just saw that for every element $f \in \mathcal{H}$ we have

$$\|\mathcal{B}_n f - \mathcal{B} f\| = \left\| (\mathcal{B}_n - \mathcal{B}) f \right\| \le \|\mathcal{B}_n - \mathcal{B}\|_{\text{op}} \|f\|.$$

So convergence in norm implies strong convergence of the sequence of operators:

$$\|\mathcal{B}_n - \mathcal{B}\|_{\text{op}} \to 0 \implies \mathcal{B}_n f \to \mathcal{B} f.$$

Furthermore, whenever we take any element satisfying $\|f\| \le B$ for some finite positive bound B, we see that the convergence is *uniform*, that is,

$$\|\mathcal{B}_n f - \mathcal{B} f\| \le B \|\mathcal{B}_n - \mathcal{B}\|_{\text{op}},$$

which is an inequality whose upper bound is independent of f: $\|f\| \le B$.

Problems _____

4.4. Prove that a sufficient condition for weak convergence to imply strong convergence of operators is $\|\mathcal{B}_n f\| \to \|\mathcal{B} f\|$. That is,

$$\mathcal{B}_n f \rightharpoonup \mathcal{B} f \text{ and } \|\mathcal{B}_n f\| \to \|\mathcal{B} f\| \implies \mathcal{B}_n f \to \mathcal{B} f.$$

4.5. Using the definition of operator norm and properties of supremum, show that for bounded operators $\mathcal{B}_1$ and $\mathcal{B}_2$ on Hilbert space $\mathcal{H}$, the following properties hold

$$\|\mathcal{B}_1 + \mathcal{B}_2\|_{\text{op}} \le \|\mathcal{B}_1\|_{\text{op}} + \|\mathcal{B}_2\|_{\text{op}}, \tag{4.7}$$

$$\|\alpha \mathcal{B}_1\|_{\text{op}} = |\alpha| \|\mathcal{B}_1\|_{\text{op}}, \tag{4.8}$$

$$\|\mathcal{B}_2 \circ \mathcal{B}_1\|_{\text{op}} \le \|\mathcal{B}_2\|_{\text{op}} \|\mathcal{B}_1\|_{\text{op}}. \tag{4.9}$$

_____

Remark 4.2. From (3.2), (3.3), (4.7) and (4.8), we can conclude that the set of all bounded operators forms a normed linear space. Furthermore, (4.9) shows a multiplicative equivalent of triangle inequality for bounded operators. Remarkably, from the completeness of the Hilbert space domain, it can be proven that the set of all bounded operators is complete and hence is a Banach space: if a sequence of operators $\{\mathcal{B}_n\}_{n=1}^{\infty}$ is fundamental, $\lim_{m,n\to\infty}\|\mathcal{B}_m - \mathcal{B}_n\|_{\text{op}} = 0$, then it has a limit $\mathcal{B} = \lim_{n\to\infty}\mathcal{B}_n$ such that $\lim_{n\to\infty}\|\mathcal{B}f - \mathcal{B}_n f\| = 0,\ \forall f \in \mathcal{H}$ and $\lim_{n\to\infty}\|\mathcal{B} - \mathcal{B}_n\|_{\text{op}} = 0$.   □

4.5   Bounded operators as matrices

Let us start with our wish list. One important property of a separable Hilbert space $\mathcal{H}$ is the existence of a countable basis $\{\varphi_n\}_{n=1}^{\infty}$. Such basis completely characterizes each element of the space like f through Fourier series expansion $f = \sum_{n=1}^{\infty}\langle f, \varphi_n\rangle \varphi_n$. It would be great if we could completely characterize the action of an operator $\mathcal{L}$ by describing how it acts on each element of the given basis and based on that could represent the action of operator on the entire space in a succinct manner. Fortunately, this is possible for bounded (linear) operators where an infinite matrix can be associated with a general bounded operator from a separable Hilbert space $\mathcal{H}$ to itself.

Let us first see the process without proof. Imagine that when a bounded operator $\mathcal{B}$ operates on a basis element of $\mathcal{H}$ such as φ_m, on output it projects it along φ_n with *gain* $b_{n,m}^{(\varphi)}$ for all $n \in \mathbb{N}$. That is,

$$b_{n,m}^{(\varphi)} \triangleq \langle \mathcal{B}\varphi_m, \varphi_n\rangle, \tag{4.10}$$

and we can write

$$\mathcal{B}\varphi_m = \sum_{n=1}^{\infty} b_{n,m}^{(\varphi)} \varphi_n.$$

Remark 4.3 (Indexing mnemonic). It always crops up whether the indexing in the expression (4.10) is around the right way or is the most sensible convention. We offer the following mnemonic (no pun intended with the first two letters)

$$\varphi_n \longleftarrow b_{n,m}^{(\varphi)} \varphi_m,$$
$$m\text{-th input from the right, } n\text{-th output to the left}$$

as a reference to the convention.   □

Now the question is how we can express the action of $\mathcal{B}$ on a general element of $\mathcal{H}$ such as f. In particular, what is the projection of $\mathcal{B}f$ into direction φ_n? For bounded operators we can break this into the following two steps:

Step 1: Project f along the direction φ_m resulting in $f_m = \langle f, \varphi_m\rangle \varphi_m$. Then because of linearity of the operator $\mathcal{B}$, it projects f_m into direction φ_n with a *combined gain* $b_{n,m}^{(\varphi)}\langle f, \varphi_m\rangle$. Basically, this tells us how much of the m-th component of f is projected along the n-th basis φ_n after going through operator $\mathcal{B}$.

Step 2: Repeat the above step for all directions m and add up the combined gains as $\sum_{m=1}^{\infty} b_{n,m}^{(\varphi)} \langle f, \varphi_m \rangle$, which gives how $\mathcal{B}f$ is projected along φ_n or $\langle \mathcal{B}f, \varphi_n \rangle$. These steps can only be done due to boundedness (continuity) of the linear operator.

Once we determine $\langle \mathcal{B}f, \varphi_n \rangle$ we can write the overall effect of operator $\mathcal{B}$ on f as

$$\mathcal{B}f = \sum_{n=1}^{\infty} \langle \mathcal{B}f, \varphi_n \rangle \varphi_n$$

$$= \sum_{n=1}^{\infty} \left(\sum_{m=1}^{\infty} \langle f, \varphi_m \rangle b_{n,m}^{(\varphi)} \right) \varphi_n.$$

We can express the above operation with a countably infinite matrix of the form

$$\mathbf{B}^{(\varphi)} = \begin{pmatrix} b_{1,1}^{(\varphi)} & b_{1,2}^{(\varphi)} & b_{1,3}^{(\varphi)} & \cdots \\ b_{2,1}^{(\varphi)} & b_{2,2}^{(\varphi)} & b_{2,3}^{(\varphi)} & \cdots \\ b_{3,1}^{(\varphi)} & b_{3,2}^{(\varphi)} & b_{3,3}^{(\varphi)} & \cdots \\ \vdots & \vdots & \vdots & \ddots \end{pmatrix},$$

where the n-th row vector denoted by $(\mathbf{b}_n^{(\varphi)})'$ represents how input directions along φ_m are contributing to the output direction n along φ_n and each column m denoted by $\mathbf{b}_m^{(\varphi)}$ represents how a single input direction m contributes to different output directions along φ_n. Representing Fourier coefficients of f by a countably infinite column vector $\mathbf{f}$, we can write

$$\langle \mathcal{B}f, \varphi_n \rangle = (\mathbf{b}_n^{(\varphi)})' \mathbf{f}.$$

Example 4.3. In ℓ^2, a complete countable basis can be defined as $\{e_n\}_{n=1}^{\infty}$ where

$$e_n \triangleq \{\delta_{n,m}\}_{m=1}^{\infty},$$

and $\delta_{n,m}$ is the Kronecker delta function.

It is clear that for the right shift operator introduced in Example 4.1, $\mathcal{S}_R e_m = e_{m+1}$. The countably infinite matrix associated with the $\{e_n\}_{n=1}^{\infty}$ basis then has elements

$$s_{n,m} \triangleq \langle \mathcal{S}_R e_m, e_n \rangle = \delta_{m+1,n}$$

and takes the form

$$\mathbf{S}_R = \begin{pmatrix} 0 & 0 & 0 & 0 & \cdots \\ 1 & 0 & 0 & 0 & \cdots \\ 0 & 1 & 0 & 0 & \cdots \\ 0 & 0 & 1 & 0 & \cdots \\ \vdots & \vdots & \vdots & \vdots & \ddots \end{pmatrix}.$$

Now for each element $a = \sum_{m=1}^{\infty} \alpha_m e_m \in \ell^2$, we can write $\langle \mathcal{S}_R a, e_n \rangle$ as

$$\left\langle \mathcal{S}_R \sum_{m=1}^{\infty} \alpha_m e_m, e_n \right\rangle = \sum_{m=1}^{\infty} \alpha_m \langle \mathcal{S}_R e_m, e_n \rangle = \sum_{m=1}^{\infty} s_{n,m} \alpha_m = \alpha_{m+1} \delta_{m+1,n},$$

as expected.   □

Now we are in a position to formalize the results through the following theorem.

Theorem 4.1. *Let $\{\varphi_n\}_{n=1}^{\infty}$ be a complete orthonormal sequence used to represent the input and output of a bounded operator $\mathcal{B}$ on a separable Hilbert space $\mathcal{H}$. Then*

$$\boxed{b_{n,m}^{(\varphi)} \triangleq \langle \mathcal{B}\varphi_m, \varphi_n \rangle,} \qquad (4.11)$$

for $m, n \in \{1, 2, 3, \dots\}$, define the coefficients of an infinite matrix $\mathbf{B}^{(\varphi)}$ that characterizes the operator.   □

Proof.   Apply the bounded operator, $\mathcal{B}$, to an arbitrary element, f, of the separable Hilbert space

$$\mathcal{B}f = \mathcal{B}\Big(\lim_{M \to \infty} \sum_{m=1}^{M} \langle f, \varphi_m \rangle \varphi_m \Big) \qquad (4.12a)$$

$$= \lim_{M \to \infty} \mathcal{B}\Big(\sum_{m=1}^{M} \langle f, \varphi_m \rangle \varphi_m \Big) \qquad (4.12b)$$

$$= \lim_{M \to \infty} \sum_{m=1}^{M} \langle f, \varphi_m \rangle \mathcal{B}\varphi_m \qquad (4.12c)$$

$$= \sum_{m=1}^{\infty} \langle f, \varphi_m \rangle \mathcal{B}\varphi_m,$$

where, crucially, in going from (4.12a) to (4.12b) we have used *continuity* of the operator (since the operator is bounded) and, in going from (4.12b) to (4.12c) we have used linearity. Now we project $\mathcal{B}f$ onto φ_n

$$\langle \mathcal{B}f, \varphi_n \rangle = \sum_{m=1}^{\infty} \langle f, \varphi_m \rangle \langle \mathcal{B}\varphi_m, \varphi_n \rangle$$
$$= \sum_{m=1}^{\infty} \langle f, \varphi_m \rangle b_{n,m}^{(\varphi)}, \qquad (4.13)$$

where the coefficients $b_{n,m}^{(\varphi)}$ are independent of f and given by

$$b_{n,m}^{(\varphi)} \triangleq \langle \mathcal{B}\varphi_m, \varphi_n \rangle.$$

Then given $\mathcal{B}f \in \mathcal{H}$, it has the expansion (in the sense of convergence in the mean)

$$\mathcal{B}f = \sum_{n=1}^{\infty} \langle \mathcal{B}f, \varphi_n \rangle \varphi_n.$$

Combining this with (4.13)

$$\mathcal{B}f = \sum_{n=1}^{\infty} \Big(\sum_{m=1}^{\infty} \langle f, \varphi_m \rangle \langle \mathcal{B}\varphi_m, \varphi_n \rangle \Big) \varphi_n$$

$$= \sum_{n=1}^{\infty} \Big(\sum_{m=1}^{\infty} \langle f, \varphi_m \rangle b_{n,m}^{(\varphi)} \Big) \varphi_n,$$

it can be seen that coefficients $b_{n,m}^{(\varphi)}$ are necessary and sufficient to characterize the bounded operator $\mathcal{B}$.   □

The coefficient $b_{n,m}^{(\varphi)}$ represents how much of φ_m as an input gets projected along the φ_n direction of the output under $\mathcal{B}$.

> **Remark 4.4.** As the notation suggests, the matrix depends on the orthonormal sequence $\{\varphi_n\}_{n=1}^{\infty}$ which acts as a basis. Changing the basis changes the matrix. There would be no difficulty having a different basis (complete orthonormal sequence) on the output from the input, and another matrix would result. See Problem 4.8, p. 120.   □

Thinking in terms of a matrix is very useful provided the operator is *bounded* and hence *continuous*. It is not recommended that you think of operators in infinite dimensions *only* as matrices with an infinite number of entries since it is not universally valid for unbounded operators. The archetype unbounded operator is the differential operator, and there exist unbounded operators which cannot be simply identified with a matrix.

> **Remark 4.5.** The other key element in the formulation is *separability* of the space and in the end the matrix just tells you how much one element of the complete and countable input orthonormal sequence (which is dense in the input space) gets mapped to one element of the complete and countable output orthonormal sequence (which is dense in the output space) under the action of the operator. Linearity in this context is equivalent to this element-to-element mapping being given by a simple complex scalar.   □

Problems

4.6. A bounded operator $\mathcal{B}$ for a separable Hilbert space $\mathcal{H}$ with the basis $\{\varphi_n\}_{n=1}^{\infty}$ is characterized by $b_{n,m}^{(\varphi)}$ for all $m, n = 1, 2, \ldots$, with the additional property that

$$B^2 = \sum_{n=1}^{\infty} \sum_{m=1}^{\infty} \left| b_{n,m}^{(\varphi)} \right|^2 < \infty.$$

What can you say about the norm of the operator?

4.7. In the finite N-dimensional case show that the operator matrix simplifies to a regular $N \times N$ matrix under a suitable choice of basis.

4.8. Let $\{\varphi_n\}_{n=1}^{\infty}$ and $\{\zeta_n\}_{n=1}^{\infty}$ be two complete orthonormal sequences used to represent the input and output, respectively, of a bounded linear operator $\mathcal{B}$ on a separable Hilbert space. Find the coefficients of an infinite matrix that completely characterizes the operator.

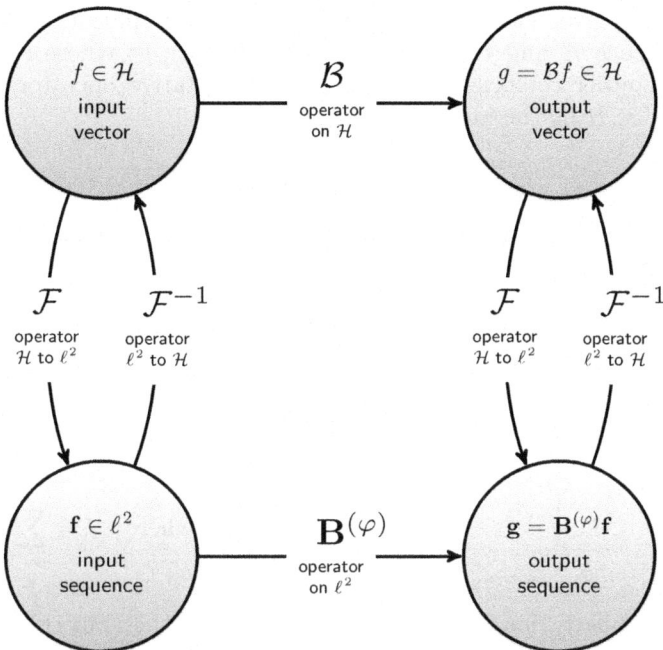

Figure 4.1 Commutative operator diagram for a bounded operator $\mathcal{B}$ on a separable Hilbert space $\mathcal{H}$ viewed under two equivalent pictures. (i) Canonical Hilbert space ℓ^2 case, which maps input element $\mathbf{f} = ((f)_1, (f)_2, (f)_3, \dots)' \in \ell^2$ to output element $\mathbf{g} = ((g)_1, (g)_2, (g)_3, \dots)' \in \ell^2$ under a matrix $\mathbf{B}^{(\varphi)}$ representation of operator $\mathcal{B}$. (ii) Abstract Hilbert space $\mathcal{H}$ case, which maps an input $f \in \mathcal{H}$ to an output $g \in \mathcal{H}$. Operator $\mathcal{F}$, which maps between the two spaces, is an invertible operator from $\mathcal{H}$ to ℓ^2.

4.6   Completing the picture: bounded operator identities

Both *boundedness* and *separability* have been invoked to establish that a bounded operator can be thought of as an infinite matrix through the identity

$$b_{n,m}^{(\varphi)} \triangleq \langle \mathcal{B}\varphi_m, \varphi_n \rangle, \tag{4.14}$$

which characterizes how the operator maps input to outputs, with respect to the given complete orthonormal sequence, $\{\varphi_n\}_{n=1}^{\infty}$, that is, $\mathcal{B} \longmapsto \mathbf{B}^{(\varphi)}$. Equation (4.14) is a prescription for generating a matrix operator $\mathbf{B}^{(\varphi)}$ from the operator $\mathcal{B}$ relative to the given complete orthonormal sequence $\{\varphi_n\}_{n=1}^{\infty}$. However, if we are first given the matrix operator $\mathbf{B}^{(\varphi)}$ then it is of interest to determine the operator $\mathcal{B}$ in the abstract Hilbert space $\mathcal{H}$, that is, $\mathbf{B}^{(\varphi)} \longmapsto \mathcal{B}$. Furthermore, it is important to understand in which sense the matrix itself is an operator and determine how its inputs and outputs (sequences) relate to the original inputs and output elements (which may be functions). This leads to consideration of the "commutative operator diagram" in Figure 4.1.

We define an invertible operator, labeled $\mathcal{F}$, which depends only on a complete orthonormal sequence, $\{\varphi_n\}_{n=1}^{\infty}$. This operator generates the Fourier coefficients of a given element and establishes an isomorphism between our abstract Hilbert space $\mathcal{H}$ and the canonical Hilbert space ℓ^2 whose elements are sequences of Fourier coefficients. As in the commutative operator diagram Figure 4.1, let $f \in \mathcal{H}$ be the input element and

$$\mathbf{f} = \big((f)_1, (f)_2, (f)_3, \ldots\big)'$$

be the column vector containing its Fourier coefficients. These are related through:

$$\mathcal{F}: \quad f \in \mathcal{H} \longmapsto \mathbf{f} \in \ell^2 \quad \text{where} \quad (f)_n = \langle f, \varphi_n \rangle,$$

$$\mathcal{F}^{-1}: \quad \mathbf{f} \in \ell^2 \longmapsto f \in \mathcal{H} \quad \text{where} \quad f = \sum_{n=1}^{\infty} (f)_n \varphi_n.$$

Similarly, for the operator output g with the Fourier coefficients vector

$$\mathbf{g} = \big((g)_1, (g)_1, (g)_3, \ldots\big)',$$

we have

$$\mathcal{F}: \quad g \in \mathcal{H} \longmapsto \mathbf{g} \in \ell^2 \quad \text{where} \quad (g)_n = \langle g, \varphi_n \rangle,$$

$$\mathcal{F}^{-1}: \quad \mathbf{g} \in \ell^2 \longmapsto g \in \mathcal{H} \quad \text{where} \quad g = \sum_{n=1}^{\infty} (g)_n \varphi_n.$$

The operators in their respective spaces lead to the expressions:

$$\mathcal{B}: \quad f \longmapsto g \in \mathcal{H} \quad \text{where} \quad g = \mathcal{B}f, \tag{4.15}$$

$$\mathbf{B}^{(\varphi)}: \quad \mathbf{f} \longmapsto \mathbf{g} \in \ell^2 \quad \text{where} \quad \mathbf{g} = \mathbf{B}^{(\varphi)}\mathbf{f}. \tag{4.16}$$

We are going to present many identities, but essentially the nub of the problem should be visualized/memorized as shown in Figure 4.1 with component identities (4.14)–(4.16).

Composing two operators leads to four expressions, which we pair off for reasons that should soon become apparent

$$\mathbf{B}^{(\varphi)} \circ \mathcal{F}:$$
$$f \in \mathcal{H} \longmapsto \mathbf{g} \in \ell^2 \quad \text{where} \quad (g)_n = \sum_{m=1}^{\infty} \langle f, \varphi_m \rangle b_{n,m}^{(\varphi)}, \qquad (4.17)$$

$$\mathcal{F} \circ \mathcal{B}:$$
$$f \in \mathcal{H} \longmapsto \mathbf{g} \in \ell^2 \quad \text{where} \quad (g)_n = \langle \mathcal{B}f, \varphi_n \rangle, \qquad (4.18)$$

and

$$\mathcal{B} \circ \mathcal{F}^{-1}:$$
$$\mathbf{f} \in \ell^2 \longmapsto g \in \mathcal{H} \quad \text{where} \quad g = \sum_{m=1}^{\infty} (f)_m (\mathcal{B}\varphi_m), \qquad (4.19)$$

$$\mathcal{F}^{-1} \circ \mathbf{B}^{(\varphi)}:$$
$$\mathbf{f} \in \ell^2 \longmapsto g \in \mathcal{H} \quad \text{where} \quad g = \sum_{n=1}^{\infty} \varphi_n \sum_{m=1}^{\infty} (f)_m b_{n,m}^{(\varphi)}. \qquad (4.20)$$

In the above equations, we have only two distinct compositions since both $\mathbf{B}^{(\varphi)} \circ \mathcal{F}$ and $\mathcal{F} \circ \mathcal{B}$ perform the same action (sending $f \longmapsto \mathbf{g}$), and both $\mathcal{B} \circ \mathcal{F}^{-1}$ and $\mathcal{F}^{-1} \circ \mathbf{B}^{(\varphi)}$ perform the same action (sending $\mathbf{f} \longmapsto g$). This implies the corresponding expressions are equivalent and we show this directly using algebraic methods.

From (4.14) and (4.17) we can write

$$(g)_n = \sum_{m=1}^{\infty} \langle f, \varphi_m \rangle b_{n,m}^{(\varphi)} = \sum_{m=1}^{\infty} \langle f, \varphi_m \rangle \langle \mathcal{B}\varphi_m, \varphi_n \rangle \qquad (4.21a)$$

$$= \sum_{m=1}^{\infty} \langle f, \varphi_m \rangle \langle \varphi_m, \mathcal{B}^{\star} \varphi_n \rangle$$

$$= \langle f, \mathcal{B}^{\star} \varphi_n \rangle$$

$$= \langle \mathcal{B}f, \varphi_n \rangle, \qquad (4.21b)$$

which is indeed (4.18). In the above calculation we have used the existence of the unique adjoint of operator $\mathcal{B}$, which is denoted by $\mathcal{B}^{\star}$, and this will be covered in Section 4.9, p. 131. Also we have used the generalized Parseval relation (2.57), p. 72, to collapse the summation.

From (4.14) and (4.20) we can write

$$g = \sum_{n=1}^{\infty} \varphi_n \sum_{m=1}^{\infty} (f)_m b_{n,m}^{(\varphi)} = \sum_{m=1}^{\infty} (f)_m \sum_{n=1}^{\infty} \langle \mathcal{B}\varphi_m, \varphi_n \rangle \varphi_n \qquad (4.22a)$$

$$= \sum_{m=1}^{\infty} (f)_m (\mathcal{B}\varphi_m), \qquad (4.22b)$$

which is indeed (4.19). Equations (4.22a) and (4.22b) show that the operator can be interpreted as transforming the orthonormal sequence, as a "basis," so in (4.22b) we see

$$\{\varphi_n\}_{n=1}^{\infty} \longmapsto \{\mathcal{B}\varphi_n\}_{n=1}^{\infty},$$

where the coefficients, $\mathbf{f}$, are invariant. That is, the input, in terms of orthonormal sequence $\{\varphi_n\}_{n=1}^{\infty}$, has Fourier coefficients $\mathbf{f}$, and the output, in terms of orthonormal sequence $\{\mathcal{B}\varphi_n\}_{n=1}^{\infty}$, has the very same Fourier coefficients $\mathbf{f}$. Equivalently, in (4.22a), the operator can be viewed as keeping the basis, $\{\varphi_n\}_{n=1}^{\infty}$, invariant and transforming the coefficients:

$$(f)_m \longmapsto (g)_n = \sum_{m=1}^{\infty} (f)_m b_{n,m}^{(\varphi)}.$$

These two perspectives on the action of the operator are, of course, equivalent.

We can now compose three operators for taking $f \in \mathcal{H}$ to $g \in \mathcal{H}$ as follows:

$$\mathcal{F}^{-1} \circ \mathbf{B}^{(\varphi)} \circ \mathcal{F}: \qquad \text{where} \quad g = \sum_{n=1}^{\infty}\sum_{m=1}^{\infty} \langle f, \varphi_m \rangle b_{n,m}^{(\varphi)} \varphi_n. \qquad (4.23)$$
$$f \longmapsto g \in \mathcal{H}$$

This composition of three operators can be shown to reduce to a single operator as follows. Using the identity $b_{n,m}^{(\varphi)} \triangleq \langle \mathcal{B}\varphi_m, \varphi_n \rangle$ in (4.23) we can write

$$\sum_{n=1}^{\infty}\sum_{m=1}^{\infty} \langle f, \varphi_m \rangle \langle \mathcal{B}\varphi_m, \varphi_n \rangle \varphi_n = \sum_{n=1}^{\infty} \langle \mathcal{B}f, \varphi_n \rangle \varphi_n \qquad (4.24a)$$

$$= \mathcal{B}f, \qquad (4.24b)$$

where we have used (4.21a) and (4.21b). This is indeed (4.15).

Furthermore, we have

$$\mathcal{F} \circ \mathcal{B} \circ \mathcal{F}^{-1}: \qquad \text{where} \quad (g)_n = \langle \mathcal{B} \sum_{m=1}^{\infty} (f)_m \varphi_m, \varphi_n \rangle. \qquad (4.25)$$
$$\mathbf{f} \longmapsto \mathbf{g} \in \ell^2$$

From (4.25) we can write

$$\langle \mathcal{B} \sum_{m=1}^{\infty} (f)_m \varphi_m, \varphi_n \rangle = \sum_{m=1}^{\infty} (f)_m \langle \mathcal{B}\varphi_m, \varphi_n \rangle$$

$$= \sum_{m=1}^{\infty} (f)_m b_{n,m}^{(\varphi)},$$

which is (4.16).

Finally, we are in a position to say how the two operators, $\mathcal{B}$ in $\mathcal{H}$, and $\mathbf{B}^{(\varphi)}$ in ℓ^2, can be inferred from one another. We have seen from (4.11), repeated here,

$$b_{n,m}^{(\varphi)} \triangleq \langle \mathcal{B}\varphi_m, \varphi_n \rangle,$$

and so we expect that we can recover operator $\mathcal{B}$ from matrix operator $\mathbf{B}^{(\varphi)}$ and we now determine this.

From (4.24) we see that

$$\mathcal{B}f = \sum_{n=1}^{\infty} \sum_{m=1}^{\infty} \langle f, \varphi_m \rangle \langle \mathcal{B}\varphi_m, \varphi_n \rangle \varphi_n \qquad (4.26a)$$

$$= \sum_{n=1}^{\infty} \sum_{m=1}^{\infty} \langle f, \varphi_m \rangle b_{n,m}^{(\varphi)} \varphi_n \qquad (4.26b)$$

$$= \sum_{n=1}^{\infty} \langle f, \sum_{m=1}^{\infty} \overline{b}_{n,m}^{(\varphi)} \varphi_m \rangle_{\mathcal{H}} \varphi_n. \qquad (4.26c)$$

So as an operator, we may write

$$\mathcal{B}(\bullet) = \sum_{n=1}^{\infty} \langle \bullet, \sum_{m=1}^{\infty} \overline{b}_{n,m}^{(\varphi)} \varphi_m \rangle_{\mathcal{H}} \varphi_n, \qquad (4.27)$$

where the $\bullet$ indicates a placeholder for the input element from the Hilbert space.

In short, in dealing with a bounded operator it may be advantageous, depending on the context, to work with either $\mathcal{B}$ or its matrix representation $\mathbf{B}^{(\varphi)}$.

Remark 4.6. In the derived identities, there is an implicit isomorphism that connects the two Hilbert spaces, $\mathcal{H}$ and ℓ^2, through the complete orthonormal sequence $\{\varphi_n\}_{n=1}^{\infty}$. This connection between the spaces is also captured through the generalized Parseval relation (2.57), p. 72, as follows:

$$
\begin{aligned}
\langle \mathbf{f}, \mathbf{g} \rangle_{\ell^2} &= \sum_{n=1}^{\infty} (f)_n \overline{(g)_n} \\
&= \sum_{n=1}^{\infty} \langle f, \varphi_n \rangle \overline{\langle g, \varphi_n \rangle} \\
&= \sum_{n=1}^{\infty} \langle f, \varphi_n \rangle \langle \varphi_n, g \rangle \\
&= \langle f, g \rangle_{\mathcal{H}},
\end{aligned}
\tag{4.28}
$$

where, of course, the left-hand inner product is in ℓ^2 and the right-hand one is the inner product in $\mathcal{H}$.   □

4.7   Archetype case: Hilbert-Schmidt integral operator

After seeing many operator identities, it is high time to ask ourselves what a bounded operator looks like in an archetype Hilbert space $L^2(\Omega)$ and how one is able to use the derived identities. The Hilbert-Schmidt integral operator is an important example that we introduce now to illustrate the concepts and to show how the developed techniques work. This type of operator has a rich history and a robust set of results accessible with limited levels of pain and ennui. We will keep revisiting this operator as we come across new concepts. Finally in Chapter 6, p. 160, we will systematically summarize the results and add new insights.

4.7.1   Some history and context

Historically, mathematicians had, over an extended period of time, attempted to sharpen their ideas on what we might call functional analysis and one significant culmination of this was the work by the Swede, Fredholm, on integral equations around 1900. In fact, it was the work by Fredholm that got people like Hilbert thinking. Hilbert had an appreciation of Cantor's transfinite cardinals, which opened up a way to think about the infinite. Inspired by Fredholm's work Hilbert showed how orthonormal sequences could be used to "reduce" functions to an infinite sequence of variables and reduce integral operators, such as the type considered by Fredholm, to matrices — that is, an array of complex numbers with one index pertaining to an input orthonormal sequence and a second index pertaining to an output orthonormal sequence. This theory developed with what we now call separable spaces, which ensures that a countably infinite sequence of functions can act to represent a larger infinity of functions in the space.

So functional analysis became accessible and familiar under Hilbert — results from finite-dimensional linear algebra could be generalized. It was left to others to refine, back-fill and more rigorously develop what we now know as Hilbert space. Whilst these approaches led to a more modern treatment, this style of presentation does not suit everyone. In fact, understanding how and why the theory developed can make more sense and provide greater insight. Often very deep ideas are treated as elementary in the modern treatment, such as understanding the convergence of Fourier series, and this is a significant misrepresentation (since it took 75 years to get close to what is understood here).

The class of integral operators considered herewith includes amongst its instantiations classical integral transform including the Fourier transform, the convolution integral, time-frequency concentration eigen-equations and Green's functions. The integral operator also naturally emerges from the representation of functionals of two variables arising from a diagonalization process which is a generalization of the singular value decomposition (SVD). A caveat in the above discussion is that various subtle technicalities may need to be employed to make the theory amenable.

4.7.2   Hilbert-Schmidt integral operator definition

Let Hilbert space $L^2(\Omega)$ be defined. Now let a function of two variables on the $\Omega \times \Omega$ be given by $K(x, y)$ with the property

$$\kappa^2 = \int_{\Omega \times \Omega} |K(x, y)|^2 dx\, dy < \infty. \tag{4.29}$$

That is, $K(x, y)$ belongs to $L^2(\Omega \times \Omega)$. For every $f(x) \in L^2(\Omega)$, we define the Hilbert-Schmidt integral operator $(\mathcal{L}_K f)(x)$ as

$$(\mathcal{L}_K f)(x) = \int_{\Omega} K(x, y) f(y)\, dy. \tag{4.30}$$

$K(x, y)$ is referred to as the Hilbert-Schmidt kernel. We can see that $(\mathcal{L}_K f)(x)$ is a linear mapping on $L^2(\Omega)$ since

$$\begin{aligned}
\|\mathcal{L}_K f\|^2 &= \int_{\Omega} |(\mathcal{L}_K f)(x)|^2 dx \\
&= \int_{\Omega} \left| \int_{\Omega} K(x, y) f(y)\, dy \right|^2 dx \\
&\leq \int_{\Omega \times \Omega} |K(x, y)|^2 dx\, dy \int_{\Omega} |f(y)|^2 dy = \kappa^2 \|f\|^2.
\end{aligned} \tag{4.31}$$

Since $\kappa^2 < \infty$, we conclude that $\mathcal{L}_K$ is a bounded operator on $L^2(\Omega)$ with its norm upper bounded by κ.

4.7.3   Matrix presentation and relations

Now, we are interested to see how this operator in $L^2(\Omega)$ is related to the matrix representation in ℓ^2. Let $f(x)$ be written in terms of a basis in $L^2(\Omega)$ as

$$f(x) = \sum_{m=1}^{\infty} (f)_m \varphi_m(x).$$

Applying $\mathcal{L}_K$, we obtain

$$(\mathcal{L}_K f)(x) = \mathcal{L}_K \Big(\sum_{m=1}^{\infty} (f)_m \varphi_m(x) \Big)$$

$$= \sum_{m=1}^{\infty} (f)_m \big(\mathcal{L}_K \varphi_m \big)(x)$$

$$= \sum_{m=1}^{\infty} (f)_m \Big(\sum_{n=1}^{\infty} \langle \mathcal{L}_K \varphi_m, \varphi_n \rangle \varphi_n(x) \Big)$$

$$= \sum_{n=1}^{\infty} \Big(\sum_{m=1}^{\infty} (f)_m k_{n,m}^{(\varphi)} \Big) \varphi_n(x),$$

where $k_{n,m}^{(\varphi)}$ signifies how much of $\varphi_m(x)$ is projected along $\varphi_n(x)$ and is given by

$$k_{n,m}^{(\varphi)} = \langle \mathcal{L}_K \varphi_m, \varphi_n \rangle = \int_{\Omega} \Big(\int_{\Omega} K(x,y)\varphi_m(y)\, dy \Big) \overline{\varphi_n(x)} dx \qquad (4.32)$$

$$= \int_{\Omega} \int_{\Omega} K(x,y)\varphi_m(y)\overline{\varphi_n(x)}\, dx\, dy$$

$$= \int_{\Omega} \int_{\Omega} K(x,y)\overline{\xi_{n,m}^{(\varphi)}(x,y)}\, dx\, dy$$

$$= \langle K, \varphi_{n,m} \rangle,$$

where $\xi_{n,m}^{(\varphi)}(x,y) \triangleq \varphi_n(x)\overline{\varphi_m(y)}$ and the inner product is defined in $L^2(\Omega \times \Omega)$.

So the key identity is

$$\boxed{ k_{n,m}^{(\varphi)} = \int_{\Omega} \int_{\Omega} K(x,y)\overline{\varphi_n(x)}\varphi_m(y)\, dx\, dy. }$$

We have just shown how we can go from the operator in $L^2(\Omega)$, $\mathcal{L}_K$, to the matrix operator in ℓ^2, $\mathbf{K}^{(\varphi)}$ with elements $k_{n,m}^{(\varphi)}$. If we are given an operator matrix $\mathbf{K}^{(\varphi)}$ in ℓ^2, we can manipulate (4.27) to obtain

$$(\mathcal{L}_K f)(x) = \sum_{n=1}^{\infty} \langle f, \sum_{m=1}^{\infty} \overline{k}_{n,m}^{(\varphi)} \varphi_m \rangle_{\Omega} \varphi_n(x)$$

$$= \sum_{n=1}^{\infty} \Big(\int_{\Omega} f(y) \Big(\sum_{m=1}^{\infty} k_{n,m}^{(\varphi)} \overline{\varphi_m(y)} \Big)\, dy \Big)\, \varphi_n(x)$$

$$= \int_{\Omega} f(y) \sum_{n=1}^{\infty} \sum_{m=1}^{\infty} k_{n,m}^{(\varphi)} \overline{\varphi_m(y)} \varphi_n(x)\, dy \qquad (4.33)$$

$$= \int_{\Omega} K(x,y)f(y)\, dy,$$

where

$$\boxed{ K(x,y) \triangleq \sum_{n=1}^{\infty} \sum_{m=1}^{\infty} k_{n,m}^{(\varphi)} \overline{\varphi_m(y)} \varphi_n(x) } \qquad (4.34)$$

is the kernel and $\langle \cdot, \cdot \rangle_\Omega$ makes explicit that the inner product is in $L^2(\Omega)$. So in the archetype case, in going from the operator matrix on ℓ^2 to the operator on $L^2(\Omega)$, the operator matrix $\mathbf{K}^{(\varphi)}$ directly forms the inner core of the kernel in an integral equation, third expression in (4.33), which realizes the operator $\mathcal{L}_K$. In the abstract case, we can regard the operator matrix as forming the inner core of a type of abstract kernel implicit in (4.27).

Take-home messages – bounded operators

M1: If a linear operator is continuous, we can switch the order of operator and limit of a converging sequence. That is, if $f_n \to f_0 \in \mathcal{H}$, then there is no difference between operating on the limit f_0 or taking the limit of operations on f_n, $\mathcal{B}f_n \to \mathcal{B}f_0$.

M2: The operator norm, $\|\mathcal{B}\|_{\text{op}}$, is the maximum gain of the operator; that is, the maximum increase in the norm of the output relative to the norm of the input. If an operator is bounded, then its norm is finite and $\|\mathcal{B}f\| \leq \|\mathcal{B}\|_{\text{op}}\|f\|$ for all $f \in \mathcal{H}$.

M3: For linear operators, boundedness and linearity imply each other. A bounded linear operator is continuous and a continuous linear operator is bounded.

M4: There are three different types of convergence for a sequence of operators $\{\mathcal{B}_n\}_{n=1}^\infty$.

- Weak convergence means that $\langle \mathcal{B}_n f, g \rangle \to \langle \mathcal{B}f, g \rangle$ for all $f, g \in \mathcal{H}$. Symbolically, $\mathcal{B}_n f \rightharpoonup \mathcal{B}f$.
- Strong convergence means that $\|\mathcal{B}_n f - \mathcal{B}f\| \to 0$ for all $f \in \mathcal{H}$. Symbolically, $\mathcal{B}_n f \to \mathcal{B}f$.
- Convergence in norm means that $\|\mathcal{B}_n - \mathcal{B}\|_{\text{op}} \to 0$

Convergence in norm implies strong convergence, which in turn implies weak convergence for operators.

M5: For bounded linear operators, the action of operator is characterized by an operator matrix. The matrix is denoted by $\mathbf{B}^{(\varphi)}$ and depends on the choice of basis $\{\varphi_n\}_{n=1}^\infty$. Its element at row n and column m is $b_{n,m}^{(\varphi)} = \langle \mathcal{B}\varphi_m, \varphi_n \rangle$, which signifies how much of the input basis φ_m is mapped along output basis φ_n as a result of operation.

M6: Instead of performing $\mathcal{B}$ on f to get $g = \mathcal{B}f$, we can first obtain the vector of Fourier coefficients of f denoted by $\mathbf{f}$ containing $(f)_n = \langle f, \varphi_n \rangle$, apply the operator in ℓ^2 by simply performing $\mathbf{g} = \mathbf{B}^{(\varphi)}\mathbf{f}$ and finally obtain g from $\mathbf{g}$ as $g = \sum_n (g)_n \varphi_n$.

Take-home messages – bounded operators, continued

M7: It is also possible to obtain the operator $\mathcal{B}$ from its operator matrix, $\mathbf{B}^{(\varphi)}$, as follows

$$\mathcal{B}(\bullet) = \sum_{n=1}^{\infty} \langle \bullet, \sum_{m=1}^{\infty} \overline{b}_{n,m}^{(\varphi)} \varphi_m \rangle_{\mathcal{H}} \varphi_n.$$

M8: An example of putting the above into action is the Hilbert-Schmidt integral operator with kernel $K(x,y)$ in $L^2(\Omega)$, defined as

$$(\mathcal{L}_K f)(x) = \int_{\Omega} K(x,y) f(y)\, dy,$$

where the operator matrix elements are given by

$$k_{n,m}^{(\varphi)} = \int_{\Omega} \int_{\Omega} K(x,y)\overline{\varphi_n(x)}\varphi_m(y)\, dx\, dy,$$

and the kernel can be retrieved as

$$K(x,y) = \sum_{n=1}^{\infty} \sum_{m=1}^{\infty} k_{n,m}^{(\varphi)} \varphi_n(x)\overline{\varphi_m(y)}.$$

4.8   Road map

It is time to take a break from mathematics and look at our map again. We have become familiar with the concept of operators, some important classes of operators and their algebraic and topological properties. We also learned how to represent bounded operators as infinite-dimensional matrices in ℓ^2 and work back and forth between abstract operators and their matrix equivalents.

Yet still the operators in their general form are somewhat hard to work with or are not intuitive enough. For example, a bounded operator with a general infinite-dimensional matrix representation $\mathbf{B}^{(\varphi)}$ can project a basis function φ_m into all possible directions φ_n where the gain is determined by $b_{n,m}^{(\varphi)}$. But, this is not so desirable. In a way, we do not have any control over what in the input goes where in the output.

Ideally, we wish to work with *special* inputs and *simple and contained* operators, such that they do not spill the input into all directions in the output. Using such simple operators we hope to represent or very well approximate a more complicated operator in a meaningful way.

In the previous part of the book, we saw how we can split a Hilbert space into smaller orthogonal subspaces, leading to Fourier expansions and projections into basis elements. In this part, we will see that projection operators are the important class of simple and contained operators we are looking for and eigenvectors or eigenspaces are the appropriate class of inputs that do not spill out into any other direction in the output, but themselves. This leads to spectral analysis of bounded operators. However, for this to make sense, we need to go through some more basic definitions and theorems. The first stop is adjoint and self-adjoint operators.

4.9   Adjoint operators

The focus has been and shall continue to be with bounded linear operators on $\mathcal{H}$. Linear operators on $\mathcal{H}$ satisfy the algebraic conditions (3.2), p. 109, and the boundedness condition is (4.2); see Example 4.4 below for these conditions in action. It is a gentle generalization to consider bounded linear mappings which cater for the situation where the output space need not be the same as the input space. There is one apparently degenerate case, where the output space is just $\mathbb{C}$, which turns out to be very useful and essential for some theoretical extensions (particularly generalizing from Hilbert spaces to Banach spaces).

Definition 4.8 (Linear functional). *A linear mapping from $\mathcal{H}$ into $\mathbb{C}$ is called a linear functional on $\mathcal{H}$.* □

This is superficially a bit like the behavior of the norm which takes in a vector and returns a number. However, the norm is not linear and can only return a non-negative real number. The norm could be regarded as a functional, but not a linear functional. So in the back of our minds we have the sequence of subsets: functionals, linear functionals and bounded linear functionals.

It is very easy to cook up a large class of linear functionals, as the example below shows.

Example 4.4 (Linear functional). Define
$$\mathcal{A}f = \langle f, h \rangle,$$
where $h \in \mathcal{H}$ is fixed (and $f \in \mathcal{H}$ can vary). We can confirm
$$\begin{aligned}
\mathcal{A}(\alpha_1 f_1 + \alpha_2 f_2) &= \langle \alpha_1 f_1 + \alpha_2 f_2, h \rangle \\
&= \langle \alpha_1 f_1, h \rangle + \langle \alpha_2 f_2, h \rangle \\
&= \alpha_1 \langle f_1, h \rangle + \alpha_2 \langle f_2, h \rangle \\
&= \alpha_1 \mathcal{A}f_1 + \alpha_2 \mathcal{A}f_2,
\end{aligned}$$
for all $f_1, f_2 \in \mathcal{H}$, which shows $\mathcal{A}$ is a *linear functional*. Furthermore, the norm of the functional, denoted by $\|\mathcal{A}\|_{\mathrm{fn}}$, is defined in a similar manner as the operator norm as
$$\|\mathcal{A}\|_{\mathrm{fn}} \triangleq \sup_{f \in \mathcal{H}, \|f\|=1} \|\mathcal{A}f\| = \sup_{f \in \mathcal{H}, \|f\|=1} |\mathcal{A}f| = \sup_{f \in \mathcal{H}, \|f\|=1} |\langle f, h \rangle| \le \|h\|,$$
where the second equality follows from the fact that $\mathcal{A}f \in \mathbb{C}$, the third equality follows from the definition of functional $\mathcal{A}$ and the inequality is the Cauchy-Schwarz inequality. Hence $\mathcal{A}$ is a *bounded linear functional*. □

The large class of linear functionals referred to above is obtained by varying $h \in \mathcal{H}$. So it is natural to ask what other linear functionals look like. Surprisingly, in the case of *bounded* linear functionals there are no others, that is, Example 4.4 exhausts all possibilities by varying $h \in \mathcal{H}$, and this is an interesting result captured through the following theorem.

Theorem 4.2 (Riesz representation theorem). *If $\mathcal{A}$ is a bounded, hence continuous, linear functional on $\mathcal{H}$ then there exists a unique vector $h \in \mathcal{H}$ such*

that
$$\mathcal{A}f = \langle f, h \rangle,$$
for all $f \in \mathcal{H}$. Further $\|\mathcal{A}\| = \|h\|$.   □

Proof. The proof is not difficult, but does not shed much light either; see (Helmberg, 1969, pp. 76–77), or (Debnath and Mikusiński, 1999, pp. 126–127). We note that from Example 4.4, $\|\mathcal{A}\| \le \|h\|$. Therefore, with the choice $f = \alpha h$ (for a suitable $\alpha \in \mathbb{R}$ such that $\|f\| = 1$) the Cauchy-Schwarz inequality is achieved and $\|\mathcal{A}\| = \|h\|$ can be inferred.   □

> **Remark 4.7.** As for the case of bounded linear operators, it is a simple matter to establish that boundedness and continuity are equivalent, that is, a bounded linear functional is also continuous.   □

We are now ready to define an adjoint operator.

Definition 4.9 (Adjoint operator). *The adjoint operator, denoted $\mathcal{B}^\star$, for any bounded (linear) operator $\mathcal{B}$ on $\mathcal{H}$ is defined as the unique operator on $\mathcal{H}$ such that*
$$\langle \mathcal{B}f, g \rangle = \langle f, \mathcal{B}^\star g \rangle, \tag{4.35}$$
for every pair of elements $f, g \in \mathcal{H}$.   □

It is well and good to define an adjoint operator as above, but it does need some justification. The real issue is whether such a $\mathcal{B}^\star$ operator actually exists and is unique. This is in contrast to the inverse of a bounded operator, which may or may not exist. This is where the Theorem 4.2 comes in. The expression $\langle \mathcal{B}f, g \rangle$ for a fixed $g \in \mathcal{H}$ is a bounded linear functional on $\mathcal{H}$ and, therefore, by the Riesz representation theorem, there exists a unique $h \in \mathcal{H}$ so that

$$\langle \mathcal{B}f, g \rangle = \langle f, h \rangle,$$

for all $f \in \mathcal{H}$. Clearly, h depends on $\mathcal{B}$ and g. The mapping which takes each $g \in \mathcal{H}$ to a unique $h \in \mathcal{H}$ is easily seen to be linear and is a linear operator. Adjoint $\mathcal{B}^\star$ is this operator and $h = \mathcal{B}^\star g$.

Finite-dimensional case

In finite-dimensional linear algebra, we can develop a convenient analog, which leads to a simple interpretation for the adjoint. Let the finite-dimensional Hilbert space be $\mathbb{C}^N$ with inner product given by the conventional dot product. Furthermore, linear operators mapping a vector in $\mathbb{C}^N$ to another vector in $\mathbb{C}^N$ are simply an $N \times N$ matrix of complex coefficients. So letting $\mathbf{a}, \mathbf{b}$ be two $N \times 1$ vectors in $\mathbb{C}^N$ and $\mathbf{M} \in \mathbb{C}^{N \times N}$

$$\langle \mathbf{Ma}, \mathbf{b} \rangle = (\mathbf{Ma})' \overline{\mathbf{b}} = \mathbf{a}' \mathbf{M}' \overline{\mathbf{b}} = \mathbf{a}' (\mathbf{M}' \overline{\mathbf{b}})$$
$$= \langle \mathbf{a}, \mathbf{M}^H \mathbf{b} \rangle,$$

where $(\cdot)^H = \overline{(\cdot)'}$ denotes the Hermitian transpose. So if a linear operator has matrix $\mathbf{M}$ then apparently its adjoint operator has matrix $\mathbf{M}^H$.

As a plausibility argument, for bounded infinite-dimensional operators which have an infinite-dimensional matrix representation, say $\mathbf{B}^{(\varphi)}$ with respect to some orthonormal sequence, their adjoint operator should be characterized by the infinite-dimensional Hermitian transpose matrix $\left(\mathbf{B}^{(\varphi)}\right)^H$ with respect to the same orthonormal sequence. To see this, using (4.16) we write

$$
\begin{aligned}
\langle \mathbf{B}^{(\varphi)}\mathbf{f}, \mathbf{g}\rangle &= \sum_{n=1}^{\infty}\sum_{m=1}^{\infty}(f)_m b_{n,m}^{(\varphi)}\overline{(g)}_n \\
&= \sum_{m=1}^{\infty}(f)_m \sum_{n=1}^{\infty} b_{n,m}^{(\varphi)}\overline{(g)}_n \\
&= \sum_{m=1}^{\infty}(f)_m \overline{\sum_{n=1}^{\infty}\overline{b}_{n,m}^{(\varphi)}(g)_n} = \langle \mathbf{f}, \left(\mathbf{B}^{(\varphi)}\right)^H\mathbf{g}\rangle.
\end{aligned}
$$

Example 4.5. Previously, we argued that the left shift operator $\mathcal{S}_L$ on ℓ^2 became close to being the inverse of right shift operator $\mathcal{S}_R$ on ℓ^2, but not quite ($\mathcal{S}_L \circ \mathcal{S}_R = \mathcal{I}$, but $\mathcal{S}_R \circ \mathcal{S}_L \neq \mathcal{I}$). We now see that $\mathcal{S}_L$ is in fact the adjoint of $\mathcal{S}_R$, and that $\mathcal{S}_R$ is the adjoint of $\mathcal{S}_L$:

$$
\langle \mathcal{S}_R a, b\rangle = \sum_{n=1}^{\infty}\alpha_n \overline{\beta}_{n+1} = \langle a, \mathcal{S}_L b\rangle,
$$

$$
\langle \mathcal{S}_L a, b\rangle = \sum_{n=1}^{\infty}\alpha_{n+1}\overline{\beta}_n = \langle a, \mathcal{S}_R b\rangle.
$$

The matrix form of the left shift operator is the transpose of right shift operator:

$$
\mathbf{S}_R = \begin{pmatrix} 0 & 0 & 0 & 0 & \cdots \\ 1 & 0 & 0 & 0 & \cdots \\ 0 & 1 & 0 & 0 & \cdots \\ 0 & 0 & 1 & 0 & \cdots \\ \vdots & \vdots & \vdots & \vdots & \ddots \end{pmatrix} \quad \text{and} \quad \mathbf{S}_L = \begin{pmatrix} 0 & 1 & 0 & 0 & \cdots \\ 0 & 0 & 1 & 0 & \cdots \\ 0 & 0 & 0 & 1 & \cdots \\ 0 & 0 & 0 & 0 & \cdots \\ \vdots & \vdots & \vdots & \vdots & \ddots \end{pmatrix}.
$$

$\square$

Remark 4.8. Because the input and output can be regarded as living in the same space $\mathcal{H}$, it gives the impression that the adjoint operator, $\mathcal{B}^\star$, has the same input and output respectively as the original operator, $\mathcal{B}$. This can be misleading (but not incorrect in this case). The adjoint operator, $\mathcal{B}^\star$, is better thought of as transforming the output to the input in terms of its direction (that is, in the same direction as would the inverse of $\mathcal{B}$ if that inverse existed). If the operator, more generally, transforms elements from a Hilbert space $\mathcal{H}_1$ to elements of a different Hilbert space $\mathcal{H}_2$, then necessarily the adjoint operator transforms elements from $\mathcal{H}_2$ to $\mathcal{H}_1$, so we can more generally define the adjoint as the unique $\mathcal{B}^\star$ such that

$$
\langle \mathcal{B}f, g\rangle_{\mathcal{H}_2} = \langle f, \mathcal{B}^\star g\rangle_{\mathcal{H}_1}. \tag{4.36}
$$

So here $\mathcal{B}: \mathcal{H}_1 \longmapsto \mathcal{H}_2$ and $\mathcal{B}^\star: \mathcal{H}_2 \longmapsto \mathcal{H}_1$. Furthermore, $f \in \mathcal{H}_1$ and $g \in \mathcal{H}_2$. This is summarized in Definition 4.10. $\square$

Remark 4.9. The above comment may be further clarified by revisiting the finite-dimensional case, but this time mapping from $\mathbb{C}^M$ to $\mathbb{C}^N$. In this case: $\mathbf{M} \in \mathbb{C}^{N \times M}$, $\mathbf{M}^H \in \mathbb{C}^{M \times N}$, $\mathbf{a}$ is a $M \times 1$ vector in $\mathbb{C}^M$ and $\mathbf{b}$ is a $N \times 1$ vector in $\mathbb{C}^N$

$$\langle \mathbf{Ma}, \mathbf{b} \rangle_{\mathbb{C}^N} = (\mathbf{Ma})' \overline{\mathbf{b}} = \mathbf{a}' \mathbf{M}' \overline{\mathbf{b}} = \mathbf{a}' (\mathbf{M}' \overline{\mathbf{b}})$$
$$= \langle \mathbf{a}, \mathbf{M}^H \mathbf{b} \rangle_{\mathbb{C}^M}.$$

□

We give the definition of the adjoint in a more general setting where the input and output spaces can be different:

Definition 4.10 (Adjoint operator generalization). *For any bounded operator* $\mathcal{B} \colon \mathcal{H}_1 \longmapsto \mathcal{H}_2$, *the adjoint operator is defined as the unique operator*

$$\mathcal{B}^\star \colon \mathcal{H}_2 \longmapsto \mathcal{H}_1,$$

such that

$$\langle \mathcal{B}f, g \rangle_{\mathcal{H}_2} = \langle f, \mathcal{B}^\star g \rangle_{\mathcal{H}_1},$$

for every pair of elements $f \in \mathcal{H}_1$ *and* $g \in \mathcal{H}_2$.

□

Problems

4.9. Let the Hilbert-Schmidt integral operator $\mathcal{L}_K$ with kernel $K(x, y)$ on $L^2(\Omega)$ be given. Detailing $\langle \mathcal{L}_K f, g \rangle$ find out the kernel for the adjoint of $\mathcal{L}_K$.

4.10. Show the following relations hold for the adjoint operator:

$$(\alpha \mathcal{B})^\star = \overline{\alpha} \mathcal{B}^\star, \quad \alpha \in \mathbb{C},$$
$$(\mathcal{B}_2 \circ \mathcal{B}_1)^\star = \mathcal{B}_1^\star \circ \mathcal{B}_2^\star,$$
$$(\mathcal{B}_1 + \mathcal{B}_2)^\star = \mathcal{B}_1^\star + \mathcal{B}_2^\star.$$

4.11. Show that the adjoint of the adjoint of an operator is the original operator itself. That is,

$$(\mathcal{B}^\star)^\star = \mathcal{B}.$$

4.12. If $\mathcal{B}$ possesses an inverse then show

$$(\mathcal{B}^\star)^{-1} = (\mathcal{B}^{-1})^\star.$$

4.13. Let $\mathcal{B}^\star$ denote the adjoint operator corresponding to a bounded operator $\mathcal{B}$. Show that the operator norm of the adjoint equals that of the operator. That is,

$$\|\mathcal{B}\|_{\mathrm{op}} = \|\mathcal{B}^\star\|_{\mathrm{op}}.$$

4.14 (Weak convergence). In a Banach space there is no inner product to employ directly to define weak convergence. Devise a way to define weak convergence in a Banach space that reduces to the inner product definition in Hilbert space. (Hint: refer back to Definition 2.25, p. 75, and use a bounded linear functional.)

4.9.1   Special forms of adjoint operators

Definition 4.11 (Self-adjoint operator). *A bounded operator on $\mathcal{H}$ is self-adjoint, if the adjoint $\mathcal{B}^\star$ equals the original operator $\mathcal{B}$. That is,*

$$\mathcal{B} = \mathcal{B}^\star, \tag{4.37}$$

which implies

$$\langle \mathcal{B}f, g \rangle = \langle f, \mathcal{B}g \rangle, \tag{4.38}$$

for every pair of elements f, g in $\mathcal{H}$.   □

A self-adjoint operator needs to be a transformation of a Hilbert space onto itself (rather than between two different Hilbert spaces). A self-adjoint operator is a generalization of a Hermitian symmetric matrix.

For self-adjoint operators we have an alternative way to compute the operator norm in terms of the inner product:

Theorem 4.3. *For a self-adjoint operator $\mathcal{B}$, we have*

$$\|\mathcal{B}\|_{\mathrm{op}} \triangleq \sup_{\|f\|=1} \|\mathcal{B}f\| = \sup_{\|f\|=1} |\langle \mathcal{B}f, f \rangle|. \tag{4.39}$$

                                                                                  □

Proof. The proof is quite standard, repeated in many books, but relies on a few tricks; see (Riesz and Sz.-Nagy, 1990, p. 230).   □

Definition 4.12 (Unitary operator). *A bounded operator on $\mathcal{H}$ is unitary, if the inverse exists and the adjoint $\mathcal{B}^\star$ equals the inverse. That is,*

$$\mathcal{B}^{-1} = \mathcal{B}^\star, \tag{4.40}$$

which implies

$$\langle \mathcal{B}f, g \rangle = \langle f, \mathcal{B}^{-1}g \rangle, \tag{4.41}$$

for every pair of elements f, g in $\mathcal{H}$.   □

Problems

4.15. Let $\mathcal{B}$ be a bounded operator on ℓ^2 which takes any element $a = \{\alpha_n\}_{n=1}^{\infty} \in \ell^2$ to $\mathcal{B}a = \{\alpha_n/n\}_{n=1}^{\infty} \in \ell^2$. Show that $\mathcal{B}$ is self-adjoint and has a norm equal to 1.

4.16. Let $\mathcal{B}_1$ and $\mathcal{B}_2$ be two self-adjoint operators. Show that the composition, $\mathcal{B}_2 \circ \mathcal{B}_1$, is self-adjoint if and only if $\mathcal{B}_1$ and $\mathcal{B}_2$ commute, that is, $\mathcal{B}_2 \circ \mathcal{B}_1 = \mathcal{B}_1 \circ \mathcal{B}_2$.

4.17. Let $\mathcal{B}_1$ and $\mathcal{B}_2$ be two self-adjoint operators. Show that the compositions $\mathcal{B}_2 \circ \mathcal{B}_1 \circ \mathcal{B}_2$, and $\mathcal{B}_1 \circ \mathcal{B}_2 \circ \mathcal{B}_1$ are self-adjoint.

4.10   Projection operators

Definition 4.13 (Projection operator). *For every subspace $\mathfrak{M}$ of a Hilbert space $\mathcal{H}$, there is a corresponding bounded linear projection operator denoted by $\mathcal{P}_{\mathfrak{M}}$, which maps every element $f \in \mathcal{H}$ into its projection in $\mathfrak{M}$, denoted $\mathcal{P}_{\mathfrak{M}} f$.*   □

Wherever there is no chance of confusion, we may drop the subspace index $\mathfrak{M}$ and represent projection operators by just $\mathcal{P}$. Projection operators have very important topological properties, which we will explore now and utilize in later sections. Perhaps the most important property of projection operators is the *control* they provide over where the input elements go in the output. No matter where the input is, the output always lands in subspace $\mathfrak{M}$. It cannot go anywhere else. Obviously, all elements that already belong to the subspace $\mathfrak{M}$ stay unchanged as a result of projection operation and this is the way one can reconstruct the corresponding subspace from a given projection operator: the closed subset of all elements of $\mathcal{H}$ which stay unchanged under the projection operator:

$$\mathfrak{M} = \{f \colon \mathcal{P}_{\mathfrak{M}} f = f\}.$$

This leads to the important *idempotent* property of projection operators: after applying projection operator once to an element f of Hilbert space, further applications do not have any extra effect. That is,

$$\mathcal{P} \circ \mathcal{P} \circ \cdots \circ \mathcal{P} f = \mathcal{P} f \quad \forall f \in \mathcal{H}.$$

From the projection theorem, Theorem 2.8, p. 67, we can write any element f as

$$f = \mathcal{P} f + \mathcal{P}^{\perp} f,$$

where $\mathcal{P}^{\perp}$ is a shorthand for the projection operator upon the complement orthogonal subspace of $\mathfrak{M}$. Applying $\mathcal{P}$ to both sides of the above equation and using the *idempotent* property of $\mathcal{P}$

$$\mathcal{P} f = \mathcal{P} \circ \mathcal{P} f + \mathcal{P} \circ \mathcal{P}^{\perp} f = \mathcal{P} f + \mathcal{P} \circ \mathcal{P}^{\perp} f,$$

meaning that $\mathcal{P} \circ \mathcal{P}^{\perp} f = o$, again confirming that the projection operator ensures that nothing from the orthogonal subspace of $\mathfrak{M}$ leaks to the output. In fact, we can reconstruct the complement orthogonal subspace corresponding to a projection operator $\mathcal{P}_{\mathfrak{M}}$ as follows

$$\mathfrak{M}^{\perp} = \{f \colon \mathcal{P}_{\mathfrak{M}} f = o\}. \tag{4.42}$$

Now we show that the projection operator is self-adjoint.

Theorem 4.4. *A projection operator $\mathcal{P}$ is self-adjoint.*   □

Proof. Since $\langle \mathcal{P} f, \mathcal{P}^{\perp} g \rangle = \langle \mathcal{P}^{\perp} f, \mathcal{P} g \rangle = 0$, we can write

$$\begin{aligned}
\langle \mathcal{P} f, g \rangle &= \langle \mathcal{P} f, \mathcal{P} g + \mathcal{P}^{\perp} g \rangle \\
&= \langle \mathcal{P} f, \mathcal{P} g \rangle \\
&= \langle \mathcal{P} f + \mathcal{P}^{\perp} f, \mathcal{P} g \rangle = \langle f, \mathcal{P} g \rangle.
\end{aligned}$$

Hence, $\mathcal{P}$ is self-adjoint.   □

In fact, it can be proven that any operator that is both self-adjoint and idempotent is a projection operator, but here we do not provide the proof. We note that identity and zero operators are trivial projection operators.

Finally, using Fourier series expansion into subspaces that we learned in the first part of the book, we can provide another definition for the projection operator, which is useful for moving onto finite rank projection operators.

Definition 4.14 (Projection operator — equivalent definition). *Let $\mathfrak{M}$ be a subspace of a Hilbert space $\mathcal{H}$ and let $\{\varphi_n\}_{n \in M}$ be a complete orthonormal sequence in $\mathfrak{M}$ where M is a set of indices, possibly of infinite cardinality. Then the projection operator onto $\mathfrak{M}$ can be defined through*

$$\mathcal{P}_{\mathfrak{M}} f = \sum_{n \in M} \langle f, \varphi_n \rangle \varphi_n. \tag{4.43}$$

□

Let us identify the projection operator matrix, based on its expansion in (4.43). Let the operator matrix corresponding to $\mathcal{P}_{\mathfrak{M}}$ be denoted by $\mathbf{P}^{(\varphi)}$, where $\{\varphi_n\}_{n=1}^{\infty}$ is the basis for the whole space $\mathcal{H}$, of which $\{\varphi_n\}_{n \in M}$ is a subset. Using the definition in (4.11) in (4.43), we obtain

$$p_{n,m}^{(\varphi)} = \langle \mathcal{P}\varphi_m, \varphi_n \rangle = \sum_{n \in M} \langle \varphi_m, \varphi_n \rangle \langle \varphi_n, \varphi_n \rangle = \begin{cases} 1 & m \in M, \\ 0 & \text{otherwise,} \end{cases}$$

which highlights that the projection operator matrix takes the very special form of a diagonal matrix, whose only non-zero diagonal elements $p_{m,m}^{(\varphi)}$ correspond to the indices $m \in M$. This can be used as an alternative proof that the operator is self-adjoint, since its operator matrix is self-Hermitian.

Problems

4.18. Show that the projection operator is self-adjoint and idempotent directly using the definition (4.43).

4.19. Show that the projection operator is a *positive operator*. That is,

$$\langle \mathcal{P}_{\mathfrak{M}} f, f \rangle \geq 0,$$

for all $f \in \mathcal{H}$.

4.10.1   Finite rank projection operators

Most of the time, we are interested in projection operators that have a *finite rank*; that is, projection operators where the underlying subspace $\mathfrak{M}$ is finite-dimensional. This is a useful key property that will allow us to break complicated operators into their smallest building blocks later on.

Definition 4.15 (Finite rank projection operator). *Given an N-dimensional orthonormal set $\{\varphi_n\}_{n=1}^N$, the finite rank projection operator $\mathcal{P}_N$ is defined by*

$$\mathcal{P}_N f = \sum_{n=1}^N \langle f, \varphi_n \rangle \varphi_n. \tag{4.44}$$

$\square$

If $\{\varphi_n\}_{n=1}^\infty$ were a complete orthonormal sequence, then (4.44) can be viewed as simply truncating the Fourier series representation of a function to a finite number of terms. That is, an orthogonal projection operator $\mathcal{P}_N$ projects an input f onto a finite N-dimensional orthonormal set $\{\varphi_n\}_{n=1}^N$ (which is inevitably incomplete in the original space $\mathcal{H}$ and only complete in the subspace spanned by $\{\varphi_n\}_{n=1}^N$).

Applying this definition to archetype Hilbert space $L^2(\Omega)$ with $\{\varphi_n\}_{n=1}^\infty$,

$$\begin{aligned}
(\mathcal{P}_N f)(x) &= \sum_{n=1}^N \langle f, \varphi_n \rangle \varphi_n(x) \\
&= \sum_{n=1}^N \Big(\int_\Omega f(y) \overline{\varphi_n(y)} \, dy \Big) \varphi_n(x) \\
&= \int_\Omega f(y) \Big(\sum_{n=1}^N \overline{\varphi_n(y)} \varphi_n(x) \Big) \, dy,
\end{aligned}$$

and comparing with (4.30), the finite rank projection kernel is readily written as

$$P_N(x, y) \triangleq \sum_{n=1}^N \varphi_n(x) \overline{\varphi_n(y)}.$$

Hermitian symmetry property or self-adjointness of the operator is verified through

$$\begin{aligned}
P_N(y, x) &= \sum_{n=1}^N \varphi_n(y) \overline{\varphi_n(x)} \\
&= \sum_{n=1}^N \overline{\varphi_n(x) \overline{\varphi_n(y)}} \\
&= \overline{P_N(x, y)}.
\end{aligned}$$

Another property worth noting is, for finite N,

$$\|P_N\|^2 \triangleq \iint_{\Omega \times \Omega} |P_N(x, y)|^2 dx \, dy < \infty.$$

And the operator norm of $\mathcal{P}_N$ is finite and using (4.31) is upper bounded by

$$\|\mathcal{P}_N\|_{\mathrm{op}} \leq \|P_N\|.$$

Problem

4.20. Show that $\|\mathcal{P}_N\|_{\text{op}} = 1$ and $\|\mathcal{I} - \mathcal{P}_N\|_{\text{op}} = 1$.

4.10.2   Further properties of projection operators

It is worthwhile to study what happens when we apply two projector operators $\mathcal{P}_{\mathfrak{M}}$ and $\mathcal{P}_{\mathfrak{N}}$ in succession defined on subspaces $\mathfrak{M}$ and $\mathfrak{N}$, respectively. If $\mathcal{P}_{\mathfrak{M}} \circ \mathcal{P}_{\mathfrak{N}}$ is a projection operator in its own right,[2] then $\mathcal{P}_{\mathfrak{M}} \circ \mathcal{P}_{\mathfrak{N}}$ projects upon the intersection $\mathfrak{M} \cap \mathfrak{N}$. Therefore, $\mathcal{P}_{\mathfrak{M}} \circ \mathcal{P}_{\mathfrak{N}}$ effectively *reduces* the range of output into a smaller subspace. If $\mathfrak{M}$ is already a subspace of $\mathfrak{N}$, applying the *bigger* operator $\mathcal{P}_{\mathfrak{N}}$ after the *smaller* operator $\mathcal{P}_{\mathfrak{M}}$ does not have any effect and $\mathcal{P}_{\mathfrak{N}} \circ \mathcal{P}_{\mathfrak{M}} = \mathcal{P}_{\mathfrak{M}} \circ \mathcal{P}_{\mathfrak{N}} = \mathcal{P}_{\mathfrak{M}}$.

If $\mathfrak{M}$ and $\mathfrak{N}$ happen to be orthogonal subspaces ($\mathfrak{N} \perp \mathfrak{M}$), we can verify using (4.42) that the succession of $\mathcal{P}_{\mathfrak{M}}$ and $\mathcal{P}_{\mathfrak{N}}$ is the zero operator $\mathcal{P}_{\mathfrak{M}} \circ \mathcal{P}_{\mathfrak{N}} = \mathcal{P}_{\mathfrak{N}} \circ \mathcal{P}_{\mathfrak{M}} = \mathcal{O}$:

$$\forall f \in \mathcal{H}: \quad \mathcal{P}_{\mathfrak{N}} f = g \in \mathfrak{N} \subset \mathfrak{M}^{\perp} \implies \mathcal{P}_{\mathfrak{M}}(\mathcal{P}_{\mathfrak{N}} f) = \mathcal{P}_{\mathfrak{M}} g = o.$$

As a result, $\mathcal{P}_{\mathfrak{M}} + \mathcal{P}_{\mathfrak{N}}$ is idempotent. Defining $\mathcal{L}^M \triangleq \underbrace{\mathcal{L} \circ \mathcal{L} \circ \cdots \circ \mathcal{L}}_{M \text{ times}}$ for an operator $\mathcal{L}$, we can write

$$(\mathcal{P}_{\mathfrak{M}} + \mathcal{P}_{\mathfrak{N}})^2 = \mathcal{P}_{\mathfrak{M}}^2 + \mathcal{P}_{\mathfrak{N}} \circ \mathcal{P}_{\mathfrak{M}} + \mathcal{P}_{\mathfrak{M}} \circ \mathcal{P}_{\mathfrak{N}} + \mathcal{P}_{\mathfrak{N}}^2$$
$$= \mathcal{P}_{\mathfrak{M}}^2 + \mathcal{O} + \mathcal{O} + \mathcal{P}_{\mathfrak{N}}^2 = \mathcal{P}_{\mathfrak{M}} + \mathcal{P}_{\mathfrak{N}}.$$

And since it is also self-adjoint, it is a projection operator.

This result is significant and allows us to break composite projection operators into smaller projection operators easily. In particular, if $\{\mathcal{P}_{\mathfrak{M}_k}\}_{k=1}^N$ is a sequence of projections into mutually orthogonal subspaces $\{\mathfrak{M}_k\}_{k=1}^N$, the projection into subspace $\sum_{k=1}^N \mathfrak{M}_k$ is nothing but $\sum_{k=1}^N \mathcal{P}_{\mathfrak{M}_k}$.

Problem

4.21. Prove that the following statements are equivalent (Helmberg, 1969, p. 107):

(a) $\mathfrak{M} \subset \mathfrak{N}$;

(b) $\mathcal{P}_{\mathfrak{N}} \circ \mathcal{P}_{\mathfrak{M}} = \mathcal{P}_{\mathfrak{M}}$;

(c) $\mathcal{P}_{\mathfrak{M}} \circ \mathcal{P}_{\mathfrak{N}} = \mathcal{P}_{\mathfrak{M}}$;

(d) $\mathcal{P}_{\mathfrak{N}} - \mathcal{P}_{\mathfrak{M}}$ is a projection operator;

(e) $\langle (\mathcal{P}_{\mathfrak{N}} - \mathcal{P}_{\mathfrak{M}}) f, f \rangle \geq 0$;

(f) $\|\mathcal{P}_{\mathfrak{M}} f\| \leq \|\mathcal{P}_{\mathfrak{N}} f\|, \quad \forall f \in \mathcal{H}$.

---

[2]   That is, if $\mathcal{P}_{\mathfrak{M}} \circ \mathcal{P}_{\mathfrak{N}}$ is idempotent and self-adjoint.

4.11   Eigenvalues, eigenvectors and more

Definition 4.16 (Eigenvalue and eigenvector of a linear operator). *A complex number λ is an eigenvalue of the linear operator $\mathcal{L}$ on $\mathcal{H}$ if there exists a non-zero vector $f \in \mathcal{H}$ such that*

$$\mathcal{L}f = \lambda f. \tag{4.45}$$

f is called the eigenvector corresponding to λ. Since the operator is linear, any multiple of f is also an eigenvector:

$$\mathcal{L}(\alpha f) = \alpha \mathcal{L}f = \alpha \lambda f = \lambda(\alpha f). \tag{4.46}$$

The set of all $f \in \mathcal{H}$ satisfying (4.45) is called the eigenspace corresponding to λ.   □

Rearranging (4.45), we can write

$$\mathcal{L}f - \lambda f = (\mathcal{L} - \lambda \mathcal{I})f = o. \tag{4.47}$$

Since $(\mathcal{L} - \lambda \mathcal{I})f = o$ is also true for the zero vector $f = o$, we conclude that if λ is an eigenvalue of operator $\mathcal{L}$ then the operator $(\mathcal{L} - \lambda \mathcal{I})$ is not one-to-one. That is, $(\mathcal{L} - \lambda \mathcal{I})f_1 = (\mathcal{L} - \lambda \mathcal{I})f_2$ for $f_1 \neq f_2$. The reverse is also true. When $(\mathcal{L} - \lambda \mathcal{I})$ is not one-to-one, λ turns out to be an eigenvalue.

When eigenvectors are fed to an operator $\mathcal{L}$, the operator acts like a scaled identity operator $\lambda \mathcal{I}$. It recognizes eigenvectors as special inputs and does not alter them, apart from a scaling factor. We cannot ask an operator to be any simpler really. As a result, eigenspace of an operator (if it exists) has the remarkable property that it stays *invariant* under the operation. Remember that for projection $\mathcal{P}_{\mathfrak{M}}$, any element f already belonging to $\mathfrak{M}$ also stayed invariant under operation. That is,

$$\mathcal{P}_{\mathfrak{M}}f = f, \quad f \in \mathfrak{M}.$$

So we see that in an invariant eigenspace, the operator reduces to a scalar multiple of projection. The invariant eigenspace does not spill out to any other subspace as a result of operation. It is self-contained.

Definition 4.17 (Regular value of a linear operator). *If $(\mathcal{L} - \alpha \mathcal{I})$ is one-to-one and the inverse of $(\mathcal{L} - \alpha \mathcal{I})$ is bounded, then α is called a regular value of $\mathcal{L}$.* □

Regular values of an operator are the least interesting scalars as far as the operator is concerned. There is nothing special about them.

Definition 4.18 (Spectrum of a linear operator). *If $(\mathcal{L} - \lambda \mathcal{I})$ is not one-to-one (λ is an eigenvalue) or if $(\mathcal{L} - \lambda \mathcal{I})$ is one-to-one, but the inverse of $(\mathcal{L} - \lambda \mathcal{I})$ is unbounded, then λ is said to belong to the spectrum of $\mathcal{L}$.*   □

What can we say about the magnitude of an eigenvalue of a bounded operator? From (4.45), if f is an eigenvector

$$\|\mathcal{L}f\| = |\lambda| \|f\| \leq \|\mathcal{L}\|_{\text{op}} \|f\| \implies |\lambda| \leq \|\mathcal{L}\|_{\text{op}}. \tag{4.48}$$

That is, the magnitude of all eigenvalues of an operator (if they exist) is upper bounded by the operator norm. It can be proven, in fact, that any positive number α with $|\alpha|$ greater than the operator norm is a regular value of the operator.

> **Example 4.6.** Let $\mathcal{B}$ be a bounded self-adjoint operator on ℓ^2 defined in Problem 4.15, p. 135, which takes any element $a = \{\alpha_n\}_{n=1}^{\infty} \in \ell^2$ to $\mathcal{B}a = \{\alpha_n/n\}_{n=1}^{\infty} \in \ell^2$. Let $\{e_n\}_{n=1}^{\infty}$ be the standard basis in ℓ^2 introduced in Example 4.3. If any vector of the form $a = \sum_{n=1}^{\infty} \alpha_n e_n$ satisfies
>
> $$\mathcal{B}a = \lambda a,$$
>
> then
>
> $$\mathcal{B}a = \sum_{n=1}^{\infty} \alpha_n \mathcal{B}e_n$$
> $$= \sum_{n=1}^{\infty} \alpha_n \frac{1}{n} e_n = \sum_{n=1}^{\infty} \alpha_n \lambda e_n,$$
>
> and we conclude that $(1/n - \lambda)\alpha_n = 0$. That is, the only possible eigenvalues are of the form $1/n$ and the only possible eigenvectors are of the form $a = \alpha_n e_n$. From $\mathcal{B}e_n = (1/n)e_n$ we confirm that the subspace spanned by e_n is invariant under this operation and does not spill to any other subspace. We also verify that $\lambda = 1/n \leq \|\mathcal{B}\|_{\mathrm{op}} = 1$.
>
> The fact that the eigenvalues of this self-adjoint operator are real and different eigenvectors are orthogonal is no coincidence. This is proven in the following two simple, but important theorems.   □

Theorem 4.5. *The eigenvalues of a self-adjoint operator $\mathcal{B}$ are real.*   □

Proof. Let λ and f be a corresponding eigenvalue and (non-zero) eigenvector for $\mathcal{B}$. Then

$$\lambda\langle f, f\rangle = \langle \lambda f, f\rangle = \langle \mathcal{B}f, f\rangle = \langle f, \mathcal{B}f\rangle = \langle f, \lambda f\rangle = \overline{\lambda}\langle f, f\rangle,$$

from which we conclude that $\lambda = \overline{\lambda}$.   □

Theorem 4.6. *Two eigenvectors corresponding to two different eigenvalues of a self-adjoint operator $\mathcal{B}$ are orthogonal.*   □

Proof. Let $\lambda_1 \neq \lambda_2$ be two different real eigenvalues of $\mathcal{B}$ and f_1 and f_2 be the corresponding eigenvectors.

$$\lambda_1\langle f_1, f_2\rangle = \langle \lambda_1 f_1, f_2\rangle = \langle \mathcal{B}f_1, f_2\rangle = \langle f_1, \mathcal{B}f_2\rangle = \langle f_1, \lambda_2 f\rangle = \lambda_2\langle f_1, f_2\rangle,$$

from which we conclude that $\langle f_1, f_2\rangle = 0$ since $(\lambda_1 - \lambda_2) \neq 0$.   □

For self-adjoint operators, eigenspaces corresponding to two different eigenvalues are orthogonal to each other, which is a stronger property than the just independence of eigenvectors. Let us see through an example how this result allows us to break a self-adjoint operator into a series of simpler operators.

Example 4.7. Let $\mathcal{P}_n$ denote the finite rank projection of ℓ^2 into subspace $\mathfrak{M}_n = \bigvee\{e_n\}$ spanned by n-th basis e_n of dimension one. Specializing (4.44), we can write

$$\mathcal{P}_n a = \langle a, e_n \rangle e_n = \alpha_n e_n.$$

Due to orthogonality of $\mathfrak{M}_n$ and $\mathfrak{M}_m$ for $m \neq n$ and according to the discussion at the end of Section 4.10.2, p. 139, we can write a as the projection into the entire ℓ^2 space spanned by $\bigvee\{e_n\}_{n=1}^{\infty}$

$$a = \mathcal{P}_{\bigvee\{e_n\}_{n=1}^{\infty}} a = \sum_{n=1}^{\infty} \mathcal{P}_n a = \sum_{n=1}^{\infty} \alpha_n e_n.$$

Then applying the operator $\mathcal{B}$ of (4.6) to both sides

$$\mathcal{B}a = \mathcal{B}\left(\sum_{n=1}^{\infty} \mathcal{P}_n a\right) = \mathcal{B}\left(\sum_{n=1}^{\infty} \alpha_n e_n\right)$$

$$= \sum_{n=1}^{\infty} \alpha_n \mathcal{B}e_n = \sum_{n=1}^{\infty} \frac{1}{n} \alpha_n e_n$$

$$= \sum_{n=1}^{\infty} \frac{1}{n} \mathcal{P}_n a = \left(\sum_{n=1}^{\infty} \frac{1}{n} \mathcal{P}_n\right) a,$$

where remarkably we have been able to break a self-adjoint operator $\mathcal{B}$ as a weighted sum of projection operators into eigenspaces where the weights are the corresponding eigenvalues. In short, we can write

$$\mathcal{B} = \sum_{n=1}^{\infty} \frac{1}{n} \mathcal{P}_n.$$

$\square$

Remark 4.10 (Diagonalizing eigenfunctions). We have already established that a bounded operator $\mathcal{B}$ on $\mathcal{H}$ can be thought of as an infinite matrix through the identity

$$b_{n,m}^{(\varphi)} \triangleq \langle \mathcal{B}\varphi_m, \varphi_n \rangle.$$

Furthermore, this matrix representation depends on the orthonormal sequence $\{\varphi_n\}_{n=1}^{\infty}$ which acts as a basis and changing the basis changes the matrix.

Let us assume that a bounded self-adjoint operator $\mathcal{B}$ on $\mathcal{H}$ has a countable set of distinct eigenvalues $\{\lambda_n\}_{n=1}^{\infty}$ and the corresponding set of unit norm eigenfunctions is denoted by $\{\phi_n\}_{n=1}^{\infty}$. Using Theorem 4.6, p. 141, we see that

$$b_{n,m}^{(\phi)} \triangleq \langle \mathcal{B}\phi_m, \phi_n \rangle$$

$$= \lambda_m \langle \phi_m, \phi_n \rangle$$

$$= \begin{cases} \lambda_m, & m = n, \\ 0, & \text{otherwise.} \end{cases}$$

That is, $\mathbf{B}^{(\phi)}$ is a diagonal matrix. If $\{\phi_n\}_{n=1}^{\infty}$ happens to be a basis for $\mathcal{H}$, then we can invoke (4.23), p. 124, to establish

$$g = \sum_{n=1}^{\infty} \sum_{m=1}^{\infty} \langle f, \phi_m \rangle b_{n,m}^{(\phi)} \phi_n$$

$$= \sum_{n=1}^{\infty} \lambda_n \langle f, \phi_n \rangle \phi_n.$$

And comparing with (4.44) shows that in fact $\mathcal{B}$ is broken into a weighted superposition of finite rank one-dimensional projection operators $\mathcal{P}_n$ each of the form

$$\mathcal{P}_n f = \langle f, \phi_n \rangle \phi_n.$$

This is possible only if the orthonormal sequence corresponding to eigenvectors forms a basis for $\mathcal{H}$ and we pick them for representation of the operator in the "commutative operator diagram" in Figure 4.1. No other basis results in such clean representation of the operator.   □

Problem _____

4.22. Show that if we truncate the operator $\mathcal{B}$ of Example 4.7

$$\mathcal{B} = \sum_{n=1}^{\infty} \frac{1}{n} \mathcal{P}_n$$

into the first N terms as

$$\mathcal{B}_N = \sum_{n=1}^{N} \frac{1}{n} \mathcal{P}_n,$$

the norm of the *remaining* operator

$$\mathcal{B} - \mathcal{B}_N = \sum_{n=N+1}^{\infty} \frac{1}{n} \mathcal{P}_n$$

is at most $1/(N + 1)$. So not only have we been able to break the original operator into projection operators into simple mutually orthogonal one-dimensional subspaces, but also we can work with only a finite number of most significant projection operators which converge to the actual operator (in the sense of operator norm) as $n \to \infty$. This is possible because the eigenvalues of the operator under study take a special form, as we will see in the following chapter.

Take-home messages – bounded operators, continued

M1: The adjoint of a bounded linear operator is a unique operator such that $\langle \mathcal{B}f, g \rangle = \langle f, \mathcal{B}^\star g \rangle$ for all $f, g \in \mathcal{H}$. If the matrix representation of $\mathcal{B}$ is $\mathbf{B}^{(\varphi)}$, the matrix of its adjoint is the Hermitian transpose of $\mathbf{B}^{(\varphi)}$, denoted by $\left(\mathbf{B}^{(\varphi)}\right)^H$.

M2: An operator is self-adjoint if $\mathcal{B}^\star = \mathcal{B}$. This happens only when $\mathbf{B}^{(\varphi)}$ is Hermitian symmetric.

M3: The projection operator $\mathcal{P}_{\mathfrak{M}}$ corresponding to a subspace $\mathfrak{M}$ of $\mathcal{H}$ maps every input f into its projection into $\mathfrak{M}$.

M4: The projection operator is idempotent. Once an element $f \in \mathcal{H}$ is projected into $\mathfrak{M}$, further applications of the projection operator do not have any effect.

M5: The projection operator is self-adjoint.

M6: An alternative definition of the projection operator into $\mathfrak{M}$ is facilitated through the expansion of f into a basis for $\mathfrak{M}$. That is, if $\{\varphi_n\}_{n \in M}$ is a basis in $\mathfrak{M}$, then $\mathcal{P}_{\mathfrak{M}} f = \sum_{n \in M} \langle f, \varphi_n \rangle \varphi_n$.

M7: The projection into the sum of mutually orthogonal subspaces $\{\mathfrak{M}_k\}_{k=1}^N$ can be accomplished by projecting into individual subspaces through $\mathcal{P}_1, \mathcal{P}_2, \ldots, \mathcal{P}_N$ and summing the output. This is only valid for mutually orthogonal subspaces.

M8: λ and f are called an eigenvalue and the corresponding eigenvector of a linear operator $\mathcal{L}$ if $\mathcal{L}f = \lambda f$.

M9: An operator does not necessarily have any eigenvalues and corresponding eigenvectors, but if it does then all eigenvalues are bounded from above by the operator norm $|\lambda| \leq \|\mathcal{L}\|_{\mathrm{op}}$.

M10: Eigenvalues of a self-adjoint operator are real and different eigenvectors corresponding to different eigenvalues are orthogonal to each other. As a result, we may break a self-adjoint operator into a series of simpler operators.

5   Compact operators

5.1   Definition

One of the most important types of operators is the class of compact operators. They have nice properties without being too trivial. Although not obvious until later, they form a natural class of operators within the class of bounded operators. We start with an obscure definition.

Definition 5.1 (Compact operator). *An operator $\mathcal{C}$ is compact or completely continuous if for every sequence $\{f_n\}_{n=1}^{\infty}$ of elements in $\mathcal{H}$ such that $\|f_n\| \leq B$, the sequence $\{\mathcal{C}f_n\}_{n=1}^{\infty}$ contains at least one subsequence which converges in the mean to an element of $\mathcal{H}$.*   □

A subsequence of $\{f_n\}_{n=1}^{\infty}$ is a sequence in its own right, which can be written as $\{f_{n_1}, f_{n_2}, f_{n_3}, \ldots\}$, where $n_1, n_2, n_3, \ldots$ is a strictly increasing sequence of natural numbers: $1 \leq n_1 < n_2 < n_3 < \cdots$. We can write such a subsequence as $\{f_{n_k}\}_{k=1}^{\infty}$. A set of points, S, is *compact* if each sequence of points in S contains a subsequence that converges to a point in S.

5.2   Compact operator: some explanation

And what exactly does it mean to talk about convergence subsequences and who cares? To make sense of it we note that a *bounded* operator transforms a bounded element, as in the definition, to an output element which is bounded. This is a notion of continuity (not of convergence). So we can infer from the definition that a bounded sequence of elements in $\mathcal{H}$ need not have any convergent subsequences. This is a result which is at odds with what we observe in finite dimensions. For example, in one dimension, if we consider the (bounded) finite interval $[0, 1]$ on the real line we can imagine that there is an infinite sequence of real numbers from this interval, say $\{0.20938, 0.99344, 0.47828, \ldots\}$. Then the claim is that it is always possible to select a subsequence that converges to at least one point on the interval $[0, 1]$. Before we show how this is possible, by a constructive process, we reflect on what this means. If there were a malicious agent who was trying to pick a sequence that avoided having a convergent subsequence then that agent has no strategy that could work. There is no way to move the input sequence around in $[0, 1]$ to avoid at least one output subsequence converging.

Finite dimensions case

In the end, with any infinite sequence, the interval $[0,1]$ will contain an ordered infinite number of points, and if a convergent subsequence exists then, by definition, it will have an infinite number of points. Partition $[0,1]$ into two subintervals, say $[0,0.5]$ and $[0.5,1]$. If either subinterval has only a finite number of points discard it and retain the other. Otherwise if both contain an infinite number of points then pick either and discard the other. Repeat the process on the new subinterval of half the previous length (here 0.5 long) to select a sub-subinterval that is 0.25 long, etc. Each retained subinterval retains an infinite number of points and halves in length at every step. Clearly there is a converging subsequence being corralled through this process. There is at least one such subsequence and there may be more (indicated by the case at any step where we have two subintervals of an infinite number of sequence points).

Generalizing to two dimensions, we can imagine picking a subsequence on a (bounded) square domain $[0,1] \times [0,1]$. Generalizing the construction for the 1-D interval $[0,1]$, one could subdivide the $[0,1] \times [0,1]$ square into 4 subsquares and iterate accordingly. For N dimensions, that is, a finite number of dimensions, the story is no different.

Countably infinite dimensions case

However, things do come unstuck if we have a countably infinite Cartesian product of $[0,1]$ intervals (a sequence whose elements belong to $[0,1]$). For example, the sequence of bounded sequences (bounded and equal in norm to 0.2, or any other value less than or equal to 1):

$$\{\{0.2, 0, 0, 0, \ldots\}, \{0, 0.2, 0, 0, \ldots\}, \{0, 0, 0.2, 0, \ldots\}, \ldots\}$$

clearly does not have a convergent subsequence. So whilst we can handle the real interval with a countably dense set, having a countable number of them without constraint on those intervals means it is possible to never have any subsequence converge. In a separable Hilbert space we can represent an element of the space by denumerable sequence of scalars (the Fourier coefficients) and so there is an analogous issue.

Let us reflect on this a bit more before moving on. What is it that the malicious agent is able to do in an infinite-dimensional space to avoid being trapped that he or she could not do in finite-dimensional spaces? If the agent has an infinite energy to spend, then an infinite-dimensional space is like heaven as there are infinite options (escape routes) to spend the energy on. The agent may choose sequences to move through each coordinate axis successively, as we saw above; whereas in finite dimensions, we know where to look for the agent. There are only a finite number of escape routes to check and being trapped is inevitable.

In infinite dimensions, we can trap a malicious agent with a *finite* total energy. Inevitably, a finite-energy agent cannot invest equal non-negligible effort in all dimensions and will have to spend smaller and smaller effort as he or she moves along the coordinate axis. As a result, the agent gets closer and closer to the origin in those coordinates, making it easy for us to capture him or her. This is exactly what a compact operator does. A compact operator can capture the agent by imposing a behavior like we see in finite dimensions without being itself finite-dimensional, by which we really mean finite rank. The key is that there is only finite energy and when dispersed over infinite dimensions then in the bulk of those dimensions it needs to be close to the origin.

5.3   Compact or not compact regions

From the reasoning above, in infinite dimensions the unit cube

$$\mathcal{C}_u = \left\{\{\alpha_n\}_{n=1}^{\infty} \colon |\alpha_n| \leq 1\right\}$$

is not compact, whereas a finite-dimensional cube is compact.[1] So a good question is: how cube-like can we make an infinite region for it to be compact? In a sense, a finite cube is an infinite cube truncated to a finite number of dimensions. So an idea is to reduce an infinite number of dimensions to something small, but not zero.

Not all points in an infinite cube are square summable. Square summable sequences $\{\alpha_n\}_{n=1}^{\infty}$ define the Hilbert space ℓ^2

$$a = \{\alpha_n\}_{n=1}^{\infty} \in \ell^2 \iff \sum_{n=1}^{\infty} |\alpha_n|^2 < \infty.$$

The Hilbert "cube," call this $\mathcal{Q}$, is a subset of ℓ^2 composed of sequences $a \in \ell^2$, for which we have

$$|\alpha_n| \leq 1/n, \quad \forall n \in \mathbb{N}. \tag{5.1}$$

It can be proven that the Hilbert cube is compact, and so any sequence $a \in \mathcal{Q} \subset \ell^2$ has a convergent subsequence. For $a \in \mathcal{Q}$, $\|a\| \leq \pi/\sqrt{6}$.[2] Imagine that in choosing a sequence of elements in ℓ^2, our malicious agent is constrained to dampen his or her effort in moving along dimension n with a factor of $1/n$. Then effectively, his or her escape routes are limited. We can capture the agent "as closely as we wish" by only searching over a finite number of dimensions.

More generally, if $a \in \ell^2$, satisfying

$$|\alpha_n| \leq \delta_n, \quad \forall n \in \mathbb{N}, \ \delta_n \geq 0,$$

is compact if and only if $\sum_{n=1}^{\infty} \delta_n^2 < \infty$, which implies $d \triangleq \{\delta_n\}_{n=1}^{\infty} \in \ell^2$.

5.4   Examples of compact and not compact operators

Bounded operators are not in general compact. Consider the identity operator and let $\{\varphi_n\}_{n=1}^{\infty}$ be an orthonormal sequence. Consider the sequence of functions $f_n = \varphi_n$ as input and, therefore, as output (and so $\|f_n\| = 1$, $\forall n$). Then clearly it has no convergent subsequence; in fact $\|f_n - f_m\|^2 = 2$, $\forall m \neq n$ and so the sequence is not Cauchy and so no subsequence can converge. We conclude that the identity operator is *not* a compact operator, although it is probably the most innocent-looking bounded operator there is.

---

[1]   The same statement can be made about an infinite unit ball and a finite unit ball.
[2] $1 + 1/2^2 + 1/3^2 + 1/4^2 + \cdots = \pi^2/6$.

We have already considered finite rank projection operators in Section 4.10.1, p. 137, which in light of the discussion so far must be compact. Finite rank projection operators are a special class of finite rank operators that we formally define below.

If an operator transforms all inputs to a finite-dimensional output subspace, say spanned by a finite orthonormal set $\{\varphi_1, \varphi_2, \ldots, \varphi_N\}$ whose elements belong to $\mathcal{H}$, then it is referred to as a finite rank operator.

Definition 5.2 (Finite rank operator). *A finite rank operator takes the form*

$$\mathcal{L}_N f = \sum_{n=1}^{N} \langle f, \zeta_n \rangle \varphi_n, \tag{5.2}$$

where $N < \infty$ is the finite rank, and $\{\varphi_1, \varphi_2, \ldots, \varphi_N\}$ and $\{\zeta_1, \zeta_2, \ldots, \zeta_N\}$ are finite orthonormal sets whose elements belong to $\mathcal{H}$. □

Applying this definition to the archetype Hilbert space $L^2(\Omega)$ with orthonormal sets $\{\varphi_n\}_{n=1}^{\infty}$ and $\{\zeta_n\}_{n=1}^{\infty}$, we obtain

$$\begin{aligned}
(\mathcal{L}_N f)(x) &= \sum_{n=1}^{N} \langle f, \zeta_n \rangle \varphi_n(x) \\
&= \sum_{n=1}^{N} \left(\int_{\Omega} f(y) \overline{\zeta_n(y)} \, dy \right) \varphi_n(x) \\
&= \int_{\Omega} f(y) \left(\sum_{n=1}^{N} \overline{\zeta_n(y)} \varphi_n(x) \right) dy,
\end{aligned}$$

and comparing with (4.30), p. 127, the operator kernel is readily written as

$$K_N(x, y) \triangleq \sum_{n=1}^{N} \varphi_n(x) \overline{\zeta_n(y)},$$

where $N < \infty$ is the finite rank, and $\{\varphi_1, \varphi_2, \ldots, \varphi_N\}$ and $\{\zeta_1, \zeta_2, \ldots, \zeta_N\}$ are finite orthonormal sets whose elements belong to $L^2(\Omega)$. The finite orthonormal set $\{\varphi_1, \varphi_2, \ldots, \varphi_N\}$ defines the spanning set of the finite-dimensional subspace whereas $\{\zeta_1, \zeta_2, \ldots, \zeta_N\}$ characterizes the relevant linear operator action. So the above development indicates that a finite rank operator is a special instance of an integral operator.

We can regard finite rank operators as "trivial" examples of compact operators. The main question is if there are non-trivial (infinite-dimensional) compact operators. The answer is yes and below we give a sufficient condition, without proof, under which a bounded operator on a separable Hilbert space is compact.

Theorem 5.1. *Let a bounded operator $\mathcal{B}$ on a separable Hilbert space $\mathcal{H}$ with basis $\{\varphi_n\}_{n=1}^{\infty}$ be defined through infinite matrix $\mathbf{B}^{(\varphi)}$ with elements $b_{n,m}^{(\varphi)}$ for all $m, n \in \{1, 2, 3, \ldots\}$. If $\mathcal{B}$ has the additional property that*

$$B^2 = \sum_{n=1}^{\infty} \sum_{m=1}^{\infty} |b_{n,m}^{(\varphi)}|^2 < \infty,$$

then $\mathcal{B}$ is compact.   □

Proof. See (Helmberg, 1969, pp. 181–182).   □

> **Remark 5.1.** This infinite matrix sufficient condition means that the matrix elements, when treated as a sequence $\{b_{n,m}\}_{n,m=1}^{\infty}$, must belong to ℓ^2. This infinite matrix condition is not a necessary condition   □

> **Remark 5.2 (Hilbert-Schmidt norm).** The finite-dimensional version of this infinite sum of squares of the matrix entries is the square of the *Frobenius norm*. The infinite-dimensional version of the Frobenius norm, when used in the context of operators here, is called the *Hilbert-Schmidt norm*.   □

We will study *compact integral operators* more comprehensively in the next chapter. Historically, around 1900, it was the study of integral equations involving integral operators that led to the genesis of compact operators, then referred to as "completely continuous transformations," an expression still used (Riesz and Sz.-Nagy, 1990, p. 204).

> **Example 5.1.** Referring to the self-adjoint operator $\mathcal{B}$ in Example 4.6, p. 141, we can verify that its matrix representation is diagonal with diagonal elements given by
> $$b_{n,n}^{(\varphi)} = 1/n.$$
> So the condition
> $$B^2 = \sum_{n=1}^{\infty} \sum_{m=1}^{\infty} \left| b_{n,m}^{(\varphi)} \right|^2 = \pi^2/6 < \infty$$
> is satisfied and we conclude that $\mathcal{B}$ is a compact operator.   □

Problem

5.1. Referring to the derivations of Hilbert-Schmidt integral operator in Section 4.7, p. 126, and specifically to (4.29) and (4.32) show that a Hilbert-Schmidt integral operator with kernel $K(x, y)$ on $L^2(\Omega)$ is compact.

5.5   Limit of finite rank operators

As a transformation in a Hilbert space, a compact operator is closely related to finite rank operators. The following theorem shows that for any compact operator it is possible to construct a (non-unique) sequence of finite rank operators (of increasing rank) that converges in norm to the compact operator:

Theorem 5.2. *Let* $\{\zeta_n\}_{n=1}^{\infty}$ *be a complete orthonormal sequence in a separable Hilbert space* $\mathcal{H}$. *If* $\mathcal{C}$ *is a compact operator on* $\mathcal{H}$, *then the sequence of operators* $\{\mathcal{C}_n\}_{n=1}^{\infty}$ *defined by*

$$\mathcal{C}_n f \triangleq \sum_{m=1}^{n} \langle f, \zeta_m \rangle \sum_{p=1}^{n} \langle \mathcal{C}\zeta_m, \zeta_p \rangle \zeta_p \qquad (5.3)$$

converges in norm to $\mathcal{C}$ *as* $n \to \infty$, *that is,* $\|\mathcal{C}_n - \mathcal{C}\|_{\mathrm{op}} \to 0$.   □

Proof. See (Riesz and Sz.-Nagy, 1990, p. 204).   □

Remark 5.3. Note that (5.3) has the appropriate structure with respect to operator $\mathcal{C}$. That is, letting $n \to \infty$ we have

$$\sum_{m=1}^{\infty} \langle f, \zeta_m \rangle \sum_{p=1}^{\infty} \langle \mathcal{C}\zeta_m, \zeta_p \rangle \zeta_p = \sum_{p=1}^{\infty} \zeta_p \sum_{m=1}^{\infty} \langle f, \zeta_m \rangle \langle \zeta_m, \mathcal{C}^{\star}\zeta_p \rangle$$

$$= \sum_{p=1}^{\infty} \langle f, \mathcal{C}^{\star}\zeta_p \rangle \zeta_p$$

$$= \sum_{p=1}^{\infty} \langle \mathcal{C}f, \zeta_p \rangle \zeta_p$$

$$= \mathcal{C}f,$$

where we have used the notions of boundedness (from compactness) and hence continuity, the generalized Parseval relation and the existence of adjoint.   □

Remark 5.4. Note that (5.3) is in the standard finite rank form, (5.2),

$$\mathcal{C}_n f = \sum_{m=1}^{n} \langle f, \zeta_m \rangle \varphi_m \quad \text{with} \quad \varphi_m \triangleq \sum_{p=1}^{n} \langle \mathcal{C}\zeta_m, \zeta_p \rangle \zeta_p,$$

where n is finite.   □

Definition 5.1, p. 145, characterizes a compact operator as mapping a bounded and generally non-converging sequence to an output sequence which is "partially convergent" but still generally non-converging. But the output is more convergent than the input in some sense. So if we restrict the class of input sequences so that they are a bit more convergent-like, then we expect the compact operator to strengthen the convergence further. Now if we have as input a convergent sequence (by which we mean strongly convergent) then *unfortunately* any bounded operator will map this to an output sequence which is also strongly convergent. It is unfortunate because this does not differentiate the compact operator from a run-of-the-mill bounded operator. What we need is something stronger than bounded (generally non-convergent) but weaker than strong convergence. This motivates an alternative way to characterize a compact operator.

5.6   Weak and strong convergent sequences

Compactness of an operator can be defined in many equivalent ways, which are useful in different contexts. One particularly useful definition is as follows.

Definition 5.3 (Compact operator). *A linear operator $\mathcal{C}$ is compact or completely continuous if it transforms every weakly convergent sequence of elements into a strongly convergent sequence, that is,*

$$\boxed{\mathcal{C} \; Compact: \quad f_n \rightharpoonup f \implies \mathcal{C}f_n \to \mathcal{C}f.} \tag{5.4}$$

□

The equivalence of this definition with the previous Definition 5.1 can be proven, but we do not go into such detail here.

From this definition it is clear that a compact operator, $\mathcal{C}$, also transforms every weakly convergent sequence of elements into a weakly convergent output sequence since the output is strongly convergent, that is, since

$$\mathcal{C}f_n \to \mathcal{C}f \implies \mathcal{C}f_n \rightharpoonup \mathcal{C}f.$$

We have the not so interesting result, which will be shown to be superfluous (since it is a property of bounded operators; see Theorem 5.3 below):

$$\boxed{\mathcal{C} \; Compact: \quad f_n \rightharpoonup f \implies \mathcal{C}f_n \rightharpoonup \mathcal{C}f.} \tag{5.5}$$

It is revealing to contrast this with what a *bounded* operator does to *weakly convergent sequences* and, for good measure, *strongly convergent sequences*. In this latter case, we have an alternative definition for a bounded operator.

Definition 5.4 (Bounded operator). *A linear operator $\mathcal{B}$ is bounded if it transforms every strongly convergent sequence of elements into a strongly convergent sequence, that is,*

$$\boxed{\mathcal{B} \; Bounded: \quad f_n \to f \implies \mathcal{B}f_n \to \mathcal{B}f.} \tag{5.6}$$

(This is a statement about the continuity of the bounded operator.)   □

Of course compact operators are bounded so it is pertinent to see what additional things a compact operator provides beyond being a bounded operator.

Theorem 5.3. *A bounded operator, $\mathcal{B}$, transforms every weakly convergent sequence of elements into a weakly convergent sequence, that is,*

$$\boxed{\mathcal{B} \; Bounded: \quad f_n \rightharpoonup f \implies \mathcal{B}f_n \rightharpoonup \mathcal{B}f.} \tag{5.7}$$

□

Proof. If $f_n \rightharpoonup f$, this implies

$$\langle f_n, g \rangle \to \langle f, g \rangle, \quad \forall g \in \mathcal{H},$$

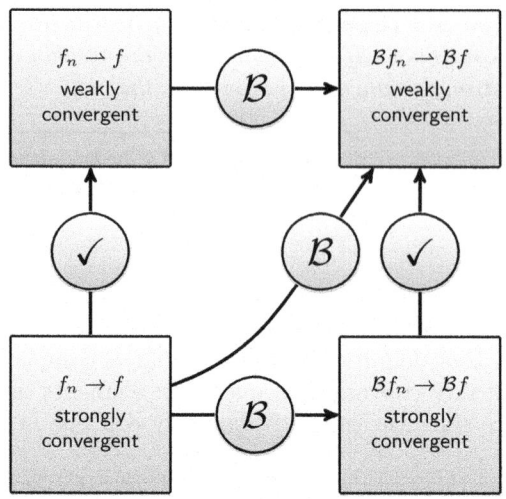

Figure 5.1 Action of a bounded operator, $\mathcal{B}$. Arrows labeled with operator $\mathcal{B}$ each show an implication, sometimes elementary. For example, $f_n \rightharpoonup f$ implies $\mathcal{B}f_n \rightharpoonup \mathcal{B}f$ for bounded $\mathcal{B}$. Arrows with a check mark ($\checkmark$) are trivial implications.

from which it follows, introducing the adjoint $\mathcal{B}^\star$ of the operator $\mathcal{B}$,

$$\langle f_n, \mathcal{B}^\star g \rangle \to \langle f, \mathcal{B}^\star g \rangle, \quad \forall g \in \mathcal{H},$$

since $\mathcal{B}^\star g \in \mathcal{H}$. Then, the adjoint relation applied to both sides implies

$$\langle \mathcal{B}f_n, g \rangle \to \langle \mathcal{B}f, g \rangle, \quad \forall g \in \mathcal{H}.$$

That is, $\mathcal{B}f_n \rightharpoonup \mathcal{B}f$. □

So a compact operator provides a stronger property, (5.4), than that provided in Theorem 5.3, (5.6), for bounded operators. This feature is captured in the additional edge in the graph in Figure 5.2 relative to the graph in Figure 5.1.

5.7   Operator compositions involving compact operators

Recall, from Theorem 2.10, p. 76, that an orthonormal sequence $\{\varphi_n\}_{n=1}^{\infty}$ is weakly convergent to the zero vector o, that is,

$$\varphi_n \rightharpoonup o.$$

Thus we can see that a compact operator $\mathcal{C}$ converts orthonormal sequences to sequences that converge in the mean to zero, which is quite interesting and mildly profound:

$$\varphi_n \rightharpoonup o \implies \mathcal{C}\varphi_n \to o.$$

This equation supports the idea that compact operators squeeze the output. This result is one element of defining the properties of the compact operators, which is taken up shortly in Section 5.8 and Section 5.9.

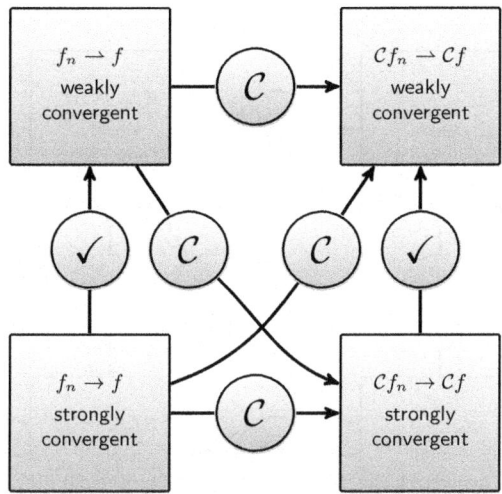

Figure 5.2 Action of a compact operator, $\mathcal{C}$. Arrows labeled with operator $\mathcal{C}$ each show an implication, sometimes elementary. For example, $f_n \rightharpoonup f$ implies $\mathcal{C}f_n \to \mathcal{C}f$ for compact operator $\mathcal{C}$ (this is non-trivial and an important property of compact operators). Arrows with a check mark ($\checkmark$) are trivial implications.

Given that compact operators have a "squeezing" property, then it is reasonable to propose that composing a compact operator, $\mathcal{C}$, with a bounded operator, $\mathcal{B}$, gives an overall compact operator. Operators, like shoes and socks, do not commute, so we need to consider two cases: $\mathcal{C} \circ \mathcal{B}$ and $\mathcal{B} \circ \mathcal{C}$.

Theorem 5.4. *Let $\mathcal{C}$ be a compact operator and $\mathcal{B}$ be a bounded operator on a Hilbert space $\mathcal{H}$. Then composition $\mathcal{C} \circ \mathcal{B}$ is compact, that is,*

$$f_n \rightharpoonup f \implies (\mathcal{C} \circ \mathcal{B})f_n \to (\mathcal{C} \circ \mathcal{B})f$$

and $\mathcal{B} \circ \mathcal{C}$ is also compact, that is,

$$f_n \rightharpoonup f \implies (\mathcal{B} \circ \mathcal{C})f_n \to (\mathcal{B} \circ \mathcal{C})f.$$

$\square$

Proof (sketch). See Figure 5.3 for a symbolic representations of the operator compositions, from which it is clear that such compositions are compact. $\square$

5.8   Spectral theory of compact operators

One important consequence of an operator being compact is that it can be very easily characterized by its spectrum, because the spectrum of a compact operator takes a very special form. In particular, it can be proven that the spectrum of a compact operator consists only of the special point 0 and eigenvalues. That is, any non-zero value such as λ is either a regular (uninteresting) value or is

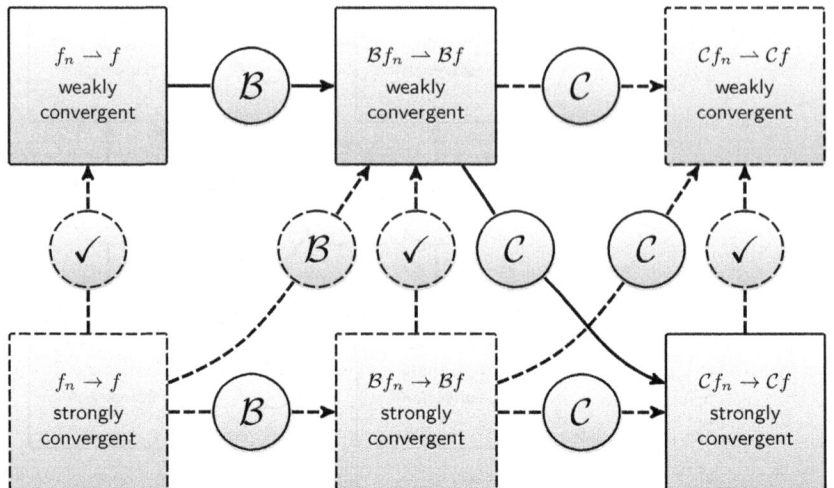

(a) Composition $\mathcal{C} \circ \mathcal{B}$. Under the first bounded operator, $\mathcal{B}$, a weakly convergent sequence is mapped to a weakly convergent sequence. Then under the second compact operator, $\mathcal{C}$, the second weakly convergent sequence is mapped to a strongly convergent sequence.

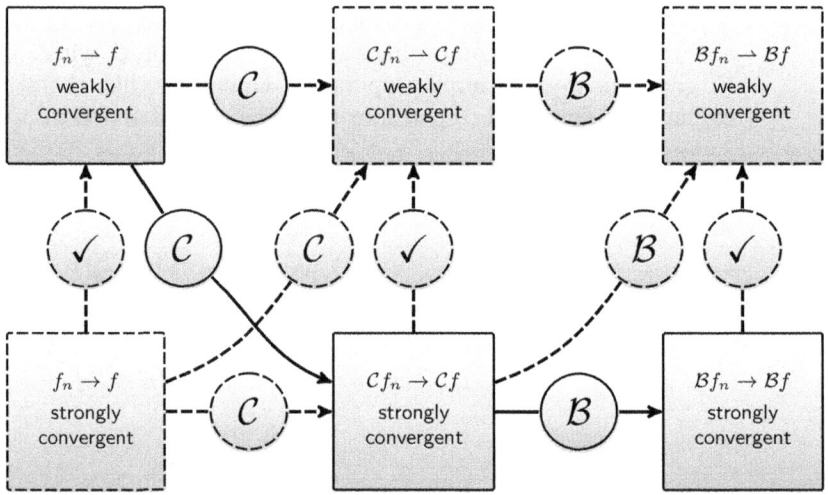

(b) Composition $\mathcal{B} \circ \mathcal{C}$. Under the first compact operator, $\mathcal{C}$, a weakly convergent sequence is mapped to a strongly convergent sequence. Then under the second bounded operator, $\mathcal{B}$, this strongly convergent sequence is mapped to a strongly convergent sequence.

Figure 5.3 Compositions $\mathcal{C} \circ \mathcal{B}$ and $\mathcal{B} \circ \mathcal{C}$, where $\mathcal{B}$ is bounded and $\mathcal{C}$ is compact (and bounded), showing the composition is compact, since weakly convergent input sequences get mapped in two operations to strongly convergent sequences via the emphasized paths (Theorem 5.4). Trivially, strongly convergent input sequences get mapped under either composition to strongly convergent sequences.

an eigenvalue for a compact operator. In addition, if there are infinitely many eigenvalues for a compact operator, they inevitably converge to zero (their only possible accumulation point is zero). The latter result about the convergence of eigenvalues is an important consequence of the following theorem.

Theorem 5.5. *Let C be a compact operator on a Hilbert space $\mathcal{H}$. Then only a finite number of linearly independent eigenvectors can exist that correspond to eigenvalues above any given positive threshold τ.* $\qquad\square$

Proof. See (Helmberg, 1969, p. 191). $\qquad\square$

Note that the above statements do not mean that a compact operator necessarily has eigenvalues. The example below shows that we can construct compact operators that do not have any eigenvalues, but here we are not specifically interested in such operators.

Example 5.2. Consider the squeezing version of the right shift operator, $\mathcal{S}_R^S$ on ℓ^2 for all $a = \{\alpha_n\}_{n=1}^{\infty} \in \ell^2$ which turns a sequence $\{\alpha_1, \alpha_2, \alpha_3, \ldots\}$ into $\{0, \alpha_1, \alpha_2/2, \alpha_3/3, \ldots\}$. Using Theorem 5.1, p. 148, we can verify that $\mathcal{S}_R^S$ is compact. But we cannot find any non-zero vector in ℓ^2 such that $\mathcal{S}_R^S a = \lambda a$. Hence, $\mathcal{S}_R^S$ does not have any eigenvalues. $\qquad\square$

The following example motivates what will be presented in the next section.

Example 5.3. We have already seen that the self-adjoint operator $\mathcal{B}$ in Example 4.6, p. 141, which takes any element $a = \{\alpha_n\}_{n=1}^{\infty} \in \ell^2$ to $\mathcal{B}a = \{\alpha_n/n\}_{n=1}^{\infty} \in \ell^2$, is in fact compact and there are infinitely many eigenvalues $\lambda_n = 1/n$, which clearly converge to zero; see also Example 5.1, Problem 4.15, p. 135, and Problem 4.22, p. 143. We also saw that this operator could be represented as a sum of projection operators weighted by eigenvalues:

$$\mathcal{B} = \sum_{n=1}^{\infty} \frac{1}{n} \mathcal{P}_n,$$

which can be approximated to any desired accuracy by only considering a finite sum of such weighted projection operators. This is no coincidence for compact self-adjoint operators and the following section aims to formalize this fundamental result. $\qquad\square$

5.9   Spectral theory of compact self-adjoint operators

Equation (4.48) indicates that the operator norm is bounded from above by the magnitude of its largest eigenvalue (if it exists). For general operators the bound may not be realized. But for compact and self-adjoint operators it does. This result, although not difficult to prove, is the breakthrough result that Hilbert generated. He did take the results further but the most tractable result is for compact self-adjoint operators. The theorem and proof can also be found in (Riesz and Sz.-Nagy, 1990, pp. 231–232).

Theorem 5.6. *For a compact, self-adjoint operator* $\mathcal{C}$,

$$\|\mathcal{C}\|_{\mathrm{op}} \triangleq \sup_{\|f\|=1} |\langle \mathcal{C}f, f \rangle| = |\lambda_{\max}|, \tag{5.8}$$

where $\lambda_{\max}$ *is the (real, possibly negative) eigenvalue with the largest absolute value.* $\square$

Proof. Equation (4.39) implies that there exists a sequence $\{f_n\}_{n=1}^{\infty}$ with $\|f_n\| = 1$, $\forall n$, such that[3]

$$|\langle \mathcal{C}f_n, f_n \rangle| \to \|\mathcal{C}\|_{\mathrm{op}}, \quad n \to \infty. \tag{5.9}$$

In (5.9) we can replace $\{f_n\}_{n=1}^{\infty}$ by a subsequence (if necessary) to strip the absolute value

$$\langle \mathcal{C}f_n, f_n \rangle \to \lambda, \quad n \to \infty, \tag{5.10}$$

where

$$\lambda = \|\mathcal{C}\|_{\mathrm{op}} \text{ or } \lambda = -\|\mathcal{C}\|_{\mathrm{op}}.$$

Note that since $\mathcal{C}$ is self-adjoint, $\langle \mathcal{C}f_n, f_n \rangle$ is a sequence of real numbers.

Next we show that

$$\mathcal{C}f_n - \lambda f_n \to 0. \tag{5.11}$$

For this purpose, we calculate

$$\begin{aligned}
\|\mathcal{C}f_n - \lambda f_n\|^2 &= \langle \mathcal{C}f_n - \lambda f_n, \mathcal{C}f_n - \lambda f_n \rangle \\
&= \|\mathcal{C}f_n\|^2 - 2\lambda \langle \mathcal{C}f_n, f_n \rangle + \lambda^2 \|f_n\|^2 \\
&\le \|\mathcal{C}\|_{\mathrm{op}}^2 \|f_n\|^2 - 2\lambda \langle \mathcal{C}f_n, f_n \rangle + \lambda^2 \|f_n\|^2 \\
&= 2\lambda^2 - 2\lambda \langle \mathcal{C}f_n, f_n \rangle.
\end{aligned} \tag{5.12}$$

However, by (5.10), this upper bound, (5.12), on $\|\mathcal{C}f_n - \lambda f_n\|^2$ tends to zero, which establishes (5.11).

To this point we do not know if $\mathcal{C}f_n$ converges (only that $\mathcal{C}f_n - \lambda f_n$ converges to zero) and this is where we invoke compactness of the operator $\mathcal{C}$. Compactness means that there is a subsequence of $\{f_n\}_{n=1}^{\infty}$, denote this by $\{f_{n_k}\}_{k=1}^{\infty}$, such that $\{\mathcal{C}f_{n_k}\}_{k=1}^{\infty}$ is a convergent sequence. Let g be this limit, that is, $\mathcal{C}f_{n_k} \to g$, or equivalently

$$\lambda \mathcal{C}f_{n_k} \to \lambda g, \tag{5.13}$$

which implies, by (5.11) which also applies to the subsequence, that $\lambda f_{n_k} \to g$, whence

$$\lambda \mathcal{C}f_{n_k} \to \mathcal{C}g, \tag{5.14}$$

by continuity, as $k \to \infty$. That is, (5.13) and (5.14) yield

$$\mathcal{C}g = \lambda g, \tag{5.15}$$

which, since $\|g\| = |\lambda| > 0$, shows that λ is an eigenvalue. If there were any other eigenvalue of greater absolute value then this would contradict the supremum in (5.9) and so $\lambda = \lambda_{\max}$, the (real, possibly negative) eigenvalue with the largest absolute value. $\square$

---

[3] Here, we have defined such a sequence since (4.39) is expressed in terms of the supremum.

This is a key result because it induces a raft of profound properties which we shall gather together and refer to as *spectral theory of compact self-adjoint operators*. It is fair to say that the following theorem is the pinnacle of the second part of the book on operators.

Theorem 5.7 (Spectral theory of compact self-adjoint operators). *Let $\mathcal{C}$ be a compact self-adjoint operator on Hilbert space $\mathcal{H}$ and let $\{\lambda_n\}_{n=1}^N$ be its different (non-zero) eigenvalues which can be finite ($N \in \mathbb{N}$) or infinite ($N = \infty$) and are indexed in descending order as*

$$|\lambda_1| > |\lambda_2| > \cdots > |\lambda_n| > \cdots .$$

Let $\mathcal{P}_n$ be the projection operator on the eigenspace $\mathfrak{M}_n$ associated with eigenvalue λ_n. The following results hold:

P1: *$\mathcal{P}_n$ is finite rank;*

P2: *$\mathcal{P}_n$ and $\mathcal{P}_m$ are mutually orthogonal for $n \neq m$. That is, for every $f \in \mathcal{H}$, $\langle \mathcal{P}_n f, \mathcal{P}_m f \rangle = 0$;*

P3: *The operator $\mathcal{C}$ is identical to the weighted sum of $\mathcal{P}_n$ as*

$$\mathcal{C} = \sum_{n=1}^N \lambda_n \mathcal{P}_n, \tag{5.16}$$

where if N is finite then $\mathcal{C}$ is of finite rank. If $N = \infty$ then

$$\mathcal{B} = \sum_{n=1}^M \lambda_n \mathcal{P}_n \tag{5.17}$$

converges to $\mathcal{C}$ in norm as $M \to \infty$;

P4: *Finally, for any truncation of the operator to the first $M < N$ terms, the norm of the remaining operator is equal to $|\lambda_{M+1}|$. That is,*

$$\left\| \mathcal{C} - \sum_{n=1}^M \lambda_n \mathcal{P}_n \right\|_{\text{op}} = |\lambda_{M+1}|. \tag{5.18}$$

□

Proof. See (Helmberg, 1969, p. 202). □

Example 5.4. If we truncate the self-adjoint operator $\mathcal{B}$ in Example 4.6, Example 5.1 and Problem 4.22 to the first N terms,

$$\mathcal{B}_N = \sum_{n=1}^N \frac{1}{n} \mathcal{P}_n,$$

then the norm of the remaining operator is $1/(N+1)$:

$$\|\mathcal{B} - \mathcal{B}_N\|_{\text{op}} = \frac{1}{N+1}.$$

Therefore, we have a clear quantitative criterion to approximate $\mathcal{B}$ to any desired accuracy. □

The spectral analysis of compact operators leads to yet another definition for the compact operator, which is given below without proving its equivalence to previous definitions.

Definition 5.5 (Compact operator). *A bounded operator $\mathcal{C}$ is compact if for every given ε there exists a finite rank bounded operator $\mathcal{B}_\varepsilon$ such that*

$$\|\mathcal{C} - \mathcal{B}_\varepsilon\|_{\mathrm{op}} < \varepsilon.$$

$\square$

Take-home messages – compact operators

M1: A compact operator mimics and extends the action of finite-dimensional operators to infinite dimensions. By putting some restrictions on how the operator output is distributed across infinite dimensions, a compact operator guarantees that any bounded input sequence, after going through the compact operator, contains a convergent subsequence. A compact operator beats a malicious agent who tries to use the infinite dimensions as escape routes for non-convergence. A compact operator makes the output "more or equally convergent" than the input in some sense.

M2: There are different equivalent definitions for a compact operator, some of which are listed below.

- An operator $\mathcal{C}$ is compact if for every bounded sequence $\{f_n\}_{n=1}^{\infty}$ of elements in $\mathcal{H}$ such that $\|f_n\| \leq B$, the sequence $\{\mathcal{C}f_n\}_{n=1}^{\infty}$ contains at least one subsequence which converges in the mean to an element of $\mathcal{H}$.

- A bounded linear operator $\mathcal{B}$ on $\mathcal{H}$ is compact if its operator matrix $\mathbf{B}^{(\varphi)}$ with elements $b_{n,m}^{(\varphi)}$ is square summable:

$$B^2 = \sum_{n=1}^{\infty} \sum_{m=1}^{\infty} |b_{n,m}^{(\varphi)}|^2 < \infty.$$

- A linear operator $\mathcal{C}$ is compact if it transforms every weakly convergent sequence of elements into a strongly convergent sequence. That is, $f_n \rightharpoonup f \implies \mathcal{C}f_n \to \mathcal{C}f$.

- A bounded operator $\mathcal{C}$ is compact if for every given ε there exists a finite rank bounded operator $\mathcal{B}_\varepsilon$ such that $\|\mathcal{C} - \mathcal{B}_\varepsilon\|_{\mathrm{op}} < \varepsilon$.

M3: Finite rank operators are trivial examples of compact operators.

M4: A necessary and sufficient condition for an operator to be compact is that it is possible to construct a sequence of finite rank operators that converges in norm to the compact operator.

M5: A compact self-adjoint operator with N different eigenvalues and associated orthogonal eigenspaces can be exactly decomposed into the weighted sum of N projection operators into the N eigenspaces. Furthermore, the operator can be arbitrarily well approximated (in operator norm sense) by a finite weighted sum of M projections into the most significant eigenspaces, where the norm of the remaining operator is $|\lambda_{M+1}|$.

6 Integral operators and their kernels

6.1 Kernel Fourier expansion and operator matrix representation

The results here extend those derived in Section 4.7, p. 126, which looked briefly at Hilbert-Schmidt kernels and their associated integral operators.

Recall that a Hilbert-Schmidt integral operator on $L^2(\Omega)$ is defined as

$$(\mathcal{L}_K f)(x) = \int_\Omega K(x, y) f(y) \, dy, \tag{6.1}$$

where the kernel $K(x, y)$ satisfies

$$\iint_{\Omega \times \Omega} |K(x, y)|^2 dx \, dy < \infty. \tag{6.2}$$

That is, $K(x, y)$ belongs to $L^2(\Omega \times \Omega)$.

Now let

$$\{\varphi_n(x)\}_{n=1}^\infty \tag{6.3}$$

be any complete orthonormal sequence in $L^2(\Omega)$, then it can be shown that

$$\left\{\xi_{n,m}^{(\varphi)}(x, y)\right\}_{n,m=1}^\infty \triangleq \left\{\varphi_n(x)\overline{\varphi_m(y)}\right\}_{n,m=1}^\infty \tag{6.4}$$

is a complete orthonormal sequence in the space of square integrable functions $L^2(\Omega \times \Omega)$; see (Helmberg, 1969, p. 63). This close connection between the expansions in $L^2(\Omega)$ and $L^2(\Omega \times \Omega)$ will be exploited in revealing the relationship between the kernel $K(x, y)$ and its associated integral operator $\mathcal{L}_K$.

> **Remark 6.1.** In the above the complete orthonormal sequence, (6.3), is arbitrary and has no special relationship to the operator $\mathcal{L}_K$ and is a free choice. Later we will find the best choice for the basis which is tuned to the $\mathcal{L}_K$. Before then surprising mileage and generality can be obtained with an arbitrary choice. □

We can expand $K(x, y) \in L^2(\Omega \times \Omega)$ in the basis (6.4), which results in the Fourier coefficients[1]

$$\boxed{k_{n,m}^{(\varphi)} = \left\langle K, \xi_{n,m}^{(\varphi)} \right\rangle_{\Omega \times \Omega},} \tag{6.5}$$

---

[1] Earlier, in Section 4.5, we used precisely this notation to represent the matrix coefficients of a bounded linear operator. We are preempting the result that the operator matrix and kernel Fourier matrix are identical.

which are square summable. The notation $\langle \cdot, \cdot \rangle_{\Omega \times \Omega}$ makes explicit the space is $L^2(\Omega \times \Omega)$. Not only are they square summable, but by Parseval relation we have

$$\sum_{n=1}^{\infty} \sum_{m=1}^{\infty} \left| k_{n,m}^{(\varphi)} \right|^2 = \iint_{\Omega \times \Omega} |K(x,y)|^2 dx \, dy < \infty. \tag{6.6}$$

This is an alternative expression for the Hilbert-Schmidt condition, (6.1).

Bringing this together we have the representation for $K(x,y)$, convergent in the mean,

$$\begin{aligned}
K(x,y) &= \sum_{m=1}^{\infty} \sum_{n=1}^{\infty} \langle K, \xi_{n,m}^{(\varphi)} \rangle_{\Omega \times \Omega} \xi_{n,m}^{(\varphi)}(x,y) \\
&= \sum_{m=1}^{\infty} \sum_{n=1}^{\infty} k_{n,m}^{(\varphi)} \xi_{n,m}^{(\varphi)}(x,y) \\
&= \sum_{m=1}^{\infty} \sum_{n=1}^{\infty} k_{n,m}^{(\varphi)} \varphi_n(x) \overline{\varphi_m(y)}.
\end{aligned} \tag{6.7}$$

The expansion in (6.7) and Fourier coefficients in (6.5) form the Fourier series in $L^2(\Omega \times \Omega)$. Also, even though we are motivating the development by appealing to the kernel, it is valid for any finite energy function in two variables on domain $\Omega \times \Omega$.

The Fourier coefficients can also be written

$$\begin{aligned}
k_{n,m}^{(\varphi)} &= \langle K, \xi_{n,m}^{(\varphi)} \rangle_{\Omega \times \Omega} \\
&= \iint_{\Omega \times \Omega} K(x,y) \overline{\varphi_n(x) \overline{\varphi_m(y)}} \, dx \, dy \\
&= \iint_{\Omega \times \Omega} K(x,y) \varphi_m(y) \overline{\varphi_n(x)} \, dx \, dy,
\end{aligned}$$

which can be interpreted in another way which is quite profound

$$\begin{aligned}
k_{n,m}^{(\varphi)} &= \int_{\Omega} \left(\int_{\Omega} K(x,y) \varphi_m(y) \, dy \right) \overline{\varphi_n(x)} \, dx \\
&= \int_{\Omega} (\mathcal{L}_K \varphi_m)(x) \overline{\varphi_n(x)} \, dx \\
&= \langle \mathcal{L}_K \varphi_m, \varphi_n \rangle_{\Omega} \equiv \langle \mathcal{L}_K \varphi_m, \varphi_n \rangle,
\end{aligned} \tag{6.8}$$

where we have re-introduced the integral operator $\mathcal{L}_K$ from (6.1) and augmented the notation on the inner product to emphasize its domain of definition.

We thus have the notable "matrix" identity

$$\langle K, \xi_{n,m}^{(\varphi)} \rangle_{\Omega \times \Omega} = \langle \mathcal{L}_K \varphi_m, \varphi_n \rangle_{\Omega} \equiv k_{n,m}^{(\varphi)}, \tag{6.9}$$

which holds for all complete orthonormal sequences

$$\{\varphi_n(x)\}_{n=1}^{\infty}$$

in $L^2(\Omega)$ and induces the complete orthonormal sequence (6.4),

$$\left\{\xi_{n,m}^{(\varphi)}(x,y)\right\}_{n,m=1}^{\infty} \triangleq \left\{\varphi_n(x)\overline{\varphi_m(y)}\right\}_{n,m=1}^{\infty}$$

in $L^2(\Omega \times \Omega)$.

We emphasize that the two inner products in (6.9) are defined on two different spaces and there is no direct way to manipulate or use this expression.

> **Remark 6.2.** In words, the Fourier coefficient matrix (or more properly vector) with respect to the complete orthonormal sequence $\{\varphi_n(x)\overline{\varphi_m(y)}\}$ of the kernel $K(x,y)$ defined on $\Omega \times \Omega$ equals the matrix of the integral operator $\mathcal{L}_K$ with respect to the arbitrary complete orthonormal sequence $\{\varphi_n(x)\}$ defined on Ω. That is, in short, the kernel "matrix" (Fourier coefficients) equals the associated integral operator matrix. $\qquad\square$

Problem _____

6.1. By substituting

$$K(x,y) = \sum_{m=1}^{\infty}\sum_{n=1}^{\infty} k_{n,m}^{(\varphi)} \varphi_n(x)\overline{\varphi_m(y)},$$

or otherwise, show that

$$\langle \mathcal{L}_K\varphi_m, \varphi_n\rangle_\Omega = \iint_{\Omega\times\Omega} K(x,y)\varphi_m(y)\,dy\,\overline{\varphi_n(x)}\,dx$$

is equal to $k_{n,m}^{(\varphi)}$.

_____

6.2   Compactness

Since (6.6) holds we conclude from Theorem 5.1, p. 148, that the Hilbert-Schmidt integral operator is indeed compact. Compactness can also be understood in terms of the limit of a sequence of kernels of increasing finite rank, which we explore next.

6.2.1   Approximating kernels with kernels of finite rank

Recall a finite rank operator on $L^2(\Omega)$, for some finite $N < \infty$, can be expressed as an integral operator

$$(\mathcal{L}_{K_N}f)(x) = \int_\Omega K_N(x,y)f(y)\,dy$$

with kernel

$$K_N(x,y) \triangleq \sum_{n=1}^{N} \varphi_n(x)\overline{\zeta_n(y)},$$

where $\{\varphi_1, \varphi_2, \ldots, \varphi_N\}$ and $\{\zeta_1, \zeta_2, \ldots, \zeta_N\}$ are finite orthonormal sets whose elements belong to $L^2(\Omega)$. So it is implicit that as a function on the region $\Omega \times \Omega$, the kernel K_N belongs to $L^2(\Omega \times \Omega)$.

Theorem 6.1. *Every Hilbert-Schmidt kernel, $K(x, y) \in L^2(\Omega \times \Omega)$, can be arbitrarily closely approximated in the mean by a kernel of finite rank, $K_N(x, y) \in L^2(\Omega \times \Omega)$.* □

Proof. This is a result of the Hilbert-Schmidt integral operator being compact. For details see (Riesz and Sz.-Nagy, 1990, pp. 158–159). □

Convergence in the mean of kernels implies convergence in the norm of the corresponding transformations, since (Riesz and Sz.-Nagy, 1990, p. 156)

$$\|\mathcal{L}_K - \mathcal{L}_{K_N}\|_{\mathrm{op}} \leq \|K - K_N\|.$$

6.3 Self-adjoint integral operators

In this section, we deal with self-adjoint integral operators and see how they can be decomposed into a weighted sum of projections into their corresponding eigenspaces. The material in here was partly related to finite rank projection operators covered in Section 4.10.1, p. 137, and in Section 4.11, p. 140, for example Remark 4.10, and can also be deduced from the spectral theorem of compact self-adjoint operators, Theorem 5.7, p. 157. Here we make the derivations explicit and self-contained.

6.3.1 Establishing self-adjointness

In the development above, no assumption was made on the compact operator $\mathcal{L}_K$, with respect to being Hermitian symmetric or not. If the operator matrix coefficients satisfy

$$k_{n,m}^{(\varphi)} = \overline{k_{m,n}^{(\varphi)}},$$

then the Hilbert-Schmidt kernel is Hermitian symmetric

$$
\begin{aligned}
K(x, y) &= \sum_{m=1}^{\infty} \sum_{n=1}^{\infty} k_{n,m}^{(\varphi)} \varphi_n(x) \overline{\varphi_m(y)} \\
&= \sum_{m=1}^{\infty} \sum_{n=1}^{\infty} \overline{k_{m,n}^{(\varphi)}} \varphi_n(x) \overline{\varphi_m(y)} = \overline{K(y, x)}.
\end{aligned}
\tag{6.10}
$$

And the operator $\mathcal{L}_K$ is self-adjoint

$$
\begin{aligned}
\langle \mathcal{L}_K f, g \rangle &= \int_\Omega \int_\Omega K(x, y) f(y) \, dy \, \overline{g(x)} \, dx \\
&= \int_\Omega \int_\Omega \overline{K(y, x)} f(y) \, dy \, \overline{g(x)} \, dx \\
&= \int_\Omega f(y) \overline{\int_\Omega K(y, x) g(x) \, dx} \, dy \\
&= \langle f, \mathcal{L}_K g \rangle.
\end{aligned}
$$

6.3.2   Spectral theory

From the spectral theory of compact self-adjoint operators, there exists an eigen-decomposition given by

$$
\begin{aligned}
\left(\mathcal{L}_K \phi_n\right)(x) &= \int_\Omega K(x,y)\phi_n(y)\,dy \\
&= \lambda_n \phi_n(x), \quad \lambda_n \in \mathbb{R},
\end{aligned}
\tag{6.11}
$$

where $\{\phi_n(x)\}_{n=1}^\infty$ is a complete orthonormal sequence in $L^2(\Omega)$ with respect to inner product (2.17), p. 47, and the $\{\lambda_n\}_{n=1}^\infty$ are real.

Problem _____

6.2. Using the spectral representation, show that
$$
\left\langle \mathcal{L}_K \phi_m, \phi_n \right\rangle = \lambda_n \delta_{m,n} \quad \text{and} \quad \lambda_n = \left\langle \mathcal{L}_K \phi_n, \phi_n \right\rangle.
$$

_____

6.3.3   Operator decomposition

The decomposition into eigenfunctions and eigenvalues completely characterizes the operator $\mathcal{L}_K$ so it is of interest to see how the associated kernel can be expressed in terms of the eigenfunctions and eigenvalues. This is given by the following theorem first published by Schmidt in 1907 (Riesz and Sz.-Nagy, 1990, p. 243).

Theorem 6.2 (Schmidt, eigenfunction kernel expansion). *Let the self-adjoint Hilbert-Schmidt integral operator*

$$
\left(\mathcal{L}_K f\right)(x) = \int_\Omega K(x,y) f(y)\,dy
$$

be given with the Hermitian symmetric Hilbert-Schmidt kernel $K(x,y) \in L^2(\Omega \times \Omega)$ satisfying
$$
K(x,y) = \overline{K(y,x)}.
$$

Then the symmetric function $K(x,y)$ can be developed, in the sense of convergence in the mean, into the expansion

$$
\boxed{K(x,y) = \sum_{n=1}^\infty \lambda_n \phi_n(x)\overline{\phi_n(y)},}
\tag{6.12}
$$

where $\{\phi_n(x)\}_{n=1}^\infty$ is the orthonormal sequence of eigenfunctions of the operator $\mathcal{L}_K$ and $\{\lambda_n\}$ is the sequence of corresponding real eigenvalues. That is,

$$
\left(\mathcal{L}_K \phi_n\right)(x) = \lambda_n \phi_n(x), \quad \lambda_n \in \mathbb{R}, \quad \forall n \in \mathbb{N}.
\tag{6.13}
$$

□

Proof. Substituting (6.13) into (6.1) with $f(x) = \phi_m(x)$ leads to

$$(\mathcal{L}_K\phi_m)(x) = \int_\Omega \Big(\sum_{n=1}^\infty \lambda_n\phi_n(x)\overline{\phi_n(y)}\Big)\phi_m(y)\,dy, \quad \forall m$$

$$= \lambda_m\phi_m(x), \quad \forall m,$$

showing (6.13) implies (6.11).

Conversely, fix ν and define

$$K_\nu(x) \triangleq K(x,\nu),$$

noting $K_\nu(x) \in L^2(\Omega)$ since the kernel (6.2) is Hilbert-Schmidt. Therefore, by completeness, we expand $f(x) = K(x,\nu)$ using a general orthonormal sequence $\{\varphi_n(x)\}_{n=1}^\infty$ in $L^2(\Omega)$ as

$$K(x,\nu) = \sum_{n=1}^\infty \langle K_\nu, \varphi_n\rangle\varphi_n(x) = \sum_{n=1}^\infty \int_\Omega K(y,\nu)\overline{\varphi_n(y)}\,dy\,\varphi_n(x)$$

$$= \sum_{n=1}^\infty \int_\Omega \overline{K(\nu,y)\varphi_n(y)}\,dy\,\varphi_n(x) \quad \text{by symmetry of the kernel}$$

$$= \sum_{n=1}^\infty \overline{\int_\Omega K(\nu,y)\varphi_n(y)\,dy}\,\varphi_n(x)$$

$$= \sum_{n=1}^\infty \overline{(\mathcal{L}_K\varphi_n)(\nu)}\,\varphi_n(x),$$

which can be written, replacing ν with y,

$$\boxed{K(x,y) = \sum_{n=1}^\infty \overline{(\mathcal{L}_K\varphi_n)(y)}\varphi_n(x) \equiv \sum_{n=1}^\infty (\mathcal{L}_K\varphi_n)(x)\overline{\varphi_n(y)} = \overline{K(y,x)}} \quad (6.15)$$

by symmetry of the kernel. This expression is valid for any complete orthonormal basis $\{\varphi_n(x)\}_{n=1}^\infty$ and simplifies with the choice of eigenfunctions $\{\phi_n\}$

$$K(x,y) = \sum_{n=1}^\infty (\mathcal{L}_K\phi_n)(x)\overline{\phi_n(y)}$$

$$= \sum_{n=1}^\infty \lambda_n\phi_n(x)\overline{\phi_n(y)}.$$

This establishes (6.12). □

Remark 6.3 (Dirac kernel). Note the kernel eigen-decomposition (6.12) resembles the expansion for the delta function using the completeness relation in any orthonormal basis $\{\varphi_n(x)\}_{n=1}^\infty$, (2.77), p. 93, transposed here

$$\delta(x-y) = \sum_{n=1}^\infty \varphi_n(x)\overline{\varphi_n(y)}. \quad (6.17)$$

However, it is not a Hilbert-Schmidt kernel. □

6.3.4   Diagonal representation

Now we compute the matrix coefficients, (6.8), for the *eigenfunction* sequence, which we denote as

$$\{\phi_n(x)\}_{n=1}^{\infty},$$

to show the operator matrix is then diagonal. Substituting for the eigenfunctions

$$
\begin{aligned}
k_{n,m}^{(\phi)} &= \int_{\Omega}\Big(\int_{\Omega} K(x,y)\phi_m(y)\,dy\Big)\overline{\phi_n(x)}\,dx \\
&= \int_{\Omega}(\mathcal{L}_K\phi_m)(x)\overline{\phi_n(x)}\,dx \\
&= \int_{\Omega}\lambda_m\phi_m(x)\overline{\phi_n(x)}\,dx \\
&= \lambda_m\int_{\Omega}\phi_m(x)\overline{\phi_n(x)}\,dx \equiv \lambda_m\langle\phi_m,\phi_n\rangle, \quad \lambda_m\in\mathbb{R}.
\end{aligned}
$$

And from the orthogonality of eigenfunctions of the self-adjoint operator $\mathcal{L}_K$, Theorem 4.6, p. 141,

$$\boxed{k_{n,m}^{(\phi)} = \lambda_n\delta_{n,m},}$$

which indeed shows the matrix becomes diagonal.

Now let us expand every $f\in L^2(\Omega)$ in terms of eigenfunctions as

$$f = \sum_{n=1}^{\infty}\langle f,\phi_n\rangle\phi_n,$$

where $\langle f,\phi_n\rangle$ are Fourier coefficients. This makes a compelling representation for any function input, f, in the integral equation (6.1), p. 160, since $\mathcal{L}_K$ acts on each eigenfunction in a straightforward manner

$$
\begin{aligned}
g = \mathcal{L}_K f &= \sum_{n=1}^{\infty}\sum_{m=1}^{\infty}\langle f,\phi_m\rangle k_{n,m}^{(\phi)}\phi_n \\
&= \sum_{n=1}^{\infty}\lambda_n\langle f,\phi_n\rangle\phi_n, \quad\quad\quad\quad (6.18)
\end{aligned}
$$

given $\mathcal{L}_K\phi_n = \lambda_n\phi_n$, for all n.

In terms of the Fourier coefficients, a function $f(x)$ as an input to the operator $\mathcal{L}_K$ with Fourier coefficients $\langle f,\phi_n\rangle$ and satisfying

$$f(x) = \sum_{n=1}^{\infty}\langle f,\phi_n\rangle\phi_n(x), \quad \forall f\in L^2(\Omega),$$

generates an output function, (6.18), whose Fourier coefficients are $\lambda_n\langle f,\phi_n\rangle$. This shows how effectively the eigen-structure characterizes the effects of the compact self-adjoint operator.

> **Remark 6.4.** Recommended reading in relation to the past few sections is (Riesz and Sz.-Nagy, 1990, pp. 242–244). □

Problems

6.3. By reconciling (6.15) with (6.7) for a symmetric kernel with $k_{n,m}^{(\varphi)} = \overline{k}_{m,n}^{(\varphi)}$, show that

$$\left(\mathcal{L}_K \varphi_m\right)(x) = \sum_{n=1}^{\infty} k_{n,m}^{(\varphi)} \varphi_n(x).$$

Interpret this result geometrically.

6.4. Show that for an integral operator with $k_{n,m}^{(\varphi)} = \overline{k}_{m,n}^{(\varphi)}$

$$\langle \mathcal{L}_K f, \varphi_n \rangle = \sum_{m=1}^{\infty} \langle f, \varphi_m \rangle k_{n,m}^{(\varphi)}.$$

Interpret the result geometrically.

6.3.5   Spectral analysis of self-adjoint integral operators

Theorem 6.3. *For a symmetric function $K(x, y) = \overline{K(y, x)} \in L^2(\Omega \times \Omega)$, the norm of the associated integral operator, $\mathcal{L}_K$, is bounded from above by the norm of the associated kernel:*

$$\|\mathcal{L}_K\|_{\mathrm{op}} \triangleq \sup_{\|f\|=1} \|\mathcal{L}_K f\| \le \|K\| = \Big(\sum_{n=1}^{\infty} \lambda_n^2\Big)^{1/2},$$

where $\{\lambda_n\}_{n=1}^{\infty}$ are the eigenvalues of $\mathcal{L}_K$. □

Proof. That

$$\|K\|^2 = \iint_{\Omega \times \Omega} |K(x, y)|^2 dx\, dy = \sum_{n=1}^{\infty} \lambda_n^2$$

follows from (6.12) and orthonormality of $\{\phi_n\}$. By the Cauchy-Schwarz inequality

$$\|\mathcal{L}_K f\|^2 = \int_\Omega \Big|\int_\Omega K(x, y) f(y)\, dy\Big|^2 dx \le \int_\Omega \Big(\int_\Omega |K(x, y)|^2 dy \int_\Omega |f(y)|^2 dy\Big) dx$$
$$= \|K\|^2 \|f\|^2,$$

from which we infer

$$\|\mathcal{L}_K f\| \le \sup_{\|f\|=1} \|\mathcal{L}_K f\| \le \|K\|, \quad \forall \|f\| = 1,$$

which shows the operator norm is upper bounded by the kernel norm $\|K\|$. □

We can sharpen Theorem 6.3 using Theorem 5.6. Since an integral operator, $\mathcal{L}_K$, with a symmetric Hilbert-Schmidt kernel is compact and self-adjoint, then the norm bound looks (and is) trivial.

Corollary 6.4. *For an integral operator,* $\mathcal{L}_K$, *with a symmetric Hilbert-Schmidt kernel, the eigenvalues are real and satisfy*

$$\lambda_{\max}^2 = \|\mathcal{L}_K\|_{\mathrm{op}}^2 \le \|K\|^2 = \sum_{n=1}^{\infty} \lambda_n^2 < \infty.$$

□

Finally, under the eigen-basis, $\{\phi_n(x)\}_{n=1}^{\infty}$, the kernel Fourier coefficient matrix and the integral operator matrix reduce to only the diagonal terms being non-zero. The diagonal elements are given by

$$\langle K, \xi_{n,n}^{(\phi)} \rangle_{\Omega \times \Omega} = \langle \mathcal{L}_K \phi_n, \phi_n \rangle_{\Omega} \equiv \lambda_n. \qquad (6.19)$$

As before, for the inner product on the left-hand side of (6.19), the orthonormal basis function notation means $\phi_{n,m}(x,y) \triangleq \phi_n(x)\overline{\phi_m(y)}$, here with $n = m$.

Using (6.13) we can easily evaluate the "trace" of the operator (sum of the eigenvalues) in terms of the kernel since

$$\mathrm{trace}(\mathcal{L}_K) = \sum_{n=1}^{\infty} \lambda_n = \int_{\Omega} K(x,x)\,dx. \qquad (6.20)$$

6.3.6   Synopsis

In Figure 6.1 and Figure 6.2 we have summarized some of the findings regarding the properties of compact self-adjoint integral operators with associated kernels. Figure 6.1 develops the properties in relation to the preferred eigenfunction representations which has the effect of diagonalizing the operator matrix representation. Figure 6.2 develops the properties in relation to an arbitrary complete orthonormal expansion, which may be viewed as a generalization of Figure 6.1.

Problems ──

6.5. If our integral operator is not Hilbert-Schmidt or symmetric what changes in the above theorems?

6.6. Show that the trace of the operator, given by (6.20), can be written as

$$\mathrm{trace}(\mathcal{L}_K) = \sum_{n=1}^{\infty} \lambda_n = \sum_{n=1}^{\infty} k_{n,n}^{(\varphi)}. \qquad (6.21)$$

6.7. In Remark 6.3, the $\delta(x-y)$ expansion, (6.17), was observed to be in the same form as the kernel eigenfunction expansion, (6.12).

   (a) Take $K(x,y) = \delta(x-y)$, and using (6.2), p. 160, determine if such a kernel is Hilbert-Schmidt.

   (b) Now formally substitute this kernel into (6.1), p. 160. What operator does this correspond to?

(c) Why is this operator not compact?

(d) The expansion of $\delta(x - y)$ is valid for any orthonormal basis and this would imply that every function is an eigenfunction and every eigenvalue is 1. Does all this make sense?

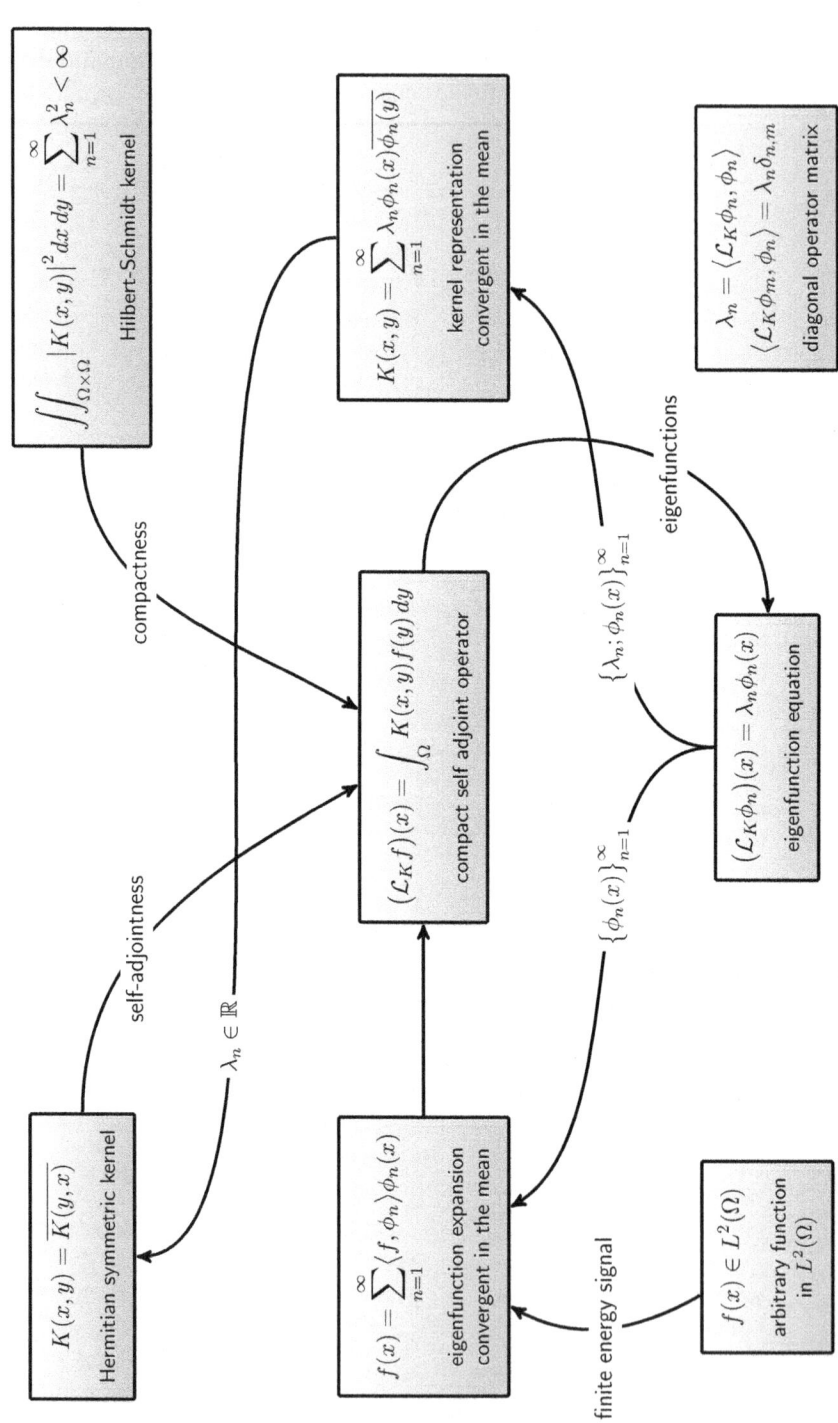

Figure 6.1 Kernel representations for compact self-adjoint integral operators with representation using the eigenvalues λ_n and eigenfunctions $\phi_n(x)$ of the operator which yields a "diagonal" kernel representation.

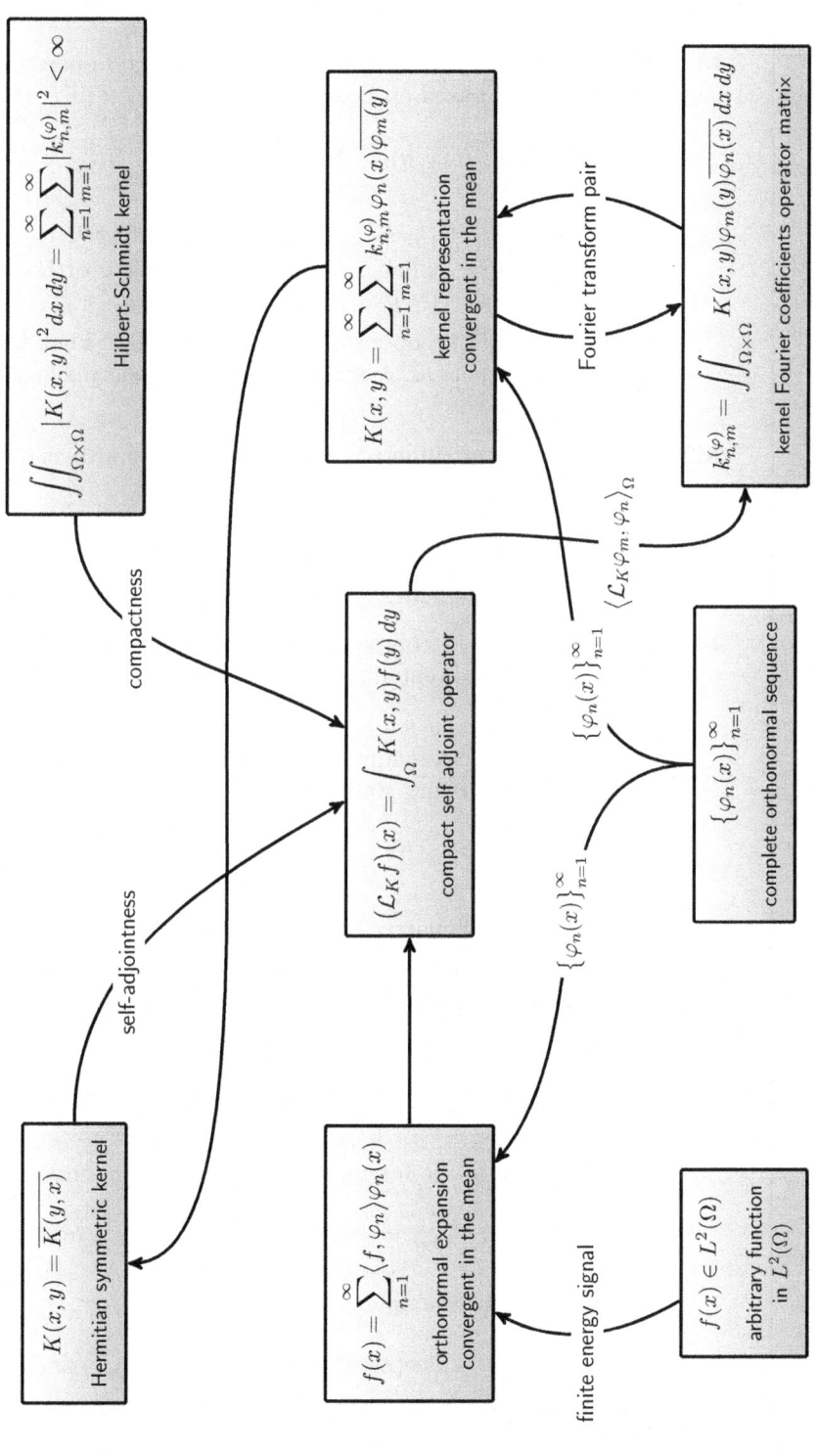

Figure 6.2 Kernel representations for compact self-adjoint integral operators using an arbitrary orthonormal expansion $\{\varphi_n(x)\}_{n=1}^{\infty}$ which yields, in general, a non-diagonal kernel Fourier representation and equivalent operator matrix with coefficients $k_{n,m}^{(\varphi)}$.

Take-home messages – integral operators

M1: The kernel $K(x, y)$ of a Hilbert-Schmidt integral operator $\mathcal{L}_K$ can be represented based on Fourier series expansion in $L^2(\Omega \times \Omega)$ as

$$K(x, y) = \sum_{m=1}^{\infty} \sum_{n=1}^{\infty} k_{n,m}^{(\varphi)} \xi_{n,m}^{(\varphi)}(x, y),$$

where $k_{n,m}^{(\varphi)} = \langle K, \xi_{n,m}^{(\varphi)} \rangle_{\Omega \times \Omega}$ are the Fourier series coefficients and $\xi_{n,m}^{(\varphi)}(x, y) = \varphi_n(x)\overline{\varphi_m(y)}$ forms a complete orthonormal sequence in $L^2(\Omega \times \Omega)$. Fourier series coefficients $k_{n,m}^{(\varphi)}$ can be written in terms of the operator matrix elements. That is, $k_{n,m}^{(\varphi)} = \langle \mathcal{L}_K \varphi_m, \varphi_n \rangle$ where φ_m is a complete orthonormal sequence in $L^2(\Omega)$.

M2: If the elements of Hilbert-Schmidt integral operator matrix satisfy $k_{n,m}^{(\varphi)} = \overline{k_{m,n}^{(\varphi)}}$ then

- the kernel is Hermitian symmetric $K(x, y) = \overline{K(y, x)}$;
- the integral operator is self-adjoint, $\langle \mathcal{L}_K f, g \rangle = \langle f, \mathcal{L}_K g \rangle$.

M3: If $\{\phi_n(x)\}_{n=1}^{\infty}$ is a complete orthonormal sequence of eigenfunctions for a self-adjoint integral operator such that $(\mathcal{L}_K \phi_n)(x) = \lambda_n \phi_n(x)$, then

- the Hermitian symmetric Hilbert-Schmidt kernel is decomposed into a very simple form as

$$K(x, y) = \sum_{n=1}^{\infty} \lambda_n \phi_n(x)\overline{\phi_n(y)};$$

- the operator matrix becomes diagonal, $k_{n,m}^{(\phi)} = \lambda_n \delta_{n,m}$;
- the Fourier coefficients of the operator output $g = \mathcal{L}_K f$ are related to those of the input through $\langle g, \phi_n \rangle = \lambda_n \langle f, \phi_n \rangle$;
- the square norm of the kernel is given by

$$\|K\|^2 = \iint_{\Omega \times \Omega} |K(x, y)|^2 dx \, dy = \sum_{n=1}^{\infty} \lambda_n^2;$$

- the norm of the integral operator, $\mathcal{L}_K$, is bounded as

$$\lambda_{\max} = \|\mathcal{L}_K\|_{\text{op}} \leq \left(\sum_{n=1}^{\infty} \lambda_n^2 \right)^{1/2}; \quad and$$

- the trace of the operator is given by

$$\text{trace}(\mathcal{L}_K) = \sum_{n=1}^{\infty} \lambda_n = \int_{\Omega} K(x, x) \, dx.$$

PART III
Applications

7   Signals and systems on 2-sphere

7.1   Introduction

The processing of signals whose domain is the 2-sphere or unit sphere[1] has been an ongoing area of research in the past few decades and is becoming increasingly more active. Such signals are widely used in geodesy and planetary studies (Simons et al., 1997; Wieczorek and Simons, 2005; Simons et al., 2006; Audet, 2011). In many cases of interest flat Euclidean modeling of planetary and heavenly data does not work. Planetary curvature should be taken into account especially for small heavenly bodies such as the Earth, Venus, Mars, and the Moon (Wieczorek, 2007). Other applications, for the processing of signals on the 2-sphere, include the study of cosmic microwave background in cosmology (Wiaux et al., 2005; Starck et al., 2006; Spergel et al., 2007), 3D beamforming/sensing (Simons et al., 2006; Górski et al., 2005; Armitage and Wandelt, 2004; Ng, 2005; Wandelt and Górski, 2001; Rafaely, 2004; Wiaux et al., 2006), computer graphics and computer vision (Brechbühler et al., 1995; Schröder and Sweldens, 2000; Han et al., 2007), electromagnetic inverse problems (Colton and Kress, 1998), brain cortical surface analysis in medical imaging (Yu et al., 2007; Yeo et al., 2008), and channel modeling for wireless communication systems (Pollock et al., 2003; Abhayapala et al., 2003). This type of processing exhibits important differences from the processing of signals on Euclidean domains — such as time-based signals whose domain is the real line $\mathbb{R}$, or 2D or 3D signals and images, whose domain is multi-dimensional, but still Euclidean. Extending well-known and useful signal processing techniques in the Euclidean domain such as convolution, filtering, smoothing, estimation, and prediction to the 2-sphere domain is a natural, but often non-trivial, way to analyze signals inherently defined on the 2-sphere.

This chapter is aimed to introduce important concepts related to signals and systems defined on the 2-sphere and lay the foundation for better understanding of some recent 2-sphere signal processing techniques that we will present in Chapter 8, p. 251, and Chapter 9, p. 293. We will spend some time elaborating different presentations and relations of spherical harmonics that are available in the literature. This can aid people wishing to start research in the area, as marrying various definitions used in different papers in different fields can be confusing. We will also discuss important subspaces and operators on the 2-sphere.

---

[1] In nomenclature, the unit sphere, $\mathbb{S}^2$, 2-sphere or simply "sphere" refer to the same thing.

7.2   Preliminaries

7.2.1   2-sphere and spherical coordinates

The 2-sphere can be regarded as being embedded[2] in the 3D Euclidean or Cartesian space, $\mathbb{R}^3$. Every point in 3D Cartesian space, $\mathbb{R}^3$, can be represented as a Cartesian 3-vector

$$\boldsymbol{u} \equiv (u_x, u_y, u_z)' \in \mathbb{R}^3,$$

where we have used $'$ to denote transpose. The components u_x, u_y, and u_z are the coordinates along the x, y, and z axes, respectively. The Euclidean norm of $\boldsymbol{u}$ is

$$|\boldsymbol{u}| \triangleq \sqrt{u_x^2 + u_y^2 + u_z^2}.$$

Note that in this chapter and the following two chapters, we use bold-face notation, such as $\boldsymbol{u}$, to refer to vectors in 3D space and have avoided using the conventional norm symbol $\|\cdot\|$ for Euclidean norm, as $\|\cdot\|$ is needed later for denoting the norm of functions on the 2-sphere.

The 2-sphere refers to the sphere of common experience, but of a nominal radius of unity. Specifically, the set containing the points on 2-sphere is defined as

$$\mathbb{S}^2 \triangleq \left\{ \boldsymbol{u} \in \mathbb{R}^3 : |\boldsymbol{u}| = 1 \right\}. \tag{7.1}$$

Note that we are only dealing with points on the "surface," that is, the sphere is the boundary of a unit ball. The unit radius is a way of saying that the radial part is unimportant and not part of the formulation. Of course in real life, if there were only one radius sphere then golf, soccer/football, cricket and baseball would all be much more interesting.[3] When we write $\mathbb{S}^2$ it is the mathematical notation for the 2-sphere and used in equations. Although the points on the 2-sphere belong to $\mathbb{R}^3$, this surface is fundamentally different from $\mathbb{R}^3$. It has a finite support and has a constant positive Gaussian curvature. In contrast, Euclidean spaces have zero curvature or are flat.

As mathematicians feel compelled to do, much to the annoyance of non-mathematicians, they generalize. A sphere can be well defined in any dimension. We want to be boring and stick with dimension 2, which relates to the two in $\mathbb{S}^2$ and the two in 2-sphere. The 2-sphere is a two-dimensional surface and can be described by two parameters. However, there is a preferred coordinate system to parameterize the 2-sphere, which we will consider next.

Vectors representing points on the 2-sphere, (7.1), are unit vectors. Unit vectors shall be denoted by $\widehat{\boldsymbol{u}}$, and can be obtained by normalizing any vector $\boldsymbol{u} \neq \boldsymbol{0}$

$$\widehat{\boldsymbol{u}} = \frac{\boldsymbol{u}}{|\boldsymbol{u}|}.$$

We can interpret $\widehat{\boldsymbol{u}}$ as being the direction and $|\boldsymbol{u}|$ as being the magnitude.

When we want a coordinate description then spherical coordinates are preferred. That is, to every point on the 2-sphere we can associate two angles:

---

[2]   This notion of embedding is an aid to interpretation and plays no role in the actual results that follow.

[3]   Some people assert cricket can never be more interesting.

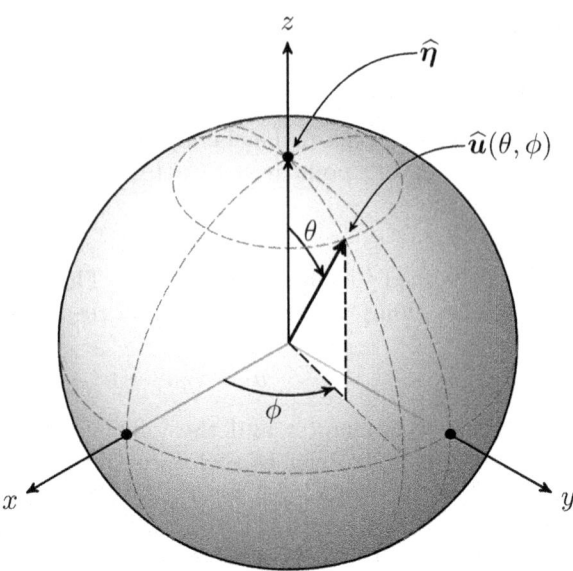

Figure 7.1 Spherical coordinates at point $\widehat{\boldsymbol{u}}(\theta, \phi) \triangleq (\sin\theta\cos\phi, \sin\theta\sin\phi, \cos\theta)'$. Here $\theta \in [0, \pi]$ is the co-latitude ($\theta = 0$ for the north pole $\widehat{\boldsymbol{\eta}}$), and $\phi \in [0, 2\pi)$ is the longitude.

$\theta \in [0, \pi]$ denoting the co-latitude measured with respect to the positive z-axis ($\theta = 0$ corresponds to the north pole, denoted by $\widehat{\boldsymbol{\eta}} = (0, 0, 1)'$), and $\phi \in [0, 2\pi)$, denoting the azimuth or longitude measured with respect to the positive x-axis in the x-y plane. This is depicted in Figure 7.1. Bringing this together we can write

$$\widehat{\boldsymbol{u}} \equiv \widehat{\boldsymbol{u}}(\theta, \phi) \triangleq (\sin\theta\cos\phi, \sin\theta\sin\phi, \cos\theta)',$$

which gives the Cartesian coordinates of points on the 2-sphere. Conversely, given a vector $\widehat{\boldsymbol{u}}$, we obtain the co-latitude and longitude angles as

$$\cos\phi = \frac{u_x}{\sqrt{u_x^2 + u_y^2}}, \quad \sin\phi = \frac{u_y}{\sqrt{u_x^2 + u_y^2}},$$

$$\cos\theta = u_z,$$

where for finding ϕ, four-quadrant inverse tangent function should be used since $\phi \in [0, 2\pi)$. So we can see that (θ, ϕ) fully parameterizes a point on the 2-sphere and, hence, can be used to interchangeably refer to a point $\widehat{\boldsymbol{u}}$ on the 2-sphere.

The inner/dot product between two points:

$$\widehat{\boldsymbol{u}} = (\sin\theta\cos\phi, \sin\theta\sin\phi, \cos\theta)',$$

$$\widehat{\boldsymbol{v}} = (\sin\vartheta\cos\varphi, \sin\vartheta\sin\varphi, \cos\vartheta)'$$

is denoted by $\widehat{\boldsymbol{u}} \cdot \widehat{\boldsymbol{v}}$ and is related to the *angular distance* Δ between the two points, which is shown in Figure 7.2. In particular,

$$\cos\Delta = \widehat{\boldsymbol{u}} \cdot \widehat{\boldsymbol{v}} = \sin\theta\sin\vartheta\cos(\phi - \varphi) + \cos\theta\cos\vartheta. \tag{7.2}$$

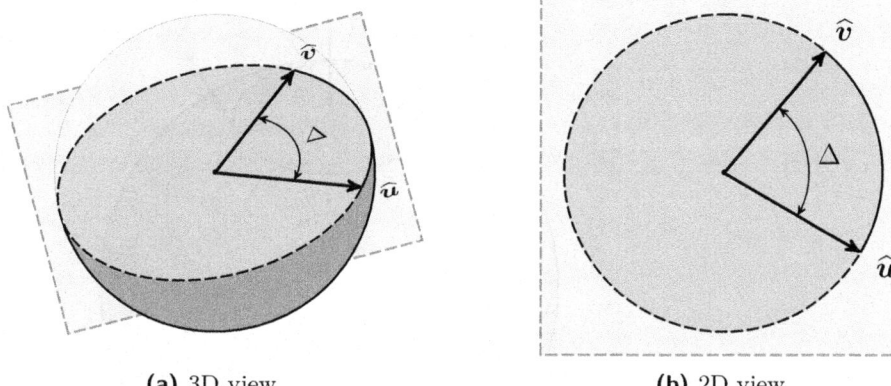

(a) 3D view.   **(b)** 2D view.

Figure 7.2 Angular distance Δ — arc length along the great circle joining two points $\widehat{u}$ and $\widehat{v}$ on the 2-sphere: (a) shows the 3D view of the the points and the great circle section and plane containing them; and (b) shows the 2D view perpendicular to the plane. Since this is a unit radius sphere the arc length, Δ, is also equal to the angle in radians. In this illustration $\Delta = 4\pi/9$ and $\cos\Delta = \widehat{u}\cdot\widehat{v} \approx 0.1736$.

Problems

7.1. Show that the 2-sphere, $\mathbb{S}^2$, is not a subspace of $\mathbb{R}^3$.

7.2. Derive the expression (7.2) for the angular distance Δ between the two unit vectors, $\widehat{u}$ and $\widehat{v}$, on the 2-sphere.

7.2.2   Regions on 2-sphere

A sufficiently regular region on the sphere $\mathbb{S}^2$ is denoted by R with positive area

$$A = \int_R ds(\widehat{u})$$

where

$$ds(\widehat{u}) = \sin\theta\, d\theta\, d\phi$$

is the differential area element. This region may be a connected region or be a union of unconnected subregions $R = R_1 \cup R_2 \cup \cdots$. The region or subregions may be irregular in shape and need not be convex. The complement region of R on $\mathbb{S}^2$ is denoted by $\bar{R} \triangleq \mathbb{S}^2 \setminus R$.

A very special and useful region on the 2-sphere is the polar cap region R_{θ_0} around the north pole, $\widehat{\eta}$, which is azimuthally symmetric and is parameterized by a maximum co-latitude angle θ_0. This is shown in Figure 7.3 and mathematically defined as

$$R_{\theta_0} \triangleq \big\{(\theta,\phi)\colon 0 \le \theta \le \theta_0,\ 0 \le \phi < 2\pi\big\}. \tag{7.3}$$

The area under the polar cap region R_{θ_0} is $2\pi(1 - \cos\theta_0)$.

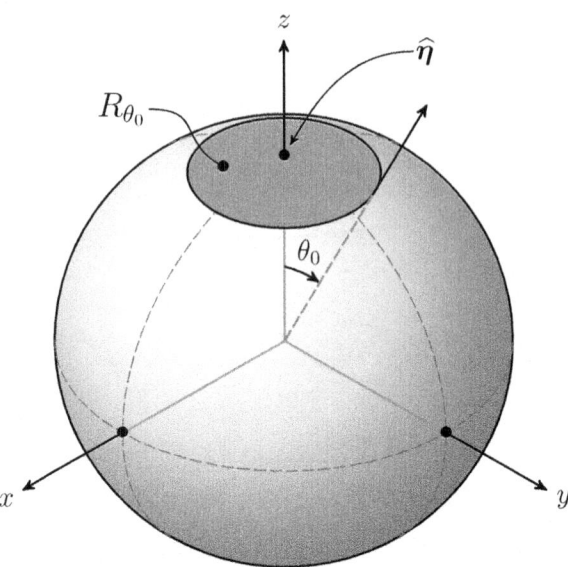

Figure 7.3 Polar cap region centered on the north pole, $\widehat{\boldsymbol{\eta}}$, of the 2-sphere. In this illustration $\theta_0 = 5\pi/36$ with area $A \approx 0.5887$.

7.2.3 Understanding rotation

Since rotation on 2-sphere will appear in a variety of applications, it is important that we elaborate on different aspects and presentations of rotation.

Understanding and visualizing rotation in three dimensions can be tricky. One main reason is that there are many different conventions and rules that are used for describing and applying rotations. There are *proper Euler angles* and *Tait-Bryan angles*. Rotations can be either *intrinsic compositions* (around moving axes) or *extrinsic compositions* (around fixed axes). Each convention can be useful for a particular application or a physical system. We do not claim to provide a comprehensive treatment of this subject, but aim to present some basic relations that aid our understanding and are necessary for the topics covered in this chapter and the following two chapters. Some interesting and useful interpretations and examples of rotation are provided in (Koks, 2006).

We first start by some discussion on orientation. Orientation of a (rigid) body in 3D can be specified in a relative manner. This is best described in terms of standard basis vectors

$$\widehat{\boldsymbol{u}}_x = (1, 0, 0)',$$

$$\widehat{\boldsymbol{u}}_y = (0, 1, 0)',$$

$$\widehat{\boldsymbol{u}}_z = (0, 0, 1)',$$

which are fixed in space. Then imagine that the body has some arbitrary initial orientation and as it changes its orientation, it takes a copy of the basis vectors with itself. By specifying these moved copies of bases in terms of the fixed bases, we can describe the new orientation of the body relative to its previous one. That is, if the moved bases are denoted by $\widehat{\boldsymbol{v}}_x$, $\widehat{\boldsymbol{v}}_y$, $\widehat{\boldsymbol{v}}_z$, then the *orientation matrix* $\boldsymbol{R}$ takes $(\widehat{\boldsymbol{u}}_x, \widehat{\boldsymbol{u}}_y, \widehat{\boldsymbol{u}}_z)$ to $(\widehat{\boldsymbol{v}}_x, \widehat{\boldsymbol{v}}_y, \widehat{\boldsymbol{v}}_z)$

$$(\widehat{\boldsymbol{v}}_x, \widehat{\boldsymbol{v}}_y, \widehat{\boldsymbol{v}}_z) = \boldsymbol{R} \times (\widehat{\boldsymbol{u}}_x, \widehat{\boldsymbol{u}}_y, \widehat{\boldsymbol{u}}_z) = \boldsymbol{R} \begin{pmatrix} 1 & 0 & 0 \\ 0 & 1 & 0 \\ 0 & 0 & 1 \end{pmatrix} = \boldsymbol{R},$$

and has the elements

$$\boldsymbol{R} \triangleq \begin{pmatrix} r_{1,1} & r_{1,2} & r_{1,3} \\ r_{2,1} & r_{2,2} & r_{2,3} \\ r_{3,1} & r_{3,2} & r_{3,3} \end{pmatrix} = (\widehat{\boldsymbol{v}}_x, \widehat{\boldsymbol{v}}_y, \widehat{\boldsymbol{v}}_z). \tag{7.4}$$

Example 7.1. An example of the moved basis vectors, represented in the fixed bases, is as follows

$$\widehat{\boldsymbol{v}}_x = (0.2803, 0.7392, -0.6124)',$$

$$\widehat{\boldsymbol{v}}_y = (-0.7392, 0.5732, 0.3536)',$$

$$\widehat{\boldsymbol{v}}_z = (0.6124, 0.3536, 0.7071)'.$$

We will revisit this in Example 7.2. □

In theory, this is all that is needed for fully describing changes in orientation in three dimensions. But this is somewhat obscure and misleading. It is obscure as we would often like to work with parametric changes in orientation and we may not be given the final position of moved basis vectors. It can be misleading because one may be tempted to think that there are nine degrees of freedom in the 3×3 orientation matrix $\boldsymbol{R}$, which is certainly not true.

There are only three degrees of freedom, where three is the dimension of the "special orthogonal group" in 3D, also known as SO(3). These three independent parameters can be described in a variety of approaches and we identify three main approaches here.

The first approach is through specifying three rotation angles around the three main x, y, and z axes. But there are 24 conventions for doing this! The conventions eventually come down to which order of rotation is taken about which axis and whether we use right-hand or left-hand rule for rotating.

The second (possibly more fundamental) approach is to describe rotation by a single angle around a unit-norm vector. This vector is in general different from the standard x, y, z axes. The equivalence between the two approaches is powered by *Euler rotation theorem* (which we do not prove here).

The third approach for describing changes in orientation is through *quaternions*, which can be advantageous in terms of computation cost, but is beyond the scope of this book.

So the message is that without understanding the first principles, applying the correct rotation can be tricky. The first step is to understand the single rotation along an arbitrary axis, which will guide our thinking about rotation and help transform one convention to another.

7.2.4   Single matrix representation of rotation

Any change in orientation of a (rigid) body can be described by a single rotation
around an appropriately chosen direction. Let us denote the unit vector along the
chosen direction by $\widehat{\boldsymbol{w}} \triangleq (w_x, w_y, w_z)'$ and the rotation angle by $\varphi \in [0, 2\pi)$ which
transforms an arbitrary initial vector $\boldsymbol{u}$ to $\boldsymbol{v}$.[4] Mathematically, the rotation is
represented by the transformation matrix $\boldsymbol{R}_\varphi^{(\widehat{\boldsymbol{w}})}$, that is,

$$\boxed{\boldsymbol{v} = \boldsymbol{R}_\varphi^{(\widehat{\boldsymbol{w}})} \boldsymbol{u}.}$$

We now present without proof how the rotation matrix is written in terms of
w_x, w_y, w_z and φ and that, in turn, shows how the change in orientation of a
rigid body is parameterized by three degrees of freedom (note that $|\widehat{\boldsymbol{w}}| = \widehat{\boldsymbol{w}}'\widehat{\boldsymbol{w}} =$
$w_x^2 + w_y^2 + w_z^2 = 1$ and hence there are only three degrees of freedom). The
rotation matrix is given by

$$\boxed{\boldsymbol{R}_\varphi^{(\widehat{\boldsymbol{w}})} \triangleq (1 - \cos\varphi)\widehat{\boldsymbol{w}}\widehat{\boldsymbol{w}}' + \cos\varphi\mathbf{I}_3 + \sin\varphi\widehat{\boldsymbol{w}}^\times,} \qquad (7.5)$$

where $\mathbf{I}_3$ is the identity matrix of size 3, $\widehat{\boldsymbol{w}}\widehat{\boldsymbol{w}}'$ is simply

$$\widehat{\boldsymbol{w}}\widehat{\boldsymbol{w}}' = \begin{pmatrix} w_x \\ w_y \\ w_z \end{pmatrix} \begin{pmatrix} w_x & w_y & w_z \end{pmatrix} = \begin{pmatrix} w_x^2 & w_x w_y & w_x w_z \\ w_y w_x & w_y^2 & w_y w_z \\ w_z w_x & w_z w_y & w_z^2 \end{pmatrix},$$

and $\widehat{\boldsymbol{w}}^\times$ is defined as

$$\widehat{\boldsymbol{w}}^\times \triangleq \begin{pmatrix} 0 & -w_z & w_y \\ w_z & 0 & -w_x \\ -w_y & w_x & 0 \end{pmatrix},$$

with the property $\widehat{\boldsymbol{w}}^\times \widehat{\boldsymbol{w}} = \widehat{\boldsymbol{w}}'\widehat{\boldsymbol{w}}^\times = 0$.

We can verify the correctness of (7.5) in two ways. First, we know that the
rotation of any multiple of $\widehat{\boldsymbol{w}}$, such as $\alpha\widehat{\boldsymbol{w}}$, around $\widehat{\boldsymbol{w}}$ should result in the same
vector $\alpha\widehat{\boldsymbol{w}}$. To see this and using $\widehat{\boldsymbol{w}}'\widehat{\boldsymbol{w}} = 1$ and $\widehat{\boldsymbol{w}}^\times \widehat{\boldsymbol{w}} = 0$, we write

$$\boldsymbol{R}_\varphi^{(\widehat{\boldsymbol{w}})}\alpha\widehat{\boldsymbol{w}} = \left((1 - \cos\varphi)\widehat{\boldsymbol{w}}\widehat{\boldsymbol{w}}' + \cos\varphi\mathbf{I}_3 + \sin\varphi\widehat{\boldsymbol{w}}^\times\right)\alpha\widehat{\boldsymbol{w}}$$

$$= \alpha\left((1 - \cos\varphi)\widehat{\boldsymbol{w}}\widehat{\boldsymbol{w}}'\widehat{\boldsymbol{w}} + \cos\varphi\mathbf{I}_3\widehat{\boldsymbol{w}} + \sin\varphi\widehat{\boldsymbol{w}}^\times\widehat{\boldsymbol{w}}\right)$$

$$= \alpha\left((1 - \cos\varphi)\widehat{\boldsymbol{w}} + \cos\varphi\widehat{\boldsymbol{w}} + 0\right) = \alpha\widehat{\boldsymbol{w}}.$$

As a result of this observation and given a non-parametric rotation matrix $\boldsymbol{R}$,
such as (7.4), we are able to compute the axis of rotation $\widehat{\boldsymbol{w}}$ and rotation angle
φ. We simply write

$$\widehat{\boldsymbol{w}} = \boldsymbol{R}\widehat{\boldsymbol{w}}.$$

---

[4] The Euclidean norm of the vector $\boldsymbol{w}$ does not affect the outcome of rotation and hence we
can simply use a unit norm vector $\widehat{\boldsymbol{w}}$ with $|\widehat{\boldsymbol{w}}|^2 = \widehat{\boldsymbol{w}}'\widehat{\boldsymbol{w}} = w_x^2 + w_y^2 + w_z^2 = 1$.

Thus, $\widehat{w}$ is the eigenvector for R corresponding to eigenvalue 1.

Determining φ is easy by noting that the trace of $R_\varphi^{(\widehat{w})}$ in (7.5) is equal to

$$\text{trace}(R_\varphi^{(\widehat{w})}) = (1 - \cos\varphi)|\widehat{w}| + 3\cos\varphi = 1 + 2\cos\varphi,$$

and hence

$$\varphi = \cos^{-1}((\text{trace}(R) - 1)/2).$$

Note that $\cos^{-1}$ returns an angle between 0 and π, but this together with the direction of the found eigenvector $\widehat{w}$ completely define the full range of rotations.

Example 7.2. In Example 7.1, R was given by

$$R = \begin{pmatrix} 0.2803 & -0.7392 & 0.6124 \\ 0.7392 & 0.5732 & 0.3536 \\ -0.6124 & 0.3536 & 0.7071 \end{pmatrix}.$$

The eigenvector corresponding to eigenvalue 1 for R is

$$\widehat{w} = (0, 0.6380, 0.7701)'$$

and the angle of rotation is

$$\varphi = \cos^{-1}((\text{trace}(R) - 1)/2) = 73.72°$$

(in degrees). □

The second way to verify the correctness of (7.5) is through rotating a vector u which is perpendicular to $\widehat{w}$ ($\widehat{w}'u = 0$) and checking that the resulting vector remains normal to u ($\widehat{w}'R_\varphi^{(\widehat{w})}u = 0$). To see this we write

$$R_\varphi^{(\widehat{w})}u = (1 - \cos\varphi)\widehat{w}\widehat{w}'u + \cos\varphi I_3 u + \sin\varphi \widehat{w}^\times u$$
$$= (\cos\varphi u + \sin\varphi \widehat{w}^\times u),$$

from which we conclude that

$$\widehat{w}'R_\varphi^{(\widehat{w})}u = \cos\varphi\widehat{w}'u + \sin\varphi\widehat{w}'\widehat{w}^\times u = 0.$$

Rotation is an invertible operation and is achieved by negating the rotation angle. That is,

$$(R_\varphi^{(\widehat{w})})^{-1} = R_{-\varphi}^{(\widehat{w})}.$$

This is one fundamental property of rotation group SO(3).

Rotations can be applied in succession and the outcome is yet another rotation around some axis by some rotation angle. This is another fundamental property of rotation group SO(3). For example, when we say that rotations $(\varphi_3, \varphi_2, \varphi_1)$ are applied around axes $\widehat{w}_3, \widehat{w}_2, \widehat{w}_1$, we mean that the overall rotation matrix is

$$R = R_{\varphi_3}^{(\widehat{w}_3)} R_{\varphi_2}^{(\widehat{w}_2)} R_{\varphi_1}^{(\widehat{w}_1)}.$$

That is, the rightmost rotation is applied first and the leftmost rotation last.

In the case of multiple rotations, the inverse of rotation becomes

$$
\begin{aligned}
\boldsymbol{R}^{-1} &= \left(\boldsymbol{R}_{\varphi_3}^{(\widehat{\boldsymbol{w}}_3)}\boldsymbol{R}_{\varphi_2}^{(\widehat{\boldsymbol{w}}_2)}\boldsymbol{R}_{\varphi_1}^{(\widehat{\boldsymbol{w}}_1)}\right)^{-1} \\
&= (\boldsymbol{R}_{\varphi_1}^{(\widehat{\boldsymbol{w}}_1)})^{-1}(\boldsymbol{R}_{\varphi_2}^{(\widehat{\boldsymbol{w}}_2)})^{-1}(\boldsymbol{R}_{\varphi_3}^{(\widehat{\boldsymbol{w}}_3)})^{-1} \\
&= \boldsymbol{R}_{-\varphi_1}^{(\widehat{\boldsymbol{w}}_1)}\boldsymbol{R}_{-\varphi_2}^{(\widehat{\boldsymbol{w}}_2)}\boldsymbol{R}_{-\varphi_3}^{(\widehat{\boldsymbol{w}}_3)}.
\end{aligned}
\tag{7.6}
$$

That is, we undo the effect of the last rotation first and so on.

7.2.5   Single rotation along the x, y or z axes

Instead of rotating an object around a single obscure direction $\widehat{\boldsymbol{w}}$, another way to arrive at the same new orientation relative to the old one is to apply three appropriately chosen successive rotations along three appropriately chosen axes among x, y, or z in some pre-specified order. This is the nub of Euler's rotation theorem.

Specializing (7.5) for the three main axes, we can write for $\widehat{\boldsymbol{u}}_x = (1,0,0)'$

$$
\boldsymbol{R}_{\varphi_x}^{(x)} \triangleq \boldsymbol{R}_{\varphi_x}^{(\widehat{\boldsymbol{u}}_x)}\Big|_{\substack{u_x=1 \\ u_y=0 \\ u_z=0}} = \begin{pmatrix} 1 & 0 & 0 \\ 0 & \cos\varphi_x & -\sin\varphi_x \\ 0 & \sin\varphi_x & \cos\varphi_x \end{pmatrix},
\tag{7.7}
$$

from which it is clear that the x component of any vector $\boldsymbol{u}$ remains unchanged under rotation. Similarly, for $\widehat{\boldsymbol{u}}_y = (0,1,0)'$

$$
\boldsymbol{R}_{\varphi_y}^{(y)} \triangleq \boldsymbol{R}_{\varphi_y}^{(\widehat{\boldsymbol{u}}_y)}\Big|_{\substack{u_x=0 \\ u_y=1 \\ u_z=0}} = \begin{pmatrix} \cos\varphi_y & 0 & \sin\varphi_y \\ 0 & 1 & 0 \\ -\sin\varphi_y & 0 & \cos\varphi_y \end{pmatrix},
\tag{7.8}
$$

which keeps the y component unchanged. And finally for $\widehat{\boldsymbol{u}}_z = (0,0,1)'$

$$
\boldsymbol{R}_{\varphi_z}^{(z)} \triangleq \boldsymbol{R}_{\varphi_z}^{(\widehat{\boldsymbol{u}}_z)}\Big|_{\substack{u_x=0 \\ u_y=0 \\ u_z=1}} = \begin{pmatrix} \cos\varphi_z & -\sin\varphi_z & 0 \\ \sin\varphi_z & \cos\varphi_z & 0 \\ 0 & 0 & 1 \end{pmatrix},
\tag{7.9}
$$

which keeps the z component unchanged and is in fact the extension of plain old 2D rotation in the x-y plane to the 3D case.

So from the above, it is gleaned that there can also be three degrees of freedom in selecting φ_x, φ_y, and φ_z for rotations around the x, y, and z axes and this is an easy-to-remember basis for specifying rotations in 3D space. But before elaborating on the possibilities, we should clarify intrinsic and extrinsic rotation compositions.

7.2.6   Intrinsic and extrinsic successive rotations

Imagine that we pick three rotation angles $\varphi \in [0, 2\pi)$, $\vartheta \in [0, \pi]$ and $\omega \in [0, 2\pi)$ and apply them using the right-hand rule around the xyz-axes (remember from right to left). But we need to be more specific about what mean by the axes. When we first rotate the object by $\omega \in [0, 2\pi)$ about the z-axis, the orientation of y-axis changes as a result. So we need to specify whether we mean to rotate by ϑ around the old or new y-axis. The same thing applies to the last rotation: is it around the newly moved x-axis or the original one?

In physical systems, a new rotation takes place around the moved (local) axis of the object. For example, an airplane has three degrees of freedom for orientation. Yaw refers to the change in the heading of the nose (around the plane's z-axis), pitch means tilting the nose up and down (around the plane's y-axis) and finally roll is rotation around the long axis that connects the plane's tail to the nose (x-axis). Of course, the rotations are specified and applied by the pilot on top of each other and based on the new orientation of the plane. This type of rotation around the moving axes is called *intrinsic* rotation.

Theoretically, we like to stick to the original fixed axes for specifying the rotations. This is desirable because we can specify the overall rotation operation beforehand as a product of three rotations around fixed axes and we need not to worry about what happens to the moved axes. In the above example, what would be meant is to rotate about the z-axis by $\omega \in [0, 2\pi)$, then around the original y-axis by $\vartheta \in [0, \pi]$ and then around the original x-axis by $\varphi \in [0, 2\pi)$. This method is called *extrinsic* rotation. Which way we use to specify the rotation should not become a source of confusion if we stick to the first principles.

From the perspective of the pilot and using (7.5), starting with an initial orthonormal set of vectors $\widehat{\boldsymbol{u}}_x$, $\widehat{\boldsymbol{u}}_y$, $\widehat{\boldsymbol{u}}_z$ along the plane's x, y, and z axes, the change in yaw results in

$$\widehat{\boldsymbol{u}}_x, \widehat{\boldsymbol{u}}_y, \widehat{\boldsymbol{u}}_z \xrightarrow{\boldsymbol{R}_\omega^{(\widehat{\boldsymbol{u}}_z)}} \widehat{\boldsymbol{v}}_x, \widehat{\boldsymbol{v}}_y, \widehat{\boldsymbol{v}}_z,$$

where, of course, $\widehat{\boldsymbol{u}}_z = \widehat{\boldsymbol{v}}_z$. The change in pitch (around plane's *new* y-axis) results in

$$\widehat{\boldsymbol{v}}_x, \widehat{\boldsymbol{v}}_y, \widehat{\boldsymbol{v}}_z \xrightarrow{\boldsymbol{R}_\vartheta^{(\widehat{\boldsymbol{v}}_y)}} \widehat{\boldsymbol{w}}_x, \widehat{\boldsymbol{w}}_y, \widehat{\boldsymbol{w}}_z,$$

where $\widehat{\boldsymbol{v}}_y = \boldsymbol{R}_\omega^{(\widehat{\boldsymbol{u}}_z)} \widehat{\boldsymbol{u}}_y$. And, finally, rolling around plane's *latest* x-axis takes the plane to its final orientation

$$\widehat{\boldsymbol{w}}_x, \widehat{\boldsymbol{w}}_y, \widehat{\boldsymbol{w}}_z \xrightarrow{\boldsymbol{R}_\varphi^{(\widehat{\boldsymbol{w}}_x)}} \widehat{\boldsymbol{t}}_x, \widehat{\boldsymbol{t}}_y, \widehat{\boldsymbol{t}}_z,$$

where $\widehat{\boldsymbol{w}}_x = \boldsymbol{R}_\vartheta^{(\widehat{\boldsymbol{v}}_y)} \widehat{\boldsymbol{v}}_x = \boldsymbol{R}_\vartheta^{(\widehat{\boldsymbol{v}}_y)} \boldsymbol{R}_\omega^{(\widehat{\boldsymbol{u}}_z)} \widehat{\boldsymbol{u}}_x$. In the end, the resulting single rotation matrix is

$$\boldsymbol{R}_{\mathrm{int}} = \boldsymbol{R}_\varphi^{(\widehat{\boldsymbol{w}}_x)} \boldsymbol{R}_\vartheta^{(\widehat{\boldsymbol{v}}_y)} \boldsymbol{R}_\omega^{(\widehat{\boldsymbol{u}}_z)}.$$

The moving axes are sometimes referred to using capital letters X, Y, and Z.

If we use the fixed axes for specifying rotations, then the single rotation matrix is

$$\boldsymbol{R}_{\mathrm{ext}} = \boldsymbol{R}_\varphi^{(x)} \boldsymbol{R}_\vartheta^{(y)} \boldsymbol{R}_\omega^{(z)} \triangleq \boldsymbol{R}_\varphi^{(\widehat{\boldsymbol{u}}_x)} \boldsymbol{R}_\vartheta^{(\widehat{\boldsymbol{u}}_y)} \boldsymbol{R}_\omega^{(\widehat{\boldsymbol{u}}_z)}.$$

Imagine that we apply the same amount of rotations $(\varphi, \vartheta, \omega)$ to the moving axes XYZ in the intrinsic method and to fixed axes xyz the in extrinsic method. We expect to arrive at completely different orientations.

But the question is how can we arrive at the same relative orientation using intrinsic and extrinsic rotations? The answer is strikingly simple. For example, if $(\varphi, \vartheta, \omega)$ is extrinsically applied to fixed axes in the order xyz, then the same orientation can be achieved if we apply $(\omega, \vartheta, \varphi)$ intrinsically to the moving axes ZYX. That is,

$$R_\varphi^{(x)} R_\vartheta^{(y)} R_\omega^{(z)} = R_\omega^{(\widehat{w}_z)} R_\vartheta^{(\widehat{v}_y)} R_\varphi^{(\widehat{u}_x)},$$

where $\widehat{v}_y = R_\varphi^{(\widehat{u}_x)} \widehat{u}_y$ and $\widehat{w}_z = R_\vartheta^{(\widehat{v}_y)} \widehat{v}_z = R_\vartheta^{(\widehat{v}_y)} R_\varphi^{(\widehat{u}_x)} \widehat{u}_z$. The above statement is generally true for any choice of rotation axes, not just xyz.

Example 7.3. A pilot applies (heading, pitch, roll) $= (45°, 20°, 30°)$, in the moving ZYX-axes.[5] The intrinsic rotation matrix is given by $R_{\text{int}} = R_{45}^{(\widehat{w}_z)} R_{20}^{(\widehat{v}_y)} R_{30}^{(\widehat{u}_x)}$, which upon using (7.5) becomes

$$R_{\text{int}} = \begin{pmatrix} 0.7414 & -0.6225 & 0.2507 \\ 0.5284 & 0.7718 & 0.3538 \\ -0.4138 & 0.1299 & 0.9011 \end{pmatrix}$$

$$\times \begin{pmatrix} 0.9397 & -0.1710 & 0.2962 \\ 0.1710 & 0.9849 & 0.1710 \\ -0.2962 & 0.0261 & 0.9548 \end{pmatrix}$$

$$\times \begin{pmatrix} 1 & 0 & 0 \\ 0 & 0.8660 & -0.5000 \\ 0 & 0.5000 & 0.8660 \end{pmatrix}.$$

To arrive at the same relative orientation, a sequence of rotations given by $(30°, 20°, 45°)$ should be applied in the fixed xyz-axes system, which results in the extrinsic rotation matrix $R_{\text{ext}} = R_{30}^{(x)} R_{20}^{(y)} R_{45}^{(z)}$

$$R_{\text{ext}} = \begin{pmatrix} 1 & 0 & 0 \\ 0 & 0.8660 & -0.5000 \\ 0 & 0.5000 & 0.8660 \end{pmatrix}$$

$$\times \begin{pmatrix} 0.9397 & 0 & 0.3420 \\ 0 & 1 & 0 \\ -0.3420 & 0 & 0.9397 \end{pmatrix}$$

$$\times \begin{pmatrix} 0.7071 & -0.7071 & 0 \\ 0.7071 & 0.7071 & 0 \\ 0 & 0 & 1 \end{pmatrix}.$$

And it is easy to check that $R_{\text{ext}} = R_{\text{int}}$. Of course, remembering and working with the extrinsic matrix is much easier and neater.   □

In terms of choosing the axes for applying rotations, all *independent* possibilities are acceptable. For example, two successive rotations around the z-axis are not independent. So for choosing the first axis we have three possibilities, then eliminating the first choice there are two remaining possibilities. For the final rotation, the first chosen axis can be used again. So in total there are

---

[5] The angles are exaggerated.

$2 \times 2 \times 3 = 12$ possibilities. Then there are the right-hand and left-hand rules for applying rotations, which brings the total number of possibilities to 24.

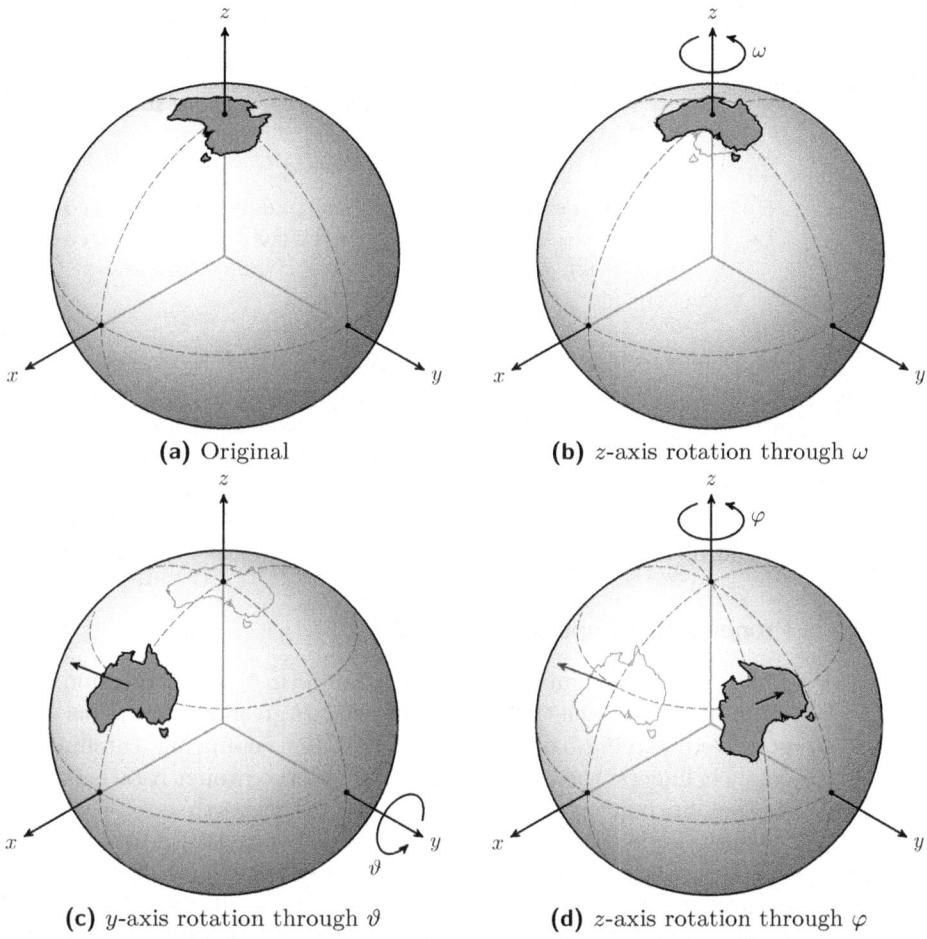

(a) Original **(b)** z-axis rotation through ω

(c) y-axis rotation through ϑ **(d)** z-axis rotation through φ

Figure 7.4 Standard zyz rotation convention. From an original position (a), we first rotate by ω about the z-axis (b), followed by ϑ about the y-axis (c), and then ϑ about the z-axis (d). All proper rotations can be expressed this way. The axes and rotations follow a right-handed convention.

7.2.7 Rotation convention used in this book

In this book, we use the right-hand rule zyz convention and specify the rotations $(\varphi, \vartheta, \omega)$ around these fixed axes, knowing that if we wanted to rotate around moving axes, we would need to swap the order of applying ω and φ to the moving Z axes to get the same result. The zyz and angle convention is shown in Figure 7.4. Our single rotation matrix is of the form

$$\boldsymbol{R} \equiv \boldsymbol{R}^{(zyz)}_{\varphi\vartheta\omega} \triangleq \boldsymbol{R}^{(z)}_{\varphi} \boldsymbol{R}^{(y)}_{\vartheta} \boldsymbol{R}^{(z)}_{\omega}, \tag{7.10}$$

where $\boldsymbol{R}^{(y)}$ and $\boldsymbol{R}^{(z)}$ are given in (7.8) and (7.9). By expanding the right-hand side of (7.10) we obtain a single rotation matrix,

$$
\boldsymbol{R} = \begin{pmatrix} \cos\varphi & -\sin\varphi & 0 \\ \sin\varphi & \cos\varphi & 0 \\ 0 & 0 & 1 \end{pmatrix} \begin{pmatrix} \cos\vartheta & 0 & \sin\vartheta \\ 0 & 1 & 0 \\ -\sin\vartheta & 0 & \cos\vartheta \end{pmatrix} \begin{pmatrix} \cos\omega & -\sin\omega & 0 \\ \sin\omega & \cos\omega & 0 \\ 0 & 0 & 1 \end{pmatrix}
$$

$$
= \begin{pmatrix} \cos\varphi & -\sin\varphi & 0 \\ \sin\varphi & \cos\varphi & 0 \\ 0 & 0 & 1 \end{pmatrix} \begin{pmatrix} \cos\vartheta\cos\omega & -\cos\vartheta\sin\omega & \sin\vartheta \\ \sin\omega & \cos\omega & 0 \\ -\sin\vartheta\cos\omega & \sin\vartheta\sin\omega & \cos\vartheta \end{pmatrix}
$$

$$
= \begin{pmatrix} \cos\varphi\cos\vartheta\cos\omega - \sin\varphi\sin\omega & -\cos\varphi\cos\vartheta\sin\omega - \sin\varphi\cos\omega & \cos\varphi\sin\vartheta \\ \sin\varphi\cos\vartheta\cos\omega + \cos\varphi\sin\omega & -\sin\varphi\cos\vartheta\sin\omega + \cos\varphi\cos\omega & \sin\varphi\sin\vartheta \\ -\sin\vartheta\cos\omega & \sin\vartheta\sin\omega & \cos\vartheta \end{pmatrix}.
$$

$$(7.11)$$

For example, looking at the first column of the rightmost matrix on the first line of (7.11) we see that the rotated unit vector along x, $(1,0,0)'$, will project into $(\cos\omega, \sin\omega, 0)'$ under the first rotation ω around z. Now if we rotate this projected vector around the fixed axis y, its y component will remain unchanged while the other components around x and z will become $\cos\vartheta\cos\omega$ and $-\sin\vartheta\cos\omega$ as can be seen from the first column of the rightmost matrix on the second line of (7.11), and so on.

Proper versus improper rotations

The caption in Figure 7.4 mentions the term "proper rotation." A proper rotation is one for which the rotation matrix, $\boldsymbol{R}$ given in (7.10), has determinant 1. If gods were playing basketball with the Earth, bouncing, catching and spinning it on their fingers, then they would only observe proper rotations. So in Figure 7.5 if Alice Springs[6] started as indicated on the north pole, Figure 7.5(a), then a feasible proper rotation could lead to a relocation indicated in Figure 7.5(b), which actually corresponds to the case given in Figure 7.4. In contrast, there is no proper rotation of Figure 7.5(a) to yield the situation shown in Figure 7.5(c), where the Australia-shaped region is rotated but mirror reversed. This transformation is unitary with determinant -1 and called an "improper rotation."

7.2.8   Three rotation angles from a single rotation matrix

We have seen so far that it is very easy to get a single rotation matrix from any consecutive rotations around some axes. We simply need to multiply the individual rotation matrices from right to left in the order they are applied.

The question is how to get the three basic rotation parameters from a given rotation matrix $\boldsymbol{R}$ where

$$
\boldsymbol{R} = \begin{pmatrix} r_{1,1} & r_{1,2} & r_{1,3} \\ r_{2,1} & r_{2,2} & r_{2,3} \\ r_{3,1} & r_{3,2} & r_{3,3} \end{pmatrix}.
\tag{7.12}
$$

---

[6]   The major central Australian town.

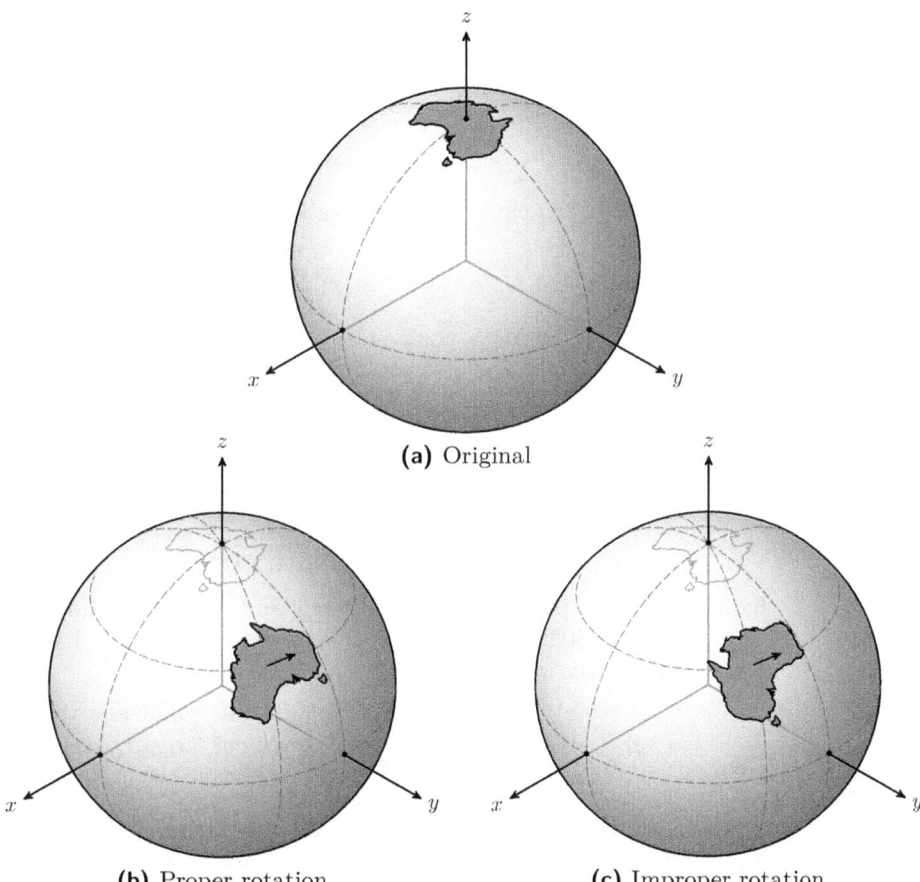

(a) Original

(b) Proper rotation

(c) Improper rotation

Figure 7.5 Proper rotation: there exists a rotation matrix with determinant of 1 which maps the north pole region (a) to the indicated position (b). Improper rotation: there is a unitary matrix of determinant −1 which maps the north pole region (a) to the indicated position (c), but no proper rotation (resulting in a rotated, but mirrored image of the original region).

This, of course, would depend on what orthonormal axes are used for rotation and in what order. For example, suppose we have used the zyz convention. By element-wise comparison of (7.12) and the last line of (7.11), we can start by solving for the easiest parameter, which seems to be ϑ.

$$\vartheta = \cos^{-1}(r_{3,3}),$$
$$\sin \vartheta = \sqrt{1 - (r_{3,3})^2}. \tag{7.13}$$

Note that since $\vartheta \in [0, \pi)$, the $\sin^{-1}$ operation is not ambiguous. Next, we note that

$$r_{1,3} = \cos \varphi \sin \vartheta,$$
$$r_{2,3} = \sin \varphi \sin \vartheta. \tag{7.14}$$

And hence $\varphi \in [0, 2\pi)$ can be found by a four-quadrant search satisfying

$$\sin \varphi = \frac{r_{2,3}}{\sqrt{1 - (r_{3,3})^2}},$$
$$\cos \varphi = \frac{r_{1,3}}{\sqrt{1 - (r_{3,3})^2}}. \tag{7.15}$$

Similarly, $\omega \in [0, 2\pi)$ can be found by a four-quadrant search satisfying

$$\sin \omega = \frac{r_{3,2}}{\sqrt{1 - (r_{3,3})^2}},$$
$$\cos \omega = \frac{-r_{3,1}}{\sqrt{1 - (r_{3,3})^2}}. \tag{7.16}$$

Example 7.4. A pilot is in an emergency and wishes to apply (heading, pitch, roll) = $(45°, 20°, 30°)$. This is the same as Example 7.3 where the overall rotation matrix is given by

$$\boldsymbol{R}_{\text{int}} = \boldsymbol{R}_{45}^{(\widehat{\boldsymbol{w}}_z)} \boldsymbol{R}_{20}^{(\widehat{\boldsymbol{v}}_y)} \boldsymbol{R}_{30}^{(\widehat{\boldsymbol{u}}_x)} = \begin{pmatrix} 0.6645 & -0.6645 & 0.3420 \\ 0.7333 & 0.4915 & -0.4698 \\ 0.1441 & 0.5630 & 0.8138 \end{pmatrix}.$$

But for some weird reason the plane does not respond to the rolling command around the X-axis. What should a cool-headed pilot do to arrive at the same orientation by changing the yaw and pitch only (using the ZYZ convention)?

We first find the angles $(\varphi, \vartheta, \omega)$ in the extrinsic zyz system. From (7.13)–(7.16), we work out $\varphi = 306.0524°$, $\vartheta = 35.5313°$ and $\omega = 104.3577°$. Therefore, the pilot should first change the yaw or heading by $\varphi = 306.0524°$ (or 53.9476 in the clockwise direction), followed by a pitch change of $\vartheta = 35.5313°$ and another yaw change of $\omega = 104.3577°$. Of course, the angles are exaggerated and may not correspond to a physical system. □

Problem

7.3. Convert a sequence of rotations $(\varphi = 15°, \vartheta = 30°, \omega = 20°)$ in the zyz system to the xyz system.

7.3   Hilbert space $L^2(\mathbb{S}^2)$

7.3.1   Definition of Hilbert space $L^2(\mathbb{S}^2)$

Let $f(\theta, \phi) \equiv f(\widehat{\boldsymbol{u}})$ and $g(\theta, \phi) \equiv g(\widehat{\boldsymbol{u}})$ be two complex-valued functions,[7] defined on the 2-sphere and define the inner product

$$
\begin{aligned}
\langle f, g \rangle &\triangleq \int_{\mathbb{S}^2} f(\widehat{\boldsymbol{u}}) \overline{g(\widehat{\boldsymbol{u}})}\, ds(\widehat{\boldsymbol{u}}) \\
&= \int_0^{2\pi} \int_0^{\pi} f(\theta, \phi) \overline{g(\theta, \phi)}\, \sin\theta\, d\theta\, d\phi,
\end{aligned}
\tag{7.17}
$$

which induces the norm

$$
\|f\| \triangleq \langle f, f \rangle^{1/2}.
\tag{7.18}
$$

7.3.2   Signals on 2-sphere

Finite energy functions whose domain is the 2-sphere are referred to as "signals on 2-sphere." Mathematically, f is a signal on the 2-sphere if and only if it has finite induced norm $\|f\| < \infty$. All such finite energy signals under inner product (7.17) form the complex separable Hilbert space $L^2(\mathbb{S}^2)$. In the following, any reference to a signal is understood to be the same as a "signal on 2-sphere."

7.3.3   Definition of spherical harmonics

The archetype complete orthonormal sequence for $L^2(\mathbb{S}^2)$ is the set of spherical harmonic functions, or spherical harmonics for short, $Y_\ell^m(\theta, \phi) \equiv Y_\ell^m(\widehat{\boldsymbol{u}})$, for degree $\ell \in \{0, 1, 2, \ldots\}$ and order $m \in \{-\ell, -\ell + 1, \ldots, \ell\}$, which are defined as[8]

$$
\boxed{
Y_\ell^m(\theta, \phi) \triangleq \sqrt{\frac{2\ell + 1}{4\pi} \frac{(\ell - m)!}{(\ell + m)!}} P_\ell^m(\cos\theta) e^{im\phi},
}
\tag{7.19}
$$

where the first term is a normalization factor to ensure Y_ℓ^m has a unit induced norm, (7.18), $\|Y_\ell^m\| = 1$, and will be denoted by

$$
\boxed{
N_\ell^m \triangleq \sqrt{\frac{2\ell + 1}{4\pi} \frac{(\ell - m)!}{(\ell + m)!}}.
}
\tag{7.20}
$$

The second (co-latitude) term involves the associated Legendre functions (see below) in the indeterminate $x = \cos\theta$, and the third (longitude) term is the complex exponential

$$
e^{im\phi} = \cos m\phi + i \sin m\phi.
\tag{7.21}
$$

---

[7]   Notations like f, $f(\theta, \phi)$, $f(\widehat{\boldsymbol{u}})$ will be used interchangeably depending on the context.

[8]   Although spherical harmonics and (associated) Legendre functions were defined in the first part of the book as examples of orthonormal basis functions, we will introduce them again to be self-contained and elaborate on them much further, as it will be needed for in-depth understanding of the material in this chapter and the following two chapters.

Both the first and second terms are real-valued, meaning the complex nature of the spherical harmonics derives from the complex exponential in ϕ only.

With $x = \cos\theta$ there is a monotonically decreasing one-to-one correspondence between $0 \leq \theta \leq \pi$ and $-1 \leq x \leq 1$. The functions $P_\ell^m(x)$ on domain $-1 \leq x \leq 1$ are the associated Legendre functions, which we now define. For order $m = 0$, they reduce to the Legendre polynomials and are given by the Rodrigues formula, (Lebedev, 1972; Riley et al., 2006), in degree $\ell \in \{0, 1, 2, \ldots\}$

$$P_\ell(x) \triangleq \frac{1}{2^\ell \ell!} \frac{d^\ell}{dx^\ell} (x^2 - 1)^\ell, \quad \ell \in \{0, 1, 2, \ldots\}, \tag{7.22}$$

and for positive order m, satisfying $m \in \{0, 1, \ldots, \ell\}$, they are given by

$$P_\ell^m(x) \triangleq (-1)^m (1 - x^2)^{m/2} \frac{d^m}{dx^m} P_\ell(x),$$

which by the Rodrigues formula can be written

$$P_\ell^m(x) \triangleq \frac{(-1)^m}{2^\ell \ell!} (1 - x^2)^{m/2} \frac{d^{\ell+m}}{dx^{\ell+m}} (x^2 - 1)^\ell, \quad m \in \{0, 1, \ldots, \ell\}, \tag{7.23}$$

for $-1 \leq x \leq 1$ (Lebedev, 1972). Because of the Rodrigues formula when $\ell + m \geq 0$ the derivatives in (7.23) remain well-defined. For $m \in \{-\ell, \ldots, 0\}$ we have the identity (see Problem 7.8)

$$P_\ell^{-m}(x) = (-1)^m \frac{(\ell - m)!}{(\ell + m)!} P_\ell^m(x), \quad m \in \{0, 1, \ldots, \ell\}, \tag{7.24}$$

for $-1 \leq x \leq 1$ (Riley et al., 2006).

Remark 7.1. The term $(-1)^m$ in the definition of associated Legendre functions is referred to as the *Condon-Shortley phase* factor (which seems an excessively gratuitous attribution for such a limited contribution to knowledge). It is somewhat esoteric, but the presence of $(-1)^m$ in (7.24) is not because $(-1)^m$ appears in (7.23); see (Riley et al., 2006, pp. 588–589) or Problem 7.8.   □

Remark 7.2. It is clear from (7.22) that the Legendre polynomials $P_\ell(x)$ are true polynomials of degree ℓ in the indeterminate x. This is a useful aide-mémoire for not confusing the spherical harmonic degree and spherical harmonic order. For odd m, $(1 - x^2)^{m/2}$ introduces a non-polynomial term into odd-order associated Legendre functions and so they are not polynomials in x in general. Furthermore, when we take $x = \cos\theta$ then $P_\ell^m(\cos\theta)$ is certainly not a polynomial in θ in general. Nevertheless, in the literature, the use of "associated Legendre polynomials" is not uncommon.   □

In summary, for the associated Legendre function portion of the spherical harmonics, they are given by equations (7.23) and (7.24) for $m \in \{0, 1, \ldots, \ell\}$, although technically we can take (7.23) to be the definition of the associated Legendre functions for all $m \in \{-\ell, -\ell+1, \ldots, \ell\}$. Equation (7.24) is useful because it shows that $P_\ell^{-m}(x)$ only differs from $P_\ell^m(x)$ by a scale factor. Furthermore, when $m = 0$ then the (un-associated) Legendre polynomial $P_\ell(x)$ in (7.22) can replace $P_\ell^0(x)$ in (7.23).

Example 7.5. With $\ell = 1$ and $m = 1$ the associated Legendre function is
$$P_1^1(x) = -(1 - x^2)^{1/2},$$
$$P_1^1(\cos\theta) = -(1 - \cos^2\theta)^{1/2} = -\sin\theta.$$
Using (7.19), the spherical harmonic for degree $\ell = 1$ and order $m = 1$ is
$$Y_1^1(\theta, \phi) = \sqrt{\frac{3}{8\pi}} P_1^1(\cos\theta) e^{i\phi}$$
$$= -\frac{1}{2}\sqrt{\frac{3}{2\pi}} \sin\theta e^{i\phi}.$$
$\square$

As in the example above, the identification
$$\left.(1 - x^2)^{m/2}\right|_{x=\cos\theta} = \sin^m\theta, \quad 0 \le \theta \le \pi$$
is uniquely defined because there is a monotonically decreasing one-to-one correspondence between the restricted range $0 \le \theta \le \pi$ and $-1 \le x \le 1$.

Example 7.6. With $\ell = 3$ and $m = 2$ the associated Legendre function is
$$P_3^2(x) = 15x(1 - x^2),$$
$$P_3^2(\cos\theta) = 15\cos\theta\sin^2\theta.$$
Using (7.19), the spherical harmonic for degree $\ell = 3$ and order $m = 2$ is
$$Y_3^2(\theta, \phi) = \frac{1}{4}\sqrt{\frac{7}{30\pi}} P_3^2(\cos\theta) e^{i2\phi}$$
$$= \frac{1}{4}\sqrt{\frac{105}{2\pi}} \sin^2\theta\cos\theta e^{i2\phi}.$$
$\square$

We defer giving a longer list of the spherical harmonics (see Example 7.9, p. 192) until we flesh out a few properties that reveal the underlying structure.

7.3.4   Orthonormality and other properties

Spherical harmonics satisfy the orthonormality condition

$$\boxed{\langle Y_\ell^m, Y_p^q \rangle = \delta_{\ell,p}^{m,q} \triangleq \delta_{\ell,p}\delta_{m,q}} \qquad (7.25)$$

with respect to the inner product (7.17). In some sense this is their raison d'être and proven in many references (Lebedev, 1972). That the spherical harmonics are complete is more important and we cover this aspect later.

In addition, the spherical harmonics have the conjugation property

$$\overline{Y_\ell^m(\theta, \phi)} = (-1)^m Y_\ell^{-m}(\theta, \phi), \tag{7.26}$$

which is a convenient expression to determine the negative-order spherical harmonics, those for $m < 0$, tainted only by the pesky Condon-Shortley phase factor.

Example 7.7. The spherical harmonic for $\ell = 1$ and $m = -1$ is given by

$$Y_1^{-1}(\theta, \phi) = (-1)^1 \overline{Y_1^1(\theta, \phi)}$$

$$= \frac{1}{2}\sqrt{\frac{3}{2\pi}} \sin\theta\, e^{-i\phi}.$$

Here we have exercised (7.26) with $m = 1$.   □

Example 7.8. The spherical harmonic expression in (7.19) is valid for all m. However, for $m < 0$ we would need to invoke (7.24) to convert the negative-order associated Legendre function to the positive-order one to enable computation from (7.23). So it is desirable to have an expression for the negative-order spherical harmonics in terms of the positive-order associated Legendre functions for evaluation (and conversely). We can make the necessary substitutions to yield an alternative expression

$$Y_\ell^{-m}(\theta, \phi) = (-1)^m \overline{Y_\ell^m(\theta, \phi)}$$

$$= (-1)^m \overline{N_\ell^m P_\ell^m(\cos\theta) e^{im\phi}}$$

$$= (-1)^m N_\ell^m P_\ell^m(\cos\theta) e^{-im\phi}, \tag{7.27}$$

which, despite the motivation to use it for $m \in \{0, 1, \dots, \ell\}$, is valid for all $m \in \{-\ell, -\ell+1, \dots, \ell\}$.   □

Equation (7.27) shows that the spherical harmonics for negative orders are very closely related to the spherical harmonics for positive orders and this allows us to introduce a slightly more compact notation

$$Y_\ell^{\mp m}(\theta, \phi) = (-1)^{\frac{m}{0}} N_\ell^m P_\ell^m(\cos\theta) e^{\mp im\phi}, \quad m \geq 0, \tag{7.28}$$

where $-m$ is associated with the upper part involving multiplication by $(-1)^m$ and $e^{-im\phi}$, and $+m$ is associated with the lower part involving multiplication by 1 and $e^{im\phi}$. The example below draws on similar notation.

Example 7.9. Here is a collection of complex spherical harmonics for degrees $\ell \in \{0, 1, 2, 3, 4\}$:

$$Y_0^0(\theta, \phi) = \frac{1}{2}\sqrt{\frac{1}{\pi}}, \qquad\qquad Y_2^{\mp 2}(\theta, \phi) = \frac{1}{4}\sqrt{\frac{15}{2\pi}} \sin^2\theta\, e^{\mp i2\phi},$$

$$Y_1^{\mp 1}(\theta, \phi) = \pm\frac{1}{2}\sqrt{\frac{3}{2\pi}} \sin\theta\, e^{\mp i\phi}, \quad Y_2^{\mp 1}(\theta, \phi) = \pm\frac{1}{2}\sqrt{\frac{15}{2\pi}} \sin\theta\cos\theta\, e^{\mp i\phi},$$

$$Y_1^0(\theta, \phi) = \frac{1}{2}\sqrt{\frac{3}{\pi}} \cos\theta, \qquad\qquad Y_2^0(\theta, \phi) = \frac{1}{4}\sqrt{\frac{5}{\pi}} (3\cos^2\theta - 1),$$

$$Y_3^{\mp 3}(\theta, \phi) = \pm\frac{1}{8}\sqrt{\frac{35}{\pi}}\sin^3\theta e^{\mp i3\phi},$$

$$Y_3^{\mp 2}(\theta, \phi) = \frac{1}{4}\sqrt{\frac{105}{2\pi}}\sin^2\theta\cos\theta e^{\mp i2\phi},$$

$$Y_3^{\mp 1}(\theta, \phi) = \pm\frac{1}{8}\sqrt{\frac{21}{\pi}}\sin\theta\left(5\cos^2\theta - 1\right)e^{\mp i\phi},$$

$$Y_3^{0}(\theta, \phi) = \frac{1}{4}\sqrt{\frac{7}{\pi}}\left(5\cos^3\theta - 3\cos\theta\right),$$

$$Y_4^{\mp 4}(\theta, \phi) = \frac{3}{16}\sqrt{\frac{35}{2\pi}}\sin^4\theta e^{\mp i4\phi},$$

$$Y_4^{\mp 3}(\theta, \phi) = \pm\frac{3}{8}\sqrt{\frac{35}{\pi}}\sin^3\theta\cos\theta e^{\mp i3\phi},$$

$$Y_4^{\mp 2}(\theta, \phi) = \frac{3}{8}\sqrt{\frac{5}{2\pi}}\sin^2\theta\left(7\cos^2\theta - 1\right)e^{\mp i2\phi},$$

$$Y_4^{\mp 1}(\theta, \phi) = \pm\frac{3}{8}\sqrt{\frac{5}{\pi}}\sin\theta\left(7\cos^3\theta - 3\cos\theta\right)e^{\mp i\phi},$$

$$Y_4^{0}(\theta, \phi) = \frac{3}{16}\sqrt{\frac{1}{\pi}}\left(35\cos^4\theta - 30\cos^2\theta + 3\right).$$

$\square$

When the order m is zero, the spherical harmonics become independent of longitude angle ϕ and are given by

$$Y_\ell^0(\theta) \equiv Y_\ell^0(\theta, \phi) = N_\ell^0 P_\ell^0(\cos\theta) = \sqrt{\frac{2\ell + 1}{4\pi}}P_\ell^0(\cos\theta).$$

For $m = 0$ in (7.23), $P_\ell^0(\cos\theta)$ reduces to the Legendre polynomial (7.22) and so we have

$$Y_\ell^0(\theta) \equiv Y_\ell^0(\theta, \phi) = \sqrt{\frac{2\ell + 1}{4\pi}}P_\ell(\cos\theta). \qquad (7.29)$$

Note that the Legendre polynomials $\{P_\ell(x)\}_{\ell=0}^\infty$ are complete in the interval $\Omega = [-1, +1]$. This will be used when dealing with a special subspace of $L^2(\mathbb{S}^2)$, which has $\{Y_\ell^0(\theta)\}_{\ell=0}^\infty$ as its basis.

An important property of spherical harmonics is the addition theorem (Colton and Kress, 1998), which we require later. It states that

$$\sum_{m=-\ell}^{\ell} Y_\ell^m(\widehat{\boldsymbol{u}})\overline{Y_\ell^m(\widehat{\boldsymbol{v}})} = \frac{(2\ell + 1)}{4\pi}P_\ell(\widehat{\boldsymbol{u}} \cdot \widehat{\boldsymbol{v}}) = \frac{(2\ell + 1)}{4\pi}P_\ell(\cos\Delta), \qquad (7.30)$$

where $\widehat{\boldsymbol{u}} \cdot \widehat{\boldsymbol{v}} = \cos\Delta$ is the dot/inner product between the unit vectors $\widehat{\boldsymbol{u}}$ and $\widehat{\boldsymbol{v}}$ in $\mathbb{R}^3$ defined in (7.2), and $P_\ell(x)$ is the Legendre polynomial of degree ℓ.

Finally, the weighted completeness relation on the 2-sphere is, as we had given earlier in the book (2.80), p. 95,

$$\sin\theta \sum_{\ell=0}^{\infty} \sum_{m=-\ell}^{\ell} Y_\ell^m(\theta,\phi)\overline{Y_\ell^m(\vartheta,\varphi)} = \delta(\theta-\vartheta)\delta(\phi-\varphi). \qquad (7.31)$$

Problems

7.4. Prove the identity

$$N_\ell^{-m}\frac{(\ell-m)!}{(\ell+m)!} = N_\ell^m, \qquad (7.32)$$

where the normalization factor is given in (7.20).

7.5. Using (7.24) prove the identity

$$N_\ell^{-m}P_\ell^{-m}(\cos\theta) = (-1)^m N_\ell^m P_\ell^m(\cos\theta), \qquad (7.33)$$

where the normalization factor is given in (7.20).

7.6. By starting with $Y_\ell^{-m}(\theta,\phi)$ and using (7.24) or otherwise prove the conjugation property (7.26) or equivalently:

$$Y_\ell^{-m}(\theta,\phi) = (-1)^m\overline{Y_\ell^m(\theta,\phi)}. \qquad (7.34)$$

7.7. From the product $(x^2-1)^\ell = (x+1)^\ell(x-1)^\ell$, show that (7.23) can be written in the alternative form

$$P_\ell^m(x) \triangleq \frac{(-1)^m}{2^\ell\ell!}(1-x^2)^{m/2}\sum_{s=0}^{\ell+m}\frac{(\ell+m)!}{s!(\ell+m-s)!}\frac{d^s}{dx^s}(x+1)^\ell\frac{d^{\ell+m-s}}{dx^{\ell+m-s}}(x-1)^\ell,$$

which is valid for $m \in \{-\ell,-\ell+1,\ldots,\ell\}$.

7.8. Prove the identity (7.24) repeated here

$$P_\ell^{-m}(x) = (-1)^m\frac{(\ell-m)!}{(\ell+m)!}P_\ell^m(x),$$

by comparing the expression for $P_\ell^m(x)$ and $P_\ell^{-m}(x)$ in (7.23) using the expression in Problem 7.7 or otherwise.

7.9. Show that (7.28) is valid for all m, that is, $m \in \{-\ell,-\ell+1,\ldots,\ell\}$.

7.10. Show that

$$Y_\ell^{-m}(\theta,\phi) = e^{im(\pi-2\phi)}Y_\ell^m(\theta,\phi), \quad m \in \{-\ell,-\ell+1,\ldots,\ell\}.$$

7.3.5   Spherical harmonic coefficients

By completeness of the spherical harmonic functions (Colton and Kress, 1998), any signal $f \in L^2(\mathbb{S}^2)$ can be expanded as

$$f(\theta,\phi) = \sum_{\ell=0}^{\infty}\sum_{m=-\ell}^{\ell}(f)_\ell^m Y_\ell^m(\theta,\phi), \qquad (7.35)$$

where

$$
\begin{aligned}
(f)_\ell^m \triangleq \langle f, Y_\ell^m \rangle &= \int_{\mathbb{S}^2} f(\widehat{\boldsymbol{u}}) \overline{Y_\ell^m(\widehat{\boldsymbol{u}})} \, ds(\widehat{\boldsymbol{u}}) \\
&= \int_0^{2\pi} \int_0^\pi f(\theta, \phi) \overline{Y_\ell^m(\theta, \phi)} \sin\theta \, d\theta \, d\phi
\end{aligned} \tag{7.36}
$$

are the spherical harmonic Fourier coefficients or spherical harmonic coefficients for short. The equality is understood in terms of convergence in the mean

$$
\lim_{L \to \infty} \left\| f - \sum_{\ell=0}^L \sum_{m=-\ell}^\ell \langle f, Y_\ell^m \rangle Y_\ell^m(\theta, \phi) \right\|^2 = 0, \quad \forall f \in L^2(\mathbb{S}^2). \tag{7.37}
$$

7.3.6   Shorthand notation

In the following, we will use the shorthand notation introduced back in the first part of the book in (2.71), p. 88, repeated here

$$
\sum_{\ell,m} \triangleq \sum_{\ell=0}^\infty \sum_{m=-\ell}^\ell
$$

and its truncated form

$$
\sum_{\ell,m}^L \triangleq \sum_{\ell=0}^L \sum_{m=-\ell}^\ell \tag{7.38}
$$

for brevity. After a short reflection, one can see there are precisely $(L+1)^2$ terms in (7.38). The other esoteric detail here is that truncating in a different way is not forbidden, but not useful. There is a precedence of degree (ℓ) over order (m). The truncated form is a truncation in the degree ℓ.

7.3.7   Enumeration

As was also described in the first part of the book, we can recast double-indexed spherical harmonics Y_ℓ^m into a single-indexed version such as Y_n for notational brevity and also compatibility with the standard convention of orthonormal sequences. This single indexing will be also useful in representing all non-zero spherical harmonic coefficients of a bandlimited function on the 2-sphere in vector form, and representing operator matrices in the spherical harmonics basis.

The bijection given in (2.73), p. 88, and shown in Figure 2.11, p. 89, does the trick. That is, given degree ℓ and order m we generate the single index n:

$$
n = \ell(\ell+1) + m. \tag{7.39}
$$

Inversely, given a single index n we generate the degree ℓ and order m,

$$
\begin{aligned}
\ell &= \lfloor \sqrt{n} \rfloor, \\
m &= n - \lfloor \sqrt{n} \rfloor (\lfloor \sqrt{n} \rfloor + 1),
\end{aligned} \tag{7.40}
$$

where $\lfloor \cdot \rfloor$ is the floor function.

With re-enumerated spherical harmonics, $\{Y_n\}_{n=0}^\infty$, the corresponding spherical harmonic coefficient can be written $(f)_n = \langle f, Y_n \rangle$.

Example 7.10. Consider $n = 12$ in the single index notation. Using (7.40), it yields $\ell = \lfloor \sqrt{12} \rfloor = 3$ and $m = 12 - 3 \times 4 = 0$, that is,

$$Y_{12}(\theta, \phi) = Y_3^0(\theta, \phi) = \frac{1}{4}\sqrt{\frac{7}{\pi}}\left(5\cos^3\theta - 3\cos\theta\right),$$

where we used the table in Example 7.9.   □

7.3.8   Spherical harmonic Parseval relation

From (7.35) and the orthonormality of spherical harmonics we can verify Parseval relation on 2-sphere

$$\int_{\mathbb{S}^2} |f(\widehat{\boldsymbol{u}})|^2 ds(\widehat{\boldsymbol{u}}) = \sum_{\ell,m} |(f)_\ell^m|^2. \tag{7.41}$$

Remark 7.3. Parseval relation follows in a direct way from the properties of complete orthonormal sequences, Definition 2.22, p. 70,

$$\langle f, f \rangle = \sum_{n=0}^{\infty} \langle f, Y_n \rangle \langle Y_n, f \rangle = \sum_{n=0}^{\infty} |(f)_n|^2 \equiv \sum_{\ell,m} |(f)_\ell^m|^2.$$

That is, we identify the re-enumerated spherical harmonic, Y_n, with the generic φ_n used in Part I and Part II in the book.   □

Problem
_____

7.11 (Generalized Parseval relation). Using the observation in Remark 7.3, show that the generalized Parseval relation for the spherical harmonics is given by

$$\int_{\mathbb{S}^2} f(\widehat{\boldsymbol{u}})\overline{g(\widehat{\boldsymbol{u}})}\, ds(\widehat{\boldsymbol{u}}) = \sum_{\ell,m} (f)_\ell^m \overline{(g)_\ell^m}. \tag{7.42}$$

_____

7.3.9   Dirac delta function on 2-sphere

For two points

$$\widehat{\boldsymbol{u}} = \left(\sin\theta\cos\phi, \sin\theta\sin\phi, \cos\theta\right)',$$

$$\widehat{\boldsymbol{v}} = \left(\sin\vartheta\cos\varphi, \sin\vartheta\sin\varphi, \cos\vartheta\right)',$$

defined on the 2-sphere, the 2-sphere Dirac delta function is denoted by $\delta(\widehat{\boldsymbol{u}}, \widehat{\boldsymbol{v}})$. The spatial representation is

$$\delta(\widehat{\boldsymbol{u}}, \widehat{\boldsymbol{v}}) = (\sin\theta)^{-1}\delta(\theta - \vartheta)\delta(\phi - \varphi).$$

It has the expected "sifting property"

$$f(\widehat{\boldsymbol{v}}) = \int_{\mathbb{S}^2} \delta(\widehat{\boldsymbol{u}}, \widehat{\boldsymbol{v}}) f(\widehat{\boldsymbol{u}})\, ds(\widehat{\boldsymbol{u}}). \tag{7.43}$$

From the completeness relation (7.31) we can write

$$\delta(\widehat{\boldsymbol{u}}, \widehat{\boldsymbol{v}}) = (\sin\theta)^{-1}\delta(\theta - \vartheta)\delta(\phi - \varphi) = \sum_{\ell,m} Y_\ell^m(\theta, \phi)\overline{Y_\ell^m(\vartheta, \varphi)},$$

or, more compactly,

$$\delta(\widehat{\boldsymbol{u}}, \widehat{\boldsymbol{v}}) = \sum_{\ell,m} Y_\ell^m(\widehat{\boldsymbol{u}})\overline{Y_\ell^m(\widehat{\boldsymbol{v}})}, \tag{7.44}$$

which means that the Dirac delta function does not belong to $L^2(\mathbb{S}^2)$ as it has an infinite energy. To see this, let us first assume that $\delta(\widehat{\boldsymbol{u}}, \widehat{\boldsymbol{v}}) \in L^2(\mathbb{S}^2)$ and compute $\langle \delta, Y_p^q \rangle$ for a fixed $\widehat{\boldsymbol{v}}$. From (7.44), we infer

$$\langle \delta, Y_p^q \rangle = \sum_{\ell,m} \langle Y_\ell^m, Y_p^q \rangle \overline{Y_\ell^m(\widehat{\boldsymbol{v}})} = \overline{Y_p^q(\widehat{\boldsymbol{v}})}.$$

And the energy of the Dirac delta function using addition theorem (7.30) would become

$$\sum_{p,q} \left| \langle \delta, Y_p^q \rangle \right|^2 = \sum_{p,q} Y_p^q(\widehat{\boldsymbol{v}})\overline{Y_p^q(\widehat{\boldsymbol{v}})} \tag{7.45}$$

$$= \sum_{p=0}^{\infty} \frac{(2p+1)}{4\pi} P_p(\widehat{\boldsymbol{v}} \cdot \widehat{\boldsymbol{v}})$$

$$= \sum_{p=0}^{\infty} \frac{(2p+1)}{4\pi} P_p(1) = \sum_{p=0}^{\infty} \frac{(2p+1)}{4\pi},$$

which is unbounded.

Problem _____

7.12. By computing

$$\int_{\mathbb{S}^2}\int_{\mathbb{S}^2} f(\widehat{\boldsymbol{u}})\delta(\widehat{\boldsymbol{u}}, \widehat{\boldsymbol{v}})\overline{g(\widehat{\boldsymbol{v}})}\, ds(\widehat{\boldsymbol{u}})\, ds(\widehat{\boldsymbol{v}})$$

two ways, that is, using the sifting property (7.43) and the delta function expansion (7.44), show the generalized Parseval relation given in (7.42).

_____

7.3.10   Energy per degree

In some applications, it is useful to work with the *energy per degree*,[9] denoted by E_ℓ. For a signal f, energy per degree ℓ is the sum of square magnitude of its spherical harmonic coefficients $(f)_\ell^m$ from $m = -\ell$ to $m = \ell$

$$E_\ell \triangleq \sum_{m=-\ell}^{\ell} |(f)_\ell^m|^2. \tag{7.46}$$

And hence using Parseval relation (7.41), total signal energy is the sum of all its energies per degree

$$\int_{\mathbb{S}^2} |f(\widehat{\boldsymbol{u}})|^2 ds(\widehat{\boldsymbol{u}}) = \sum_{\ell=0}^{\infty} E_\ell.$$

7.3.11   Vector spectral representation

In the following discussions, we may deal with all spherical harmonic coefficients of the signal f in vector form. We call this the (column) *vector spectral representation* or spectral representation for short for f and generically denote it by

$$\mathbf{f} = (\ldots, (f)_\ell^m, \ldots)',$$

where a certain ordering is assumed for arranging the coefficients. Note that since the signal f has finite energy, $\mathbf{f}$ belongs to the space of square summable sequences ℓ^2. In this book, we use the single indexing convention (7.39) to write

$$\mathbf{f} = ((f)_0, (f)_1, (f)_2, \ldots)' = ((f)_0^0, (f)_1^{-1}, (f)_1^0, \ldots)'. \tag{7.47}$$

Note that other ordering conventions are used in some works such as (Simons et al., 2006) where coefficients with the same order m come first. That is

$$\mathbf{f} = ((f)_0, (f)_1, (f)_2, \ldots)' = ((f)_0^0, (f)_1^0, (f)_2^0, \ldots)'. \tag{7.48}$$

As we will see in later chapters, this representation may be useful for visualizing some special cases of signal concentration problem on the 2-sphere. However, it has the possible disadvantage that there can be a never-ending (or at least very long) sequence of zero-order coefficients in (7.48), whereas in (7.47) all $(2\ell + 1)$ terms corresponding to degree ℓ are organized in sequence.

7.4   More on spherical harmonics

7.4.1   Alternative definition of complex spherical harmonics

We have given one form of the spherical harmonics in the previous sections, but there is not a universal agreement in the literature as to their specific mathematical definition. This is flagged here because care needs to be taken when reading papers and books as to what specific formulation of the spherical harmonics is used. In this and the following section we review and define a variety of forms of

---

[9]   This means spherical harmonic degree not angular degree.

spherical harmonics given in the literature and relate them to our preferred formulation in this book. Also our intention is not to direct credit to the inventors of the different forms and so our attributions to follow do not correspond to the originators of the alternative definitions, but are rather arbitrary.

We now introduce the spherical harmonics defined in (Colton and Kress, 1998). These are denoted with a breve and their degree is denoted with the symbol n, as

$$\breve{Y}_n^m(\theta,\phi)$$

for the purposes of reducing confusion with the spherical harmonics, which we write as $Y_\ell^m(\theta,\phi)$. We refer to them as the "Colton-Kress spherical harmonic" of order m and degree n and they are related to the spherical harmonics given in Section 7.3.3 as follows:

$$\breve{Y}_n^m(\theta,\phi) = \begin{cases} Y_\ell^m(\theta,\phi)\Big|_{\ell=n} & \text{if } m \in \{0,-1,\ldots,-\ell\} \\ (-1)^m Y_\ell^m(\theta,\phi)\Big|_{\ell=n} & \text{if } m \in \{1,2,\ldots,\ell\}, \end{cases} \tag{7.49}$$

the proof of which can be inferred from the development below. So the Colton-Kress spherical harmonics differ in the multiplication by -1 from the spherical harmonics only when the order is positive and odd. Otherwise, they are identical (for non-positive or positive even orders).

Example 7.11. Below are the spherical harmonics of degree $\ell \in \{0,1,2\}$ to compare the above relation (7.49):

$$Y_0^0(\theta,\phi) = \frac{1}{2}\sqrt{\frac{1}{\pi}} = \breve{Y}_0^0(\theta,\phi),$$

$$Y_1^{\mp 1}(\theta,\phi) = \pm\frac{1}{2}\sqrt{\frac{3}{2\pi}}\sin\theta e^{\mp i\phi} = \pm\breve{Y}_1^{\mp 1}(\theta,\phi),$$

$$Y_1^0(\theta,\phi) = \frac{1}{2}\sqrt{\frac{3}{\pi}}\cos\theta = \breve{Y}_0^0(\theta,\phi),$$

$$Y_2^{\mp 2}(\theta,\phi) = \frac{1}{4}\sqrt{\frac{15}{2\pi}}\sin^2\theta e^{\mp i2\phi} = \breve{Y}_2^{\mp 2}(\theta,\phi),$$

$$Y_2^{\mp 1}(\theta,\phi) = \pm\frac{1}{2}\sqrt{\frac{15}{2\pi}}\sin\theta\cos\theta e^{\mp i\phi} = \pm\breve{Y}_2^{\mp 1}(\theta,\phi),$$

$$Y_2^0(\theta,\phi) = \frac{1}{4}\sqrt{\frac{5}{\pi}}\left(3\cos^2\theta - 1\right) = \breve{Y}_2^0(\theta,\phi).$$

The Colton-Kress spherical harmonic is the negative of the spherical harmonic for the instance of an odd positive order (superscript) — here the case when $\ell = 1$ and $m = 1$, and $\ell = 2$ and $m = 1$.   □

We can immediately infer from (7.49) that the Colton-Kress spherical harmonics have unit norm and orthogonal because the spherical harmonics are orthonormal, (7.25). That is, the Colton-Kress spherical harmonics satisfy the orthonormality condition

$$\langle Y_\ell^m, Y_p^q \rangle = \delta_{\ell,p}^{m,q} \implies \langle \breve{Y}_n^m, \breve{Y}_p^q \rangle = \delta_{n,p}^{m,q}.$$

Furthermore, in comparison with (7.26) and corroborated by the examples in Example 7.11, they have the simpler conjugation property

$$\overline{\breve{Y}_n^m(\theta, \phi)} = \breve{Y}_n^{-m}(\theta, \phi). \tag{7.50}$$

And this well emulates the analogous property of the complex exponentials

$$\overline{e^{im\phi}} = e^{-im\phi}. \tag{7.51}$$

Of course, $e^{im\phi}$ appears directly in the various definitions of the spherical harmonics as the longitude ϕ portion; see (7.19). The co-latitude θ portion is characterized by the associated Legendre functions, which are purely real. In fact, this is the source of the difference in definitions. Colton and Kress (Colton and Kress, 1998) have used the following definitions. For the Legendre polynomials they use the standard Rodrigues formula, (7.22), but this time using the symbol n for the degree

$$P_n(x) \triangleq \frac{1}{2^n n!} \frac{d^n}{dx^n} (x^2 - 1)^n, \quad n \in \{0, 1, 2, \ldots\}. \tag{7.52}$$

Their version of the associated Legendre functions (Colton and Kress, 1998) is

$$\begin{aligned}
\breve{P}_n^m(x) &\triangleq (1 - x^2)^{m/2} \frac{d^m}{dx^m} P_n(x) \\
&= \frac{1}{2^n n!} (1 - x^2)^{m/2} \frac{d^{n+m}}{dx^{n+m}} (x^2 - 1)^n, \quad m \in \{0, 1, \ldots, n\},
\end{aligned} \tag{7.53}$$

for $-1 \le x \le 1$, and the same definition is used in (Riley et al., 2006). For negative-order m Colton and Kress (Colton and Kress, 1998) buck convention and define[10]

$$\breve{P}_n^{-m}(x) \triangleq \frac{(n-m)!}{(n+m)!} \breve{P}_n^m(x), \quad m \in \{0, 1, \ldots, n\}, \tag{7.54}$$

for $-1 \le x \le 1$. Equations (7.53)–(7.54) differ subtly from (7.23)–(7.24). In contrast to (7.23)–(7.24), the Condon-Shortley phase factor $(-1)^m$ is not present in either expression (7.53)–(7.54). One possible motivation for defining things this way, apart from the desire to be annoying, is to ensure that the negative-order associated Legendre functions are only a positive real scale factor different from the positive-order associated Legendre functions. This positive real scaling factor defined for normalizing them on the interval $[-1, +1]$ (see Section 2.5.4, p. 61, in Part I) is irrelevant once incorporated into the spherical harmonics which override the normalization.

In a technical sense (7.54) is mathematically inconsistent with (7.53) were we to want to employ (7.53) with $m < 0$ such that $n + m \ge 0$. That is, taking (7.53) to be valid for $n + m \ge 0$, and indeed all $m \in \{-n, -n+1, \ldots, n\}$ and not just $m \in \{0, 1, \ldots, n\}$, then the Condon-Shortley phase factor $(-1)^m$ would

---

[10] Actually, they do not define this equation and they do not need negative-order m associated Legendre functions, and they even call them associated Legendre functions, but this equation can be inferred and is useful for comparison.

appear as multiplicative factor on the right side of (7.54) as it does it in (7.24). This result is derived in (Riley et al., 2006, pp. 588–589) and is equivalent to Problem 7.8, p. 194.[11]

Armed with this formulation of the associated Legendre functions, the Colton-Kress spherical harmonics for degree n satisfying $n \geq 0$ and order m satisfying $-n \leq m \leq n$ are defined as

$$\breve{Y}_n^m(\theta, \phi) = \sqrt{\frac{2n+1}{4\pi} \frac{(n-|m|)!}{(n+|m|)!}} \breve{P}_n^{|m|}(\cos\theta) e^{im\phi}, \qquad (7.55)$$

which can be compared with the arguably less tidy (7.19) and (7.27). When expressed this way then the conjugate property in (7.50) is straightforward to see because it reduces to (7.51). Also noteworthy is that this definition does not require (7.54), because only positive-order associated Legendre functions are needed.

Example 7.12. For $m > 0$, and using (7.54),

$$\sqrt{\frac{(n-m)!}{(n+m)!}} \breve{P}_n^m(\cos\theta) = \sqrt{\frac{(n-m)!}{(n+m)!} \frac{(n+m)!}{(n-m)!}} \breve{P}_n^{-m}(\cos\theta)$$

$$= \sqrt{\frac{(n+m)!}{(n-m)!}} \breve{P}_n^{-m}(\cos\theta)$$

$$= \sqrt{\frac{(n-(-m))!}{(n+(-m))!}} \breve{P}_n^{-m}(\cos\theta),$$

and so this expression is invariant to the sign of m. This is why we can use $|m|$ in all but the $e^{im\phi}$ part of (7.55). $\square$

Our final comments on the spherical harmonics given in (Colton and Kress, 1998): our notation, such as the inner product in (7.17), is close to that given there. And further, because of the equivalence between the spherical harmonics the proof of completeness given in Colton and Kress (Colton and Kress, 1998) is particularly useful as a reference.

7.4.2   Complex spherical harmonics: a synopsis

Complex spherical harmonics are composed of complex exponentials, associated Legendre functions and a normalization. The normalized can always be derived as a positive factor such that the spherical harmonics have unit norm induced from the inner product (7.17). The complex exponential is never in dispute (although people fond of real spherical harmonics, Section 7.4.3, may think otherwise). So the variation and inconsistency comes down to the definition of the associated Legendre functions. And then there can be variation in how the negative-order associated Legendre functions are defined (or derived). It is conventionally the case that a generalization of the Rodrigues formula means the

---

[11] Consequently, the formulation in (Riley et al., 2006, pp. 588–589) represents a third alternative definition of the associated Legendre functions with (7.53) being valid for all $|m| \leq \ell$ and a negative-order identity looking like (7.24), but with $\ell = n$.

negative-order associated Legendre functions can be deduced from the positive-order associated Legendre functions and this identity holds independently of the specific associated Legendre functions adopted.

Stripped of normalization and possible multiplication by -1, the spherical harmonics are simply functions in co-latitude θ and longitude ϕ proportional to

$$P_\ell^{|m|}(\cos\theta)e^{im\phi},$$

with the degree $\ell \in \{0, 1, 2, \ldots\}$ and order $m \in \{-\ell, -\ell+1, \ldots, \ell\}$. This is consistent with (7.19) and (7.55). We recall that $P_\ell^{|m|}(\cos\theta)$ is real-valued and it is only the complex exponential, $e^{im\phi}$, that can contribute to the imaginary part. One is immediately drawn to view this situation as being very reminiscent of the Fourier series representation of periodic time functions (Oppenheim et al., 1996). Once comfortable with the complex case, it is compelling to regard the complex exponential form of classical Fourier series as more elegant and self-contained than the trigonometric form, in which Fourier gave his initial result. In the spirit of nostalgia, we can now explore a different definition of spherical harmonics which is popular in some fields.

7.4.3   Real spherical harmonics

Another important variety of spherical harmonics consists of the set of real spherical harmonics. The motivation this time is implicit. Suppose the functions we wish to represent are real-valued rather than complex-valued. Then one really only needs to have orthonormal sequences of real functions. However, the complex spherical harmonics are more than powerful enough to form a basis to represent real functions as a special case, because we can combine spherical harmonics with opposite orders (complex linear combinations), such as m with $-m$, to yield a purely real result. We shall defer further discussion until after we have defined our real spherical harmonics.

If we are only interested in real functions on $\mathbb{S}^2$ and their real linear combinations, then the Hilbert space that we seek to find is one in which the scalar field is real, that is, it is a real separable Hilbert space. In comparison with the complex inner product, (7.17), we define the real inner product between two real-valued functions on the sphere, $f(\widehat{\boldsymbol{u}})$ and $g(\widehat{\boldsymbol{u}})$,

$$\langle f, g \rangle \triangleq \int_{\mathbb{S}^2} f(\widehat{\boldsymbol{u}})g(\widehat{\boldsymbol{u}})\,ds(\widehat{\boldsymbol{u}}) \tag{7.56}$$
$$= \int_0^{2\pi}\int_0^\pi f(\theta,\phi)g(\theta,\phi)\,\sin\theta\,d\theta\,d\phi,$$

which induces the norm

$$\|f\| \triangleq \langle f, f \rangle^{1/2} = \Big(\int_0^{2\pi}\int_0^\pi \big(f(\theta,\phi)\big)^2 \sin\theta\,d\theta\,d\phi\Big)^{1/2}.$$

Now we motivate our real spherical harmonics by looking at their lower-dimensional analogy. In classical Fourier series, the role of the normalized complex exponential (see Section 2.5.3, p. 59)

$$\Big\{\frac{1}{\sqrt{2\pi}}e^{im\phi}\Big\}_{m\in\mathbb{Z}},$$

(in the complex case) is played by the normalized DC ($m = 0$), normalized sine terms and normalized cosine terms (in the real case), that is,

$$\frac{1}{\sqrt{2\pi}}, \ \left\{\frac{1}{\sqrt{\pi}}\sin m\phi\right\}_{m=1}^{\infty} \text{ and } \left\{\frac{1}{\sqrt{\pi}}\cos m\phi\right\}_{m=1}^{\infty}.$$

Taking due note of the relative weighting to keep things normalized properly we can infer that the $e^{im\phi}$ part of the complex spherical harmonics, (7.58), can be mapped as follows to give a real form:

$$e^{im\phi} \longmapsto \begin{cases} \sqrt{2}\cos m\phi = \dfrac{1}{\sqrt{2}}\left(e^{im\phi} + e^{-im\phi}\right) & \text{if } m \in \{-1,-2,-3,\ldots\} \\ 1 & \text{if } m = 0, \\ \sqrt{2}\sin m\phi = \dfrac{1}{i\sqrt{2}}\left(e^{im\phi} - e^{-im\phi}\right) & \text{if } m \in \{1,2,3,\ldots\}, \end{cases} \tag{7.57}$$

where, of course, the DC ($m = 0$) term is unchanged. Note that we have mapped the cosine portion to the negative m (and because they are even functions the sign of m does not matter), sines to positive m and DC to $m = 0$, and in this way can use a single index $m \in \mathbb{Z}$ in preparation for the real spherical harmonic generalization.

To distinguish them in notation from the complex spherical harmonics, the real spherical harmonics are denoted by $Y_{\ell,m}(\theta,\phi)$. Following the mapping convention in (7.57), we can infer the real spherical harmonics

$$Y_\ell^m(\theta,\phi) \equiv N_\ell^m P_\ell^m(\cos\theta)e^{im\phi} \longmapsto Y_{\ell,m}(\theta,\phi).$$

That is,

$$Y_{\ell,m}(\theta,\phi) \triangleq \begin{cases} \sqrt{2}N_\ell^m P_\ell^m(\cos\theta)\cos m\phi & \text{if } m \in \{-1,-2,\ldots,-\ell\}, \\ N_\ell^0 P_\ell(\cos\theta) & \text{if } m = 0, \\ \sqrt{2}N_\ell^m P_\ell^m(\cos\theta)\sin m\phi & \text{if } m \in \{1,2,\ldots,\ell\}, \end{cases} \tag{7.58}$$

where the normalization, N_ℓ^m, is the normalization for the complex case given in (7.20). Defined this way, the real spherical harmonics are orthonormal with respect to the inner product (7.56).

The real spherical harmonics can equivalently be deduced from the complex spherical harmonics by extracting their real and imaginary parts and renormalizing, as follows:

$$Y_{\ell,m}(\theta,\phi) \triangleq \begin{cases} \sqrt{2}\,\mathfrak{Re}\{Y_\ell^m(\theta,\phi)\} & \text{if } m \in \{-1,-2,\ldots,-\ell\}, \\ Y_\ell^0(\theta,\phi) & \text{if } m = 0, \\ \sqrt{2}\,\mathfrak{Im}\{Y_\ell^m(\theta,\phi)\} & \text{if } m \in \{1,2,\ldots,\ell\}, \end{cases} \tag{7.59}$$

where $\Im m(\cdot)$ returns the imaginary part of its argument. This is a useful way to obtain their explicit expressions given the more widely embraced complex spherical harmonics.

Given the complex portion only involves the longitude part, $e^{im\phi}$, we introduce a slightly more compact notation as an alternative to (7.59)

$$
\begin{aligned}
Y_{\ell,0}(\theta,\phi) &= N_\ell^0 P_\ell(\cos\theta), \quad m = 0, \\
Y_{\ell,\mp m}(\theta,\phi) &= \sqrt{2} N_\ell^m P_\ell^m(\cos\theta)_{\sin}^{\cos} m\phi, \quad m \in \{1, 2, \ldots, \ell\},
\end{aligned}
\tag{7.60}
$$

where $-m$ is associated with the upper part involving $\cos m\phi$, and $+m$ is associated with the lower part involving $\sin m\phi$, and $\ell \in \{0, 1, 2, \ldots\}$. The example below draws on similar notation.

Example 7.13. Here is a collection of real spherical harmonics for degrees $\ell \in \{0, 1, 2, 3, 4\}$, best determined from (7.59) or (7.60) and a table of complex spherical harmonics:

$$
Y_{0,0}(\theta,\phi) = \frac{1}{2}\sqrt{\frac{1}{\pi}}, \qquad\qquad Y_{2,\mp 2}(\theta,\phi) = \frac{1}{4}\sqrt{\frac{15}{\pi}}\sin^2\theta_{\sin}^{\cos}2\phi,
$$

$$
Y_{1,\mp 1}(\theta,\phi) = \pm\frac{1}{2}\sqrt{\frac{3}{\pi}}\sin\theta_{\sin}^{\cos}\phi, \quad Y_{2,\mp 1}(\theta,\phi) = \pm\frac{1}{2}\sqrt{\frac{15}{\pi}}\sin\theta\cos\theta_{\sin}^{\cos}\phi,
$$

$$
Y_{1,0}(\theta,\phi) = \frac{1}{2}\sqrt{\frac{3}{\pi}}\cos\theta, \qquad\quad Y_{2,0}(\theta,\phi) = \frac{1}{4}\sqrt{\frac{5}{\pi}}\left(3\cos^2\theta - 1\right),
$$

$$
Y_{3,\mp 3}(\theta,\phi) = \pm\frac{1}{4}\sqrt{\frac{35}{2\pi}}\sin^3\theta_{\sin}^{\cos}3\phi,
$$

$$
Y_{3,\mp 2}(\theta,\phi) = \frac{1}{4}\sqrt{\frac{105}{\pi}}\sin^2\theta\cos\theta_{\sin}^{\cos}2\phi,
$$

$$
Y_{3,\mp 1}(\theta,\phi) = \pm\frac{1}{4}\sqrt{\frac{21}{2\pi}}\sin\theta\left(5\cos^2\theta - 1\right)_{\sin}^{\cos}\phi,
$$

$$
Y_{3,0}(\theta,\phi) = \frac{1}{4}\sqrt{\frac{7}{\pi}}\left(5\cos^3\theta - 3\cos\theta\right),
$$

$$
Y_{4,\mp 4}(\theta,\phi) = \frac{3}{16}\sqrt{\frac{35}{\pi}}\sin^4\theta_{\sin}^{\cos}4\phi,
$$

$$
Y_{4,\mp 3}(\theta,\phi) = \pm\frac{3}{4}\sqrt{\frac{35}{2\pi}}\sin^3\theta\cos\theta_{\sin}^{\cos}3\phi,
$$

$$
Y_{4,\mp 2}(\theta,\phi) = \frac{3}{8}\sqrt{\frac{5}{\pi}}\sin^2\theta\left(7\cos^2\theta - 1\right)_{\sin}^{\cos}2\phi,
$$

$$
Y_{4,\mp 1}(\theta,\phi) = \pm\frac{3}{4}\sqrt{\frac{5}{2\pi}}\sin\theta\left(7\cos^3\theta - 3\cos\theta\right)_{\sin}^{\cos}\phi,
$$

$$
Y_{4,0}(\theta,\phi) = \frac{3}{16}\sqrt{\frac{1}{\pi}}\left(35\cos^4\theta - 30\cos^2\theta + 3\right).
$$

$\square$

7.4.4   Unnormalized real spherical harmonics

In some fields of science, the use of unnormalized real spherical harmonics such that

$$\langle Y_{\ell,m}, Y_{p,q} \rangle = 4\pi \delta_{\ell,p} \delta_{m,q} \tag{7.61}$$

is not uncommon (Wieczorek and Simons, 2005), where 4π is indeed the area of 2-sphere. Then the inner product between two (real-valued) functions f and g is explicitly normalized with the 2-sphere area and given by

$$\langle f, g \rangle \triangleq \frac{1}{4\pi} \int_{\mathbb{S}^2} f(\widehat{u}) g(\widehat{u}) \, ds(\widehat{u}), \tag{7.62}$$

which induces the norm

$$\|f\| \triangleq \left(\frac{1}{4\pi} \int_{\mathbb{S}^2} \left(f(\widehat{u}) \right)^2 ds(\widehat{u}) \right)^{1/2}. \tag{7.63}$$

Hence, $\|f\|$ is understood as the energy of the signal f per unit area or simply the "signal power."

Were we to use the unnormalized spherical harmonics, the spherical harmonic coefficients would need to be explicitly normalized as

$$(f)_{\ell,m} \triangleq \langle f, Y_{\ell,m} \rangle = \frac{1}{4\pi} \int_{\mathbb{S}^2} f(\widehat{u}) Y_{\ell,m}(\widehat{u}) \, ds(\widehat{u}). \tag{7.64}$$

And the Parseval relation would take the form

$$\frac{1}{4\pi} \int_{\mathbb{S}^2} \left(f(\widehat{u}) \right)^2 ds(\widehat{u}) = \sum_{\ell,m} \left((f)_{\ell,m} \right)^2. \tag{7.65}$$

Therefore, $(f)_{\ell,m}$ would in fact describe the power of the signal for degree ℓ and order m.

> **Remark 7.4.** In some works such as (Wieczorek and Simons, 2005), energy per degree E_ℓ is referred to as *power spectrum*, which is consistent with their use of unnormalized spherical harmonics and defining spherical harmonic coefficients for describing the signal power in (7.65). In some other works (Simons et al., 2006), the authors prefer to work with *power spectral density* (PSD)
>
> $$\widehat{E}_\ell = \frac{E_\ell}{(2\ell + 1)},$$
>
> which normalizes E_ℓ by the $2\ell + 1$ terms that make it. Using the notion of power spectral density and referring to (7.45), Dirac delta function on the sphere will have a *white* PSD, $\widehat{E}_\ell = 1/(4\pi)$ for all degrees $\ell \in \{0, 1, 2, \ldots\}$. □

We briefly discussed this alternative for the sake of completeness. Our preference, however, in this book is to use normalized versions of spherical harmonics and directly work with signal energy rather than with energy per area or power.

7.4.5   Complex Hilbert space with real spherical harmonics

We motivated real spherical harmonics for the case of real-valued functions on the 2-sphere and implicitly sought a real Hilbert space. But the real spherical harmonics can be used with complex scalars to represent complex-valued functions on the 2-sphere leading to a complex Hilbert space.[12] This apparently really complex situation is not complex really. All we need to show is the complex spherical harmonics can be expressed in terms of the real spherical harmonics using complex scalars. This is somewhat the converse of what is captured in (7.59).

When $m = 0$ we have the identity

$$Y_\ell^0(\theta, \phi) = Y_{\ell,m}(\theta, \phi)\Big|_{m=0}.$$

Also for all m, that is, $-\ell \le m \le \ell$,

$$
\begin{aligned}
Y_\ell^m(\theta, \phi) &= N_\ell^m P_\ell^m(\cos\theta) e^{im\phi} \\
&= N_\ell^m P_\ell^m(\cos\theta)\big(\cos m\phi + i\sin m\phi\big) \\
&= \frac{1}{\sqrt{2}}\Big(\sqrt{2}N_\ell^m P_\ell^m(\cos\theta)\cos m\phi\Big) + \frac{i}{\sqrt{2}}\Big(\sqrt{2}N_\ell^m P_\ell^m(\cos\theta)\sin m\phi\Big),
\end{aligned}
$$

and we consider two cases of this expression.

For $m \in \{1, 2, \dots, \ell\}$ we can obtain

$$
\begin{aligned}
Y_\ell^m(\theta, \phi) &= \frac{1}{\sqrt{2}}\Big(\sqrt{2}(-1)^m N_\ell^{-m} P_\ell^{-m}(\cos\theta)\cos m\phi\Big) + \frac{i}{\sqrt{2}}Y_{\ell,m}(\theta, \phi) \\
&= \frac{(-1)^m}{\sqrt{2}}Y_{\ell,-m}(\theta, \phi) + \frac{i}{\sqrt{2}}Y_{\ell,m}(\theta, \phi),
\end{aligned}
$$

where we have used $(-1)^m = (-1)^{-m}$ and $\cos(-m\phi) = \cos(m\phi)$. That is,

$$Y_\ell^m(\theta, \phi) = \frac{(-1)^m}{\sqrt{2}}Y_{\ell,-m}(\theta, \phi) + i\frac{1}{\sqrt{2}}Y_{\ell,m}(\theta, \phi), \quad m \in \{1, 2, \dots, \ell\}. \qquad (7.66)$$

For $m \in \{-1, -2, \dots, -\ell\}$ we can obtain

$$
\begin{aligned}
Y_\ell^m(\theta, \phi) &= \frac{1}{\sqrt{2}}Y_{\ell,m}(\theta, \phi) - \frac{i}{\sqrt{2}}\Big(\sqrt{2}(-1)^m N_\ell^{-m} P_\ell^{-m}(\cos\theta)\sin(-m\phi)\Big) \\
&= \frac{1}{\sqrt{2}}Y_{\ell,m}(\theta, \phi) - i\frac{(-1)^m}{\sqrt{2}}Y_{\ell,-m}(\theta, \phi).
\end{aligned}
$$

That is,

$$Y_\ell^m(\theta, \phi) = \frac{1}{\sqrt{2}}Y_{\ell,m}(\theta, \phi) - i\frac{(-1)^m}{\sqrt{2}}Y_{\ell,-m}(\theta, \phi), \quad m \in \{-1, -2, \dots, -\ell\}.$$

$$(7.67)$$

---

[12] And indeed the complex spherical harmonics can be used to represent real functions.

This implies the real spherical harmonics can be used to completely represent complex-valued functions on the 2-sphere.

Remark 7.5. We could have used the conjugation property (7.26), p. 192, in (7.66) to derive (7.67). □

7.4.6   Real spherical harmonics: a synopsis

Most simply seen from (7.60), stripped of normalization and possible multiplication by -1, the real spherical harmonics are simply functions in co-latitude θ and longitude ϕ proportional to one of

$$P_\ell(\cos\theta) \quad \text{or} \quad P_\ell^m(\cos\theta){\textstyle\genfrac{}{}{0pt}{}{\cos}{\sin}} m\phi, \quad m \in \{1, 2, \ldots, \ell\},$$

with the degree ℓ satisfying $\ell \in \{0, 1, 2, \ldots\}$.

Problems

7.13. In the norm

$$\|f\| = \left(\int_0^{2\pi} \big(f(\phi)\big)^2 d\phi \right)^{1/2},$$

show that the functions on the right-hand side of (7.57) satisfy

$$\|\sqrt{2}\cos m\phi\| = \|1\| = \|\sqrt{2}\sin m\phi\| = \sqrt{2\pi}, \quad m > 0.$$

7.14. Show that the real spherical harmonics in (7.58) can be expressed in terms of the complex spherical harmonics in the following way

$$Y_{\ell,m}(\theta,\phi) = \begin{cases} \dfrac{1}{\sqrt{2}}\big(Y_\ell^m(\theta,\phi) + (-1)^m Y_\ell^{-m}(\theta,\phi)\big) & \text{if } m \in \{-1,-2,\ldots,-\ell\}, \\[2mm] Y_\ell^0(\theta,\phi) & \text{if } m = 0, \\[2mm] \dfrac{1}{i\sqrt{2}}\big(Y_\ell^m(\theta,\phi) - (-1)^m Y_\ell^{-m}(\theta,\phi)\big) & \text{if } m \in \{1,2,\ldots,\ell\}. \end{cases}$$

$$(7.68)$$

7.15. In (Simons et al., 2006) they gave the following set of definitions for what they called the "real surface spherical harmonics,"

$$Y_{\ell,m}(\theta,\phi) = \begin{cases} \sqrt{2}X_{\ell,m}(\theta)\cos m\phi & \text{if } m \in \{-1,-2,\ldots,-\ell\}, \\[2mm] X_{\ell,0}(\theta) & \text{if } m = 0, \\[2mm] \sqrt{2}X_{\ell,m}(\theta)\sin m\phi & \text{if } m \in \{1,2,\ldots,\ell\}, \end{cases} \tag{7.69a}$$

$$X_{\ell,m}(\theta) = (-1)^m \left(\frac{(2\ell+1)}{4\pi} \right)^{1/2} \left[\frac{(\ell-m)!}{(\ell+m)!} \right]^{1/2} P_{\ell,m}(\cos\theta), \tag{7.69b}$$

$$P_{\ell,m}(\mu) = \frac{1}{2^\ell \ell!}(1-\mu^2)^{m/2}\frac{d^{\ell+m}}{d\mu^{\ell+m}}(\mu^2-1)^\ell. \tag{7.69c}$$

How does this compare with the real spherical harmonics in (7.58)?

7.16. Prove the assertion in Remark 7.5.

7.4.7   Visualization of spherical harmonics

Algebraic expressions fall short of giving the sense of what the spherical harmonics really look like. We are interested in both the real and imaginary parts of the spherical harmonics and apart from annoying constants, these coincide precisely with the real spherical harmonics. That is, the real/imaginary parts of the spherical harmonics are proportional to the negative/positive order, $\cos/\sin$, real spherical harmonics.

The spherical harmonics are functions of the two variables, θ the co-latitude and ϕ the longitude. For purposes of presentation, the function value can be mapped to a radial term and the resulting function displayed as a surface in spherical polar coordinates. For example, to visualize the real part of (complex) spherical harmonic $Y_\ell^m(\theta, \phi)$, which is proportional to the real spherical harmonic $Y_{\ell,-m}(\theta, \phi)$ for $m > 0$, we can use the non-negative spherical polar function:

$$r(\theta, \phi) = \left| \gamma_0 + \gamma_1 \widehat{Y}_{\ell,-m}(\theta, \phi) \right| \in [0, \gamma_0 + \gamma_1], \tag{7.70}$$

where $\gamma_0 \geq 0$ is a non-negative offset parameter, $\gamma_1 \geq 0$ is a non-negative scaling parameter, and the spherical harmonic is normalized such that its maximum absolute real value is 1, that is, $\widehat{Y}_{\ell,-m}(\theta, \phi) \in [-1, +1]$. Similarly the imaginary part of (complex) spherical harmonic $Y_\ell^m(\theta, \phi)$ is proportional to the real spherical harmonic $Y_{\ell,m}(\theta, \phi)$ for $m > 0$ and can be visualized with

$$r(\theta, \phi) = \left| \gamma_0 + \gamma_1 \widehat{Y}_{\ell,m}(\theta, \phi) \right| \in [0, \gamma_0 + \gamma_1]. \tag{7.71}$$

In addition, it is conventional/optional to color-map the surface as a function and for illustration we can set

$$c(\theta, \phi) = \widehat{Y}_{\ell,-m}(\theta, \phi) \in [-1, +1],$$

which takes values in the relevant color-space. When a color-map is used, $c(\theta, \phi)$ holds the information about the signal and then γ_1 in (7.70)–(7.71) can be chosen as zero (meaning $r(\theta, \phi) = |\gamma_0|$).

> **Example 7.14.** As an example, we plot the normalized real and imaginary parts of $\widehat{Y}_4^3(\theta, \phi)$ and $\widehat{Y}_{11}^7(\theta, \phi)$ according to three sets of parameter values as shown in Table 7.1 according to (7.70) and (7.71).
>
> The results for $\widehat{Y}_4^3(\theta, \phi)$ are shown in Figure 7.6. Figure 7.6a and Figure 7.6b correspond to using parameters from set 1: $\gamma_0 = 0$ and $\gamma_1 = 1.25$; Figure 7.6c and Figure 7.6d correspond to using parameters from set 2: $\gamma_0 = 1$ and $\gamma_1 = 0.25$; and Figure 7.6e and Figure 7.6f correspond to using parameters from set 3: $\gamma_0 = 1$ and $\gamma_1 = 0$. The results for $\widehat{Y}_{11}^7(\theta, \phi)$ are shown in Figure 7.7 using the same ordering with the three sets.   □

All three styles of plots obtained by varying γ_0 and γ_1 have their own advantages and disadvantages. In order:

1: The first combination is good for low degrees and relatively useless for higher degrees. This is the balloon puppeteer's favorite — a tommy gun is always a challenge.

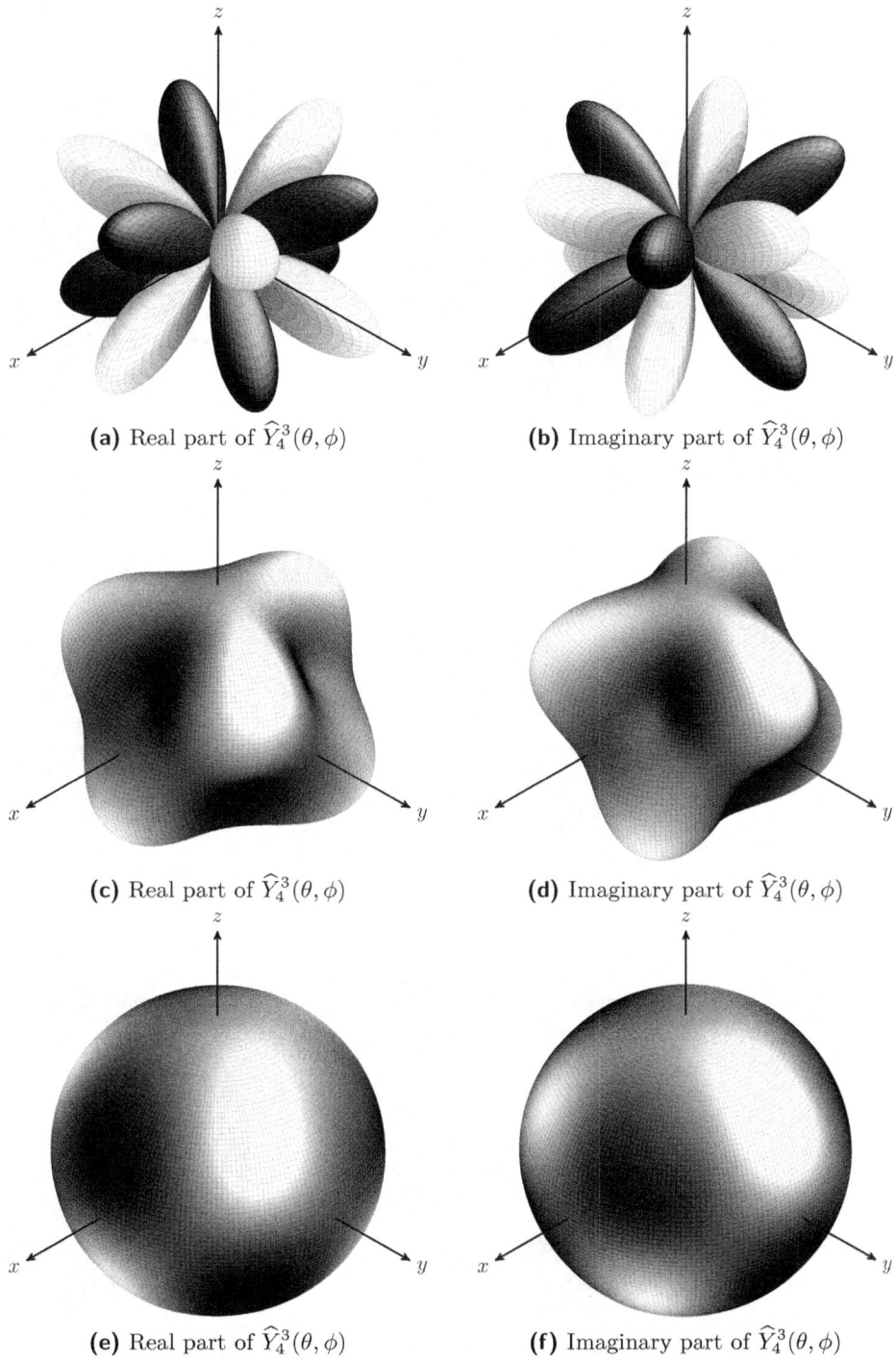

Figure 7.6 Real part (left) and imaginary part (right) of $\widehat{Y}_4^3(\theta, \phi)$, following Table 7.1. Plots use a gray-scale color-map with black (minimum) to white (maximum).

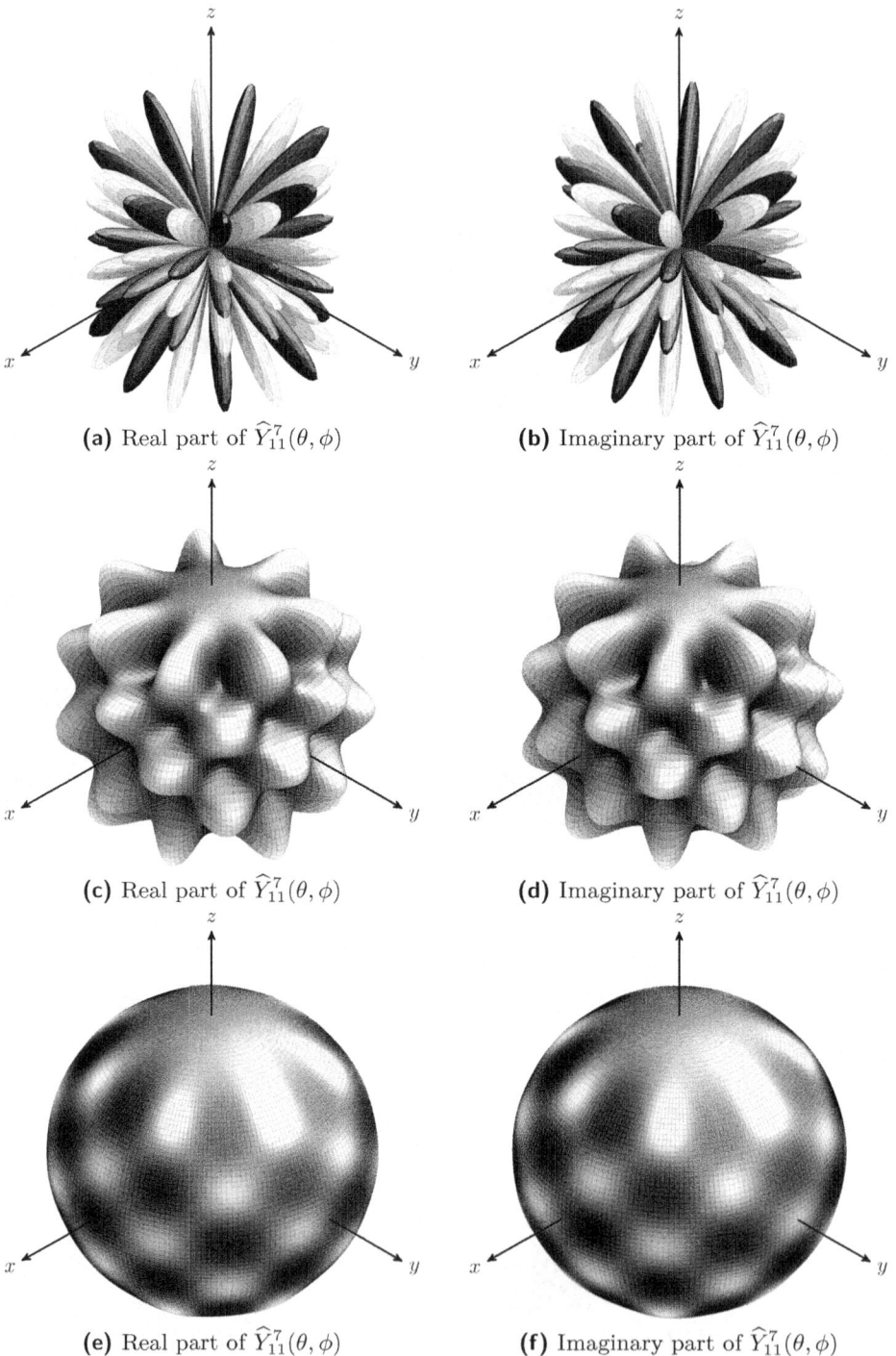

(a) Real part of $\widehat{Y}_{11}^{7}(\theta, \phi)$

(b) Imaginary part of $\widehat{Y}_{11}^{7}(\theta, \phi)$

(c) Real part of $\widehat{Y}_{11}^{7}(\theta, \phi)$

(d) Imaginary part of $\widehat{Y}_{11}^{7}(\theta, \phi)$

(e) Real part of $\widehat{Y}_{11}^{7}(\theta, \phi)$

(f) Imaginary part of $\widehat{Y}_{11}^{7}(\theta, \phi)$

Figure 7.7 Real part (left) and imaginary part (right) of $\widehat{Y}_{11}^{7}(\theta, \phi)$, following Table 7.1. Plots use a gray-scale color-map with black (minimum) to white (maximum).

Table 7.1 Plot parameters for representing the spherical harmonics and, more generally, any function on the 2-sphere.

Parameter set				
	γ_0	γ_1	$r(\theta,\phi)$ **range**	**Value encoding**
set 1	0	1.25	$[0, 1.25]$	Large variations about origin + color-map
set 2	1	0.25	$[0.75, 1.25]$	Small variations about 2-sphere + color-map
set 3	1	0	1	Only in color-map, plotted on 2-sphere

2: The second combination corresponds to what might be best used to model the topography on a planet for example (with more subdued γ_1), as in the famous Mars Orbiter Laser Altimeter (MOLA) image data (Zuber et al., 1992). As displayed here, it is a true hybrid between information carried by the bumps and by the color-map.

3: The third combination is intellectually the most honest with the color-map carrying all the information. In this case you do not need to think of the sphere as being embedded in a 3D Euclidean space. The color-map could be replaced by contour lines and the like.

7.4.8   Visual catalog of spherical harmonics

Having tested our wares on visualization we now provide a visual catalog of the spherical harmonics. There are infinitely many spherical harmonics and we can glimpse just a few. In Figure 7.8, the real spherical harmonics for degrees $\ell \in \{0, 1, 2, 3, 4\}$ and orders $m \in \{-\ell, -\ell + 1, \ldots, \ell\}$ are depicted. As these are the lowest-order spherical harmonics the balloon puppeteer representation is used. The figure shows the degrees horizontally and the orders vertically. Various symmetries and rotations can be discerned, as can the relationship between the number of blobs and degrees and orders. But proving such properties isn't worth a pair of fetid dingo's kidneys.

7.5   Useful subspaces of $L^2(\mathbb{S}^2)$

7.5.1   Subspace of bandlimited signals

An important subspace of $L^2(\mathbb{S}^2)$ is the subspace spanned by orthonormal spherical harmonics Y_ℓ^m for degrees smaller than or equal to a finite value L. This subspace is $(L+1)^2$-dimensional and is denoted by $\mathcal{H}_L(\mathbb{S}^2)$ defined as

$$\mathcal{H}_L(\mathbb{S}^2) = \Big\{ f \in L^2(\mathbb{S}^2) \colon f(\theta,\phi) = \sum_{\ell,m}^{L} (f)_\ell^m Y_\ell^m(\theta,\phi) \Big\}. \qquad (7.72)$$

The signals belonging to $\mathcal{H}_L(\mathbb{S}^2)$ are *bandlimited* with bandwidth L because their maximum spectral degree is L and their spectral representation is a finite vector

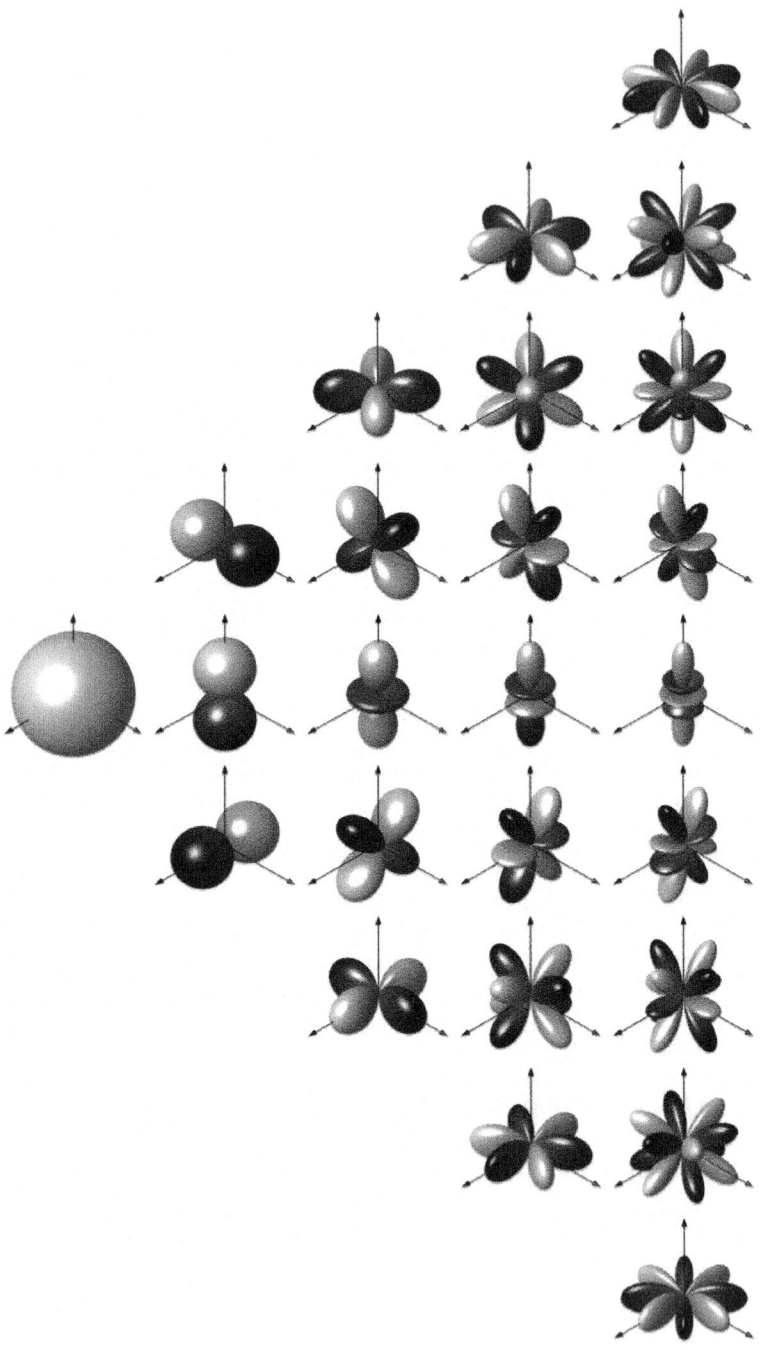

Figure 7.8 The normalized real spherical harmonics, $Y_{\ell,m}$, for degrees $0 \leq \ell \leq 4$ (horizontal) and orders $-\ell \leq m \leq \ell$ (vertical).

with $(L+1)^2$ elements

$$\mathbf{f} = \left((f)_0, (f)_1, (f)_2, \ldots, (f)_{L^2+2L}\right)' = \left((f)_0^0, (f)_1^{-1}, (f)_1^0, \ldots, (f)_L^L\right)'.$$

Using the vector spectral representation of a signal in $\mathcal{H}_L(\mathbb{S}^2)$, Parseval relation takes a simple form as

$$\int_{\mathbb{S}^2} \left|f(\widehat{\boldsymbol{u}})\right|^2 ds(\widehat{\boldsymbol{u}}) = \mathbf{f}'\overline{\mathbf{f}} = \mathbf{f}^H\mathbf{f}, \quad \forall f \in \mathcal{H}_L(\mathbb{S}^2). \tag{7.73}$$

7.5.2   Subspace of spacelimited signals

Another important subspace of $L^2(\mathbb{S}^2)$ is the subspace of *spacelimited* signals in a region R denoted by $\mathcal{H}_R(\mathbb{S}^2)$ defined as[13]

$$\mathcal{H}_R(\mathbb{S}^2) \triangleq \{f \in L^2(\mathbb{S}^2) : f(\theta, \phi) = 0, \text{ for } \{\theta, \phi\} \in \bar{R} \triangleq \mathbb{S}^2 \setminus R\}. \tag{7.74}$$

This space is by nature infinite-dimensional, because no signal can have a finite representation in both spectral and spatial domains. In other words, no non-trivial signal can simultaneously belong to $\mathcal{H}_R(\mathbb{S}^2)$ and $\mathcal{H}_L(\mathbb{S}^2)$ for a finite L and for R being a strict subset of $\mathbb{S}^2$. If the finite region of interest is azimuthally symmetric around the north pole, the infinite-dimensional basis for $\mathcal{H}_R(\mathbb{S}^2)$ will be of the form Y_ℓ^0, as we shall see next.

7.5.3   Subspace of azimuthally symmetric signals

As it is important in the sequel, we define the linear subspace of $L^2(\mathbb{S}^2)$, denoted $\mathcal{H}^0(\mathbb{S}^2)$, as

$$\mathcal{H}^0(\mathbb{S}^2) \triangleq \{f \in L^2(\mathbb{S}^2) : f(\theta, \phi) = f(\theta)\}, \tag{7.75}$$

which is the subspace of finite energy signals with azimuthal/rotational symmetry about the north pole $\widehat{\boldsymbol{\eta}} = (0,0,1)'$. According to the discussion following (7.29), the completeness of the Legendre polynomials implies completeness of the $m = 0$ spherical harmonics $\{Y_\ell^0(\theta)\}_{\ell=0}^\infty$. So any $f \in \mathcal{H}^0(\mathbb{S}^2)$ can be written

$$f(\theta) = \sum_{\ell=0}^\infty (f)_\ell^0 Y_\ell^0(\theta), \quad \forall f \in \mathcal{H}^0(\mathbb{S}^2), \tag{7.76}$$

where the spherical harmonic coefficients are

$$(f)_\ell^0 \triangleq \langle f, Y_\ell^0 \rangle = \int_0^{2\pi}\int_0^\pi f(\theta)\overline{Y_\ell^0(\theta)} \sin\theta\, d\theta\, d\phi$$

$$= \sqrt{\pi(2\ell+1)} \int_0^\pi f(\theta)P_\ell(\cos\theta) \sin\theta\, d\theta, \tag{7.77}$$

using (7.29).

---

[13] Note that $f(\theta, \phi) = 0$ is understood in the context of Lebesgue measure theory. The function does not have to be identically zero outside R in a pointwise sense. See Section 2.4.6, p. 47, for more details.

The space $\mathcal{H}^0(\mathbb{S}^2)$ contains azimuthally symmetric signals whose domain is the entire 2-sphere, as well as those that are spacelimited. A valid subspace of $\mathcal{H}^0(\mathbb{S}^2)$ is the set of azimuthally symmetric spacelimited signals that are non-zero over a *polar cap* region R_{θ_0} parameterized by θ_0. That is,

$$\mathcal{H}^0_{\theta_0}(\mathbb{S}^2) \triangleq \{f \in \mathcal{H}^0(\mathbb{S}^2) \colon f(\theta) = 0, \text{ for } \theta > \theta_0\}.$$

By the same token, $\mathcal{H}^0(\mathbb{S}^2)$ contains the subspace of azimuthally symmetric signals whose spectral degree ℓ is unbounded, as well as those which are band-limited with a maximum degree L:

$$\mathcal{H}^0_L(\mathbb{S}^2) \triangleq \{f \in \mathcal{H}^0(\mathbb{S}^2) \colon \langle f, Y_\ell^0 \rangle = 0, \text{ for } \ell > L\}.$$

7.6   Sampling on 2-sphere

Imagine that we wish to numerically compute the spherical harmonic coefficient of a function $f \in L^2(\mathbb{S}^2)$, $(f)_\ell^m$, and we have "sufficient" samples of f at points (θ_j, ϕ_k) over the entire 2-sphere to write

$$(f)_\ell^m \approx d\theta\, d\phi \sum_j \sum_k f(\theta_j, \phi_k) \overline{Y_\ell^m(\theta_j, \phi_k)} \sin \theta_j, \qquad (7.78)$$

where $d\theta$, $d\phi$ are the sampling resolutions. This is fine in principle, but there may be some computational issues with it. First, we cannot be sure about the numerical accuracy and stability of this technique for a given number of samples. This is especially relevant if the results are going to be fed to other sophisticated algorithms and error accumulation may occur. Second, if we attempt to improve accuracy by oversampling, the computational and storage costs increase.

So it is desirable to have more efficient and provably accurate techniques for signal processing and relevant computations on 2-sphere. This has been an active and practically relevant area of research; see (Driscoll and Healy, 1994; Yeo et al., 2008; Huang et al., 2011) and the references therein.

Specifically, given a signal f is uniformly sampled with N points along co-latitude θ and longitude ϕ, a widely-used method has been proposed in (Driscoll and Healy, 1994) that turns the above approximation into an exact computation for all degrees smaller than $N/2$, where N is the number of samples along the co-latitude and longitude as described below. This is accomplished using a "quadrature" rule with appropriately chosen weights. We briefly review the results, without proof, as they can be useful in applications involving signal processing on 2-sphere.

7.6.1   Sampling distribution

Assume that we have a signal $f \in L^2(\mathbb{S}^2)$ such that $(f)_\ell^m = 0$ for all $\ell > L_f$. That is, there are a maximum of $(L_f + 1)^2$ non-zero spherical harmonic coefficients for f. Now any $(L_f + 1)^2$ independent samples as our sampling distribution (that is, $(L_f + 1)^2$ independent co-latitude and longitude pairs) lead to a linear system in the unknown spherical harmonic coefficients, which can be solved exactly via matrix inversion. The real issue here, though, is having a sampling distribution that leads to a numerically well-conditioned system, matches real-world sampling schemes and more importantly leads to reasonable computational complexity.

We uniformly sample both the co-latitude and longitude angles with N samples, satisfying $N \geq 2(L_f + 1)$, at points

$$
\begin{aligned}
\theta_j &= \frac{\pi j}{N}, \quad j \in \{0, 1, \dots, N - 1\}, \\
\phi_k &= \frac{2\pi k}{N}, \quad k \in \{0, 1, \dots, N - 1\}.
\end{aligned}
\tag{7.79}
$$

Obviously, there are more samples around the poles than the equator and this needs to be compensated for with some proper weighting. The specific weighted sampling distribution function was proposed in (Driscoll and Healy, 1994), and corrected for a minor scaling error (an additional $2\sqrt{\pi}$ needed below), as

$$
s(\theta, \phi) = \frac{2\pi\sqrt{2}}{N} \sum_{j=0}^{N-1} \sum_{k=0}^{N-1} \alpha_j^{(N)} \delta(\theta - \theta_j) \delta(\phi - \phi_k),
\tag{7.80}
$$

where the coefficients $\alpha_j^{(N)}$ must satisfy the following set of N equations

$$
\sum_{j=0}^{N-1} \alpha_j^{(N)} P_\ell(\cos\theta_j) = \delta_{\ell,0}, \quad \ell \in \{0, 1, \dots, N - 1\},
\tag{7.81}
$$

where $P_\ell(\cdot)$ is the Legendre polynomial of degree ℓ. If N is chosen to be a power of 2, then the coefficients have a closed-form solution given by

$$
\alpha_j^{(N)} = \frac{2\sqrt{2}}{N} \sin\frac{\pi j}{N} \sum_{k=0}^{N/2-1} \frac{1}{2k+1} \sin\left((2k+1)\frac{\pi k}{N}\right), \quad j \in \{0, 1, \dots, N-1\}. \tag{7.82}
$$

7.6.2   Sampling theorem on 2-sphere

The following theorem was proven in (Driscoll and Healy, 1994).

Theorem 7.1. *Let a function $f(\theta, \phi)$ be bandlimited on 2-sphere such that $(f)_\ell^m = 0$ for all degrees $\ell \in \{L_f + 1, L_f + 2, L_f + 3, \dots\}$. Let the signal be sampled with at least $N = 2(L_f + 1)$ uniform points on the sphere in both co-latitude and longitude according to (7.79) and the signal values be denoted by $f(\theta_j, \phi_k)$. The spherical harmonic coefficients of signal f up to degree ℓ can be exactly recovered using*

$$
(f)_\ell^m = \frac{2\pi\sqrt{2}}{N} \sum_{j=0}^{N-1} \sum_{k=0}^{N-1} \alpha_j^{(N)} f(\theta_j, \phi_k) \overline{Y_\ell^m(\theta_j, \phi_k)},
\tag{7.83}
$$

for degree $\ell \in \{0, 1, \dots, L_f\}$ and order $m \in \{-\ell, -\ell + 1, \dots, \ell\}$, where $\alpha_j^{(N)}$ satisfy (7.81). Any possible aliasing will occur for degrees beyond L_f.   □

Example 7.15. Let us compare the exact method (7.83) and the approximate method (7.78) for a reference signal, whose spherical harmonic coefficient is known. We pick $f(\theta, \phi) = Y_2^{-1}(\theta, \phi)$ and hence $(f)_\ell^m = \delta_{2,\ell}\delta_{-1,m}$. We picked $N = 8$ to evaluate $Y(\theta_j, \phi_k)$. From (7.82)

$$\alpha = \big(\, 0,\ 0.1258,\ 0.1751,\ 0.2782,\ 0.2559,\ 0.2782,\ 0.1751,\ 0.1258 \,\big)'.$$

In simulation, the error, defined as

$$\varepsilon \triangleq \left(\sum_{\ell=0}^{4} \sum_{m=-\ell}^{\ell} \left| (f)_\ell^m - \delta_{2,\ell}\delta_{-1,m} \right|^2 \right)^{1/2},$$

was computed to be 2.1314×10^{-16} using the exact method (7.83), whereas the error using (7.78) with the same number of samples $N = 8$ was found to be 1.7000×10^{-3}. Increasing the number of samples to $N' = 32$ reduced the error to 5.8463×10^{-6}. □

Example 7.16. Samples of a signal f at points $\theta_j = j\pi/N$ and $\phi_k = k2\pi/N$ for $j, k = 0, 1, \ldots, N - 1$ and $N = 4$ are given by

$$f(\theta_j, \phi_k) = \begin{pmatrix} 0.5481 & 0.4606 & 0.9703 & 0.4558 \\ 0.7807 & 0.9209 & 0.7523 & 0.5205 \\ 0.5589 & 0.8862 & 0.0155 & 0.0889 \\ 0.3017 & 0.8795 & 0.9923 & 0.7851 \end{pmatrix},$$

where the horizontal direction represents change in co-latitude θ and vertical direction corresponds to change in longitude ϕ. Using the exact method (7.83) the vector of spherical coefficients is found as

$$\mathbf{f} = \big(\, 2.2829,\ 0.3304 + i\,0.1439,\ 0.4692,\ -0.3304 + i\,0.1439 \,\big)'.$$

Whereas using (7.78) we find

$$\mathbf{f} = \big(\, 2.1801,\ 0.3945 + i\,0.1499,\ 0.3909,\ -0.3945 + i\,0.1499 \,\big)'.$$

The difference in values is non-negligible. The normalized error, defined as

$$\hat{\varepsilon} \triangleq \frac{\left(\sum_{\ell=0}^{2} \sum_{m=-\ell}^{\ell} \left| (f)_\ell^m - (\widetilde{f})_\ell^m \right|^2 \right)^{1/2}}{\left(\sum_{\ell=0}^{2} \sum_{m=-\ell}^{\ell} \left| (f)_\ell^m \right|^2 \right)^{1/2}},$$

is $\hat{\varepsilon} = 0.0662$.

Now we use $\mathbf{f}$ as our reference to oversample the signal with $N = 8$ as

$$f_2(\theta_j, \phi_k) = \sum_{\ell=0}^{1} \sum_{m=-\ell}^{\ell} (f)_\ell^m Y(\theta_j, \phi_k).$$

As expected, we get the same spherical harmonic coefficients as in $\mathbf{f}$, if we feed f_2 in (7.83) with the normalized error $\hat{\varepsilon} = 2.332 \times 10^{-16}$. We can also improve the accuracy of direct calculation

$$\mathbf{f} = \big(\, 2.2535,\ 0.3305 + i\,0.1440,\ 0.4508,\ -0.3305 + i\,0.1440 \,\big)'.$$

where the normalized error is reduced to $\hat{\varepsilon} = 0.0145$. □

In summary, the exact method of (Driscoll and Healy, 1994) is superior to direct computation method in terms of reliability and computation cost.

7.7   Bounded linear operators on 2-sphere

A bounded linear operator, $\mathcal{B}$, mapping $f \in L^2(\mathbb{S}^2)$ to $g \in L^2(\mathbb{S}^2)$ is denoted by

$$g(\widehat{\boldsymbol{u}}) = (\mathcal{B}f)(\widehat{\boldsymbol{u}}), \tag{7.84}$$

or $g = \mathcal{B}f$ for short. For the most part, $\mathcal{B}$ is used to denote a generic bounded operator or used as the stem in the notation of a specific bounded operator. The major exception is for rotation operators where the stem $\mathcal{D}$ is used.

7.7.1   Systems on 2-sphere

The terminology "systems on 2-sphere" can be used when describing a linear operator on $L^2(\mathbb{S}^2)$. A system under this interpretation maps signals (finite energy functions on the 2-sphere) to other such signals. It is clear that such a definition of "system" could be narrowed or broadened depending on the context and application. We focus on the narrower notation of "bounded linear operators" for which we have much theory. This class is rich enough to include important classes of spatially invariant and spatially varying operators. Finally, we remark that the "systems" terminology is most useful when primarily in the broad setting, as we have done in the title of this chapter. For the remainder of this chapter we use the "operator" terminology.

7.7.2   Matrix representation

As we saw in the second part of the book, any bounded linear operator, $\mathcal{B}$, on a separable Hilbert space admits an infinite matrix representation $\mathbf{B}^{(\varphi)}$ with respect to a given complete orthonormal sequence $\{\varphi_n\}_{n=1}^{\infty}$. The element $b_{n,m}^{(\varphi)}$ at row n and column m of this matrix represents how input direction along φ_m is contributing to the output direction n along φ_n.

Since $L^2(\mathbb{S}^2)$ is separable and the set of spherical harmonics is a complete orthonormal sequence, any bounded linear operator, $\mathcal{B}$ on $L^2(\mathbb{S}^2)$, admits an infinite matrix representation. We almost exclusively work with spherical harmonics as our complete orthonormal sequence and hence can drop the superscript dependence on the basis function and simply write $\mathbf{B}$ as matrix representation of $\mathcal{B}$. The only point to care for such operators in $L^2(\mathbb{S}^2)$ is that their matrix elements have double indices. That is, the true elements of $\mathbf{B}$ are of the form $b_{\ell,p}^{m,q}$ specifying the gain of operator in mapping input Y_p^q along Y_ℓ^m or specifying how much of Y_p^q as an input gets projected along the Y_ℓ^m direction of the output under $\mathcal{B}$:

$$b_{\ell,p}^{m,q} \triangleq \langle \mathcal{B}Y_p^q, Y_\ell^m \rangle, \tag{7.85}$$

for all $\ell, p \in \{0, 1, 2, \ldots\}$ and $m, q \in \{-\ell, -\ell+1, \ldots, \ell\}$. In (7.85) we should think of the pair ℓ, m specifying a single output row index, and the pair p, q a single input column index. The single index conversion can be accomplished using (7.39). Using this convention, as we do in the rest of this chapter, the operator matrix will look like:[14]

$$
\mathbf{B} = \left(\begin{array}{ccccccc}
b_{0,0} & b_{0,1} & b_{0,2} & b_{0,3} & b_{0,4} & \cdots \\
b_{1,0} & b_{1,1} & b_{1,2} & b_{1,3} & b_{1,4} & \cdots \\
b_{2,0} & b_{2,1} & b_{2,2} & b_{2,3} & b_{2,4} & \cdots \\
b_{3,0} & b_{3,1} & b_{3,2} & b_{3,3} & b_{3,4} & \cdots \\
b_{4,0} & b_{4,1} & b_{4,2} & b_{4,3} & b_{4,4} & \cdots \\
\vdots & \vdots & \vdots & \vdots & \vdots & \ddots
\end{array}\right)
$$

$$
= \left(\begin{array}{ccccccc}
b_{0,0}^{0,0} & b_{0,1}^{0,-1} & b_{0,1}^{0,0} & b_{0,1}^{0,1} & b_{0,2}^{0,-2} & \cdots \\
b_{1,0}^{-1,0} & b_{1,1}^{-1,-1} & b_{1,1}^{-1,0} & b_{1,1}^{-1,1} & b_{1,2}^{-1,-2} & \cdots \\
b_{1,0}^{0,0} & b_{1,1}^{0,-1} & b_{1,1}^{0,0} & b_{1,1}^{0,1} & b_{1,2}^{0,-2} & \cdots \\
b_{1,0}^{1,0} & b_{1,1}^{1,-1} & b_{1,1}^{1,0} & b_{1,1}^{1,1} & b_{1,2}^{1,-2} & \cdots \\
b_{2,0}^{-2,0} & b_{2,1}^{-2,-1} & b_{2,1}^{-2,0} & b_{2,1}^{-2,1} & b_{2,2}^{-2,-2} & \cdots \\
\vdots & \vdots & \vdots & \vdots & \vdots & \ddots
\end{array}\right).
$$

$$(7.86)$$

So a bounded linear operator, $\mathcal{B}$, mapping $f \in L^2(\mathbb{S}^2)$ with spherical harmonic coefficients $(f)_p^q = \langle f, Y_p^q \rangle$ to $g \in L^2(\mathbb{S}^2)$ with spherical harmonic coefficients $(g)_\ell^m = \langle g, Y_\ell^m \rangle = \langle \mathcal{B}f, Y_\ell^m \rangle$ can be expressed in the spherical harmonic domain in terms of (7.85) as

$$
(g)_\ell^m = \sum_{p,q} b_{\ell,p}^{m,q}(f)_p^q \tag{7.87}
$$

and

$$
g(\widehat{\boldsymbol{u}}) = \sum_{\ell,m} (g)_\ell^m Y_\ell^m(\widehat{\boldsymbol{u}}) = \sum_{\ell,m}\sum_{p,q} b_{\ell,p}^{m,q}(f)_p^q Y_\ell^m(\widehat{\boldsymbol{u}}). \tag{7.88}
$$

Finally, the operator matrix of the operator composition, say $\mathcal{P} = \mathcal{B} \circ \mathcal{D}$, is simply given by the matrix multiplication of the operator matrices of the component operators with the elements given by

$$
p_{\ell,s}^{m,t} = \sum_{p,q} b_{\ell,p}^{m,q} d_{p,s}^{q,t}.
$$

7.7.3   Kernel representation

Given an operator matrix $\mathbf{B}$ with operator matrix elements $b_{\ell,p}^{m,q}$ defined in (7.85), we wish to express the operator kernel $B(\widehat{\boldsymbol{u}}, \widehat{\boldsymbol{v}})$ such that

$$
g(\widehat{\boldsymbol{u}}) = \int_{\mathbb{S}^2} B(\widehat{\boldsymbol{u}}, \widehat{\boldsymbol{v}}) f(\widehat{\boldsymbol{v}})\, ds(\widehat{\boldsymbol{v}}).
$$

---

[14] In (7.39), p. 195, the enumeration starts at $n = 0$ rather than 1, so our matrices are indexed from 0. Also in the representation of the matrices using elements (7.85) and ordering (7.39) the dashed lines, when present, indicate the ranges over which the degree ℓ is fixed.

The approach is similar to (4.26), p. 125. Using (7.88) we write

$$g(\widehat{\boldsymbol{u}}) = (\mathcal{B}f)(\widehat{\boldsymbol{u}}) = \sum_{\ell,m}(g)_\ell^m Y_\ell^m(\widehat{\boldsymbol{u}})$$

$$= \sum_{\ell,m}\sum_{p,q} b_{\ell,p}^{m,q}(f)_p^q Y_\ell^m(\widehat{\boldsymbol{u}}) = \sum_{\ell,m}\sum_{p,q} b_{\ell,p}^{m,q}\langle f, Y_p^q\rangle Y_\ell^m(\widehat{\boldsymbol{u}})$$

$$= \sum_{\ell,m}\langle f, \sum_{p,q}\overline{b_{\ell,p}^{m,q}} Y_p^q\rangle Y_\ell^m(\widehat{\boldsymbol{u}})$$

$$= \sum_{\ell,m}\left(\int_{\mathbb{S}^2} f(\widehat{\boldsymbol{v}})\sum_{p,q} b_{\ell,p}^{m,q}\overline{Y_p^q(\widehat{\boldsymbol{v}})}\,ds(\widehat{\boldsymbol{v}})\right) Y_\ell^m(\widehat{\boldsymbol{u}})$$

$$= \int_{\mathbb{S}^2} f(\widehat{\boldsymbol{v}})\sum_{\ell,m}\sum_{p,q} b_{\ell,p}^{m,q}\overline{Y_p^q(\widehat{\boldsymbol{v}})} Y_\ell^m(\widehat{\boldsymbol{u}})\,ds(\widehat{\boldsymbol{v}}),$$

from which we infer the kernel to be

$$\boxed{B(\widehat{\boldsymbol{u}},\widehat{\boldsymbol{v}}) = \sum_{\ell,m}\sum_{p,q} b_{\ell,p}^{m,q} Y_\ell^m(\widehat{\boldsymbol{u}})\overline{Y_p^q(\widehat{\boldsymbol{v}})}.} \tag{7.89}$$

7.7.4   Obtaining matrix elements from kernel

We first operate on $Y_p^q(\widehat{\boldsymbol{u}})$ to obtain $g(\widehat{\boldsymbol{u}}) = (\mathcal{B}Y_p^q)(\widehat{\boldsymbol{u}})$ and then write it in terms of the kernel integral

$$g(\widehat{\boldsymbol{u}}) = (\mathcal{B}Y_p^q)(\widehat{\boldsymbol{u}}) = \int_{\mathbb{S}^2} B(\widehat{\boldsymbol{u}},\widehat{\boldsymbol{v}}) Y_p^q(\widehat{\boldsymbol{v}})\,ds(\widehat{\boldsymbol{v}}).$$

Then the matrix element $b_{\ell,p}^{m,q} = \langle \mathcal{B}Y_p^q, Y_\ell^m\rangle$ is obtained as

$$\boxed{b_{\ell,p}^{m,q} = \langle \mathcal{B}Y_p^q, Y_\ell^m\rangle = \int_{\mathbb{S}^2}\int_{\mathbb{S}^2} B(\widehat{\boldsymbol{u}},\widehat{\boldsymbol{v}}) Y_p^q(\widehat{\boldsymbol{v}})\overline{Y_\ell^m(\widehat{\boldsymbol{u}})}\,ds(\widehat{\boldsymbol{v}})\,ds(\widehat{\boldsymbol{u}}).} \tag{7.90}$$

7.8   Spectral truncation operator

A spectral truncation operator of degree L, $\mathcal{B}_L$, is a finite rank projection operator that projects any signal $f \in L^2(\mathbb{S}^2)$ to $\mathcal{H}_L(\mathbb{S}^2)$. It is an idempotent self-adjoint compact operator, which zeros out any spherical harmonic coefficients with order $\ell > L$ and leaves any coefficient with degree $\ell \in \{0, 1, \dots, L\}$ in the input signal f unchanged. That is,

$$g(\widehat{\boldsymbol{u}}) = (\mathcal{B}_L f)(\widehat{\boldsymbol{u}})$$

$$= \sum_{\ell,m}^{L}(f)_\ell^m Y_\ell^m(\widehat{\boldsymbol{u}}). \tag{7.91}$$

Its operator matrix elements are therefore

$$b^{m,q}_{\ell,p} = \langle \mathcal{B}_L Y^q_p, Y^m_\ell \rangle$$

$$= \begin{cases} \delta^{m,q}_{\ell,p} & \ell \in \{0, 1, \ldots, L\}, \ m \in \{-\ell, -\ell+1, \ldots, \ell\}, \\ 0 & \text{otherwise,} \end{cases} \tag{7.92}$$

using (7.25). So referring to (7.86) we observe that the operator matrix is diagonal where the first $(L+1)^2$ diagonal elements are one (shown shaded below) and the remaining diagonal elements are zero:

$$\mathbf{B}_L = \begin{pmatrix} 1 & 0 & \cdots & 0 & 0 & 0 & \cdots \\ 0 & 1 & \cdots & 0 & 0 & 0 & \cdots \\ \vdots & \vdots & \ddots & \vdots & \vdots & \vdots & \cdots \\ 0 & 0 & \cdots & 1 & 0 & 0 & \cdots \\ 0 & 0 & \cdots & 0 & 0 & 0 & \cdots \\ 0 & 0 & \cdots & 0 & 0 & 0 & \cdots \\ \vdots & \vdots & \vdots & \vdots & \vdots & \vdots & \ddots \end{pmatrix},$$

which can be written more compactly[15]

$$\mathbf{B}_L = \operatorname{diag}\big(\underbrace{1; \ 1, 1, 1; \ 1, \ldots, 1}_{(L+1)^2 \ \text{terms}}; \ \ldots \big).$$

As a result, the spectral truncation operator *does not mix degrees or orders*. That is, the input degree and order p, q either gets mapped to the same degree and order ℓ, m in the output or is zeroed out.

And then using (7.92) and (7.89), the truncation operator kernel $B_L(\widehat{\boldsymbol{u}}, \widehat{\boldsymbol{v}})$ is

$$B_L(\widehat{\boldsymbol{u}}, \widehat{\boldsymbol{v}}) = \sum_{\ell,m} Y^m_\ell(\widehat{\boldsymbol{u}}) \overline{Y^m_\ell(\widehat{\boldsymbol{v}})} = \sum_{\ell=0}^{L} \frac{(2\ell+1)}{4\pi} P_\ell(\widehat{\boldsymbol{u}} \cdot \widehat{\boldsymbol{v}}), \tag{7.93}$$

where the last equality follows from the addition theorem (7.30), p. 193.

7.9   Spatial truncation operator

A spatial truncation operator to region $R \subset \mathbb{S}^2$, $\mathcal{B}_R$, projects a signal $f \in L^2(\mathbb{S}^2)$ to $\mathcal{H}_R(\mathbb{S}^2)$. It zeros out the input signal outside the region R and keeps it unchanged inside R. That is,

$$(\mathcal{B}_R f)(\widehat{\boldsymbol{u}}) = \begin{cases} f(\widehat{\boldsymbol{u}}) & \widehat{\boldsymbol{u}} \in R, \\ 0 & \widehat{\boldsymbol{u}} \in \bar{R} \triangleq \mathbb{S}^2 \setminus R. \end{cases} \tag{7.94}$$

---

[15] In the diagonal matrix representation the semi-colons indicate the range where the degree is fixed: before the first semi-colon $\ell = 0$ (1 term), between the first and second semi-colons $\ell = 1$ (3 terms), etc.

Denoting the indicator or characteristic function of the region R by $\mathbf{1}_R$, we can write the above operation as

$$(\mathcal{B}_R f)(\widehat{\boldsymbol{u}}) = \int_{\mathbb{S}^2} \mathbf{1}_R(\widehat{\boldsymbol{v}}) \delta(\widehat{\boldsymbol{u}}, \widehat{\boldsymbol{v}}) f(\widehat{\boldsymbol{v}}) \, ds(\widehat{\boldsymbol{v}}), \qquad (7.95)$$

from which we infer that the operator kernel is given by

$$\boxed{B_R(\widehat{\boldsymbol{u}}, \widehat{\boldsymbol{v}}) = \mathbf{1}_R(\widehat{\boldsymbol{v}}) \delta(\widehat{\boldsymbol{u}}, \widehat{\boldsymbol{v}}).} \qquad (7.96)$$

Therefore, from (7.90) we can obtain the operator matrix elements as

$$\begin{aligned}
b_{\ell,p}^{m,q} &= \int_{\mathbb{S}^2} \int_{\mathbb{S}^2} \mathbf{1}_R(\widehat{\boldsymbol{v}}) \delta(\widehat{\boldsymbol{u}}, \widehat{\boldsymbol{v}}) Y_p^q(\widehat{\boldsymbol{v}}) \overline{Y_\ell^m(\widehat{\boldsymbol{u}})} \, ds(\widehat{\boldsymbol{v}}) \, ds(\widehat{\boldsymbol{u}}) \\
&= \int_R \int_R \delta(\widehat{\boldsymbol{u}}, \widehat{\boldsymbol{v}}) Y_p^q(\widehat{\boldsymbol{v}}) \overline{Y_\ell^m(\widehat{\boldsymbol{u}})} \, ds(\widehat{\boldsymbol{v}}) \, ds(\widehat{\boldsymbol{u}}) \\
&= \int_R Y_p^q(\widehat{\boldsymbol{u}}) \overline{Y_\ell^m(\widehat{\boldsymbol{u}})} \, ds(\widehat{\boldsymbol{u}}), \qquad (7.97)
\end{aligned}$$

which does not have a closed form when the region of interest is not the whole 2-sphere and has to be evaluated numerically.

In contrast to the spectral truncation operator, the spatial truncation operator *mixes degrees and orders*. In general, an input degree and order p, q can get mapped to any degree and order ℓ, m in the output and all elements in the operator matrix in (7.86) are expected to be non-zero. As we will see for other operators, such as the rotation operator, mixing of orders is not uncommon. However, mixing of degrees can occur if the operator is spatially varying and treats different spatial regions on the 2-sphere differently, as does the spatial truncation operator. The analog in the time-frequency analysis is the application of a time-varying filter or a time domain mask, which can mix different frequencies of the signal and filter.

7.10   Spatial masking of signals with a window

We can generalize the spatial truncation operator to the spatial masking operator. Masking of a signal with a window function refers to pointwise multiplication of the signal of interest $f(\widehat{\boldsymbol{u}})$ with an appropriately designed window function $h(\widehat{\boldsymbol{u}})$ in spatial domain. Let us denote the spatial masking operator by $\mathcal{B}_h$ with operator matrix elements $b_{\ell,p}^{m,q}$ and operator kernel $B_h(\widehat{\boldsymbol{u}}, \widehat{\boldsymbol{v}})$. The operator matrix elements and operator kernel can be represented in a variety of forms which can be useful in different applications and we will elaborate on this now.

7.10.1   Different operator representations

In its simplest form, the masked signal in spatial domain is given by

$$\boxed{(\mathcal{B}_h f)(\widehat{\boldsymbol{u}}) = h(\widehat{\boldsymbol{u}}) f(\widehat{\boldsymbol{u}}).} \qquad (7.98)$$

We can write this operation in integral form as

$$(\mathcal{B}_h f)(\widehat{\boldsymbol{u}}) = \int_{\mathbb{S}^2} h(\widehat{\boldsymbol{v}})\delta(\widehat{\boldsymbol{u}},\widehat{\boldsymbol{v}})f(\widehat{\boldsymbol{v}})\,ds(\widehat{\boldsymbol{v}}), \qquad (7.99)$$

from which we infer that the operator kernel is given by

$$B_h(\widehat{\boldsymbol{u}},\widehat{\boldsymbol{v}}) = h(\widehat{\boldsymbol{v}})\delta(\widehat{\boldsymbol{u}},\widehat{\boldsymbol{v}}). \qquad (7.100)$$

Based on this kernel representation and using (7.90) we obtain the operator matrix elements as

$$b_{\ell,p}^{m,q} = \int_{\mathbb{S}^2} h(\widehat{\boldsymbol{u}})Y_p^q(\widehat{\boldsymbol{u}})\overline{Y_\ell^m(\widehat{\boldsymbol{u}})}\,ds(\widehat{\boldsymbol{u}}), \qquad (7.101)$$

which is interpreted as the weighted inner product of Y_p^q and Y_ℓ^m and needs to be evaluated numerically.

Now we provide a second representation for the operator kernel and operator matrix elements. Let the spherical harmonic coefficients of the window be denoted by $(h)_s^t$. Using the expansion of the Dirac delta from (7.44), we can write the kernel in (7.100) in an alternative form as

$$B_h(\widehat{\boldsymbol{u}},\widehat{\boldsymbol{v}}) = \sum_{s,t}(h)_s^t Y_s^t(\widehat{\boldsymbol{v}})\sum_{j,k}Y_j^k(\widehat{\boldsymbol{u}})\overline{Y_j^k(\widehat{\boldsymbol{v}})}. \qquad (7.102)$$

Upon using $h(\widehat{\boldsymbol{u}}) = \sum_{s,t}(h)_s^t Y_s^t(\widehat{\boldsymbol{u}})$ in (7.101) we can find the second representation for operator matrix elements as

$$b_{\ell,p}^{m,q} = \sum_{s,t}(h)_s^t \int_{\mathbb{S}^2} Y_s^t(\widehat{\boldsymbol{u}})Y_p^q(\widehat{\boldsymbol{u}})\overline{Y_\ell^m(\widehat{\boldsymbol{u}})}\,ds(\widehat{\boldsymbol{u}}). \qquad (7.103)$$

Upon defining a shorthand notation for the integral above

$$y(s,t;p,q;\ell,m) \triangleq \int_{\mathbb{S}^2} Y_s^t(\widehat{\boldsymbol{u}})Y_p^q(\widehat{\boldsymbol{u}})\overline{Y_\ell^m(\widehat{\boldsymbol{u}})}\,ds(\widehat{\boldsymbol{u}}), \qquad (7.104)$$

the operator matrix elements are succinctly written

$$b_{\ell,p}^{m,q} = \sum_{s,t}(h)_s^t y(s,t;p,q;\ell,m). \qquad (7.105)$$

In quantum mechanics, the integral involving three spherical harmonics, precisely as in (7.104), occurs frequently and can be represented using Wigner $3j$

symbols (Wigner, 1959), which are related to Clebsch-Gordan coefficients. Such an expression for (7.104) is (Sakurai, 1994)

$$y(s,t;p,q;\ell,m) = (-1)^m \sqrt{\frac{(2s+1)(2p+1)(2\ell+1)}{4\pi}} \begin{pmatrix} s & p & \ell \\ 0 & 0 & 0 \end{pmatrix} \begin{pmatrix} s & p & \ell \\ t & q & -m \end{pmatrix},$$

where the last two terms are not matrices, but the Wigner $3j$ symbols, which will be discussed shortly.

To provide a third representations of the operator kernel, we use (7.105) and (7.89) to write

$$B_h(\widehat{\boldsymbol{u}}, \widehat{\boldsymbol{v}}) = \sum_{\ell,m} \sum_{p,q} \sum_{s,t} (h)_s^t y(s,t;p,q;\ell,m) Y_\ell^m(\widehat{\boldsymbol{u}}) \overline{Y_p^q(\widehat{\boldsymbol{v}})}. \tag{7.106}$$

The integral (7.104) is in general non-zero and similar to the spatial truncation operator and referring to (7.105) the spatial masking operator is expected to mix degrees. We again attribute this to the spatially varying treatment of the input signal f by window h.

Before proceeding, we take a little detour to investigate the definition, computation, and some of the basic properties of Wigner $3j$ symbols, as they crop up in a variety of applications.

7.10.2   Wigner $3j$ symbols

Wigner in (Wigner, 1959, p. 191, p. 290) defined the "three-j" symbols as

$$\begin{pmatrix} j_1 & j_2 & j_3 \\ m_1 & m_2 & m_3 \end{pmatrix} \triangleq \frac{(-1)^{j_1-j_2-m_3}}{\sqrt{2j_3+1}} s_{j_3,m_1,m_2}^{j_1,j_2} \delta_{m_1+m_2+m_3,0}, \tag{7.107}$$

where $s_{j_3,m_1,m_2}^{j_1,j_2}$ is given by

$$s_{j_3,m_1,m_2}^{j_1,j_2} =$$

$$\frac{\sqrt{(j_3+j_1-j_2)!\,(j_3-j_1+j_2)!\,(j_1+j_2-j_3)!\,(j_3-m_3)!\,(j_3+m_3)!}}{\sqrt{(j_1+j_2+j_3+1)!\,(j_1-m_1)!\,(j_1+m_1)!\,(j_2-m_2)!\,(j_2+m_2)!}}$$

$$\times \sum_t \frac{(-1)^{t+j_2+m_2}\sqrt{2j_3+1}(j_3+j_2+m_1-t)!\,(j_1-m_1+t)!}{(j_3-j_1+j_2-t)!\,(j_3-m_3-t)!\,(t)!\,(j_1-j_2+m_3+t)!}.$$

The sum over t is such that all the arguments inside the factorial are non-negative.

In addition to the condition of $m_1 + m_2 + m_3 = 0$, $|j_2 - j_3| \leq j_1 \leq j_2 + j_3$, $|j_1 - j_3| \leq j_2 \leq j_1 + j_3$, $|j_1 - j_2| \leq j_3 \leq j_1 + j_2$ and $|m_k| \leq j_k$ for $k = 1, 2, 3$ must also be satisfied for the Wigner $3j$ symbol to be non-zero.

There are various other ways for defining and calculating Wigner $3j$ symbols. One of the most widely used ones is the so-called Racah Formula (Messiah, 1961,

p. 1058):

$$\begin{pmatrix} j_1 & j_2 & j_3 \\ m_1 & m_2 & m_3 \end{pmatrix} = (-1)^{j_1-j_2-m_3} \sqrt{\Delta(j_1, j_2, j_3)}$$

$$\times \sqrt{(j_1 - m_1)! \, (j_1 + m_1)! \, (j_2 - m_2)! \, (j_2 + m_2)! \, (j_3 - m_3)! \, (j_3 + m_3)!}$$

$$\times \sum_t \frac{(-1)^t}{(t)! \, (j_1 + j_2 - j_3 - t)! \, (j_1 - m_1 - t)! \, (j_2 + m_2 - t)!}$$

$$\times \frac{1}{(j_3 - j_2 + m_1 + t)! \, (j_3 - j_1 - m_2 + t)!},$$

where $\Delta(j_1, j_2, j_3)$ is given by

$$\Delta(j_1, j_2, j_3) = \frac{(j_1 + j_2 - j_3)! \, (j_1 - j_2 + j_3)! \, (j_2 - j_1 + j_3)!}{(j_1 + j_2 + j_3 + 1)!}.$$

There is a multitude of symmetry, orthogonality, and recursion relations for Wigner $3j$ symbols, which is beyond the scope of this book and interested readers are referred to books on quantum mechanics. Here we present some of the most basic symmetry and orthogonality relations (Wigner, 1959; Messiah, 1961). The main symmetry relations are:

$$\begin{pmatrix} j_1 & j_2 & j_3 \\ m_1 & m_2 & m_3 \end{pmatrix} = \begin{pmatrix} j_2 & j_3 & j_1 \\ m_2 & m_3 & m_1 \end{pmatrix} = \begin{pmatrix} j_3 & j_1 & j_2 \\ m_3 & m_1 & m_2 \end{pmatrix},$$

$$(-1)^{j_1+j_2+j_3} \begin{pmatrix} j_1 & j_2 & j_3 \\ m_1 & m_2 & m_3 \end{pmatrix} = \begin{pmatrix} j_1 & j_3 & j_2 \\ m_1 & m_3 & m_2 \end{pmatrix} = \begin{pmatrix} j_3 & j_2 & j_1 \\ m_3 & m_2 & m_1 \end{pmatrix}$$

$$= \begin{pmatrix} j_2 & j_1 & j_3 \\ m_2 & m_1 & m_3 \end{pmatrix},$$

$$(-1)^{j_1+j_2+j_3} \begin{pmatrix} j_1 & j_2 & j_3 \\ m_1 & m_2 & m_3 \end{pmatrix} = \begin{pmatrix} j_1 & j_2 & j_3 \\ -m_1 & -m_2 & -m_3 \end{pmatrix},$$

and two main orthogonality relations are:

$$(2j_3 + 1) \sum_{m_1=-j_1}^{j_1} \sum_{m_2=-j_2}^{j_2} \begin{pmatrix} j_1 & j_2 & j_3 \\ m_1 & m_2 & m_3 \end{pmatrix} \begin{pmatrix} j_1 & j_2 & j_3' \\ m_1 & m_2 & m_3' \end{pmatrix} = \delta_{j_3, j_3'} \delta_{m_3, m_3'},$$

$$\sum_{j_3=|j_1-j_2|}^{j_1+j_2} \sum_{m_3=-j_3}^{j_3} (2j_3 + 1) \begin{pmatrix} j_1 & j_2 & j_3 \\ m_1 & m_2 & m_3 \end{pmatrix} \begin{pmatrix} j_1 & j_2 & j_3 \\ m_1' & m_2' & m_3 \end{pmatrix} = \delta_{m_1, m_1'} \delta_{m_2, m_2'}.$$

7.10.3   Spherical harmonic coefficients of $\mathcal{B}_h f$

Referring to (7.87) and (7.105), if the spectral harmonic coefficients of the masked signal g are denoted by $(g)_\ell^m$ and those of the original signal are denoted by $(f)_p^q$, we can write

$$\boxed{(g)_\ell^m = \sum_{s,t} (h)_s^t \sum_{p,q} (f)_p^q y(s, t; p, q; \ell, m),} \qquad (7.108)$$

Wigner, Eugene (1902–1995) — Along with John von Neumann (1903–1957), Leó Szilárd
(1898–1964), Edward Teller (1908–2003) and other talented Hungarians at the start of
the twentieth century, Eugene Wigner was referred to as one of "the Martians." Various
attributes gave the Martians away. They spoke together in a language unrelated to
Indo-European languages — to an outsider it sounded like Martian (but it probably
sounded more like Bela Lugosi (1882–1956) than a Martian). They were otherworldly
intelligent and so were thought to have been the result of some recreational stop-over
to Earth by Martians. Szilárd, who is generally held responsible for the Martian joke,
hypothesized nuclear fission and filed a patent, and realized the runaway process would
make an effective bomb. Not to be outdone, fellow Martian Teller conceived of the
more powerful hydrogen (fusion) bomb and rather unkindly is generally thought to be
character basis of Dr. Strangelove in the movie of the same name.

Wigner did seminal work on quantum mechanics and championed the introduction of group theory techniques, along with Weyl. He first published his influential
book "Gruppentheorie und ihre Anwendungen auf die Quantenmechanik der Atom-
spektren" in German 1931 and it has been reprinted many times and translated to
English (Wigner, 1959). Wigner did receive the Nobel Prize in Physics in 1963 for
"contributions to the theory of the atomic nucleus and the elementary particles, particularly through the discovery and application of fundamental symmetry principles."
On the personal side, he had the distinction of being the brother-in-law of the inspiring
Paul Dirac (1902–1984); see (Farmelo, 2009).

which can also be alternatively represented using (7.101) as

$$(g)_\ell^m = \sum_{p,q} (f)_p^q \int_{\mathbb{S}^2} h(\hat{u}) Y_p^q(\hat{u}) \overline{Y_\ell^m(\hat{u})} \, ds(\hat{u}). \tag{7.109}$$

Given (7.77), in the case where the designed window function is azimuthally symmetric, we only need to compute integrals of the form $y(s, 0; p, q; \ell, m)$. When one of the orders is zero (t here) $y(s, 0; p, q; \ell, m)$ will take non-zero values only under certain conditions. For example, it is necessary to have $q = m$ and $|\ell - s| \le p \le \ell + s$. If in addition the window is bandlimited to a maximum spectral degree of L_h and the signal of interest is also either strictly or *practically* bandlimited to a maximum spectral degree of L_f,[16] we can further simplify (7.108)

$$(g)_\ell^m = (-1)^m \sqrt{\frac{(2\ell + 1)}{4\pi}} \sum_{s=0}^{L_h} \sum_{p=|\ell-s|}^{\ell+s} (h)_s^0 (f)_p^m \sqrt{(2s+1)(2p+1)}$$
$$\times \begin{pmatrix} s & p & \ell \\ 0 & 0 & 0 \end{pmatrix} \begin{pmatrix} s & p & \ell \\ 0 & m & -m \end{pmatrix}. \tag{7.110}$$

According to (7.110), for $\ell \le L_h$ each windowed coefficient $(g)_\ell^m$ will receive contributions from the data spherical coefficients $(f)_p^m$ with degrees in the range $0 \le p \le \ell + L_h$ and for $\ell \ge L_h$ it will receive contributions from the data coefficients $(f)_p^m$ with degrees in the range $|\ell - L_h| \le p \le \ell + L_h$. Therefore, the windowed coefficients are spectrally smoother versions of the original signal. The amount of smoothing depends on the window bandwidth and the shape of its spectral response.

In addition, since we assume that the signal is bandlimited to degree L_f, the maximum possible degree of the windowed signal is $L_g = L_f + L_h$.

7.11   Rotation operator

We can define the effect of rotation on signals in $L^2(\mathbb{S}^2)$ as an operator. We denote the rotation operator as $\mathcal{D}(\varphi, \vartheta, \omega)$ parameterized in the Euler angles in the *zyz* convention, which corresponds to the rotation matrix $\boldsymbol{R} \in \mathbb{R}^{3\times3}$ defined in (7.10), p. 185, and repeated here,[17,18]

$$\boldsymbol{R} \equiv \boldsymbol{R}_{\varphi\vartheta\omega}^{(zyz)} \triangleq \boldsymbol{R}_\varphi^{(z)} \boldsymbol{R}_\vartheta^{(y)} \boldsymbol{R}_\omega^{(z)}. \tag{7.111}$$

The action of the operator $\mathcal{D}(\varphi, \vartheta, \omega)$ can be decomposed into three simpler rotation operators along the z, y, and z axes

$$\mathcal{D}(\varphi, \vartheta, \omega) = \mathcal{D}_z(\varphi) \circ \mathcal{D}_y(\vartheta) \circ \mathcal{D}_z(\omega). \tag{7.112}$$

---

[16] That is $h(\hat{u})$ belongs to $\mathcal{H}_{L_h}^0(\mathbb{S}^2)$ and $f(\hat{u})$ belongs to $\mathcal{H}_{L_f}(\mathbb{S}^2)$.

[17] $\boldsymbol{R}$ should not to be confused with "rotation operator matrix" that we define shortly.

[18] Unless otherwise stated, a general rotation over $\varphi \in [0, 2\pi)$, $\vartheta \in [0, \pi]$, $\omega \in [0, 2\pi)$ using the rotation matrix $\boldsymbol{R}$ is understood. Other notations may be used later to represent special forms of rotation.

The effect of $\mathcal{D}(\varphi, \vartheta, \omega)$ on the signal $f \in L^2(\mathbb{S}^2)$ can be realized through an inverse rotation of the coordinate system. That is,

$$(\mathcal{D}(\varphi, \vartheta, \omega)f)(\widehat{\boldsymbol{u}}) = f(\boldsymbol{R}^{-1}\widehat{\boldsymbol{u}}), \qquad (7.113)$$

which means that the value of $\mathcal{D}(\varphi, \vartheta, \omega)f$ at point $\widehat{\boldsymbol{u}}$ is equal to the value of the original function f at point $\boldsymbol{R}^{-1}\widehat{\boldsymbol{u}}$ (Wigner, 1959, p. 358).

If a signal f with spherical harmonic coefficients $(f)_\ell^m$ is rotated on the sphere under rotation operator $\mathcal{D}(\varphi, \vartheta, \omega)$, the spherical harmonic coefficient of degree ℓ and order m of the rotated signal is a linear combination of different order spherical harmonics of the original signal of the *same* degree ℓ as

$$\langle \mathcal{D}(\varphi, \vartheta, \omega)f, Y_\ell^m \rangle = \sum_{m'=-\ell}^{\ell} D_{m,m'}^\ell(\varphi, \vartheta, \omega)(f)_\ell^{m'}, \qquad (7.114)$$

where $D_{m,m'}^\ell(\varphi, \vartheta, \omega)$ is the Wigner D-matrix given by

$$D_{m,m'}^\ell(\varphi, \vartheta, \omega) \triangleq e^{-im\varphi} d_{m,m'}^\ell(\vartheta) e^{-im'\omega}, \qquad (7.115)$$

and $d_{m,m'}^\ell(\vartheta)$ is the Wigner d-matrix defined as

$$
\begin{aligned}
d_{m,m'}^\ell(\vartheta) &\triangleq \\
&\sum_n (-1)^{n-m'+m} \frac{\sqrt{(\ell+m')!\,(\ell-m')!(\ell+m)!\,(\ell-m)!}}{(\ell+m'-n)!\,(n)!\,(\ell-n-m)!\,(n-m'+m)!} \\
&\qquad \times \left(\cos\frac{\vartheta}{2}\right)^{2\ell-2n+m'-m} \left(\sin\frac{\vartheta}{2}\right)^{2n-m'+m},
\end{aligned} \qquad (7.116)
$$

where the sum is over all n such that denominator terms do not become negative (Sakurai, 1994, p. 223). That is, $\max(0, m'-m) \leq n \leq \min(\ell-m, \ell+m')$. It is clear from the definition that the Wigner d-matrix is real-valued which we prove by a different method a bit later.

7.11.1   Rotation operator matrix

Using the notation convention in this book the operator matrix corresponding to bounded operator $\mathcal{D}(\varphi, \vartheta, \omega)$ (implicitly with respect to using the spherical harmonics as the complete orthonormal sequence) is written $d_{\ell,p}^{m,q}(\varphi, \vartheta, \omega)$. This should not be confused with the Wigner d-matrix, $d_{m,m'}^\ell(\vartheta)$ defined in (7.116), although the two are closely related. These operator matrix coefficients of the rotation operator $\mathcal{D}(\varphi, \vartheta, \omega)$ can be related to the Wigner D-matrix using (7.114) as follows

$$
\begin{aligned}
d_{\ell,p}^{m,q}(\varphi, \vartheta, \omega) &\triangleq \langle \mathcal{D}(\varphi, \vartheta, \omega)Y_p^q, Y_\ell^m \rangle \\
&= \sum_{m'=-\ell}^{\ell} D_{m,m'}^\ell(\varphi, \vartheta, \omega)\langle Y_p^q, Y_\ell^{m'} \rangle \qquad (7.117) \\
&= D_{m,q}^\ell(\varphi, \vartheta, \omega)\delta_{\ell,p},
\end{aligned}
$$

where $m, q \in \{-\ell, -\ell+1, \ldots, \ell\}$.

Remark 7.6. By defining the Wigner D-matrix as follows

$$D_{m,m'}^{\ell}(\varphi, \vartheta, \omega) \triangleq \langle \mathcal{D}(\varphi, \vartheta, \omega) Y_{\ell}^{m'}, Y_{\ell}^{m} \rangle, \tag{7.118}$$

gives them a clear geometric interpretation to complement the somewhat esoteric algebraic definition given in (7.115). The Wigner D-matrix holds the non-trivial[19] coefficients of the rotation operator matrix with respect to using the spherical harmonics shown in (7.117). □

Remark 7.7. For the Wigner d-matrix, we similarly see

$$d_{m,m'}^{\ell}(\vartheta) \triangleq \langle \mathcal{D}(0, \vartheta, 0) Y_{\ell}^{m'}, Y_{\ell}^{m} \rangle = \langle \mathcal{D}_y(\vartheta) Y_{\ell}^{m'}, Y_{\ell}^{m} \rangle \tag{7.119}$$

which follows from (7.115). □

From (7.117) we observe that the operator does not mix degrees: an input degree p cannot get mapped to another degree ℓ other than p. This is because the rotation operator is spatially uniformly treating the signal. But it can mix orders. Hence the structure of the operator matrix is block diagonal where the elements are appropriately filled with Wigner D-matrices $D_{m,q}^{\ell}(\varphi, \vartheta, \omega)$. Visually, the operator matrix in the spherical harmonics will look like (see Section 7.7.2)

$$\mathbf{D}(\varphi, \vartheta, \omega) =$$

$$\begin{pmatrix}
D_{0,0}^0 & 0 & 0 & 0 & 0 & 0 & 0 & 0 & 0 & \cdots \\
0 & D_{-1,-1}^1 & D_{-1,0}^1 & D_{-1,1}^1 & 0 & 0 & 0 & 0 & 0 & \cdots \\
0 & D_{0,-1}^1 & D_{0,0}^1 & D_{0,1}^1 & 0 & 0 & 0 & 0 & 0 & \cdots \\
0 & D_{1,-1}^1 & D_{1,0}^1 & D_{1,1}^1 & 0 & 0 & 0 & 0 & 0 & \cdots \\
0 & 0 & 0 & 0 & D_{-2,-2}^2 & D_{-2,-1}^2 & D_{-2,0}^2 & D_{-2,1}^2 & D_{-2,2}^2 & \cdots \\
0 & 0 & 0 & 0 & D_{-1,-2}^2 & D_{-1,-1}^2 & D_{-1,0}^2 & D_{-1,1}^2 & D_{-1,2}^2 & \cdots \\
0 & 0 & 0 & 0 & D_{0,-2}^2 & D_{0,-1}^2 & D_{0,0}^2 & D_{0,1}^2 & D_{0,2}^2 & \cdots \\
0 & 0 & 0 & 0 & D_{1,-2}^2 & D_{1,-1}^2 & D_{1,0}^2 & D_{1,1}^2 & D_{1,2}^2 & \cdots \\
0 & 0 & 0 & 0 & D_{2,-2}^2 & D_{2,-1}^2 & D_{2,0}^2 & D_{2,1}^2 & D_{2,2}^2 & \cdots \\
\vdots & \vdots & \vdots & \vdots & \vdots & \vdots & \vdots & \vdots & \vdots & \ddots
\end{pmatrix}, \tag{7.120}$$

where we have suppressed the dependence on the Euler angles φ, ϑ, and ω in the matrix elements on the right-hand side. In the matrix, only the (shaded) diagonal blocks of successive sizes 1×1, 3×3, 5×5, etc., can be non-zero.[20]

---

[19] The trivial coefficients are those that are zero for all possible values of φ, ϑ, and ω. Non-trivial coefficients can still take the value zero for specific values of φ, ϑ, and ω, but are generally non-zero.

[20] This block diagonal structure features no less than on the front cover of Sakurai's book "Modern Quantum Mechanics" and within it (Sakurai, 1994, p. 193).

7.11.2   Rotation operator kernel

The rotation operator kernel is derived using (7.89) and substituting (7.117)

$$
D_{\varphi,\vartheta,\omega}(\widehat{\boldsymbol{u}},\widehat{\boldsymbol{v}}) = \sum_{\ell,m} \sum_{q=-\ell}^{\ell} D_{m,q}^{\ell}(\varphi,\vartheta,\omega) Y_{\ell}^{m}(\widehat{\boldsymbol{u}}) \overline{Y_{\ell}^{q}}(\widehat{\boldsymbol{v}}),
\tag{7.121}
$$

and is used in the integral operator form

$$
\big(\mathcal{D}(\varphi,\vartheta,\omega)f\big)(\widehat{\boldsymbol{u}}) = \int_{\mathbb{S}^2} D_{\varphi,\vartheta,\omega}(\widehat{\boldsymbol{u}},\widehat{\boldsymbol{v}}) f(\widehat{\boldsymbol{v}})\, ds(\widehat{\boldsymbol{v}}).
$$

7.11.3   Important relations pertaining to rotation operation

First from (7.6), p. 182, and the fact that the rotation convention zyz symmetrically uses rotations around the z-axis, we conclude that the inverse of rotation operator is given by

$$
\mathcal{D}(\varphi,\vartheta,\omega)^{-1} = \mathcal{D}(-\omega,-\vartheta,-\varphi).
\tag{7.122}
$$

It may be noted that a negative rotation about the y-axis in ϑ is outside the range of admissible $\vartheta \in [0,\pi]$ values, but algebraically and geometrically it is perfectly fine. If one wants the same effect as a negative ϑ rotation about the y-axis then one can use the positive ϑ rotation and compensate for the sign flip in the treatment of the two z-axis rotations. The key identity, which can be established using (7.8) and (7.9), is

$$
\mathcal{D}(\varphi,\vartheta,\omega) = \mathcal{D}(\pi + \varphi, -\vartheta, \pi + \omega),
\tag{7.123}
$$

where the z-axis rotations can be regarded as modulo 2π. In short, (7.123) means we can pre- and post-rotate through π around the z-axis to give the same effect as a $-\vartheta$ y-axis rotation.

Another well-known identity due to Wigner in an unpublished manuscript in 1940 but reprinted later (Wigner, 1965), is the following operator decomposition (Edmonds, 1957; Wandelt and Górski, 2001; Yeo et al., 2008):

$$
\mathcal{D}(\varphi,\vartheta,\omega) = \mathcal{D}\left(\varphi + \frac{\pi}{2}, \frac{\pi}{2}, \vartheta + \pi\right) \circ \mathcal{D}\left(0, \frac{\pi}{2}, \omega + \frac{\pi}{2}\right),
\tag{7.124}
$$

which at first glance appears complicated. At its heart it implements the y-axis ϑ rotation by transferring the operation to the z-axis via pre- and post-rotation (operator compositions), Rubik's Cube style. With this cryptic insight, we now interpret (7.124) geometrically.[21] First we observe that, (7.112),

$$
\begin{aligned}
\mathcal{D}(\varphi,\vartheta,\omega) &= \mathcal{D}_z(\varphi) \circ \mathcal{D}_y(\vartheta) \circ \mathcal{D}_z(\omega) \\
&= \mathcal{D}(\varphi,0,0) \circ \mathcal{D}_y(\vartheta) \circ \mathcal{D}(0,0,\omega)
\end{aligned}
$$

---

[21] Of course, if you love trigonometry then you can show it by multiplying out and simplifying the equivalent 3×3 rotation matrices. This misses the point somewhat because you would not have found the relation in the first place.

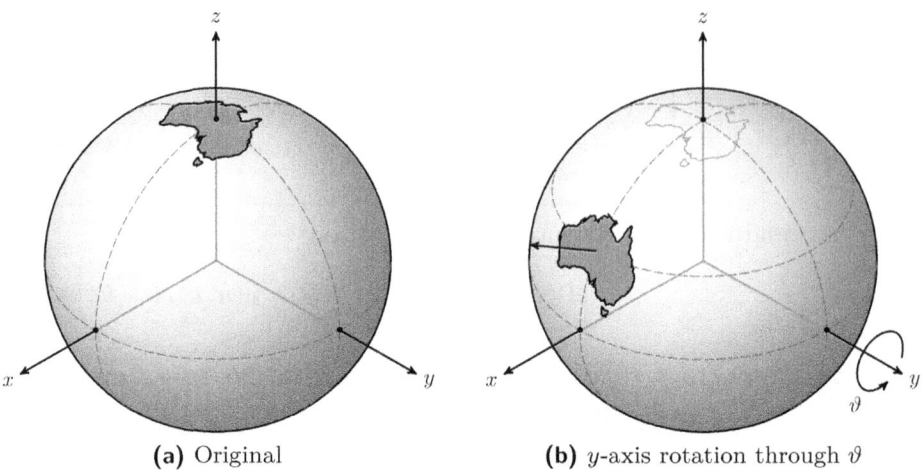

(a) Original　　　　　　　**(b)** y-axis rotation through ϑ

Figure 7.9 The effect of a rotation about the y-axis through ϑ.

So, to establish (7.124), it is sufficient to show

$$\mathcal{D}_y(\vartheta) = \mathcal{D}\left(\frac{\pi}{2}, \frac{\pi}{2}, \vartheta + \pi\right) \circ \mathcal{D}\left(0, \frac{\pi}{2}, \frac{\pi}{2}\right)$$
$$= \mathcal{D}\left(-\frac{\pi}{2}, -\frac{\pi}{2}, \vartheta\right) \circ \mathcal{D}\left(0, \frac{\pi}{2}, \frac{\pi}{2}\right)$$

where we have used (7.123). This becomes the result found in (Risbo, 1996):

$$\mathcal{D}_y(\vartheta) = \mathcal{D}\left(-\frac{\pi}{2}, -\frac{\pi}{2}, 0\right) \circ \mathcal{D}_z(\vartheta) \circ \mathcal{D}\left(0, \frac{\pi}{2}, \frac{\pi}{2}\right). \tag{7.125}$$

It is elementary from geometric considerations. The y-axis of the operand is pre-rotated first to the negative x-axis then to the z-axis. The ϑ rotation is applied about the z-axis. Then the operand is transferred back to the negative x-axis then to the y-axis, of course, through the inverse of the pre-rotations. Figure 7.9 depicts a y-axis rotation through ϑ (in this figure ϑ is taken to be $\pi/3$). Figure 7.10 depicts the equivalent five rotations given in (7.125) that yield the same result.

In identity (7.125) the only rotations about the y-axis are fixed at angles $\pm\pi/2$, as depicted in Figure 7.10c and Figure 7.10e. The more general expression (7.124) involves only the $\pi/2$ rotation about the y-axis (but applied twice). All remaining rotations are about the z-axis, and are used to affect rotations through $\varphi + \pi/2$, $\vartheta + \pi$ and $\omega + \pi/2$. A key observation is that such z-axis rotations are much simpler algebraically to characterize given they have a diagonal operator matrix representation

$$\langle \mathcal{D}_z(\varphi)Y_p^q, Y_\ell^m \rangle = e^{-im\varphi}\delta_{\ell,p}^{m,q},$$

which can be inferred from (7.115); see also (Wigner, 1959, p. 167). An explicit representation of this operator matrix follows:

$$
\mathbf{D}_z(\varphi) =
\begin{pmatrix}
1 & 0 & 0 & 0 & 0 & 0 & 0 & 0 & 0 & \cdots \\
0 & e^{i\varphi} & 0 & 0 & 0 & 0 & 0 & 0 & 0 & \cdots \\
0 & 0 & 1 & 0 & 0 & 0 & 0 & 0 & 0 & \cdots \\
0 & 0 & 0 & e^{-i\varphi} & 0 & 0 & 0 & 0 & 0 & \cdots \\
0 & 0 & 0 & 0 & e^{i2\varphi} & 0 & 0 & 0 & 0 & \cdots \\
0 & 0 & 0 & 0 & 0 & e^{i\varphi} & 0 & 0 & 0 & \cdots \\
0 & 0 & 0 & 0 & 0 & 0 & 1 & 0 & 0 & \cdots \\
0 & 0 & 0 & 0 & 0 & 0 & 0 & e^{-i\varphi} & 0 & \cdots \\
0 & 0 & 0 & 0 & 0 & 0 & 0 & 0 & e^{-i2\varphi} & \cdots \\
\vdots & \vdots & \vdots & \vdots & \vdots & \vdots & \vdots & \vdots & \vdots & \ddots
\end{pmatrix}
$$

$$
= \operatorname{diag}\left(1;\ e^{j\varphi}, 1, e^{-j\varphi};\ e^{j2\varphi}, e^{j\varphi}, 1, e^{-j\varphi}, e^{-j2\varphi};\ \ldots\right).
$$

Problems

7.17. An alternative to identity (7.125) is

$$
\mathcal{D}_y(\vartheta) = \mathcal{D}\left(\frac{\pi}{2}, \frac{\pi}{2}, 0\right) \circ \mathcal{D}_z(\vartheta) \circ \mathcal{D}\left(0, -\frac{\pi}{2}, -\frac{\pi}{2}\right).
$$

Show this result and interpret it geometrically.

7.18. Find an analogous identity to (7.125) that transfers a z-axis φ rotation to the y-axis, through pre- and post-rotations, and interpret it geometrically.

7.11.4   Wigner d-matrix and Wigner D-matrix properties

We mention some further important properties of the Wigner d-matrix and the Wigner D-matrix. For a given m and m', the Wigner d-matrix obeys the following orthogonality relation

$$
\boxed{\int_0^\pi d_{m,m'}^\ell(\vartheta) d_{m,m'}^p(\vartheta) \sin\vartheta \, d\vartheta = \frac{2}{2\ell+1}\delta_{\ell,p},}
\tag{7.126}
$$

from which, and using the definition (7.115), we conclude that Wigner D-matrix is fully orthogonal over all three rotation angles in SO(3) as

$$
\boxed{\int_0^{2\pi}\int_0^\pi\int_0^{2\pi} D_{m,m'}^\ell(\varphi,\vartheta,\omega)\overline{D_{q,q'}^p(\varphi,\vartheta,\omega)} \sin\vartheta \, d\omega \, d\vartheta \, d\varphi = \frac{8\pi^2}{2\ell+1}\delta_{\ell,p}\delta_{m,q}\delta_{m',q'}.}
$$

$$
\tag{7.127}
$$

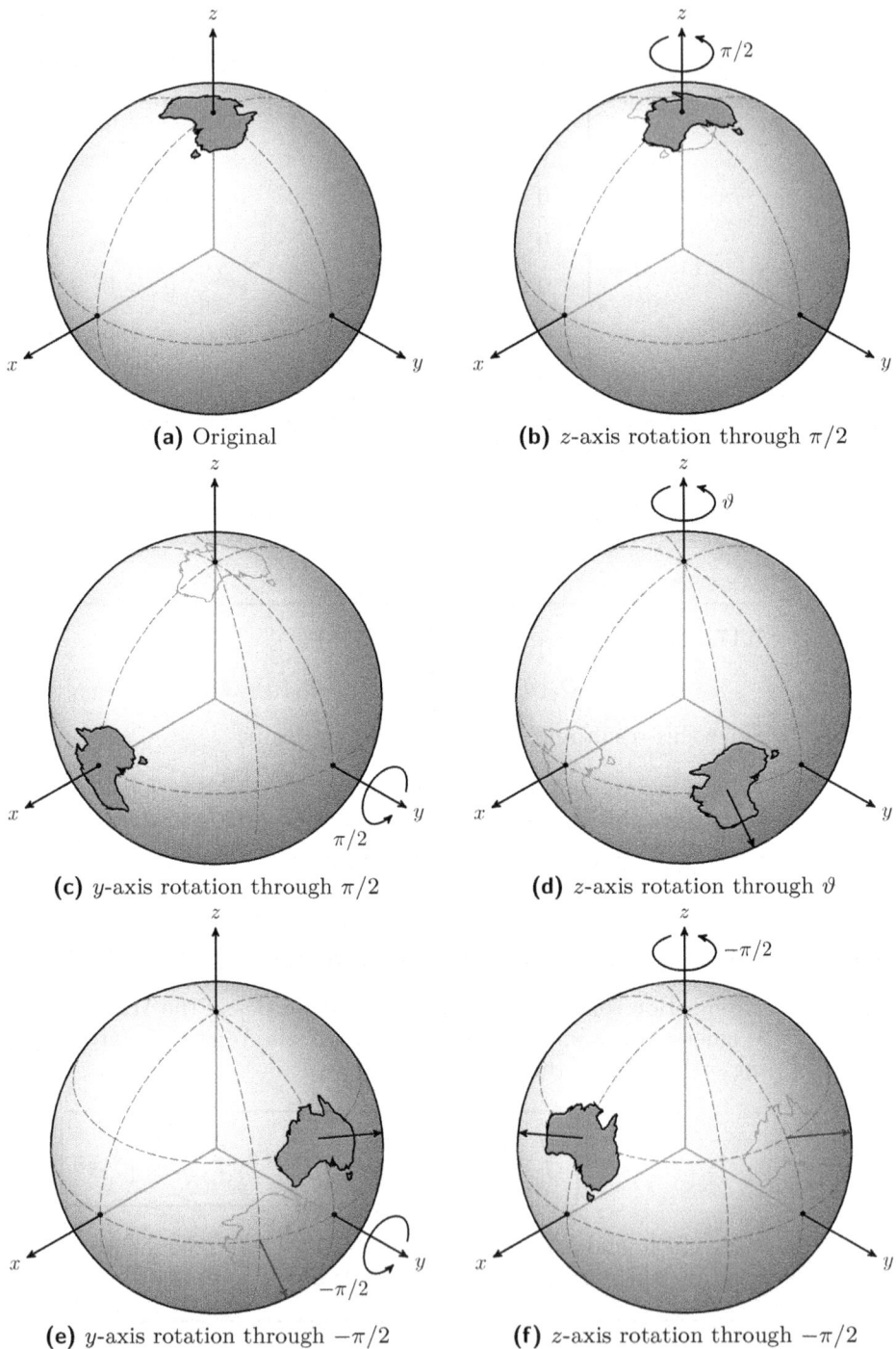

(a) Original   **(b)** z-axis rotation through $\pi/2$

(c) y-axis rotation through $\pi/2$ **(d)** z-axis rotation through ϑ

(e) y-axis rotation through $-\pi/2$ **(f)** z-axis rotation through $-\pi/2$

Figure 7.10 A demonstration that the five indicated elementary rotations realize a y-axis rotation through ϑ equivalent to that shown in Figure 7.9.

The shorthand notation will be

$$\langle D^\ell_{m,m'}, D^p_{q,q'} \rangle_{\mathrm{SO}(3)} = \frac{8\pi^2}{2\ell+1}\delta_{\ell,p}\delta_{m,q}\delta_{m',q'},$$ (7.128)

which signifies that the inner product is over all independent SO(3) rotation parameters and not over $\mathbb{S}^2$.

When $m' = 0$ the first rotation by ω around the z-axis has no effect on the Wigner D-matrix and can be taken as zero, $D^\ell_{m,0}(\varphi, \vartheta, 0)$. This $m' = 0$ Wigner D-matrix can be related to the spherical harmonic $Y^m_\ell(\vartheta, \varphi)$ through

$$D^\ell_{m,0}(\varphi, \vartheta, 0) = \sqrt{\frac{4\pi}{2\ell+1}}\overline{Y^m_\ell(\vartheta, \varphi)},$$ (7.129)

which also implies that

$$d^\ell_{m,0}(\vartheta) = \sqrt{\frac{(\ell-m)!}{(\ell+m)!}}P^m_\ell(\cos\vartheta).$$ (7.130)

Problems

7.19. Infer the orthogonality of the Wigner D-matrix (7.127) from the orthogonality identity (7.126).

7.20. Show (7.129) implies (7.130).

7.11.5   Wigner d-matrix symmetry relations

The Wigner d-matrix, $d^\ell_{m,m'}(\vartheta)$, satisfies a number of useful symmetry properties where the orders m and m' are exchanged, and the rotation about the y-axis, through ϑ, is negated $(-\vartheta)$ or supplemented $(\pi - \vartheta)$. These can be proven in a tedious manner algebraically from the expression (7.116). Doing so is somewhat against the spirit of the book where we prefer geometric methods. So the approach we take to demonstrating symmetry relations uses the inner product identity $d^\ell_{m,m'}(\vartheta) \triangleq \langle \mathcal{D}_y(\vartheta)Y^{m'}_\ell, Y^m_\ell \rangle$ given in (7.119).

Real-valuedness

One of the most useful properties of the Wigner d-matrix is that it is real-valued. This we pointed out at the time we gave without proof the algebraic definition (7.116), but this does not constitute an explanation of why they are real-valued. As a property it is somewhat unexpected. However, it is clear that Wigner regarded it as obvious, that is, obvious if you are a Martian. At its heart this is why the zyz rotation convention was preferred by Wigner. Now we explain why $d^\ell_{m,m'}(\vartheta)$ is real-valued without using (7.116).

Wigner's explanation would have gone as follows: the spherical harmonics have a "conjugate y-symmetry" about the xz-plane (or the $y = 0$ plane where either $\phi = 0$ or $\phi = \pi$) given by

$$Y_\ell^m(\theta, \phi) = \overline{Y_\ell^m(\theta, -\phi)}. \qquad (7.131)$$

Conjugate y-symmetry also holds for the complex conjugates of the spherical harmonics and for spherical harmonics rotated about the y-axis; see below.[22] Therefore, in evaluating the inner product in (7.119) the integrand is a product of conjugate y-symmetric functions, for every rotation ϑ. Then the contributions to the inner product from the negative-y hemisphere are the complex conjugates of the contributions from the positive-y hemisphere. This ensures that the contribution from the imaginary portions integrates to zero.

We can put the explanation above into action as follows. Define

$$I_\vartheta(\phi_1, \phi_2) \triangleq \int_{\phi_1}^{\phi_2} \int_0^\pi \left(\mathcal{D}_y(\vartheta) Y_\ell^{m'}\right)(\theta, \phi) \overline{Y_\ell^m(\theta, \phi)} \sin\theta \, d\theta \, d\phi.$$

Then $d_{m,m'}^\ell(\vartheta)$ is real-valued because

$$\begin{aligned}
d_{m,m'}^\ell(\vartheta) = I_\vartheta(0, 2\pi) &= I_\vartheta(0, \pi) + I_\vartheta(\pi, 2\pi) \\
&= I_\vartheta(0, \pi) + I_\vartheta(-\pi, 0) \\
&= I_\vartheta(0, \pi) + \overline{I_\vartheta(0, \pi)} = 2\,\mathfrak{Re}\big\{I_\vartheta(0, \pi)\big\}.
\end{aligned}$$

This follows because the integrand

$$\left(\mathcal{D}_y(\vartheta) Y_\ell^{m'}\right)(\theta, \phi) \overline{Y_\ell^m(\theta, \phi)}$$

is conjugate y-symmetric since

$$\begin{aligned}
\left(\mathcal{D}_y(\vartheta) Y_\ell^{m'}\right)(\theta, \phi) &= Y_\ell^{m'}(\theta - \vartheta, \phi) \\
&= \overline{Y_\ell^{m'}(\theta - \vartheta, -\phi)} = \overline{\left(\mathcal{D}_y(\vartheta) Y_\ell^{m'}\right)(\theta, -\phi)},
\end{aligned}$$

and

$$\overline{Y_\ell^m(\theta, \phi)} = Y_\ell^m(\theta, -\phi)$$

are conjugate y-symmetric using (7.131).

In short, the spherical harmonics have a conjugate y-symmetry and rotations about the y-axis preserve this property and that is why the Wigner d-matrix is real-valued. So we combine the identity (7.119) and its property of being real-valued into one equation

$$\textbf{Real-valued:} \quad d_{m,m'}^\ell(\vartheta) \triangleq \langle \mathcal{D}_y(\vartheta) Y_\ell^{m'}, Y_\ell^m \rangle = \langle Y_\ell^m, \mathcal{D}_y(\vartheta) Y_\ell^{m'} \rangle. \qquad (7.132)$$

----

[22] The conjugate y-symmetry explains why, for example, the spherical harmonics are real-valued along $y = 0$.

Symmetry relations

The operator $\mathcal{D}_y(-\vartheta)$ being the adjoint of $\mathcal{D}_y(\vartheta)$ can be used to establish the first symmetry relation (see Problem 7.22)

$$
\boxed{\textbf{Symmetry 1:}\quad d^\ell_{m,m'}(\vartheta) = d^\ell_{m',m}(-\vartheta).}
\tag{7.133}
$$

Using the conjugation property of spherical harmonics, (7.26), p. 192, compactly written $\overline{Y^m_\ell} = (-1)^m Y^{-m}_\ell$, and the identity (7.132), we see

$$
\begin{aligned}
d^\ell_{m,m'}(\vartheta) &= \langle \mathcal{D}_y(\vartheta) Y^{m'}_\ell, Y^m_\ell \rangle = \overline{\langle \mathcal{D}_y(\vartheta) Y^{m'}_\ell, Y^m_\ell \rangle} \\
&= \langle \overline{\mathcal{D}_y(\vartheta) Y^{m'}_\ell}, \overline{Y^m_\ell} \rangle = \langle \mathcal{D}_y(\vartheta) \overline{Y^{m'}_\ell}, \overline{Y^m_\ell} \rangle \\
&= \langle \mathcal{D}_y(\vartheta)(-1)^{m'} Y^{-m'}_\ell, (-1)^m Y^{-m}_\ell \rangle \\
&= (-1)^{m-m'} \langle \mathcal{D}_y(\vartheta) Y^{-m'}_\ell, Y^{-m}_\ell \rangle = (-1)^{m-m'} d^\ell_{-m,-m'}(\vartheta),
\end{aligned}
$$

where we have used the property that when applying a y-axis rotation the linear combination (7.114) only has real weights, as given in (7.132).[23] That is, we have the second symmetry relation

$$
\boxed{\textbf{Symmetry 2:}\quad d^\ell_{m,m'}(\vartheta) = (-1)^{m-m'} d^\ell_{-m,-m'}(\vartheta).}
\tag{7.134}
$$

The next symmetry property of the Wigner d-matrix comes from the following special case of the rotation identity (7.123) given by

$$
\mathcal{D}_y(\vartheta) = \mathcal{D}_z(\pi) \circ \mathcal{D}_y(-\vartheta) \circ \mathcal{D}_z(\pi)
$$

married with the following property of spherical harmonics

$$
\boxed{\left(\mathcal{D}_z(\pi) Y^m_\ell \right)(\theta, \phi) = Y^m_\ell(\theta, \phi - \pi) = e^{-im\pi} Y^m_\ell(\theta, \phi),}
$$

compactly written, $\mathcal{D}_z(\pi) Y^m_\ell = (-1)^m Y^m_\ell$. Pursuing this we have

$$
\begin{aligned}
d^\ell_{m,m'}(\vartheta) &= \langle \mathcal{D}_y(\vartheta) Y^{m'}_\ell, Y^m_\ell \rangle \\
&= \langle \mathcal{D}_z(\pi) \circ \mathcal{D}_y(-\vartheta) \circ \mathcal{D}_z(\pi) Y^{m'}_\ell, Y^m_\ell \rangle \\
&= \langle \mathcal{D}_z(\pi) Y^{m'}_\ell, \mathcal{D}_y(\vartheta) \circ \mathcal{D}_z(\pi) Y^m_\ell \rangle \\
&= (-1)^{m-m'} \langle Y^{m'}_\ell, \mathcal{D}_y(\vartheta) Y^m_\ell \rangle \\
&= (-1)^{m-m'} \langle \mathcal{D}_y(\vartheta) Y^m_\ell, Y^{m'}_\ell \rangle = (-1)^{m-m'} d^\ell_{m',m}(\vartheta)
\end{aligned}
$$

where we have employed various adjoint transformations. Therefore, we obtain our third symmetry relation

$$
\boxed{\textbf{Symmetry 3:}\quad d^\ell_{m,m'}(\vartheta) = (-1)^{m-m'} d^\ell_{m',m}(\vartheta).}
\tag{7.135}
$$

---

[23] The parity term $(-1)^{m-m'}$ can be written in different ways such as $(-1)^{m+m'}$. We choose to write it in the form that is most commonly used in the literature.

Finally, we can combine (7.134) and (7.135) to obtain

$$\boxed{\textbf{Symmetry 4:}\quad d_{m,m'}^{\ell}(\vartheta) = d_{-m',-m}^{\ell}(\vartheta).}\qquad(7.136)$$

These symmetries are graphically depicted in Figure 7.11.[24] Because of symmetry relations 2, 3 and 4 we can reduce the number of computations required when determining the $(2\ell + 1)^2$ entries of $d_{m,m'}^{\ell}(\vartheta)$ for each ℓ. The amount of reduction is easily determined by inspection. We only need to compute $(\ell + 1)^2$ of the $d_{m,m'}^{\ell}(\vartheta)$ and the remainder of the $(2\ell + 1)^2$ entries can be inferred (for $\ell = 0$ there is only one entry and therefore there is no strict reduction). This reduction is applicable for all degrees $\ell > 0$ under consideration and asymptotically approaches $1/4$ as ℓ increases.

For $\ell = 0$ we have the 1×1 matrix $\mathbf{D}^0(0, \vartheta, 0) = (1)$, and for $\ell = 1$ the 3×3 matrix

$$
\begin{array}{ccc}
m' = -1 & m' = 0 & m' = +1
\end{array}
$$

$$
\mathbf{D}^1(0, \vartheta, 0) =
\begin{pmatrix}
\dfrac{1 + \cos\vartheta}{2} & \dfrac{\sin\vartheta}{\sqrt{2}} & \dfrac{1 - \cos\vartheta}{2} \\[2.2ex]
-\dfrac{\sin\vartheta}{\sqrt{2}} & \cos\vartheta & \dfrac{\sin\vartheta}{\sqrt{2}} \\[2.2ex]
\dfrac{1 - \cos\vartheta}{2} & -\dfrac{\sin\vartheta}{\sqrt{2}} & \dfrac{1 + \cos\vartheta}{2}
\end{pmatrix}
\begin{array}{l}
m = -1 \\[2.2ex]
m = 0, \\[2.2ex]
m = +1
\end{array}
\qquad(7.137)
$$

where $(2\ell+1) \times (2\ell+1)$ matrix $\mathbf{D}^\ell(\varphi, \vartheta, \omega)$ is the ℓ-th block in the block-diagonal matrix (7.120).[25] As an example of an entry in (7.137), for $\ell = 1, m = 1, m' = 0$ we have

$$d_{1,0}^1(\vartheta) = (-1)^{(0-0+1)}\sqrt{2}\cos\frac{\vartheta}{2}\sin\frac{\vartheta}{2} = -\frac{\sin\vartheta}{\sqrt{2}},$$

where only $n = 0$ leads to a non-zero summand in (7.116). Finally in (7.137) we have highlighted four entries (being $(\ell + 1)^2$ in number) which are sufficient to determine the remainder of the nine entries by using the symmetry relations.

> **Remark 7.8.** This example goes back to Wigner's original findings, which he published in 1931 (Wigner, 1959, p. 168). The same matrix can be found in (Sakurai, 1994, p. 195), but appears as the transpose of (7.137) because Sakurai's ordering is $m = +1, 0, -1$. To emphasize the ordering in this book we have augmented the matrix (7.137) with input column indices $m' = -1, 0, +1$ and output row indices $m = -1, 0, +1$.   □

A final type of symmetry for the Wigner d-matrix arises when considering a rotation through $\pi - \vartheta$ about the y-axis and relating it to a rotation through ϑ

---

[24] There are further solid and dashed lines that could be depicted between the matrix and its adjoint, but have been omitted for clarity.

[25] $\mathbf{D}^\ell(\varphi, \vartheta, \omega)$ is referred to as the $(2\ell+1)$-dimensional irreducible representation of the operator $\mathcal{D}(\varphi, \vartheta, \omega)$, (Sakurai, 1994, p. 192).

about the y-axis (Wigner, 1959, p. 216). A representative symmetry relation is
given by

$$\boxed{\textbf{Symmetry 5:}\quad d_{m,m'}^{\ell}(\pi - \vartheta) = (-1)^{\ell-m} d_{m',-m}^{\ell}(\vartheta).} \tag{7.138}$$

To prove this relation we use the following property of spherical harmonics
(see Problem 7.25)

$$\boxed{\left(\mathcal{D}_y(\pi)Y_\ell^m\right)(\theta, \phi) = (-1)^{\ell-m} Y_\ell^{-m}(\theta, \phi),} \tag{7.139}$$

compactly written $\mathcal{D}_y(\pi)Y_\ell^m = (-1)^{\ell-m}Y_\ell^{-m}$. We expand the left-hand side of
(7.138) using various adjoint manipulations analogous to earlier derivations, as
well as (7.139), which leads to

$$
\begin{aligned}
d_{m,m'}^{\ell}(\pi - \vartheta) &= \left\langle \mathcal{D}_y(\pi - \vartheta)Y_\ell^{m'}, Y_\ell^m \right\rangle \\
&= \left\langle \mathcal{D}_y(-\vartheta)Y_\ell^{m'}, \mathcal{D}_y(\pi)Y_\ell^m \right\rangle \\
&= \left\langle \mathcal{D}_y(\vartheta) \circ \mathcal{D}_y(\pi)Y_\ell^m, Y_\ell^{m'} \right\rangle \\
&= (-1)^{\ell-m}\left\langle \mathcal{D}_y(\vartheta)Y_\ell^{-m}, Y_\ell^{m'} \right\rangle = (-1)^{\ell-m} d_{m',-m}^{\ell}(\vartheta),
\end{aligned}
$$

which is the right-hand side of (7.138). An alternative proof of (7.138) from
the algebraic definition of the Wigner d-matrix, (7.116), is left to Problem 7.24.
Further symmetries in the spirit of (7.138) are considered in Problem 7.27.

Problems

7.21. Define the subspace

$$\mathcal{H}_\ell \triangleq \left\{ f \in L^2(\mathbb{S}^2) \colon f = \sum_{m=-\ell}^{\ell} \langle f, Y_\ell^m \rangle Y_\ell^m \right\} \subset L^2(\mathbb{S}^2) \tag{7.140}$$

whose members are signals containing only a single degree ℓ. Show that the
generalized Parseval relation for spherical harmonics, (7.42), p. 196, becomes

$$\sum_{m=-\ell}^{\ell} \langle f, Y_\ell^m \rangle \langle Y_\ell^m, g \rangle = \langle f, g \rangle, \quad f, g \in \mathcal{H}_\ell \tag{7.141}$$

for signals restricted to $\mathcal{H}_\ell$.

7.22. Show that the adjoint of the operator $\mathcal{D}_y(\vartheta)$ is $\mathcal{D}_y(-\vartheta)$; that is,

$$\langle \mathcal{D}_y(\vartheta)f, g \rangle = \langle f, \mathcal{D}_y(-\vartheta)g \rangle, \quad \forall f, g \in L^2(\mathbb{S}^2). \tag{7.142}$$

Given that the Wigner d-matrix is real-valued, prove the first symmetry relation
(7.133)

$$d_{m,m'}^{\ell}(\vartheta) = d_{m',m}^{\ell}(-\vartheta).$$

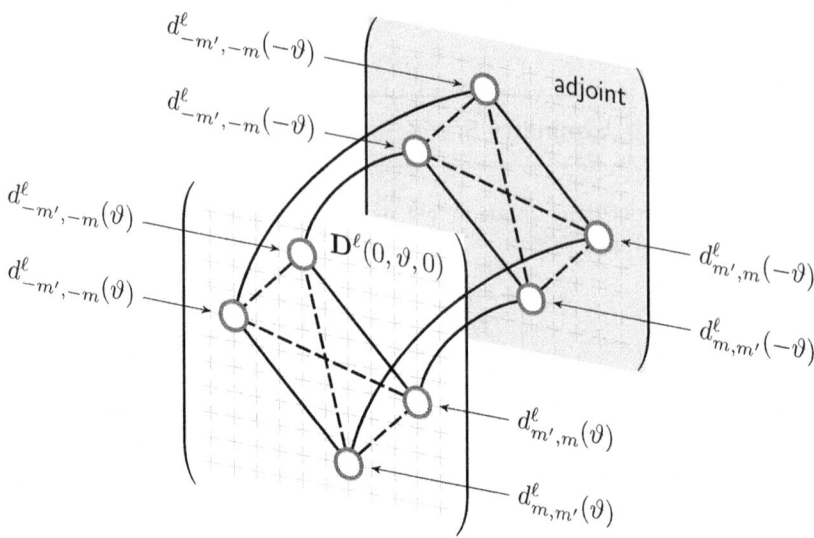

Figure 7.11 Symmetries 1, 2, 3 and 4 of the Wigner d-matrix $d_{m,m'}^\ell(\vartheta)$. In the foreground is a representation of the $(2\ell + 1) \times (2\ell + 1)$ matrix for degree ℓ, with ordering $m, m' = -\ell, \ldots, \ell$. In the background is the adjoint matrix. Equal elements (symmetries 1 and 4 of the Wigner d-matrix) are connected with a solid line and those that differ by a multiplicative factor of $(-1)^{m-m'}$ (symmetries 2 and 3 of the Wigner d-matrix) are connected with a dashed line.

7.23. Use the results from Problem 7.21 and Problem 7.22, Wigner d-matrix identity (7.119), and the property that rotations do not mix degrees, to show

$$\sum_{m'=-\ell}^{\ell} d_{m,m'}^\ell(\vartheta_1) d_{m',q}^\ell(\vartheta_2) = d_{m,q}^\ell(\vartheta_1 + \vartheta_2) \qquad (7.143)$$

and interpret the results.

7.24. Use (7.116) and (7.143) to give an algebraic proof of the fifth symmetry relation (7.138), p. 237, repeated here,

$$d_{m,m'}^\ell(\pi - \vartheta) = (-1)^{\ell-m} d_{m',-m}^\ell(\vartheta).$$

7.25. Using geometric arguments establish the identity (7.139)

$$\left(\mathcal{D}_y(\pi) Y_\ell^m\right)(\theta, \phi) = (-1)^{\ell-m} Y_\ell^{-m}(\theta, \phi).$$

(Hint: first try to prove the parity relation of the spherical harmonics

$$Y_\ell^m(\pi - \theta, \pi + \phi) = (-1)^\ell Y_\ell^m(\theta, \phi),$$

where the angular transformation corresponds to a reflection about the origin.)

7.26. Prove the following two Wigner d-matrix identities

$$d^\ell_{m,m'}(0) = \delta_{m,m'},$$

and

$$d^\ell_{m,m'}(\pi) = (-1)^{\ell+m'}\delta_{m,-m'}.$$

7.27 (Further symmetry relations).   Using methods similar to the proof of symmetry relation (7.138), show

Symmetry 6: $d^\ell_{m,m'}(\pi - \vartheta) = (-1)^{\ell-m}d^\ell_{-m,m'}(-\vartheta),$ (7.144)

Symmetry 7: $d^\ell_{m,m'}(\pi + \vartheta) = (-1)^{\ell-m}d^\ell_{-m,m'}(\vartheta),$ (7.145)

and

Symmetry 8: $d^\ell_{m,m'}(\pi + \vartheta) = (-1)^{\ell-m}d^\ell_{m',-m}(-\vartheta).$ (7.146)

7.11.6   Fourier series representation of Wigner D-matrix

Taking cognizance of the rotation operator decomposition (7.124) we seek an alternative expression for the Wigner D-matrix, (7.115), repeated here

$$D^\ell_{m,m'}(\varphi, \vartheta, \omega) \triangleq e^{-im\varphi}d^\ell_{m,m'}(\vartheta)e^{-im'\omega}.$$

Given that φ and ω are involved in an elementary way then it is the part representing the rotation about the y-axis through angle ϑ, $d^\ell_{m,m'}(\vartheta)$, that can be targeted for simplification. The advantage of working with (7.124) is that the y-axis rotations do not depend on φ, ϑ and ω. Therefore, we define

$$\Delta^\ell_{m,m'} \triangleq d^\ell_{m,m'}(\pi/2).$$ (7.147)

Using (7.124) we can write

$$\mathcal{D}(\varphi, \vartheta, \omega)f = \mathcal{D}\left(\varphi + \frac{\pi}{2}, \frac{\pi}{2}, \vartheta + \pi\right)g,$$ (7.148)

where we have defined

$$g \triangleq \mathcal{D}\left(0, \frac{\pi}{2}, \omega + \frac{\pi}{2}\right)f.$$ (7.149)

The key identity we wish to redevelop is

$$\langle \mathcal{D}(\varphi, \vartheta, \omega)f, Y^m_\ell \rangle = \sum_{m'=-\ell}^{\ell} D^\ell_{m,m'}(\varphi, \vartheta, \omega)\langle f, Y^{m'}_\ell \rangle,$$ (7.150)

which is just (7.114) rewritten.

From (7.148) we can utilize (7.114), (7.115) and (7.147) to obtain

$$\langle \mathcal{D}(\varphi, \vartheta, \omega) f, Y_\ell^m \rangle = \left\langle \mathcal{D}\left(\varphi + \frac{\pi}{2}, \frac{\pi}{2}, \vartheta + \pi\right) g, Y_\ell^m \right\rangle$$

$$= \sum_{m''=-\ell}^{\ell} D_{m,m''}^\ell \left(\varphi + \frac{\pi}{2}, \frac{\pi}{2}, \vartheta + \pi\right) \langle g, Y_\ell^{m''} \rangle$$

$$= \sum_{m''=-\ell}^{\ell} e^{-im\varphi} i^{-m} \Delta_{m,m''}^\ell e^{-im''\vartheta} (-1)^{m''} \langle g, Y_\ell^{m''} \rangle. \qquad (7.151)$$

Similarly, from (7.149), and using (7.114), (7.115) and (7.147) again, we arrive at

$$\langle g, Y_\ell^{m''} \rangle = \left\langle \mathcal{D}\left(0, \frac{\pi}{2}, \omega + \frac{\pi}{2}\right) f, Y_\ell^{m''} \right\rangle$$

$$= \sum_{m'=-\ell}^{\ell} D_{m'',m'}^\ell \left(0, \frac{\pi}{2}, \omega + \frac{\pi}{2}\right) \langle f, Y_\ell^{m'} \rangle$$

$$= \sum_{m'=-\ell}^{\ell} \Delta_{m'',m'}^\ell e^{-im'\omega} i^{-m'} \langle f, Y_\ell^{m'} \rangle. \qquad (7.152)$$

Now we can substitute (7.152) into (7.151) to yield

$$\langle \mathcal{D}(\varphi, \vartheta, \omega) f, Y_\ell^m \rangle = \sum_{m''=-\ell}^{\ell} e^{-im\varphi} i^{-m} \Delta_{m,m''}^\ell e^{-im''\vartheta} (-1)^{m''}$$

$$\times \sum_{m'=-\ell}^{\ell} \Delta_{m'',m'}^\ell e^{-im'\omega} i^{-m'} \langle f, Y_\ell^{m'} \rangle$$

$$= \sum_{m'=-\ell}^{\ell} i^{m-m'} e^{-im\varphi} e^{-im'\omega} \sum_{m''=-\ell}^{\ell} \Delta_{m'',m}^\ell \Delta_{m'',m'}^\ell e^{-im''\vartheta} \langle f, Y_\ell^{m'} \rangle,$$

where we have used the identity $\Delta_{m,m''}^\ell = (-1)^{m-m''} \Delta_{m'',m}^\ell$, which is a special case of (7.135) with $\vartheta = \pi/2$. Comparing this with (7.150) we infer

$$\boxed{D_\ell^{m,m'}(\varphi, \vartheta, \omega) = e^{-im\varphi} \left(i^{m-m'} \sum_{m''=-\ell}^{\ell} \Delta_{m'',m}^\ell \Delta_{m'',m'}^\ell e^{-im''\vartheta}\right) e^{-im'\omega},} \qquad (7.153)$$

which also means

$$\boxed{d_{m,m'}^\ell(\vartheta) = i^{m-m'} \sum_{m''=-\ell}^{\ell} \Delta_{m'',m}^\ell \Delta_{m'',m'}^\ell e^{-im''\vartheta}.} \qquad (7.154)$$

Remark 7.9. This representation of the Wigner d-matrix, (7.154), is of the form of a finite Fourier series. And similarly, the Wigner D-matrix admits a multi-dimensional Fourier series representation (7.153). □

Remark 7.10. Given that $d_{m,m'}^{\ell}(\vartheta)$ and $\Delta_{m,m'}^{\ell}$, are both real-valued then (7.154) further simplifies (Risbo, 1996). A deeper study into recursive computations involving the Wigner d-matrix can be found in (Trapani and Navaza, 2006).   □

7.11.7   Spherical harmonics revisited

We can exploit the relationship between the Wigner D-matrix and the spherical harmonics, (7.129), to establish

$$
\begin{aligned}
Y_\ell^m(\theta,\phi) &= \sqrt{\frac{2\ell+1}{4\pi}}\overline{D_{m,0}^{\ell}(\phi,\theta,0)} \\
&= \sqrt{\frac{2\ell+1}{4\pi}}e^{im\phi}d_{m,0}^{\ell}(\theta),
\end{aligned}
\tag{7.155}
$$

which, when compared with $Y_\ell^m(\theta,\phi) = N_\ell^m P_\ell^m(\cos\theta)e^{im\phi}$, implies (7.130).

From (7.154), we have a Fourier series representation, which can be conjugated, because $d_{m,0}^{\ell}(\theta)$ is real-valued

$$
\begin{aligned}
d_{m,0}^{\ell}(\theta) &= i^m \sum_{m'=-\ell}^{\ell} \Delta_{m',m}^{\ell}\Delta_{m',0}^{\ell}e^{-im'\theta} \\
&= (-i)^m \sum_{m'=-\ell}^{\ell} \Delta_{m',m}^{\ell}\Delta_{m',0}^{\ell}e^{im'\theta},
\end{aligned}
\tag{7.156}
$$

and so we can write

$$
\boxed{Y_\ell^m(\theta,\phi) = \sqrt{\frac{2\ell+1}{4\pi}}(-i)^m \sum_{m'=-\ell}^{\ell} \Delta_{m',m}^{\ell}\Delta_{m',0}^{\ell}e^{im'\theta}e^{im\phi},}
$$

where $\Delta_{m',m}^{\ell} = d_{m',m}^{\ell}(\pi/2)$, (7.147). A generalization is treated in (McEwen and Wiaux, 2011). Following (Risbo, 1996), this is in a form that computationally can be assisted by multi-dimensional Fast Fourier Transform (FFT), which we now sketch.

7.11.8   Fast computation

Consider the expansion of a bandlimited function in the spherical harmonics written in the form (7.155)

$$
f(\theta,\phi) = \sum_{\ell=0}^{L} \sum_{m=-\ell}^{\ell} (f)_\ell^m \sqrt{\frac{2\ell+1}{4\pi}}e^{im\phi}d_{m,0}^{\ell}(\theta)
$$

with $d_{m,0}^{\ell}(\theta)$ given as in (7.156). This can be written as a two-dimensional classical Fourier expansion (Risbo, 1996), amenable to efficient evaluation using the FFT,

$$
f(\theta,\phi) = \sum_{m=-L}^{L} \sum_{m'=-L}^{L} F_{m,m'}e^{im\phi}e^{im'\theta},
$$

where

$$F_{m,m'} \triangleq \sum_{\ell=\ell(m,m')}^{L} \sqrt{\frac{2\ell+1}{4\pi}} (-i)^m \Delta_{m',m}^\ell \Delta_{m',0}^\ell (f)_\ell^m,$$

and $\ell(m,m') \triangleq \max(|m|,|m'|)$, which ensures that the sum only includes terms of compatible constituent degree, ℓ, and orders, m and m'.

Fast algorithms that take advantage of multi-dimensional FFTs based on the Fourier series of the Wigner d-matrix (or special case the associated Legendre functions) whenever there are summations involving the Wigner D-matrix (or special case spherical harmonics) have been developed in a number of contexts (Risbo, 1996; Wandelt and Górski, 2001; McEwen et al., 2007; Khalid et al., 2012b, 2013a).

7.12   Projection into $\mathcal{H}^0(\mathbb{S}^2)$

Define the operator $\mathcal{B}^0$ to project a signal f, which is not necessarily azimuthally symmetric, into the subspace of azimuthally symmetric signals $\mathcal{H}^0(\mathbb{S}^2)$ (7.75). Since $\{Y_\ell^0\}_{\ell=0}^\infty$ is complete in $\mathcal{H}^0(\mathbb{S}^2)$, we can write

$$\begin{aligned}
(\mathcal{B}^0 f)(\widehat{\boldsymbol{u}}) &= (\mathcal{B}^0 f)(\theta,\phi) \\
&\triangleq \sum_{\ell=0}^\infty (f)_\ell^0 Y_\ell^0(\theta).
\end{aligned} \tag{7.157}$$

The operator matrix elements are therefore given by

$$\boxed{b_{\ell,p}^{m,q} \triangleq \langle \mathcal{B}^0 Y_p^q, Y_\ell^m \rangle = \delta_{\ell,p}\delta_{0,q}\delta_{0,m}.} \tag{7.158}$$

The matrix is sparse-diagonal and looks like

$$\mathbf{B}^0 = \begin{pmatrix}
1 & 0 & 0 & 0 & 0 & 0 & 0 & 0 & 0 & \cdots \\
0 & 0 & 0 & 0 & 0 & 0 & 0 & 0 & 0 & \cdots \\
0 & 0 & 1 & 0 & 0 & 0 & 0 & 0 & 0 & \cdots \\
0 & 0 & 0 & 0 & 0 & 0 & 0 & 0 & 0 & \cdots \\
0 & 0 & 0 & 0 & 0 & 0 & 0 & 0 & 0 & \cdots \\
0 & 0 & 0 & 0 & 0 & 0 & 0 & 0 & 0 & \cdots \\
0 & 0 & 0 & 0 & 0 & 0 & 1 & 0 & 0 & \cdots \\
0 & 0 & 0 & 0 & 0 & 0 & 0 & 0 & 0 & \cdots \\
0 & 0 & 0 & 0 & 0 & 0 & 0 & 0 & 0 & \cdots \\
\vdots & \vdots & \vdots & \vdots & \vdots & \vdots & \vdots & \vdots & \vdots & \ddots
\end{pmatrix}$$

$$= \mathrm{diag}(1;\ 0,1,0;\ 0,0,1,0,0;\ \ldots).$$

The operator kernel can be found using (7.89)

$$B^0(\widehat{\boldsymbol{u}}, \widehat{\boldsymbol{v}}) = \sum_{\ell=0}^{\infty} Y_\ell^0(\widehat{\boldsymbol{u}})\overline{Y_\ell^0}(\widehat{\boldsymbol{v}}), \tag{7.159}$$

and is used in the integral operator form

$$\big(\mathcal{B}^0 f\big)(\widehat{\boldsymbol{u}}) = \int_{\mathbb{S}^2} B^0(\widehat{\boldsymbol{u}}, \widehat{\boldsymbol{v}}) f(\widehat{\boldsymbol{v}})\, ds(\widehat{\boldsymbol{v}}).$$

Projection into $\mathcal{H}^0(\mathbb{S}^2)$ is a spatially invariant operation and, as such, does not mix degrees.

We can prove that the projection $\mathcal{B}^0 f$ into $\mathcal{H}^0(\mathbb{S}^2)$ can be interpreted as the averaging of f over all rotations about the z-axis. To see this, we first apply the rotation operator $\mathcal{D}(0, 0, \omega)$ to f and then average $\big(\mathcal{D}(0, 0, \omega)f\big)(\widehat{\boldsymbol{u}})$, at a given $\widehat{\boldsymbol{u}}$, over all values of ω to write

$$\widetilde{f}(\widehat{\boldsymbol{u}}) \triangleq \frac{1}{2\pi} \int_0^{2\pi} \big(\mathcal{D}(0, 0, \omega)f\big)(\widehat{\boldsymbol{u}})\, d\omega. \tag{7.160}$$

Now we use the spherical harmonic expansion of $\mathcal{D}(0, 0, \omega)f$, (7.114) and (7.115), which gives

$$\widetilde{f}(\widehat{\boldsymbol{u}}) = \frac{1}{2\pi} \int_0^{2\pi} \sum_{\ell,m} \Big(\sum_{m'=-\ell}^{\ell} D_{m,m'}^{\ell}(0,0,\omega)(f)_\ell^{m'} \Big) Y_\ell^m(\widehat{\boldsymbol{u}})\, d\omega$$

$$= \frac{1}{2\pi} \sum_{\ell,m} \sum_{m'=-\ell}^{\ell} (f)_\ell^{m'} Y_\ell^m(\widehat{\boldsymbol{u}}) \int_0^{2\pi} D_{m,m'}^{\ell}(0,0,\omega)\, d\omega$$

$$= \sum_{\ell,m} \sum_{m'=-\ell}^{\ell} (f)_\ell^{m'} Y_\ell^m(\widehat{\boldsymbol{u}}) d_{m,m'}^{\ell}(0) \underbrace{\frac{1}{2\pi} \int_0^{2\pi} e^{-im'\omega}\, d\omega}_{\delta_{m',0}}$$

$$= \sum_{\ell,m} (f)_\ell^0 Y_\ell^m(\widehat{\boldsymbol{u}}) \underbrace{d_{m,0}^{\ell}(0)}_{\delta_{m,0}}$$

$$= \sum_{\ell=0}^{\infty} (f)_\ell^0 Y_\ell^0(\widehat{\boldsymbol{u}}) = \big(\mathcal{B}^0 f\big)(\widehat{\boldsymbol{u}}).$$

Here we have used $d_{m,0}^{\ell}(0) = \delta_{m,0}$ (see Problem 7.26 or Problem 7.28 below). In summary, average $\widetilde{f}(\widehat{\boldsymbol{u}})$ defined in (7.160) is identical to $\big(\mathcal{B}^0 f\big)(\theta)$ in (7.157).

This interpretation will be useful when we deal with different definitions of convolution on 2-sphere in Chapter 9, p. 293.

7.28. Using (7.116) prove that

$$d^\ell_{m,0}(0) = \delta_{m,0}.$$

7.29. Show that

$$\big(\mathcal{D}(0,0,\omega)f\big)(\theta,\phi) = f(\theta,\phi-\omega)$$

and thereby give an alternative proof that

$$\big(\mathcal{B}^0 f\big)(\widehat{\boldsymbol{u}}) = \frac{1}{2\pi}\int_0^{2\pi}\big(\mathcal{D}(0,0,\omega)f\big)(\widehat{\boldsymbol{u}})\,d\omega.$$

(Hint: use the definition of the spherical harmonics (7.19), p. 189.)

7.13 Rotation of azimuthally symmetric signals

For an azimuthally symmetric signal, written as $f(\theta)$ with spherical harmonic coefficients $(f)^0_\ell$, the full SO(3) rotation operator (7.113) has the same effect as a special type of rotation operator. This is because the initial rotation around the z-axis has no effect and hence we can use operator $\mathcal{D}(\varphi,\vartheta,0)$, which has the effect of rotating a signal first through an angle of ϑ about the y-axis followed by an angle of φ about the z-axis.

Given the identity (7.129), the only relevant Wigner D-matrix when operating on azimuthally symmetric signals occurs when $m' = 0$ and we obtain

$$\big\langle\mathcal{D}(\varphi,\vartheta,0)f,Y_\ell^m\big\rangle = \sqrt{\frac{4\pi}{2\ell+1}}\,\overline{Y_\ell^m(\vartheta,\varphi)}(f)^0_\ell, \quad f \in \mathcal{H}^0(\mathbb{S}^2). \qquad (7.161)$$

To use this expression in the event that the supplied signal is not azimuthally symmetric requires a simple pre-composition by the projection operator $\mathcal{B}^0$ defined in (7.157)

$$\big\langle\mathcal{D}(\varphi,\vartheta,0)\circ\mathcal{B}^0 f,Y_\ell^m\big\rangle = \sqrt{\frac{4\pi}{2\ell+1}}\,\overline{Y_\ell^m(\vartheta,\varphi)}(f)^0_\ell, \quad f \in L^2(\mathbb{S}^2),$$

where for azimuthally symmetric signals $\mathcal{B}^0 f = f$, and this reduces to (7.161). The composite operator, $\mathcal{D}(\varphi,\vartheta,0)\circ\mathcal{B}^0$, has matrix elements given by

$$b^{m,q}_{\ell,p} \triangleq \big\langle\mathcal{D}(\varphi,\vartheta,0)\circ\mathcal{B}^0 Y_p^q,Y_\ell^m\big\rangle = \sqrt{\frac{4\pi}{2\ell+1}}\,\overline{Y_\ell^m(\vartheta,\varphi)}\delta_{\ell,p}\delta_{0,q},$$

which creates a special type of block diagonal operator matrix structure and referring to (7.120) and (7.129) only the middle column of each block diagonal part can be non-zero,

$$
\mathbf{D}(\varphi,\vartheta,0)\mathbf{B}^0 =
\begin{pmatrix}
D^0_{0,0} & 0 & 0 & 0 & 0 & 0 & 0 & 0 & 0 & \cdots \\
0 & 0 & D^1_{-1,0} & 0 & 0 & 0 & 0 & 0 & 0 & \cdots \\
0 & 0 & D^1_{0,0} & 0 & 0 & 0 & 0 & 0 & 0 & \cdots \\
0 & 0 & D^1_{1,0} & 0 & 0 & 0 & 0 & 0 & 0 & \cdots \\
0 & 0 & 0 & 0 & 0 & 0 & D^2_{-2,0} & 0 & 0 & \cdots \\
0 & 0 & 0 & 0 & 0 & 0 & D^2_{-1,0} & 0 & 0 & \cdots \\
0 & 0 & 0 & 0 & 0 & 0 & D^2_{0,0} & 0 & 0 & \cdots \\
0 & 0 & 0 & 0 & 0 & 0 & D^2_{1,0} & 0 & 0 & \cdots \\
0 & 0 & 0 & 0 & 0 & 0 & D^2_{2,0} & 0 & 0 & \cdots \\
\vdots & \vdots & \vdots & \vdots & \vdots & \vdots & \vdots & \vdots & \vdots & \ddots
\end{pmatrix}
$$

7.14   Operator classification based on operator matrix

Based on the various examples that we have seen so far, we can classify operators based on the structure of their operator matrix or equivalently based on how they act on input degrees and orders.

C1: There are operators that mix neither degrees nor orders. The operator matrix is fully diagonal and is of the form

$$
b^{m,q}_{\ell,p} = b^m_\ell \delta^{m,q}_{\ell,p}. \tag{7.162}
$$

Examples include spectral truncation operator, $\mathcal{B}_L$ (where $b^m_\ell = 1$ only when $\ell \in \{0,1,\ldots,L\}$ and $m \in \{-\ell, -\ell+1, \ldots, \ell\}$), or projection into $\mathcal{H}^0(\mathbb{S}^2)$, $\mathcal{B}^0$ (where $b^m_\ell = \delta_{m,0}$). In later chapters, we will see other examples of fully diagonal operator matrices such as isotropic convolution. These operators are classified as spatially invariant.

C2: There are operators that do not mix degrees, but do mix orders. The operator matrix is block diagonal and is of the form

$$
b^{m,q}_{\ell,p} = b^{m,q}_\ell \delta_{\ell,p}. \tag{7.163}
$$

An example is the rotation, $\mathcal{D}(\varphi,\vartheta,\omega)$, where $b^{m,q}_\ell = D^\ell_{m,q}(\varphi,\vartheta,\omega)$ is the Wigner D-matrix. These operators are also spatially invariant.

C3: There are operators that mix both degrees and orders. All the operator matrix elements $b^{m,q}_{\ell,p}$ are in general non-zero. Examples include spatial truncation and spatial masking with a window function. These operators are spatially varying.

7.15   Quadratic functionals on 2-sphere

Linear functionals were defined in Definition 4.8, p. 131, and were seen as a special case of functionals that take a vector and produce a complex number. In this section, we look at quadratic functionals (in fact *bounded* quadratic functionals), which are the next simplest type of functional and include the square of a norm, $\|f\|^2$, as an example.

For a given bounded linear operator, $\mathcal{B}$, on $L^2(\mathbb{S}^2)$ with given operator matrix $\mathbf{B}$, operator kernel $B(\widehat{\boldsymbol{u}}, \widehat{\boldsymbol{v}})$, and for a given function $f \in L^2(\mathbb{S}^2)$, a quadratic functional is defined as

$$\boxed{I_\mathcal{B}(f) \triangleq \langle \mathcal{B}f, f \rangle,} \tag{7.164}$$

which essentially gives the amount of cross energy between the original signal and its operated version. From (7.88), p. 218, we can express $\mathcal{B}f$ as

$$(\mathcal{B}f)(\widehat{\boldsymbol{u}}) = \sum_{\ell,m} \sum_{p,q} b_{\ell,p}^{m,q}(f)_p^q Y_\ell^m(\widehat{\boldsymbol{u}}). \tag{7.165}$$

And hence the quadratic functional can be expressed as

$$\begin{aligned}
I_\mathcal{B}(f) = \langle \mathcal{B}f, f \rangle &= \Big\langle \sum_{\ell,m} \sum_{p,q} b_{\ell,p}^{m,q}(f)_p^q Y_\ell^m, \sum_{s,t}(f)_s^t Y_s^t \Big\rangle \\
&= \sum_{\ell,m} \sum_{p,q} b_{\ell,p}^{m,q}(f)_p^q \overline{(f)_\ell^m} \\
&= \langle \mathbf{B}\mathbf{f}, \mathbf{f} \rangle,
\end{aligned} \tag{7.166}$$

where $\langle Y_\ell^m, Y_s^t \rangle = \delta_{s,\ell}^{t,m}$ is used and $\mathbf{f}$ is the vector spectral representation of f containing all $(f)_\ell^m$ in (7.47), p. 198, and the same spherical harmonic ordering is used for storing elements in $\mathbf{f}$ and $\mathbf{B}$. The left-hand side of (7.166) is expressed in terms of the inner product in $L^2(\mathbb{S}^2)$ and the final inner product on the right-hand side is the one for ℓ^2.

We can also express the functional based on the kernel representation of the operator. To see this, we use (7.89) to write the functional

$$\begin{aligned}
I_\mathcal{B}(f) = \langle \mathcal{B}f, f \rangle &= \Big\langle \int_{\mathbb{S}^2} B(\,\cdot\,, \widehat{\boldsymbol{v}}) f(\widehat{\boldsymbol{v}}) \, ds(\widehat{\boldsymbol{v}}), f \Big\rangle \\
&= \int_{\mathbb{S}^2} \int_{\mathbb{S}^2} B(\widehat{\boldsymbol{u}}, \widehat{\boldsymbol{v}}) f(\widehat{\boldsymbol{v}}) \, ds(\widehat{\boldsymbol{v}}) \overline{f(\widehat{\boldsymbol{u}})} \, ds(\widehat{\boldsymbol{u}}).
\end{aligned} \tag{7.167}$$

We can apply any bounded linear operator, including the basic operators of the previous sections, to obtain a bounded quadratic functional. Let us consider two simple examples.

Example 7.17 (Spectral truncation). We apply the spectral truncation operator $\mathcal{B}_L$ with operator matrix elements $b_{\ell,p}^{m,q} = \delta_{\ell,p}^{m,q}$ for $\ell \in \{0, 1, \dots, L\}$

(and zero otherwise) defined in (7.92), p. 220, to (7.166) to obtain

$$
\begin{aligned}
I_{\mathcal{B}_L}(f) &\triangleq \sum_{p,q}(f)_p^q \sum_{\ell,m} b_{\ell,p}^{m,q} \overline{(f)_\ell^m} \\
&= \sum_{\ell,m}^{L} \left| (f)_\ell^m \right|^2,
\end{aligned}
$$

(7.168)

which gives the energy of the signal f within the range of spectral degrees $\ell \in \{0, 1, \ldots, L\}$.   □

Example 7.18 (Spatial truncation). Consider spatial truncation operator $\mathcal{B}_R$ to region $R \subset \mathbb{S}^2$. In this case it is easier to use the kernel representation of the operator $B_R(\widehat{\boldsymbol{u}}, \widehat{\boldsymbol{v}}) = \mathbb{1}_R(\widehat{\boldsymbol{v}})\delta(\widehat{\boldsymbol{u}}, \widehat{\boldsymbol{v}})$ defined in (7.96), p. 221, in (7.167) to obtain

$$
\begin{aligned}
I_{\mathcal{B}_R}(f) &\triangleq \int_{\mathbb{S}^2}\int_{\mathbb{S}^2} B_R(\widehat{\boldsymbol{u}}, \widehat{\boldsymbol{v}}) f(\widehat{\boldsymbol{v}}) \, ds(\widehat{\boldsymbol{v}}) \overline{f(\widehat{\boldsymbol{u}})} \, ds(\widehat{\boldsymbol{u}}) \\
&= \int_{\mathbb{S}^2}\int_{\mathbb{S}^2} \mathbb{1}_R(\widehat{\boldsymbol{v}})\delta(\widehat{\boldsymbol{u}}, \widehat{\boldsymbol{v}}) f(\widehat{\boldsymbol{v}}) \, ds(\widehat{\boldsymbol{v}}) \overline{f(\widehat{\boldsymbol{u}})} \, ds(\widehat{\boldsymbol{u}}) \\
&= \int_R \left| f(\widehat{\boldsymbol{u}}) \right|^2 ds(\widehat{\boldsymbol{u}}),
\end{aligned}
$$

(7.169)

which gives the energy of the signal f within the spatial region R.   □

7.16   Classification of quadratic functionals

Based on the classification of operators in Section 7.14, we can identify three general forms for the functionals.

C1: If the operator matrix is fully diagonal with elements (7.162), repeated here

$$
b_{\ell,p}^{m,q} = b_\ell^m \delta_{\ell,p}^{m,q},
$$

then the functional in (7.166) simplifies to

$$
I_{\mathcal{B}_D}(f) \triangleq \langle \mathcal{B}_D f, f \rangle = \sum_{\ell,m} b_\ell^m \left| (f)_\ell^m \right|^2,
$$

(7.170)

which can be interpreted as a weighted energy of the signal f where the weights are determined by the action of the operator on different degrees and orders. We have already seen one example above for the spectral truncation.

C2: If the operator matrix is block diagonal with elements (7.163), repeated here

$$
b_{\ell,p}^{m,q} = b_\ell^{m,q} \delta_{\ell,p},
$$

then the functional in (7.166) simplifies to

$$
I_{\mathcal{B}_{BD}}(f) \triangleq \langle \mathcal{B}_{BD} f, f \rangle = \sum_{\ell,m} \sum_{q=-\ell}^{\ell} b_\ell^{m,q} (f)_\ell^q \overline{(f)_\ell^m}.
$$

C3: Finally, if the operator matrix is a general matrix with "mixing-degree" property, the functional representation in (7.166) does not simplify any further. In this case, it may be easier to represent the functional in terms of the operator kernel. For example, for the general masking operator with window function $h(\hat{\boldsymbol{u}})$ defined in (7.100), we can write

$$
\begin{aligned}
I_{\mathcal{B}_h}(f) &= \int_{\mathbb{S}^2} \int_{\mathbb{S}^2} h(\hat{\boldsymbol{v}}) \delta(\hat{\boldsymbol{u}}, \hat{\boldsymbol{v}}) f(\hat{\boldsymbol{v}}) \, ds(\hat{\boldsymbol{v}}) \overline{f(\hat{\boldsymbol{u}})} \, ds(\hat{\boldsymbol{u}}) \\
&= \int_{\mathbb{S}^2} h(\hat{\boldsymbol{u}}) \big| f(\hat{\boldsymbol{u}}) \big|^2 ds(\hat{\boldsymbol{u}}),
\end{aligned}
\tag{7.171}
$$

which can be interpreted as the weighted energy of the signal f where the weights are determined by the operator mask value $h(\hat{\boldsymbol{u}})$ at different spatial points $\hat{\boldsymbol{u}}$.

Quadratic functionals corresponding to some of the studied operators in this chapter are summarized in Table 7.2.

Problems _____

7.30. Derive the quadratic functional expressions given for the three operators,

$$
\mathcal{B}_h, \quad \big(\mathcal{D}(\varphi, \vartheta, \omega) f\big)(\hat{\boldsymbol{u}}) \quad \text{and} \quad \mathcal{B}^0,
$$

in Table 7.2, which have not been established explicitly in the text. That is, show

$$
\langle \mathcal{B}_h f, f \rangle = \int_{\mathbb{S}^2} h(\hat{\boldsymbol{u}}) \big| f(\hat{\boldsymbol{u}}) \big|^2 ds(\hat{\boldsymbol{u}}),
$$

$$
\langle \mathcal{D}(\varphi, \vartheta, \omega) f, f \rangle = \sum_{\ell, m} \sum_{q=-\ell}^{\ell} D_{m,q}^{\ell}(\varphi, \vartheta, \omega)(f)_{\ell}^{q} \overline{(f)_{\ell}^{m}},
$$

and

$$
\langle \mathcal{B}^0 f, f \rangle = \sum_{\ell=0}^{\infty} \big| (f)_{\ell}^{0} \big|^2.
$$

7.31. Consider a bounded operator, $\mathcal{B}_A$, whose output can be expressed as

$$
(\mathcal{B}_A f)(\hat{\boldsymbol{u}}) = \int_{\mathbb{S}^2} B(\hat{\boldsymbol{u}} \cdot \hat{\boldsymbol{v}}) f(\hat{\boldsymbol{v}}) \, ds(\hat{\boldsymbol{v}}),
$$

where the kernel takes the special form

$$
B_A(\hat{\boldsymbol{u}}, \hat{\boldsymbol{v}}) = B(\hat{\boldsymbol{u}} \cdot \hat{\boldsymbol{v}}),
$$

and $B(\cdot)$ is defined on the domain $[-1, +1]$. Following the pattern in Table 7.2, find the corresponding expressions for the operator matrix elements of $\mathcal{B}_A$ and quadratic functional $I_{\mathcal{B}_A}(f)$.

7.32. Consider a bounded operator, $\mathcal{B}_I$, whose matrix elements are given by

$$
b_{\ell,p}^{m,q} = b_{\ell} \delta_{\ell,p}^{m,q},
$$

which is a special case of (7.162). The matrix can be written

$$\mathbf{B}_I = \begin{pmatrix} b_0 & 0 & 0 & 0 & 0 & 0 & 0 & 0 & 0 & \cdots \\ 0 & b_1 & 0 & 0 & 0 & 0 & 0 & 0 & 0 & \cdots \\ 0 & 0 & b_1 & 0 & 0 & 0 & 0 & 0 & 0 & \cdots \\ 0 & 0 & 0 & b_1 & 0 & 0 & 0 & 0 & 0 & \cdots \\ 0 & 0 & 0 & 0 & b_2 & 0 & 0 & 0 & 0 & \cdots \\ 0 & 0 & 0 & 0 & 0 & b_2 & 0 & 0 & 0 & \cdots \\ 0 & 0 & 0 & 0 & 0 & 0 & b_2 & 0 & 0 & \cdots \\ 0 & 0 & 0 & 0 & 0 & 0 & 0 & b_2 & 0 & \cdots \\ 0 & 0 & 0 & 0 & 0 & 0 & 0 & 0 & b_2 & \cdots \\ \vdots & \vdots & \vdots & \vdots & \vdots & \vdots & \vdots & \vdots & \vdots & \ddots \end{pmatrix}$$

$$= \mathrm{diag}(b_0;\ b_1, b_1, b_1;\ b_2, b_2, b_2, b_2, b_2;\ \ldots).$$

Following the pattern in Table 7.2, find the corresponding expressions for the operator output $(\mathcal{B}_I f)(\widehat{\boldsymbol{u}})$, kernel $B_I(\widehat{\boldsymbol{u}}, \widehat{\boldsymbol{v}})$ and quadratic functional $I_{\mathcal{B}_I}(f)$.

Table 7.2 Some quadratic functionals and their spatial-domain and spectral-domain operator kernels.

Operator	Output	Matrix elements	Kernel	Quadratic functional
$\mathcal{B}$	$(\mathcal{B}f)(\widehat{u})$ (7.84)	$b^{m,q}_{\ell,p}$ $\langle \mathcal{B}Y^q_p, Y^m_\ell\rangle$ (7.85)	$B(\widehat{u},\widehat{v})$ $\sum_{\ell,m}\sum_{p,q} b^{m,q}_{\ell,p} Y^m_\ell(\widehat{u})\overline{Y^q_p(\widehat{v})}$ (7.89)	$I_B(f)=\langle \mathcal{B}f, f\rangle$ (7.164) $\sum_{\ell,m}\sum_{p,q} b^{m,q}_{\ell,p}(f)^q_p\overline{(f)^m_\ell}$ (7.166)
Spectral truncation $\mathcal{B}_L$	$\sum_{\ell,m}^{L}(f)^m_\ell Y^m_\ell(\widehat{u})$ (7.91)	$\begin{cases}\delta^{m,q}_{\ell,p} & 0\le\ell\le L\\ 0 & \ell>L\end{cases}$ (7.92)	$\sum_{\ell=0}^{L}\dfrac{(2\ell+1)}{4\pi}P_\ell(\widehat{u}\cdot\widehat{v})$ (7.93)	$\sum_{\ell,m}^{L}\left|(f)^m_\ell\right|^2$ (7.168)
Spatial truncation $\mathcal{B}_R$	$\begin{cases}f(\widehat{u}) & \widehat{u}\in R\\ 0 & \widehat{u}\in S^2\backslash R\end{cases}$ (7.94)	$\int_R Y^q_p(\widehat{u})\overline{Y^m_\ell(\widehat{u})}\,ds(\widehat{u})$ (7.97)	$1_R(\widehat{v})\delta(\widehat{u},\widehat{v})$ (7.96)	$\int_R \left|f(\widehat{u})\right|^2 ds(\widehat{u})$ (7.169)
Spatial masking $\mathcal{B}_h$	$h(\widehat{u})f(\widehat{u})$ (7.99)	$\int_{S^2} h(\widehat{u})Y^q_p(\widehat{u})\overline{Y^m_\ell(\widehat{u})}\,ds(\widehat{u})$ (7.103) $\sum_{s,t}(h)^t_s\,y(s,t;p,q;\ell,m)$ (7.105)	$h(\widehat{v})\delta(\widehat{u},\widehat{v})$ (7.100) $\sum_{\ell,m}\sum_{p,q}\sum_{s,t}(h)^t_s$ $\times y(s,t;p,q;\ell,m)Y^m_\ell(\widehat{u})\overline{Y^q_p(\widehat{v})}$ (7.106)	$\int_{S^2} h(\widehat{u})\left|f(\widehat{u})\right|^2 ds(\widehat{u})$
Rotation $\mathcal{D}(\varphi,\vartheta,\omega)$	$(\mathcal{D}(\varphi,\vartheta,\omega)f)(\widehat{u})$ (7.113) $=f(R^{-1}\widehat{u})$	$\underbrace{D^\ell_{m,q}(\varphi,\vartheta,\omega)}_{\text{Wigner }D}\delta_{\ell,p}$ (7.117)	$\sum_{\ell,m}\sum_{q=-\ell}^{\ell}D^\ell_{m,q}(\varphi,\vartheta,\omega)$ $\times Y^m_\ell(\widehat{u})\overline{Y^q_\ell(\widehat{v})}$ (7.121)	$\sum_{\ell,m}\sum_{q=-\ell}^{\ell}D^\ell_{m,q}(\varphi,\vartheta,\omega)$ $\times (f)^q_\ell\overline{(f)^m_\ell}$
Projection into $\mathcal{H}^0(S^2)$ $\mathcal{B}^0$	$\sum_{\ell=0}^{\infty}(f)^0_\ell Y^0_\ell(\theta)$ (7.157)	$\delta_{\ell,p}\delta_{0,q}\delta_{0,m}$ (7.158)	$\sum_{\ell=0}^{\infty}Y^0_\ell(\widehat{u})\overline{Y^0_\ell(\widehat{v})}$ (7.159)	$\sum_{\ell=0}^{\infty}\left|(f)^0_\ell\right|^2$

8 Advanced topics on 2-sphere

8.1 Introduction

In this chapter, we continue our study of signals on the 2-sphere. Compared to the previous chapter, which covered fundamental topics related to signals and systems on the sphere, here we will treat more advanced topics, that is, signal concentration and joint spatio-spectral analysis on the sphere.

Before we start this treatment, we shall briefly revisit some of the classical concentration problems in time and frequency domains, which were laid out in a series of papers by Slepian, Pollak, and Landau between the early 1960s and late 1970s (Slepian and Pollak, 1961; Landau et al., 1961; Landau and Pollak, 1962; Slepian, 1978). In addition to these landmark papers, there are many more recent fine works that treat the time-frequency concentration problem (Papoulis, 1977), and time-frequency analysis in general (Cohen, 1989, 1995). Therefore, our intention is not to provide an in-depth and comprehensive coverage of the subject in this classical domain, but to refresh our minds about the time-frequency problem to be able to appreciate the similarities and differences that exist in the evil twin problem on the unit sphere.

8.2 Time-frequency concentration reviewed

8.2.1 Preliminaries

Let $L^2(\mathbb{R}) \triangleq L^2(-\infty, \infty)$ be the space of square integrable functions, such as $f(t)$, on the entire real time line $t \in (-\infty, \infty)$ with the inner product defined as

$$\langle f, g \rangle = \int_{-\infty}^{\infty} f(t)\overline{g(t)}\, dt,$$

which induces the norm

$$\|f\| = \left(\int_{-\infty}^{\infty} |f(t)|^2 dt \right)^{1/2}.$$

The Fourier transform of $f(t)$ is given by

$$F(\omega) \triangleq \int_{-\infty}^{\infty} f(t)e^{-i\omega t} dt, \qquad (8.1)$$

where ω is the angular frequency and the inverse Fourier transform by

$$f(t) = \frac{1}{2\pi} \int_{-\infty}^{\infty} F(\omega)e^{i\omega t} d\omega. \tag{8.2}$$

The functions belonging to $L^2(\mathbb{R})$ satisfy the generalized Parseval relation

$$\int_{-\infty}^{\infty} f(t)\overline{g(t)}\, dt = \int_{-\infty}^{\infty} F(\omega)\overline{G(\omega)}\, d\omega. \tag{8.3}$$

By the Cauchy-Schwarz inequality, Theorem 2.1, p. 37, the *angle* between two functions f and g is defined as

$$\theta(f,g) \triangleq \cos^{-1} \frac{\mathfrak{Re}(\langle f, g \rangle)}{\|f\| \, \|g\|}, \tag{8.4}$$

where $\mathfrak{Re}(\cdot)$ takes the real part of its argument.

Let $\mathcal{B}_W$ be the (self-adjoint and idempotent) projection operator into the subspace $\mathcal{H}_W$ of bandlimited functions with the maximum angular frequency $|\omega| \le W$ such that

$$(\mathcal{B}_W F)(\omega) = \begin{cases} F(\omega) & \text{if } |\omega| \le W, \\ 0 & \text{if } |\omega| > W. \end{cases} \tag{8.5}$$

In time domain, the operator action is

$$(\mathcal{B}_W f)(t) = \frac{1}{2\pi} \int_{-W}^{W} F(\omega)e^{i\omega t} d\omega. \tag{8.6}$$

Using the definition of Fourier transform (8.1) in (8.6) and noting that

$$\frac{1}{2\pi} \int_{-W}^{W} e^{i\omega(t-s)} d\omega = \frac{\sin W(t-s)}{\pi(t-s)},$$

results in

$$\begin{aligned}
(\mathcal{B}_W f)(t) &= \frac{1}{2\pi} \int_{-W}^{W} F(\omega)e^{i\omega t}\, d\omega \\
&= \frac{1}{2\pi} \int_{-W}^{W} \int_{-\infty}^{\infty} f(s)e^{-i\omega s}\, ds\, e^{i\omega t} d\omega \\
&= \int_{-\infty}^{\infty} f(s) \frac{1}{2\pi} \int_{-W}^{W} e^{i\omega(t-s)}\, d\omega\, ds \\
&= \int_{-\infty}^{\infty} f(s) K_W(t,s)\, ds,
\end{aligned} \tag{8.7}$$

where

$$K_W(t,s) \triangleq \frac{\sin W(t-s)}{\pi(t-s)} \tag{8.8}$$

is the operator kernel. Self-adjointness of $\mathcal{B}_W$ can be easily established since the operator kernel is real and symmetric; see (6.10), p. 163.

And similarly, let $\mathcal{B}_T$ be the (self-adjoint and idempotent) projection operator into the subspace $\mathcal{H}_T$ of functions limited to time interval $|t| \leq T$ such that

$$(\mathcal{B}_W f)(t) = \begin{cases} f(t) & \text{if } |t| \leq T, \\ 0 & \text{if } |t| > T. \end{cases} \tag{8.9}$$

In frequency domain, the operator action is

$$(\mathcal{B}_T F)(\omega) = \int_{-T}^{T} f(t) e^{-i\omega t} \, dt. \tag{8.10}$$

By using the definition of inverse Fourier transform (8.2) in (8.10) and following similar procedures as in (8.7) we obtain

$$(\mathcal{B}_T F)(\omega) = \int_{-\infty}^{\infty} F(\omega') \frac{\sin T(\omega - \omega')}{\pi(\omega - \omega')} \, d\omega'. \tag{8.11}$$

8.2.2   Problem statement

For a given finite W and T, define the energy concentration ratios for signal f in time and angular frequency domains as

$$\alpha = \frac{\displaystyle\int_{-T}^{T} |f(t)|^2 dt}{\displaystyle\int_{-\infty}^{\infty} |f(t)|^2 dt}$$

and

$$\beta = \frac{\displaystyle\int_{-W}^{W} |F(\omega)|^2 d\omega}{\displaystyle\int_{-\infty}^{\infty} |F(\omega)|^2 d\omega}.$$

Some of the questions that Slepian, Pollak, and Landau set to answer in (Slepian and Pollak, 1961; Landau et al., 1961) were:

Questions (Slepian, Pollak and Landau)

Q1: If a function f is already bandlimited and belongs to $\mathcal{H}_W$ (or equivalently $\beta = 1$), what range of time concentration ratios α is achievable and what type of function can achieve the maximum concentration?

Q2: If a function f is already timelimited and belongs to $\mathcal{H}_T$ (or equivalently $\alpha = 1$), what range of frequency concentration ratios β is achievable and what type of function can achieve the maximum concentration?

Q3: Given that no function can simultaneously belong to $\mathcal{H}_W$ and $\mathcal{H}_T$, what is the minimum angle possible between subspaces $\mathcal{H}_W$ and $\mathcal{H}_T$? That is, what function or functions $f \in \mathcal{H}_W$ and $g \in \mathcal{H}_T$ attain $\min \theta(f, g) > 0$?

Q4: What range of α and β are simultaneously achievable and what type of functions can achieve them?

8.2.3   Answer to time concentration question Q1

The answer to the first question can be elaborated as follows. Assume that $\beta = 1$ and hence signal $f \in \mathcal{H}_W$ is already bandlimited. Using the self-adjoint and idempotent property of projection operator $\mathcal{B}_T$, we write the time concentration ratio as

$$\alpha = \frac{\displaystyle\int_{-T}^{T} |f(t)|^2 dt}{\displaystyle\int_{-\infty}^{\infty} |f(t)|^2 dt} = \frac{\langle \mathcal{B}_T f, \mathcal{B}_T f \rangle}{\langle f, f \rangle} = \frac{\langle \mathcal{B}_T f, f \rangle}{\langle f, f \rangle}. \tag{8.12}$$

Now using generalized Parseval relation and (8.11), the numerator can be written in the frequency domain as

$$\langle \mathcal{B}_T f, f \rangle = \int_{-T}^{T} (\mathcal{B}_T f)(t) \overline{f(t)} \, dt = \int_{-\infty}^{\infty} \int_{-\infty}^{\infty} F(\omega') \frac{\sin T(\omega - \omega')}{\pi(\omega - \omega')} \overline{F(\omega)} \, d\omega' \, d\omega$$

$$= \int_{-W}^{W} \int_{-W}^{W} F(\omega') \frac{\sin T(\omega - \omega')}{\pi(\omega - \omega')} \overline{F(\omega)} \, d\omega' \, d\omega,$$

where the last equality follows from the initial assumption that $f \in \mathcal{H}_W$ is bandlimited to W. Therefore, the time concentration ratio in (8.12) is written as

$$\alpha = \frac{\displaystyle\int_{-W}^{W} \int_{-W}^{W} F(\omega') \frac{\sin T(\omega - \omega')}{\pi(\omega - \omega')} \overline{F(\omega)} \, d\omega' \, d\omega}{\displaystyle\int_{-W}^{W} |F(\omega)|^2 d\omega}. \tag{8.13}$$

The functions satisfying the following integral eigenequation

$$\boxed{\int_{-W}^{W} F(\omega') \frac{\sin T(\omega - \omega')}{\pi(\omega - \omega')} \, d\omega' = \lambda F(\omega), \quad |\omega| \le W,} \tag{8.14}$$

where λ is the eigenvalue, achieve the stationary points of the ratio in (8.13). Taking the inverse Fourier transform of both sides of (8.14) and after a few simple manipulations of the left-hand side, we obtain

$$\boxed{\int_{-T}^{T} f(s) \frac{\sin W(t - s)}{\pi(t - s)} \, ds = \lambda f(t), \quad t \in (-\infty, \infty).} \tag{8.15}$$

Properties of eigenfunctions and eigenvalues

Here we present some important properties of eigenfunctions and eigenvalues of the time concentration problem.

P1: There is a countable set of bandlimited eigenfunctions, denoted by $\Psi_n(t)$, with their corresponding eigenvalue, denoted by λ_n, that satisfy (8.14) or equivalently (8.15).

P2: The eigenvalues are distinct, real and strictly between 0 and 1. Arranging them in descending order, we write

$$1 > \lambda_0 > \lambda_1 > \lambda_2 > \cdots > 0.$$

P3: The eigenfunctions are also real and can be found by appropriately scaling the solutions to the following differential equation

$$(1 - t^2)\frac{d^2 u}{dt^2} - 2t\frac{du}{dt} + (\Xi - T^2 W^2 t^2)u = 0, \quad |t| \le 1.$$

The "time-bandwidth" product TW is the single parameter characterizing the differential equation and Ξ is an eigenvalue of the equation.

P4: The eigenfunctions are orthonormal on the entire time line

$$\langle \Psi_n, \Psi_k \rangle = \int_{-\infty}^{\infty} \Psi_n(t)\Psi_k(t)\,dt = \delta_{n,k}.$$

P5: The eigenfunctions form a basis for the subspace $\mathcal{H}_W$. That is, any function $f \in \mathcal{H}_W$ can be written in the form

$$f(t) = \sum_{n=0}^{\infty} (f)_n \Psi_n(t) = \sum_{n=0}^{\infty} \langle f, \Psi_n \rangle \Psi_n(t), \tag{8.16}$$

where the sum converges in the mean to f.

P6: The eigenfunctions also have the "curious property" (Slepian and Pollak, 1961) that they are also orthogonal over the time interval $|t| \le T$. To see this, we write using (8.15)

$$\int_{-\infty}^{\infty} \Psi_n(t)\Psi_k(t)\,dt$$

$$= \frac{1}{\lambda_n \lambda_k} \int_{-\infty}^{\infty} \int_{-T}^{T} \Psi_n(s)\frac{\sin W(t - s)}{\pi(t - s)}\,ds \int_{-T}^{T} \Psi_k(u)\frac{\sin W(t - u)}{\pi(t - u)}\,du\,dt$$

$$= \frac{1}{\lambda_n \lambda_k} \int_{-T}^{T} \int_{-T}^{T} \Psi_n(s)\Psi_k(u) \int_{-\infty}^{\infty} \frac{\sin W(t - s)}{\pi(t - s)}\frac{\sin W(t - u)}{\pi(t - u)}\,dt\,ds\,du$$

$$= \frac{1}{\lambda_n \lambda_k} \int_{-T}^{T} \Psi_k(u) \int_{-T}^{T} \frac{\sin W(u - s)}{\pi(u - s)}\Psi_n(s)\,ds\,du$$

$$= \frac{1}{\lambda_k} \int_{-T}^{T} \Psi_k(u)\Psi_n(u)\,du = \delta_{n,k}.$$

Therefore, $\langle \mathcal{B}_T \Psi_n, \mathcal{B}_T \Psi_k \rangle = \lambda_k \delta_{n,k}$. The eigenfunctions are also orthogonal over the complementary time interval $|t| > T$.

P7: The time concentration ratio α attains its largest value equal to λ_0 when $f(t) = \Psi_0(t)$ is used in (8.12). That is, $\Psi_0(t)$ is the unique orthonormal function bandlimited to $|\omega| \le W$, which is maximally concentrated in time interval $|t| \le T$.

P8: The eigenvalues are almost binary-valued. They are either very close to one or very close to zero with a sharp transition between the two. As a result, the corresponding eigenfunctions are either almost concentrated in

$|t| \leq T$ or almost excluded from it, belonging to $|t| > T$. The number of significant eigenvalues near one is known as the "Shannon number" and is equal to $2WT$. It specifies the *degrees of freedom* in choosing functions that are bandlimited and almost timelimited to W and T.

Eigenfunctions $\Psi_n(t)$ have another important property that will be discussed below, when we review the answer to the second question.

8.2.4   Answer to frequency concentration question Q2

The answer to the second question can be elaborated as follows. Assume that $\alpha = 1$ and hence signal $f \in \mathcal{H}_T$ is already timelimited. Using the self-adjoint and idempotent property of projection operator $\mathcal{B}_W$, we write the frequency concentration ratio as

$$\beta = \frac{\int_{-W}^{W} |F(\omega)|^2 d\omega}{\int_{-\infty}^{\infty} |F(\omega)|^2 d\omega} = \frac{\langle \mathcal{B}_W F, \mathcal{B}_W F \rangle}{\langle F, F \rangle} = \frac{\langle \mathcal{B}_W F, F \rangle}{\langle F, F \rangle}. \tag{8.17}$$

Now using generalized Parseval relation and (8.7), the numerator can be written in time domain

$$\begin{aligned}
\langle \mathcal{B}_W F, F \rangle &= \int_{-W}^{W} (\mathcal{B}_W F)(\omega) \overline{F(\omega)} \, d\omega \\
&= \int_{-\infty}^{\infty} \int_{-\infty}^{\infty} f(s) \frac{\sin W(t-s)}{\pi(t-s)} \overline{f(t)} \, ds \, dt \\
&= \int_{-T}^{T} \int_{-T}^{T} f(s) \frac{\sin W(t-s)}{\pi(t-s)} \overline{f(t)} \, ds \, dt,
\end{aligned}$$

where the last equality follows from the initial assumption that $f \in \mathcal{H}_T$ is time-limited to T. Therefore, the frequency concentration ratio in (8.17) is written as

$$\beta = \frac{\int_{-T}^{T} \int_{-T}^{T} f(s) \frac{\sin W(t-s)}{\pi(t-s)} \overline{f(t)} \, ds \, dt}{\int_{-T}^{T} |f(t)|^2 dt}. \tag{8.18}$$

The functions satisfying the following integral eigenequation

$$\boxed{\int_{-T}^{T} f(s) \frac{\sin W(t-s)}{\pi(t-s)} \, ds = \lambda f(t), \quad |t| \leq T,} \tag{8.19}$$

where λ is the eigenvalue, achieve the stationary points of the ratio in (8.18).

Properties of eigenfunctions and eigenvalues

Comparison of (8.19) with (8.14) reveals a striking similarity between the two eigenequations. In fact, apart from a scaling factor, they are identical and two simple changes of variables $\omega T/W = t$ and $\omega' T/W = s$ will reveal that if $\Psi_n(\omega)$ is an eigensolution to (8.14), then $\Psi_n(\omega T/W)$ would be an eigensolution to (8.19). Therefore, we can consider the frequency concentration problem as the dual problem of time concentration. Hence, by solving one set of eigenequations and appropriate scaling we obtain the solutions to both problems.

8.2.5   Answer to minimum angle question Q3

The third question has a geometric significance. It signifies the *closeness*, measured by the relative inner product (8.4), of the elements of subspaces $\mathcal{H}_W$ and $\mathcal{H}_T$. An angle near 0 or π between two elements $f \in \mathcal{H}_W$ and $g \in \mathcal{H}_T$ means that the elements are almost proportional (or close) to each other and an angle near to $\pi/2$ means that the elements are almost orthogonal to (or far from) each other. Apart from the obvious zero element, no other function in $L^2(\mathbb{R})$ can live in both $\mathcal{H}_W$ and $\mathcal{H}_T$ and the question is what the closest non-zero elements of these two subspaces are.

Let us choose an arbitrary member f in $\mathcal{H}_W$. The first question is: for a given $f \in \mathcal{H}_W$, what function $g \in \mathcal{H}_T$ achieves the infimum angle $\theta(f, g)$? From the discussion on page 68, we know the answer to this question to be $g = \mathcal{B}_T f$ or a multiple of it. For a given f, the projection of bandlimited function f into the subspace of timelimited functions is our best bet and the infimum angle would be

$$\inf \theta(f, g) = \cos^{-1} \frac{\langle f, \mathcal{B}_T f \rangle}{\|f\| \|\mathcal{B}_T f\|} = \cos^{-1} \frac{\langle \mathcal{B}_T f, \mathcal{B}_T f \rangle}{\|f\| \|\mathcal{B}_T f\|} = \cos^{-1} \frac{\|\mathcal{B}_T f\|}{\|f\|} > 0. \quad (8.20)$$

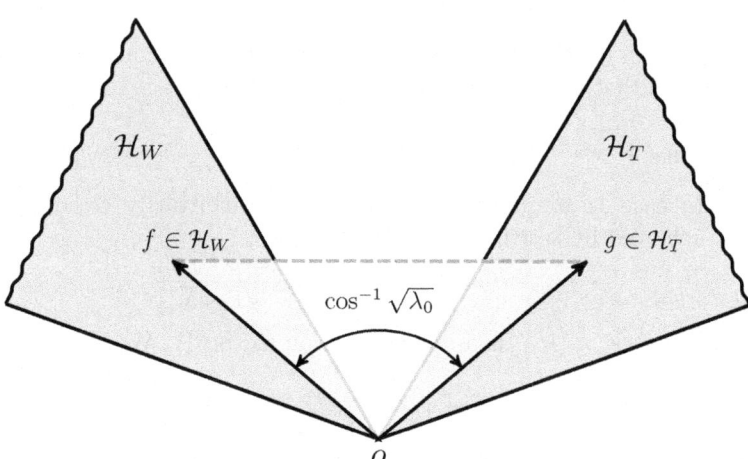

Figure 8.1 Minimum angle between the two subspaces $\mathcal{H}_W$ and $\mathcal{H}_T$. Here $\Psi_0(t)$ is the eigenfunction with eigenvalue λ_0.

Now the second question, specifically pertinent to the concentration problem, is that among all possible $f \in \mathcal{H}_W$, which function or functions can attain the least angle between the two subspaces. Intuitively, we expect this function to be $\Psi_0(t)$ or a multiple of it, which is jointly bandlimited and almost concentrated in $|t| \leq T$. In other words, as far as timelimited functions in $\mathcal{H}_T$ are concerned, $\Psi_0(t)$ is the closest bandlimited function they can befriend, which has invested almost all its energy in their subspace $\mathcal{H}_T$. It has almost nothing to hide. This can be easily proven by invoking orthonormal expansion (8.16) in (8.20), the fact that $\langle \mathcal{B}_T \Psi_n, \Psi_k \rangle = \lambda_k \delta_{n,k}$ and that λ_0 is the largest eigenvalue. When having the option of choosing the most concentrated signal $\Psi_0(t) \in \mathcal{H}_W$, why would one bother choosing a suboptimal function? In the end, by setting $f = \Psi_0$ the least angle between subspaces $\mathcal{H}_W$ and $\mathcal{H}_T$ will become

$$\theta(f,g) \geq \theta_0(f,g)\bigg|_{\substack{f=\Psi_0 \\ g=\mathcal{B}_T\Psi_0}} = \cos^{-1} \frac{\|\mathcal{B}_T \Psi_0\|}{\|\Psi_0\|} = \cos^{-1} \sqrt{\lambda_0} > 0. \qquad (8.21)$$

8.2.6   Answer to time-frequency concentration question Q4

Let $f \in L^2(\mathbb{R})$ be a unit-norm function, with $\|f\| = 1$ with time and frequency concentration ratios $\|\mathcal{B}_T f\|^2 = \alpha$ and $\|\mathcal{B}_W f\|^2 = \beta$.[1] The question is what ranges of α and β are jointly achievable. Due to the symmetry of the concentration problems, one can vary time concentration ratio α from 0 to 1, and find the admissible range of frequency concentration ratio β. As proved in (Landau et al., 1961), four cases should be considered:

Case 1: If $\alpha = 0$, then $0 \leq \beta < 1$;

Case 2: If $0 < \alpha < \lambda_0$, then $0 \leq \beta \leq 1$;

Case 3: If $\lambda_0 \leq \alpha < 1$, then $\sqrt{\beta} \leq \cos(\cos^{-1}\sqrt{\lambda_0} - \cos^{-1}\sqrt{\alpha})$;

Case 4: If $\alpha = 1$, then $0 < \beta \leq \lambda_0$.

In each case, we only provide some important comments and examples of functions that can achieve particular values of β.

Case 1

In case 1, frequency concentration β arbitrarily close to zero or one can be achieved by setting

$$f = \frac{\Psi_n - \mathcal{B}_T \Psi_n}{\sqrt{1 - \lambda_n}},$$

where $\Psi_n \in \mathcal{H}_W$ is an eigenfunction to (8.14). We can easily verify that

$$\|\mathcal{B}_T f\|^2 = \frac{1}{1 - \lambda_n} \|\mathcal{B}_T(\Psi_n - \mathcal{B}_T \Psi_n)\|^2 = 0$$

---

[1]   Note that compared to the original papers (Slepian and Pollak, 1961; Landau et al., 1961), we have used a slightly different notation for concentration ratios. We use α and not α^2 and β and not β^2 to refer to time and frequency energy concentration ratios. Hence, we have appropriately modified the answer to the fourth question.

and

$$\|f\|^2 = \frac{1}{1-\lambda_n}\|\Psi_n - \mathcal{B}_T\Psi_n\|^2 = \frac{1+\lambda_n - 2\lambda_n}{1-\lambda_n} = 1.$$

For the frequency concentration ratio we write

$$\beta = \|\mathcal{B}_W f\|^2 = \frac{1}{1-\lambda_n}\|\mathcal{B}_W(\Psi_n - \mathcal{B}_T\Psi_n)\|^2 \tag{8.22}$$

$$= \frac{1}{1-\lambda_n}\langle \mathcal{B}_W(\Psi_n - \mathcal{B}_T\Psi_n), \mathcal{B}_W(\Psi_n - \mathcal{B}_T\Psi_n)\rangle \tag{8.23}$$

$$= \frac{1}{1-\lambda_n}\left(1 + \langle \mathcal{B}_W \circ \mathcal{B}_T\Psi_n, \mathcal{B}_W \circ \mathcal{B}_T\Psi_n\rangle - 2\langle \mathcal{B}_W \Psi_n, \mathcal{B}_W \circ \mathcal{B}_T\Psi_n\rangle\right) \tag{8.24}$$

$$= \frac{1}{1-\lambda_n}\left(1 + \lambda_n^2 - 2\langle \Psi_n, \mathcal{B}_T\Psi_n\rangle\right) = \frac{1+\lambda_n^2 - 2\lambda_n}{1-\lambda_n} = 1 - \lambda_n, \tag{8.25}$$

where in going from (8.23) to (8.24) we have used the fact that $\Psi_n \in \mathcal{H}_W$ is already orthonormal and bandlimited and in going from (8.24) to (8.25), we have used a result proven in (Slepian and Pollak, 1961) that timelimiting and then bandlimiting eigenfunction $\Psi_n \in \mathcal{H}_W$ will reduce its energy to λ_n^2 and that since Ψ_n is already bandlimited, $\langle \mathcal{B}_W \Psi_n, \mathcal{B}_W \circ \mathcal{B}_T\Psi_n\rangle = \langle \mathcal{B}_W \circ \mathcal{B}_W \Psi_n, \mathcal{B}_T\Psi_n\rangle = \langle \Psi_n, \mathcal{B}_T\Psi_n\rangle = \langle \mathcal{B}_T\Psi_n, \mathcal{B}_T\Psi_n\rangle = \lambda_n$. Since eigenvalues that are arbitrarily close to one or zero exist, β can get arbitrarily close to zero or one. Existence of non-zero functions that are orthogonal to both $\mathcal{H}_W$ and $\mathcal{H}_T$ was established in (Landau et al., 1961), which proves that $\beta = 0$ can also be achieved.

Case 2

In case 2, for any $\alpha > 0$, there exists an eigenvalue such that $\lambda_n < \alpha$. By choosing

$$f = \frac{\sqrt{\alpha - \lambda_n}\Psi_0 + \sqrt{\lambda_0 - \alpha}\Psi_n}{\sqrt{\lambda_0 - \lambda_n}},$$

and noting that $f \in \mathcal{H}_W$ and $\langle \mathcal{B}_T\Psi_n, \mathcal{B}_T\Psi_k\rangle = \lambda_k\delta_{n,k}$, we can easily verify that $\beta = \|\mathcal{B}_W f\|^2 = \|f\|^2 = 1$ and $\|\mathcal{B}_T f\|^2 = \alpha$. Achievability of other values of β was shown in (Landau et al., 1961).

Case 3

The third case is interesting. Roughly speaking, it is telling us that as we increase our desired time concentration ratio beyond λ_0, less and less frequency concentration is possible. We have already seen that when $\alpha = \lambda_0$, then it is possible to achieve full frequency concentration $\beta = 1$ by choosing $f = \Psi_0$. However, as α increases beyond λ_0, $\beta = 1$ becomes impossible. In the extreme case where α approaches one, β approaches λ_0, which is strictly smaller than one. It has been proven in (Landau et al., 1961) that by choosing

$$f = p\Psi_0 + q\mathcal{B}_T\Psi_0$$

$$\triangleq \sqrt{\frac{1-\alpha}{1-\lambda_0}}\Psi_0 + \left(\sqrt{\frac{\alpha}{\lambda_0}} - \sqrt{\frac{1-\alpha}{1-\lambda_0}}\right)\mathcal{B}_T\Psi_0,$$

we can achieve $\|\mathcal{B}_T f\|^2 = \alpha$, $\|f\|^2 = 1$ and more importantly, the bound

$$\|\mathcal{B}_W f\| = \sqrt{\beta} = \cos(\cos^{-1}\sqrt{\lambda_0} - \cos^{-1}\sqrt{\alpha}).$$

To see the last assertion, we follow similar steps as in (8.22)–(8.25) and some algebraic simplifications to write

$$\begin{aligned}
\beta = \|\mathcal{B}_W f\|^2 &= \left\|\mathcal{B}_W(p\Psi_0 + q\mathcal{B}_T\Psi_o)\right\|^2 \\
&= p^2 + \lambda_0^2 q^2 + 2pq\lambda_0 \\
&= \left(\sqrt{(1-\alpha)(1-\lambda_0)} + \sqrt{\alpha\lambda_0}\right)^2.
\end{aligned}$$

Noting that $\sqrt{\alpha} < 1$, $\sqrt{\lambda_0} < 1$ and defining $\sqrt{\alpha} = \cos\phi$ and $\sqrt{\lambda_0} = \cos\theta$, we obtain

$$\beta = (\sin\phi\sin\theta + \cos\phi\cos\theta)^2 = \left(\cos(\theta - \phi)\right)^2,$$

or simply

$$\sqrt{\beta} = \cos(\cos^{-1}\sqrt{\lambda_0} - \cos^{-1}\sqrt{\alpha}).$$

Case 4

The fourth case can be considered as a limiting case of the third case, except for the extra condition that $\beta = 0$ has to be excluded.

8.3   Introduction to concentration problem on 2-sphere

The objective is to find bandlimited functions belonging to $\mathcal{H}_L(\mathbb{S}^2)$, having maximum degree L, on the 2-sphere that are *optimally concentrated* within a spatial region of interest R. Optimal concentration is a bit vague here and is yet to be defined. In fact, there may be different measures of concentration. In (Simons et al., 2006), the most natural Slepian-type concentration problem is formulated and investigated in detail where the measure of concentration is the fraction of signal energy in the region of interest. We will later discuss generalized concentration measures. The dual problem is to find spacelimited functions belonging to $\mathcal{H}_R(\mathbb{S}^2)$ on 2-sphere whose spectral representation is *optimally concentrated* within the spectral degree $\ell \in \{0, 1, \ldots, L\}$.

Having reviewed the time-frequency concentration problem, we will highlight the similarities and differences of the spatio-spectral concentration problem. While there are many expected similarities, there are some important differences. For example, the arbitrary shape of the spatial concentration region R on the sphere "enriches" the problem on 2-sphere (Simons et al., 2006).

Motivation

Finding optimally concentrated signals in spatial or spectral domains has many applications in practice, which are similar in spirit and can be as diverse as applications of the Slepian concentration problem in time-frequency domain. For example, we may need to smooth a signal on the sphere to reduce the effects of noise or high spectral components by performing local *convolutional* averaging or filtering, which will be the topic of Chapter 9, p. 293. Naturally, we would like to design a filter kernel such that it is best concentrated around the point of interest and has minimal *leakage* outside the desired neighborhood. Because otherwise, other parts of the signal might significantly contribute to the filtered output. In other examples, optimally concentrated signals are applied for spectral estimation in (Dahlen and Simons, 2008) and localized spectral analysis of two windowed (masked) functions in (Wieczorek and Simons, 2005). In Section 8.13, p. 279, we will discuss an application of optimally concentrated windows that allow accurate joint spatio-spectral analysis of signals on the sphere.

8.4   Optimal spatial concentration of bandlimited signals

We first need a measure of concentration. To start with, we take the energy of a signal as our measure, which is physically relevant in many applications. The idea is very simple. The signal we are looking for is denoted by f and is bandlimited to a maximum degree L. That is $f \in \mathcal{H}_L(\mathbb{S}^2)$. We wish to maximize the ratio of signal energy in the desired region R to the total energy over the 2-sphere. Mathematically, we wish to maximize

$$
\alpha = \frac{\displaystyle\int_R \left| f(\widehat{\boldsymbol{u}}) \right|^2 ds(\widehat{\boldsymbol{u}})}{\displaystyle\int_{\mathbb{S}^2} \left| f(\widehat{\boldsymbol{u}}) \right|^2 ds(\widehat{\boldsymbol{u}})}, \tag{8.26}
$$

where α is a measure of spatial concentration. Using the spherical harmonic representation of signal f in (7.72), p. 211, we can expand the numerator in (8.26) as

$$
\int_R \left| f(\widehat{\boldsymbol{u}}) \right|^2 ds(\widehat{\boldsymbol{u}}) = \sum_{\ell,m}^L \overline{(f)_\ell^m} \sum_{p,q}^L (f)_p^q D_{\ell,p}^{m,q},
$$

where we have defined

$$
D_{\ell,p}^{m,q} \triangleq \int_R Y_p^q(\widehat{\boldsymbol{u}}) \overline{Y_\ell^m(\widehat{\boldsymbol{u}})} \, ds(\widehat{\boldsymbol{u}}). \tag{8.27}
$$

Note that $Y_\ell^m(\widehat{\boldsymbol{u}})$ and $Y_p^q(\widehat{\boldsymbol{u}})$ are not orthogonal over a strict subregion of $\mathbb{S}^2$ and $D_{\ell,p}^{m,q}$ should be computed numerically. We will discuss some examples later. Using the Parseval relation in (7.73), p. 213, for bandlimited signals, we can rewrite the concentration ratio in (8.26) in matrix form as

$$
\alpha = \frac{\mathbf{f}^H \mathbf{D} \mathbf{f}}{\mathbf{f}^H \mathbf{f}}, \tag{8.28}
$$

where $\mathbf{f}$ is the vector containing the spectral representation of f and $(\cdot)^H$ denotes Hermitian transpose. In the above equation, matrix $\mathbf{D}$ gathers all elements such as $D_{\ell,p}^{m,q}$ and has a dimension of $(L+1)^2 \times (L+1)^2$. Since there are four indices in

$D^{m,q}_{\ell,p}$, one needs to be careful about writing them in a two-dimensional matrix format. It is important to be consistent with the way elements in the vector spectral representation $\mathbf{f}$ are arranged or interpreted. Other than this, it is somewhat arbitrary.

Here we use the single indexing convention (7.39), p. 195, which is consistent with our presentation of $\mathbf{f}$. Define

$$n \triangleq \ell(\ell+1) + m \quad \text{and} \quad r \triangleq p(p+1) + q$$

to write $D^{m,q}_{\ell,p}$ as $D_{n,r}$, which occupies $\mathbf{D}$ at row n and column r. Therefore, letting $L' \triangleq L^2 + 2L$, matrix $\mathbf{D}$ is

$$\mathbf{D} = \begin{pmatrix} D_{0,0} & D_{0,1} & D_{0,2} & \cdots & D_{0,L'} \\ D_{1,0} & D_{1,1} & D_{1,2} & \cdots & D_{1,L'} \\ D_{2,0} & D_{2,1} & D_{2,2} & \cdots & D_{2,L'} \\ \vdots & \vdots & \vdots & \ddots & \vdots \\ D_{L',0} & D_{L',1} & D_{L',2} & \cdots & D_{L',L'} \end{pmatrix}$$

$$= \begin{pmatrix} D^{0,0}_{0,0} & D^{0,1}_{0,-1} & D^{0,1}_{0,0} & \cdots & D^{0,L}_{0,L} \\ D^{1,0}_{-1,0} & D^{1,1}_{-1,-1} & D^{1,1}_{-1,0} & \cdots & D^{1,L}_{-1,L} \\ D^{1,0}_{0,0} & D^{1,1}_{0,-1} & D^{1,1}_{0,0} & \cdots & D^{1,L}_{0,L} \\ \vdots & \vdots & \vdots & \ddots & \vdots \\ D^{L,0}_{L,0} & D^{L,1}_{L,-1} & D^{L,1}_{L,0} & \cdots & D^{L,L}_{L,L} \end{pmatrix}.$$

At stationary points of the ratio (8.28), the vector $\mathbf{f}$ is a solution to the following eigenfunction problem:

$$\boxed{\mathbf{D}\mathbf{f} = \lambda\mathbf{f}.} \tag{8.29}$$

Or more explicitly for each element $(f)^m_\ell$ in $\mathbf{f}$ we have

$$\sum_{p,q}^{L} D^{m,q}_{\ell,p}(f)^q_p = \lambda(f)^m_\ell. \tag{8.30}$$

Determining matrix $\mathbf{D}$ with elements given by (8.27) is enough for (numerically) solving the eigenfunction problem (8.29).

Remark 8.1. Compared with the time concentration problem of bandlimited signals, this analogy is somewhat simpler conceptually and mathematically because the problem is finite-dimensional. □

8.4.1   Orthogonality relations

Since $\mathbf{D}$ with elements defined in (8.27) is Hermitian symmetric and positive definite, all eigenvalues are real and the eigenvectors corresponding to distinct eigenvalues are orthogonal to each other. We can sort the eigenvalues in descending order

$$1 > \lambda_0 \geq \lambda_1 \geq \cdots \geq \lambda_{L^2+2L} > 0. \tag{8.31}$$

We consider unit-norm orthonormal eigenvectors $\mathbf{f}_n$ and $\mathbf{f}_k$ corresponding to two distinct eigenvalues λ_n and λ_k such that

$$\langle \mathbf{f}_n, \mathbf{f}_k \rangle = \mathbf{f}_n' \overline{\mathbf{f}_k} = \delta_{n,k}, \tag{8.32}$$

where

$$\mathbf{f}_n = \left(\left((f)_0^0 \right)_n, \left((f)_1^{-1} \right)_n, \left((f)_1^0 \right)_n, \ldots, \left((f)_L^L \right)_n \right)'.$$

The eigenvector $\mathbf{f}_n$ in spectral domain, corresponding to eigenvalue λ_n, gives rise to a signal on the 2-sphere $f_n(\widehat{\boldsymbol{u}})$ such that

$$f_n(\widehat{\boldsymbol{u}}) = \sum_{\ell,m}^{L} \left((f)_\ell^m \right)_n Y_\ell^m(\widehat{\boldsymbol{u}}), \quad \widehat{\boldsymbol{u}} \in \mathbb{S}^2. \tag{8.33}$$

Therefore, the corresponding eigenfunctions $f_n(\widehat{\boldsymbol{u}})$ and $f_k(\widehat{\boldsymbol{u}})$ are orthonormal on the whole 2-sphere:

$$\int_{\mathbb{S}^2} f_n(\widehat{\boldsymbol{u}}) \overline{f_k(\widehat{\boldsymbol{u}})} \, ds(\widehat{\boldsymbol{u}}) = \delta_{n,k}. \tag{8.34}$$

The concentration ratio of $f_n(\widehat{\boldsymbol{u}})$ in region R is $\lambda_n < 1$. All eigenvalues are strictly smaller than one because a bandlimited signal cannot be simultaneously spacelimited and inevitably part of its energy leaks out to the region $\bar{R} \triangleq \mathbb{S}^2 \setminus R$. $f_0(\widehat{\boldsymbol{u}})$, corresponding to the largest eigenvalue, has the largest spatial concentration in R and is our "favorite" eigenfunction, while $f_{L^2+2L}(\widehat{\boldsymbol{u}})$ is least appealing as far as concentration in R goes.

Remark 8.2. If we reverse the ordering of eigenvalues in (8.31) from smallest to biggest, we obtain the corresponding eigenfunctions that are optimally *excluded* from region R. The eigenfunction $f_{L^2+2L}(\widehat{\boldsymbol{u}})$ is the most excluded function in R and hence most concentrated in complementary region $\bar{R}$. □

In addition to satisfying the orthonormality property on the whole 2-sphere, $f_n(\widehat{\boldsymbol{u}})$ and $f_k(\widehat{\boldsymbol{u}})$ have the interesting property that they are also orthogonal (but not orthonormal) in region R. To see this, we first note that

$$\mathbf{f}_n' \overline{\mathbf{D}\mathbf{f}_k} = \lambda_k \mathbf{f}_n' \overline{\mathbf{f}_k} = \lambda_k \delta_{n,k},$$

where we have used (8.29) and $\mathbf{f}_n' \overline{\mathbf{f}_k} = \delta_{n,k}$. Now using (8.33) and the definition of $D_{\ell,p}^{m,q}$ in (8.27) we can verify that

$$\int_R f_n(\widehat{\boldsymbol{u}}) \overline{f_k(\widehat{\boldsymbol{u}})} \, ds(\widehat{\boldsymbol{u}}) = \sum_{\ell,m}^{L} \left((f)_\ell^m \right)_n \sum_{p,q}^{L} \overline{D_{\ell,p}^{m,q} \left((f)_p^q \right)_k} \tag{8.35}$$

$$= \mathbf{f}_n' \overline{\mathbf{D}\mathbf{f}_k} = \lambda_k \mathbf{f}_n' \overline{\mathbf{f}_k} = \lambda_k \delta_{n,k}.$$

Below, we summarize the orthonormality and orthogonality relations

$$\boxed{\mathbf{f}_n' \overline{\mathbf{f}_k} = \delta_{n,k} \iff \int_{\mathbb{S}^2} f_n(\widehat{\boldsymbol{u}}) \overline{f_k(\widehat{\boldsymbol{u}})} \, ds(\widehat{\boldsymbol{u}}) = \delta_{n,k}}$$

and

$$\mathbf{f}'_n \overline{\mathbf{Df}_k} = \lambda_k \delta_{n,k} \iff \int_R f_n(\widehat{\boldsymbol{u}}) \overline{f_k(\widehat{\boldsymbol{u}})} \, ds(\widehat{\boldsymbol{u}}) = \lambda_k \delta_{n,k}.$$

8.4.2   Eigenfunction kernel representation

The eigenfunction problem in (8.29) has an equivalent kernel representation in $L^2(\mathbb{S}^2)$. To see this, we multiply the left-hand side of (8.30) by spherical harmonic $Y_\ell^m(\widehat{\boldsymbol{u}})$ and sum over all possible degrees and orders to obtain

$$\sum_{p,q}^L (f)_p^q \sum_{\ell,m}^L D_{\ell,p}^{m,q} Y_\ell^m(\widehat{\boldsymbol{u}}) = \sum_{p,q}^L (f)_p^q \sum_{\ell,m}^L \int_R Y_p^q(\widehat{\boldsymbol{v}}) \overline{Y_\ell^m(\widehat{\boldsymbol{v}})} \, ds(\widehat{\boldsymbol{v}}) Y_\ell^m(\widehat{\boldsymbol{u}})$$

$$= \int_R \sum_{p,q}^L (f)_p^q Y_p^q(\widehat{\boldsymbol{v}}) \sum_{\ell,m}^L Y_\ell^m(\widehat{\boldsymbol{u}}) \overline{Y_\ell^m(\widehat{\boldsymbol{v}})} \, ds(\widehat{\boldsymbol{v}})$$

$$= \int_R \sum_{\ell=0}^L \sum_{m=-\ell}^\ell Y_\ell^m(\widehat{\boldsymbol{u}}) \overline{Y_\ell^m(\widehat{\boldsymbol{v}})} f(\widehat{\boldsymbol{v}}) \, ds(\widehat{\boldsymbol{v}}).$$

This equation is in the form of an integral operator $\mathcal{D}$ on $L^2(\mathbb{S}^2)$:

$$(\mathcal{D}f)(\widehat{\boldsymbol{u}}) = \int_{\mathbb{S}^2} D(\widehat{\boldsymbol{u}}, \widehat{\boldsymbol{v}}) f(\widehat{\boldsymbol{v}}) \, ds(\widehat{\boldsymbol{v}}),$$

with kernel

$$D(\widehat{\boldsymbol{u}}, \widehat{\boldsymbol{v}}) = \mathbf{1}_R(\widehat{\boldsymbol{v}}) \sum_{\ell,m}^L Y_\ell^m(\widehat{\boldsymbol{u}}) \overline{Y_\ell^m(\widehat{\boldsymbol{v}})} = \mathbf{1}_R(\widehat{\boldsymbol{v}}) \sum_{\ell=0}^L \frac{(2\ell+1)}{4\pi} P_\ell(\widehat{\boldsymbol{u}} \cdot \widehat{\boldsymbol{v}}), \qquad (8.36)$$

where $\mathbf{1}_R(\widehat{\boldsymbol{v}})$ is the indicator function for region R, and we have used the addition theorem (7.30), p. 193.

Applying the same procedure to the right-hand side of (8.30) yields

$$\lambda \sum_{\ell,m}^L (f)_\ell^m Y_\ell^m(\widehat{\boldsymbol{u}}) = \lambda f(\widehat{\boldsymbol{u}}), \quad \widehat{\boldsymbol{u}} \in \mathbb{S}^2. \qquad (8.37)$$

So the end result is

$$\int_{\mathbb{S}^2} D(\widehat{\boldsymbol{u}}, \widehat{\boldsymbol{v}}) f(\widehat{\boldsymbol{v}}) \, ds(\widehat{\boldsymbol{v}}) = \lambda f(\widehat{\boldsymbol{u}}), \quad \widehat{\boldsymbol{u}} \in \mathbb{S}^2, \qquad (8.38)$$

where the kernel $D(\widehat{\boldsymbol{u}}, \widehat{\boldsymbol{v}})$ is given by (8.36).

8.5   Optimal spectral concentration of spacelimited signals

The dual problem is as follows. The signal we are looking for is denoted by g with spectral coefficients $(g)_\ell^m$ and is spacelimited to a region R. That is, $g \in \mathcal{H}_R(\mathbb{S}^2)$. We wish to optimally concentrate the signal in the desired spectral interval with a maximum degree L. Here, we first choose energy of the signal as the measure of concentration. Mathematically, we wish to maximize the following ratio

$$\beta = \frac{\displaystyle\sum_{\ell,m}^{L} |(g)_\ell^m|^2}{\displaystyle\sum_{\ell,m} |(g)_\ell^m|^2}. \tag{8.39}$$

Noting that g is only non-zero in R, its spectral coefficients $(g)_\ell^m$ are given by

$$(g)_\ell^m = \int_R g(\widehat{\boldsymbol{u}}) \overline{Y_\ell^m(\widehat{\boldsymbol{u}})} \, ds(\widehat{\boldsymbol{u}}). \tag{8.40}$$

Plugging this equation into the numerator in (8.39) we obtain

$$\sum_{\ell,m}^{L} |(g)_\ell^m|^2 = \int_R \int_R g(\widehat{\boldsymbol{v}}) \sum_{\ell,m}^{L} \overline{Y_\ell^m(\widehat{\boldsymbol{v}})} Y_\ell^m(\widehat{\boldsymbol{u}}) \overline{g(\widehat{\boldsymbol{u}})} \, ds(\widehat{\boldsymbol{u}}) \, ds(\widehat{\boldsymbol{v}})$$

$$= \int_R \int_R g(\widehat{\boldsymbol{v}}) D(\widehat{\boldsymbol{u}}, \widehat{\boldsymbol{v}}) \overline{g(\widehat{\boldsymbol{u}})} \, ds(\widehat{\boldsymbol{u}}) \, ds(\widehat{\boldsymbol{v}}),$$

where we have used the definition of the kernel $D(\widehat{\boldsymbol{u}}, \widehat{\boldsymbol{v}})$ in (8.36). Finally using the Parseval relation for the denominator of (8.39), we can write the concentration ratio as

$$\beta = \frac{\displaystyle\int_R \int_R g(\widehat{\boldsymbol{v}}) D(\widehat{\boldsymbol{u}}, \widehat{\boldsymbol{v}}) \overline{g(\widehat{\boldsymbol{u}})} \, ds(\widehat{\boldsymbol{u}}) \, ds(\widehat{\boldsymbol{v}})}{\displaystyle\int_R |g(\widehat{\boldsymbol{u}})|^2 ds(\widehat{\boldsymbol{u}})}. \tag{8.41}$$

At stationary points of the above ratio, function g is a solution to the following integral eigenfunction problem

$$\int_R D(\widehat{\boldsymbol{u}}, \widehat{\boldsymbol{v}}) \, g(\widehat{\boldsymbol{v}}) \, ds(\widehat{\boldsymbol{v}}) = \lambda g(\widehat{\boldsymbol{u}}), \quad \widehat{\boldsymbol{u}} \in R. \tag{8.42}$$

Comparing (8.42) and (8.38) we see that these two eigenfunction problems are the same with the only difference that the domain of (8.38) is the entire 2-sphere whereas (8.42) is only valid when $\widehat{\boldsymbol{u}} \in R$. Remarkably, within the region R, the spacelimited eigensolution to (8.42) $g(\widehat{\boldsymbol{u}})$ is identical to the eigensolution $f(\widehat{\boldsymbol{u}})$ to (8.38) in its dual spatial concentration problem. That is,

$$g(\widehat{\boldsymbol{u}}) = \begin{cases} f(\widehat{\boldsymbol{u}}) & \text{if } \widehat{\boldsymbol{u}} \in R, \\ 0 & \text{if } \widehat{\boldsymbol{u}} \in \bar{R}. \end{cases} \tag{8.43}$$

Given the spatial and spectral parameters R and L, the following summarizes the steps that should be taken to solve the optimal spectral concentration problem (8.42):

Step 1: Determine the $(L+1)^2 \times (L+1)^2$ matrix $\mathbf{D}$ with elements given by (8.27).

Step 2: (Numerically) Solve the eigenfunction problem (8.33). The eigenvector $\mathbf{f}_n$ corresponding to eigenvalue λ_n leads to eigenfunction $f_n(\widehat{\boldsymbol{u}})$ using (8.33).

Step 3: Project $f_n(\widehat{\boldsymbol{u}})$ to subspace $\mathcal{H}_R(\mathbb{S}^2)$ using (8.43) to obtain $g_n(\widehat{\boldsymbol{u}})$.

Step 4: Use (8.40) to obtain the spherical harmonic coefficients of $g_n(\widehat{\boldsymbol{u}})$ for which λ_n fraction of their energy is concentrated in spectral interval $0 \le \ell \le L$.

8.6   Area-bandwidth product or spherical Shannon number

Let us look at the sum of all eigenvalues of matrix $\mathbf{D}$ in (8.29).

$$N \triangleq \sum_{n=0}^{L^2+2L} \lambda_n = \text{trace}(\mathbf{D}) = \sum_{\ell,m} D_{\ell,\ell}^{m,m}$$

$$= \int_R \sum_{\ell,m}^{L} Y_\ell^m(\widehat{\boldsymbol{u}})\overline{Y_\ell^m(\widehat{\boldsymbol{u}})}\, ds(\widehat{\boldsymbol{u}}) \tag{8.44}$$

$$= \int_R \sum_{\ell=0}^{L} \frac{(2\ell+1)}{4\pi} P_\ell(1)\, ds(\widehat{\boldsymbol{u}}) = \frac{A}{4\pi}(L+1)^2,$$

where we have used addition theorem (7.30), p. 193, with $\widehat{\boldsymbol{u}}\cdot\widehat{\boldsymbol{u}} = 1$, $P_\ell(1) = 1$ and $A = \int_R ds(\widehat{\boldsymbol{u}})$. Obviously, this is true regardless of distribution of the eigenvalues. However, eigenvalues are almost binary-valued: they are close either to one or to zero with a narrow transition band between the two values. Therefore, to a good degree of accuracy, the *area-bandwidth product*

$$\boxed{N = \frac{A}{4\pi}(L+1)^2} \tag{8.45}$$

quantifies the number of significant eigenvalues or the number of significant functions that are simultaneously bandlimited to spectral degree L and (almost) concentrated in the region R.[2]

Remark 8.3. The *area-bandwidth product* N is the analog of *Shannon's time-bandwidth* product, $2WT$, on the 2-sphere.   □

---

[2] This value also quantifies the number of functions that are simultaneously spacelimited to spatial region R and (almost) bandlimited to the spectral degree L.

8.7 Operator formulation

We briefly outline how we can express spatial and spectral concentration problems in operator form. This will be detailed further in Section 8.12 when we discuss a generalized variational framework developed in (Franks, 1969, Ch. 6) applied to the 2-sphere. Recalling from Section 7.8, p. 219, and Section 7.9, p. 220, two projection operators into the subspace of bandlimited signals $\mathcal{H}_L(\mathbb{S}^2)$ and the subspace of spacelimited signals $\mathcal{H}_R(\mathbb{S}^2)$ were denoted by $\mathcal{B}_L$ and $\mathcal{B}_R$, respectively. Using this notation, we can write the optimal spatial concentration problem in (8.26) as

$$\lambda = \frac{\langle \mathcal{B}_R \circ \mathcal{B}_L f, \mathcal{B}_R \circ \mathcal{B}_L f \rangle}{\langle \mathcal{B}_L f, \mathcal{B}_L f \rangle}.$$

Noting that the projection operators $\mathcal{B}_R$ and $\mathcal{B}_L$ are both idempotent and self-adjoint, we can write the numerator above as

$$\langle \mathcal{B}_R \circ \mathcal{B}_L f, \mathcal{B}_R \circ \mathcal{B}_L f \rangle = \langle (\mathcal{B}_R \circ \mathcal{B}_L)^* \circ \mathcal{B}_R \circ \mathcal{B}_L f, f \rangle$$
$$= \langle \mathcal{B}_L^* \circ \mathcal{B}_R^* \circ \mathcal{B}_R \circ \mathcal{B}_L f, f \rangle$$
$$= \langle \mathcal{B}_L \circ \mathcal{B}_R \circ \mathcal{B}_L f, f \rangle.$$

Similarly, the denominator is

$$\langle \mathcal{B}_L f, \mathcal{B}_L f \rangle = \langle \mathcal{B}_L^* \circ \mathcal{B}_L f, f \rangle$$
$$= \langle \mathcal{B}_L f, f \rangle.$$

And hence

$$\lambda = \frac{\langle \mathcal{B}_L \circ \mathcal{B}_R \circ \mathcal{B}_L f, f \rangle}{\langle \mathcal{B}_L f, f \rangle}.$$

This shows that at the stationary points of the ratio, function f satisfies

$$(\mathcal{B}_L \circ \mathcal{B}_R \circ \mathcal{B}_L)(\mathcal{B}_L f) = \lambda(\mathcal{B}_L f),$$

where again we have used the idempotent property of projection operator $\mathcal{B}_L$.
Similarly for the spectral concentration problem, we can write

$$\lambda = \frac{\langle \mathcal{B}_L \circ \mathcal{B}_R g, \mathcal{B}_L \circ \mathcal{B}_R g \rangle}{\langle \mathcal{B}_R g, \mathcal{B}_R g \rangle}.$$

And following similar steps as above, two key equations are derived:

$$\lambda = \frac{\langle \mathcal{B}_R \circ \mathcal{B}_L \circ \mathcal{B}_R g, g \rangle}{\langle \mathcal{B}_R g, g \rangle}$$
$$(\mathcal{B}_R \circ \mathcal{B}_L \circ \mathcal{B}_R)(\mathcal{B}_R g) = \lambda(\mathcal{B}_R g).$$

8.8   Special case: azimuthally symmetric polar cap region

An important region of interest on the sphere is an axis-symmetric polar cap, R_{θ_0}, which is parameterized by θ_0. By a proper rotation, this cap can be positioned around any point on the 2-sphere. Therefore, the assumption of symmetry around the pole is not too restrictive. Here we wish to study bandlimited signals that are optimally concentrated in this polar cap, knowing that we can solve the dual spacelimited problem easily.

For this symmetric region and referring to the definition of spherical harmonics in (7.19), the elements $D_{\ell,p}^{m,q}$ of matrix $\mathbf{D}$ take a special form as

$$
\begin{aligned}
D_{\ell,p}^{m,q} &= \int_0^{\theta_0} \int_0^{2\pi} Y_p^q(\theta,\phi)\overline{Y_\ell^m(\theta,\phi)}\,\sin\theta\,d\phi\,d\theta \\
&= 2\pi\delta_{m,q} \int_0^{\theta_0} N_\ell^m P_\ell^m(\cos\theta) N_p^m P_p^m(\cos\theta)\,\sin\theta\,d\theta.
\end{aligned}
\tag{8.46}
$$

Therefore, the elements of $\mathbf{D}$ can only be non-zero when $m = q$. Because of this special property and by appropriate switching of rows and columns of $\mathbf{D}$, it can be converted into a block diagonal matrix, where non-zero elements such as $D_{\ell,p}^{m,m}$ with a fixed order m for $m \le \ell, p \le L$ appear next to each other in the following submatrix $\mathbf{D}^{(m)}$ as[3]

$$
\mathbf{D}^{(m)} = \begin{pmatrix}
D_{m,m}^{m,m} & D_{m,m}^{m,m+1} & D_{m,m}^{m,m+2} & \cdots & D_{m,m}^{m,L} \\
D_{m,m}^{m+1,m} & D_{m,m}^{m+1,m+1} & D_{m,m}^{m+1,m+2} & \cdots & D_{m,m}^{m+1,L} \\
D_{m,m}^{m+2,m} & D_{m,m}^{m+3,m+2} & D_{m,m}^{m+2,m+2} & \cdots & D_{m,m}^{m+2,L} \\
\vdots & \vdots & \vdots & \ddots & \vdots \\
D_{m,m}^{L,m} & D_{m,m}^{L,m+1} & D_{m,m}^{L,m+2} & \cdots & D_{m,m}^{L,L}
\end{pmatrix}.
\tag{8.47}
$$

That is, for a fixed non-negative order $m \in \{0, 1, \ldots, L\}$, there can be at most $(L - m + 1) \times (L - m + 1)$ non-zero elements of $\mathbf{D}$. As a result, for the special case of axis-symmetric polar cap region R_{θ_0}, instead of solving the full eigenvalue problem (8.29), we can solve $L + 1$ smaller problems, each for a fixed order $m \in \{0, 1, \ldots, L\}$, each with a size of $L - m + 1$. The corresponding eigenfunction problem becomes

$$
\mathbf{D}^{(m)}\mathbf{f}^{(m)} = \lambda\mathbf{f}^{(m)}.
\tag{8.48}
$$

Similar to the general case, the n-th solution $\mathbf{f}_n^{(m)}$ to the eigenfunction problem above, corresponding to eigenvalue $\lambda_n^{(m)}$, gives rise to eigenfunction $f_n^{(m)}(\widehat{\boldsymbol{u}})$ on the whole 2-sphere

$$
f_n^{(m)}(\widehat{\boldsymbol{u}}) = \sum_{\ell=m}^{L} \left((f)_\ell^m\right)_n Y_\ell^m(\widehat{\boldsymbol{u}}), \quad n \in \{0, 1, \ldots, L - m\}.
\tag{8.49}
$$

---

[3] As was alluded to in the discussion after (7.48), p. 198, in (Simons et al., 2006) spectral coefficients with a fixed order m are arranged next to each other and the same ordering is used for $\mathbf{D}$. With this ordering, the block diagonal structure of $\mathbf{D}$ would become immediately apparent without any rearranging of rows and columns.

For a given order m, they satisfy the following orthonormality and orthogonality relations

$$\left(\mathbf{f}_n^{(m)}\right)'\overline{\mathbf{f}_k^{(m)}} = \delta_{n,k} \iff \int_{\mathbb{S}^2} f_n^{(m)}(\widehat{\boldsymbol{u}})\overline{f_k^{(m)}(\widehat{\boldsymbol{u}})}\,ds(\widehat{\boldsymbol{u}}) = \delta_{n,k}$$

and

$$\mathbf{f}_n^{(m)}\overline{\mathbf{D}^{(m)}\mathbf{f}_k^{(m)}} = \lambda_k^{(m)}\delta_{n,k} \iff \int_R f_n^{(m)}(\widehat{\boldsymbol{u}})\overline{f_k^{(m)}(\widehat{\boldsymbol{u}})}\,ds(\widehat{\boldsymbol{u}}) = \lambda_k^{(m)}\delta_{n,k}.$$

Remark 8.4. Note that for a fixed order m, the size of eigenvector $\mathbf{f}^{(m)}$ corresponding to $\mathbf{D}^{(m)}$ is $(L - m + 1) \times 1$. The full-size eigenvector corresponding to the full-size matrix $\mathbf{D}$ (assuming the single-indexing convention of (7.39), p. 195) can be obtained as follows:

Step 1: Start with a zero vector $\mathbf{f}$ of size $(L + 1)^2 \times 1$.

Step 2: For a fixed m and each $\ell \in \{m, m + 1, \ldots, L\}$, find the corresponding single index $n = \ell(\ell + 1) + m$. Therefore, $n \in \{m^2 + 2m, \ldots, L^2 + L + m\}$.

Step 3: Fill the n-th position in $\mathbf{f}$ with the $(\ell - m + 1)$-th element in $\mathbf{f}^{(m)}$.

Through the above procedure, we can verify that the obtained vectors $\mathbf{f}$ have non-overlapping non-zero elements for different orders m. Since their orthonormality was already ensured through (8.8) for a fixed m, we confirm that they are in fact orthonormal over the entire spectral range $\ell \in \{0, 1, \ldots, L\}$ and $m \in \{-\ell, -\ell + 1, \ldots, \ell\}$, as expected.   □

Remark 8.5. Being able to break the original eigenfunction problem in (8.29) into $(L + 1)$ smaller subproblems is only possible for azimuthally symmetric regions. In general, $\mathbf{D}$ cannot be converted into a block diagonal matrix and therefore we cannot solve smaller eigenfunction problems corresponding to fixed orders m, because we cannot guarantee that the resulting eigenvectors maintain their orthonormality across different orders. In general, in order to ensure full orthonormality among all eigenfunctions regardless of order m and degree ℓ, one needs to solve the full eigenfunction problem (8.29).   □

Similar to (8.44), for a fixed order m, the number of significant eigenvalues is given by

$$N^{(m)} \triangleq \sum_{n=m}^{L-m+1} \lambda_n^{(m)} = \text{trace}(\mathbf{D}^{(m)}) = \sum_{\ell=m}^{L} D_{\ell,\ell}^{m,m}$$

$$= 2\pi \int_0^{\theta_0} \sum_{\ell=m}^{L} \left(N_\ell^m P_\ell^m(\cos\theta)\right)^2 \sin\theta\,d\theta,$$

$$(8.50)$$

which does not lead to a simple closed form solution.

Once all $L-m+1$ eigenvalues $\lambda^{(m)}$ are obtained for all orders $m \in \{0, 1, \ldots, L\}$, they can be globally ranked regardless of their order. Combining (8.44) and (8.50), the total number of significant eigenvalues will be $N = N^{(0)} + 2N^{(m)}$, where the factor of 2 appears due to symmetry for positive and negative orders m.

8.9   Azimuthally symmetric concentrated signals in polar cap

In Section 8.8, for every order m we obtained a family of $L - m + 1$ eigenfunctions that are optimally concentrated in the polar cap region R_{θ_0}. We note that although the region of interest is azimuthally symmetric, the family of eigenfunctions do not have to be. They can have any shape inside R_{θ_0} and can be a function of longitude angle ϕ. However, if we focus only on the zeroth-order family $m = 0$, we end up with $L + 1$ azimuthally symmetric eigenfunctions that are optimally concentrated in an azimuthally symmetric polar cap. The corresponding eigenfunction problem is

$$\mathbf{D}^{(0)}\mathbf{f}^{(0)} = \lambda\mathbf{f}^{(0)}, \tag{8.51}$$

where the sum of eigenvalues is denoted by $N^{(0)}$. Two important questions are (1) how the sum of eigenvalues $N^{(0)}$ is related to the number of significant azimuthally symmetric eigenfunctions inside R_{θ_0} and (2) how this number is related to the region size parameterized by θ_0 and bandwidth L.

Fortunately, similar to the general case, the eigenvalues corresponding to azimuthally symmetric concentrated functions are either very close to one or zero with a very sharp transition in between. Therefore, $N^{(0)}$ accurately represents the number of significant eigenvalues. In (Wieczorek and Simons, 2005), the authors have shown empirically that $N^{(0)}$ is well described by the analog of the cartesian Shannon number, that is,

$$\boxed{N^{(0)} = (L + 1)\frac{\theta_0}{\pi}.} \tag{8.52}$$

Note that $N^{(0)}$ is directly related to the maximum co-latitude angle θ_0 and not to the polar cap area $A = 2\pi(1 - \cos\theta_0)$. It was observed in (Wieczorek and Simons, 2005) that the first eigenfunction $f_0^{(0)}(\hat{\boldsymbol{u}})$ corresponding to the largest eigenvalue in (8.51) approaches the *spherical uncertainty principle*, which is described below.

Example 8.1. In this example, we reproduce some of the results in (Simons et al., 2006, Fig. 5.1). We solve the problem of spatial concentration of bandlimited signals where the spatial region of interest is a polar cap with maximum co-latitude of $\theta_0 = 40°$ and the maximum spectral degree is $L = 18$. The first four eigenvalues (arranged horizontally) for $m = \{0, 1, 2, 3\}$ (arranged vertically) are computed and drawn in Figure 8.2. Elements of matrix $\mathbf{D}$ are numerically calculated using (8.27) with 2520 samples along the longitude and co-latitudes in the range $0 \le \theta \le \theta_0$ and then used in (8.47) and (8.48) to compute the eigenvalues and eigenvectors.

It should be noted that here we are only showing the co-latitude response of the eigenfunctions in spatial domain together with the normalization constants. The longitude-dependent part of Y_ℓ^m, $\cos m\phi$ and $\sin m\phi$, are not shown. (Note that (Simons et al., 2006) use real spherical harmonics.) As expected, when the eigenvalue decreases, the leakage into co-latitudes beyond $\theta_0 = 40°$ becomes more pronounced, whereas the first eigenfunctions for each m are almost fully concentrated to the desired polar cap (grayed region) for all practical purposes.

□

8.10   Uncertainty principle for azimuthally symmetric functions

The uncertainty principle on the 2-sphere is a bound on the best joint spatial and spectral concentration of signals in $\mathbb{S}^2$. Roughly speaking, the spatial and spectral distribution of a signal cannot be simultaneously very small. A small *variance* in spatial domain results in a large *variance* in spherical harmonic domain and vice versa. Let us first quantify spatial and spectral variance. For an azimuthally symmetric function $h(\theta)$, the variance in the spatial domain is given by (Wieczorek and Simons, 2005; Narcowich and Ward, 1996)

$$\sigma_S^2 = 1 - \left(\pi \int_0^\pi \sin(2\theta)\big|h(\theta)\big|^2 d\theta\right)^2. \tag{8.53}$$

The spectral variance of h with spherical harmonic coefficients $(h)_\ell^0$ is defined as

$$\sigma_L^2 = \sum_{\ell=0}^\infty \ell(\ell+1)\big|(h)_\ell^0\big|^2. \tag{8.54}$$

The following inequality, referred to as the uncertainty principle, holds for unit energy azimuthally symmetric functions defined on the 2-sphere (Narcowich and Ward, 1996; Freeden and Michel, 1999)

$$\left(\frac{\sigma_S}{\sqrt{1-\sigma_S^2}}\right)\sigma_L \geq 1. \tag{8.55}$$

An azimuthally symmetric function is said to be optimally concentrated from the perspective of the uncertainty principle if the product of its spatial and spectral variance can achieve or get very close to the lower bound of 1 in (8.55). For example, the first eigenfunction $f_0^{(0)}(\hat{u})$ corresponding to the largest eigenvalue in (8.51) is optimal in this sense (Wieczorek and Simons, 2005).

8.11   Comparison with time-frequency concentration problem

Here we summarize some of the important similarities and differences that exist between the concentration problems on spatio-spectral and time-frequency domains. Some of these were implied or explicitly discussed in previous sections. Here we list them and also add some new properties.

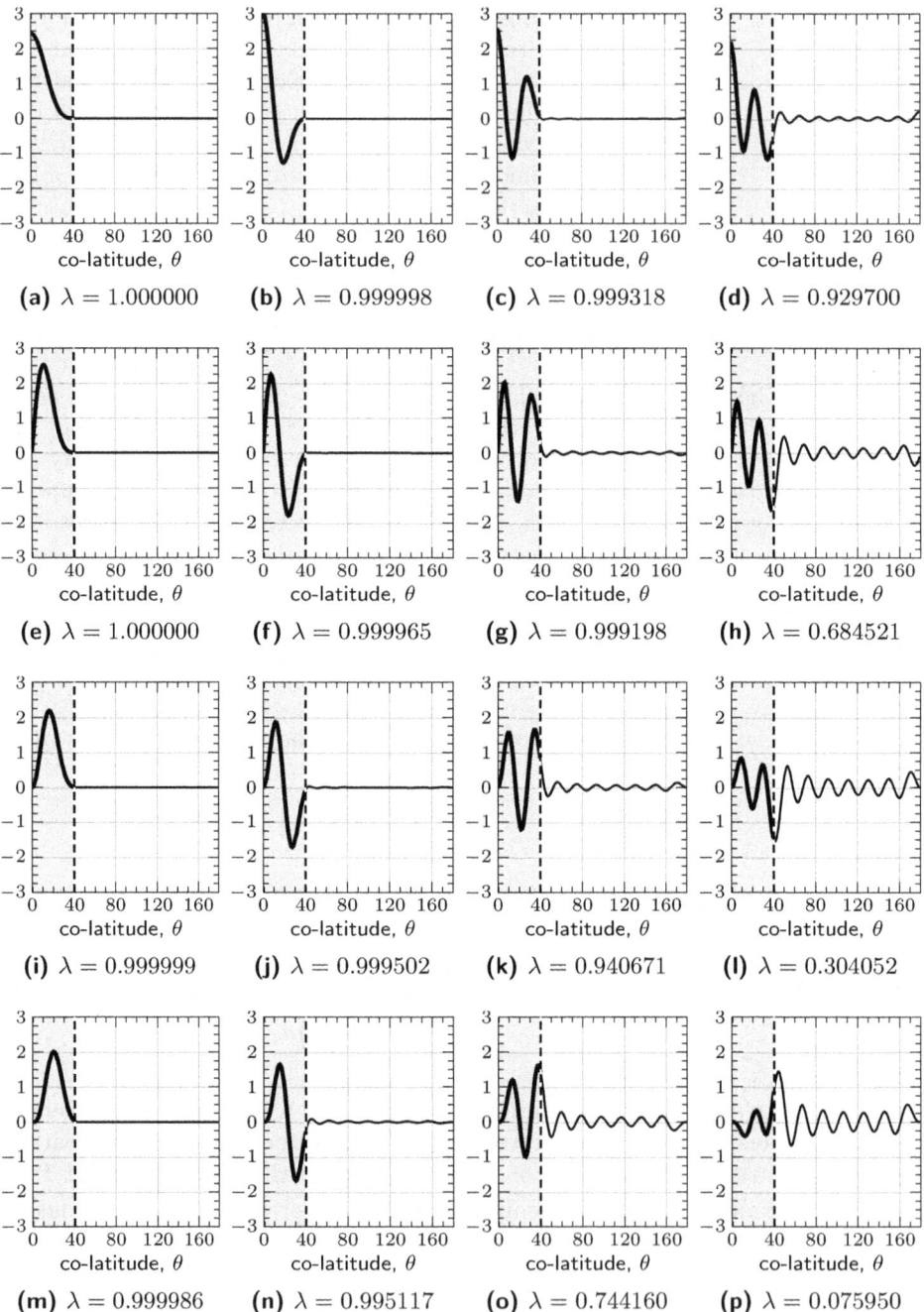

Figure 8.2 Eigenfunctions and eigenvalues for the spatial concentration problem. Left
to right are the first four most concentrated signals. Top to bottom are the results
for $m = 0$ (top row), $m = 1$, $m = 2$, and $m = 3$ (bottom row). The accuracy of
eigenvalues is 6 decimal places.

P1: For a finite L, there are only a finite number, equal to $(L+1)^2$, of band-limited spatial eigenfunctions and corresponding eigenvalues on 2-sphere. This is in contrast with the time-frequency problem, where the number of eigenfunctions is countably infinite.

P2: Unlike the time-frequency problem, finding the bandlimited eigenfunctions is a finite-dimensional problem, which can be solved using matrix formulation (8.29), p. 262.

P3: Similar to the time-frequency case, the eigenvalues are distinct, real and strictly between 0 and 1. This is because $\mathbf{D}$ in (8.29), p. 262, is Hermitian symmetric.

P4: Unlike the time-frequency case, the eigenfunctions can be in general complex-valued. This is because $\mathbf{D}$ in (8.29), p. 262, is in general complex.

P5: Similar to the time-frequency case, the eigenfunctions are orthonormal on the entire 2-sphere as stated in (8.32), p. 263.

P6: Similar to the time-frequency case, the eigenfunctions are orthogonal in region R as stated in (8.35), p. 263.

P7: Somewhat similar to the time-frequency case, the $(L+1)^2$ eigenfunctions form a basis for the subspace of bandlimited functions $\mathcal{H}_L(\mathbb{S}^2)$ on 2-sphere. The only difference is that $\mathcal{H}_L(\mathbb{S}^2)$ is finite-dimensional.

P8: Similar to the time-frequency case, the spatial concentration ratio α attains its largest value equal to λ_0 when the first eigenfunction f_0 is used in (8.26).

P9: Similar to the time-frequency case, there is an equivalent differential equation for solving the spatial concentration problem. The details are discussed in (Simons et al., 2006).

P10: Similar to the time-frequency case, the eigenvalues are almost binary-valued. They are either very close to one or very close to zero with a sharp transition between the two. Somewhat similar to the time-frequency case, the *degrees of freedom* in choosing functions that are bandlimited to degree L and almost spatially limited to region R is given by (8.45), p. 266.

P11: Unlike the time-frequency case, where time-bandwidth product $2WT$ is the only determining factor in the concentration problem, the size of the spatial region A and the bandwidth L do not uniquely specify the eigenfunctions. As we have seen through various examples, the shape of region R, whether it is symmetric or not, joint or disjoint, etc., has a critical effect on the type of eigenfunctions.

P12: Similar to the time-frequency case, the dual spectral concentration problem can be readily solved by appropriately modifying the solutions to the spatial concentration problem, as outlined in Section 8.5, p. 265.

P13: Similar to the time-frequency case, the minimum angle between the subspace of bandlimited functions $\mathcal{H}_L(\mathbb{S}^2)$ and the subspace of spacelimited functions $\mathcal{H}_R(\mathbb{S}^2)$ on the sphere is equal to $\cos^{-1}\sqrt{\lambda_0}$ and is achieved between $f_0 \in \mathcal{H}_L(\mathbb{S}^2)$ and its projection $\mathcal{B}_R f_0 \in \mathcal{H}_R(\mathbb{S}^2)$.

P14: All the results in Section 8.2.6, p. 258, for the time-frequency case carry over to the spatio-spectral concentration problem. In particular, for the third case where $\lambda_0 \le \alpha < 1$, setting

$$f = p f_0 + q \mathcal{B}_R f_0 \triangleq \sqrt{\frac{1-\alpha}{1-\lambda_0}} f_0 + \left(\sqrt{\frac{\alpha}{\lambda_0}} - \sqrt{\frac{1-\alpha}{1-\lambda_0}} \right) \mathcal{B}_R f_0$$

will achieve the spectral concentration bound

$$\sqrt{\beta} = \cos\left(\cos^{-1}\sqrt{\lambda_0} - \cos^{-1}\sqrt{\alpha}\right).$$

8.12 Franks generalized variational framework on 2-sphere

The spatial and spectral concentration problems studied in Section 8.4 and Section 8.5 are only two special examples of a general variational problem (Wei et al., 2011) that can be defined on the 2-sphere closely analogous to the Franks framework in the time-frequency domain (Franks, 1969, Ch. 6). We begin with introducing this general variational problem on 2-sphere. We then show how the previous examples fit into this framework.

> **Remark 8.6.** A goal here is to kill off the temptation to develop more specific "novel" formulations of concentration problems (based on quadratic functionals) by showing there is a general unifying framework.   □

8.12.1 Variational problem formulation

We consider three quadratic functionals as follows:

Spatial quadratic functional

The first quadratic functional is $I_{\mathcal{B}_h}(f)$, where the operator $\mathcal{B}_h$ spatially masks the signal $f(\widehat{\boldsymbol{u}})$ on $\mathbb{S}^2$ with the window function $h(\widehat{\boldsymbol{u}}) \in L^2(\mathbb{S}^2)$, as defined in Section 7.10, p. 221, with the additional condition that $h(\widehat{\boldsymbol{u}})$ is real-valued and positive a.e. That is,

$$\left(\mathcal{B}_h f\right)(\widehat{\boldsymbol{u}}) = h(\widehat{\boldsymbol{u}}) f(\widehat{\boldsymbol{u}}), \quad h(\widehat{\boldsymbol{u}}) = \overline{h(\widehat{\boldsymbol{u}})} > 0 \text{ a.e.} \tag{8.56}$$

Recalling (7.171), p. 248, we have

$$\boxed{I_{\mathcal{B}_h}(f) = \int_{\mathbb{S}^2} h(\widehat{\boldsymbol{u}}) \left| f(\widehat{\boldsymbol{u}}) \right|^2 ds(\widehat{\boldsymbol{u}}),} \tag{8.57}$$

which is the weighted energy of the signal f, where the weights are determined by the operator mask value at different spatial points. Proving that such a spatial masking operator is self-adjoint is left to Problem 8.1.

With operator $\mathcal{B}_h$ we can associate an operator matrix $\mathbf{H}$ (using the spherical harmonics as the complete orthonormal sequence) with matrix elements

$$h_{\ell,p}^{m,q} = \int_{\mathbb{S}^2} h(\widehat{\boldsymbol{u}}) Y_p^q(\widehat{\boldsymbol{u}}) \overline{Y_\ell^m(\widehat{\boldsymbol{u}})} \, ds(\widehat{\boldsymbol{u}}), \tag{8.58}$$

the proof of which is left to Problem 8.2. These matrix elements further satisfy

$$h_{\ell,p}^{m,q} = \overline{h_{p,\ell}^{q,m}}, \quad \text{or equivalently,} \quad \mathbf{H} = \mathbf{H}^H, \tag{8.59}$$

see Problem 8.3.

Spectral quadratic functional

The second quadratic functional that we consider is denoted by $I_{\mathcal{B}_s}(f)$ where the operator $\mathcal{B}_s$ is a self-adjoint, but otherwise arbitrary, bounded operator. The operator matrix is denoted $\mathbf{S}$ and should have matrix elements that satisfy

$$s_{\ell,p}^{m,q} = \overline{s_{p,\ell}^{q,m}}, \quad \text{or equivalently,} \quad \mathbf{S} = \mathbf{S}^H. \tag{8.60}$$

The operator output is

$$\left(\mathcal{B}_s f\right)(\widehat{\boldsymbol{u}}) = \sum_{\ell,m} \sum_{p,q} s_{\ell,p}^{m,q} (f)_p^q Y_\ell^m(\widehat{\boldsymbol{u}}). \tag{8.61}$$

Then according to (7.166), p. 246, the functional is given by

$$\boxed{I_{\mathcal{B}_s}(f) = \sum_{p,q} \sum_{\ell,m} s_{\ell,p}^{m,q}(f)_p^q \overline{(f)_\ell^m} = \langle \mathbf{S} \mathbf{f}, \mathbf{f} \rangle,} \tag{8.62}$$

which represents a very general *spectral energy masking* of the signal f with weights $s_{\ell,p}^{m,q}$.

Remark 8.7. An astute reader may realize that this formulation of spectral quadratic functional is almost too general and can represent any quadratic functional based on an arbitrary bounded self-adjoint operator in either spatial or spectral domain. This issue is the subject of Problem 8.4. More specific operators such as ones whose operator matrices are block diagonal and whose spectral weightings do not mix degrees prove more useful and belong to this class of spectral quadratic functionals.   □

Remark 8.8. To ensure that $I_{\mathcal{B}_s}$ is a true energy functional one should impose positivity in the form $\langle \mathbf{S} \mathbf{f}, \mathbf{f} \rangle > 0$ for non-zero $\mathbf{f}$.   □

Energy quadratic functional

The third quadratic functional, $I(f) = \langle f, f \rangle$, is very simple and basically measures the energy of the signal:

$$\boxed{I(f) = \int_{\mathbb{S}^2} \left| f(\widehat{\boldsymbol{u}}) \right|^2 ds(\widehat{\boldsymbol{u}}) = \sum_{\ell,m} \left| (f)_\ell^m \right|^2.} \tag{8.63}$$

Lagrange functional

Armed with arbitrary weighted measures of signal energy in spatial and spectral domains in (8.57) and (8.62), respectively, the general variational problem is to find a function f such that it maximizes the weighted energy $I_{\mathcal{B}_h}(f)$ (or $I_{\mathcal{B}_s}(f)$), subject to a constraint on the other weighted energy $I_{\mathcal{B}_s}(f)$ (or $I_{\mathcal{B}_h}(f)$) and the total energy of the signal being $I(f)$. This is equivalent to maximizing the following Lagrange functional,

$$G(f) = \mu_h I_{\mathcal{B}_h}(f) + \mu_s I_{\mathcal{B}_s}(f) + I(f), \qquad (8.64)$$

where μ_h and μ_s are two Lagrange multipliers. Without loss of generality, we can limit ourselves to finding unit-norm functions such that the energy quadratic functional satisfies $I(f) = 1$.

Problems

8.1. Show that the spatial masking operator

$$(\mathcal{B}_h f)(\widehat{\boldsymbol{u}}) = h(\widehat{\boldsymbol{u}}) f(\widehat{\boldsymbol{u}})$$

using a real-valued window

$$h(\widehat{\boldsymbol{u}}) = \overline{h(\widehat{\boldsymbol{u}})}$$

is self-adjoint. (There is no need to assume $h(\widehat{\boldsymbol{u}}) \geq 0$ here.)

8.2. Given the spatial masking operator $\mathcal{B}_h$ on $L^2(\mathbb{S}^2)$, which is defined in (8.56), show that the operator matrix denoted by $\mathbf{H}$ has matrix elements

$$
\begin{aligned}
h_{\ell,p}^{m,q} &\triangleq \langle \mathcal{B}_h Y_p^q, Y_\ell^m \rangle \\
&= \int_{\mathbb{S}^2} h(\widehat{\boldsymbol{u}}) Y_p^q(\widehat{\boldsymbol{u}}) \overline{Y_\ell^m(\widehat{\boldsymbol{u}})} \, ds(\widehat{\boldsymbol{u}}),
\end{aligned}
\qquad (8.65)
$$

as given in (8.58). Infer also that

$$\langle \mathcal{B}_h f, Y_\ell^m \rangle = \sum_{p,q} h_{\ell,p}^{m,q}(f)_p^q, \quad \text{where } (f)_p^q = \langle f, Y_p^q \rangle. \qquad (8.66)$$

8.3. Show that for the matrix $\mathbf{H}$ in Problem 8.2 that the matrix elements satisfy

$$h_{\ell,p}^{m,q} = \overline{h_{p,\ell}^{q,m}}.$$

8.4. Show that the general spectral quadratic functional in (8.62) with matrix elements

$$s_{\ell,p}^{m,q} = \sum_{s,t} \langle h, Y_s^t \rangle \int_{\mathbb{S}^2} Y_s^t(\widehat{\boldsymbol{u}}) Y_p^q(\widehat{\boldsymbol{u}}) \overline{Y_\ell^m(\widehat{\boldsymbol{u}})} \, ds(\widehat{\boldsymbol{u}}),$$

yields the spatial masking operator (8.56). (This shows that (8.62) is rather general and its action is not just limited to spectral shaping.)

8.12.2   Stationary points of Lagrange functional $G(f)$

To find the stationary points of $G(f)$, one needs to first find the derivative of $G(f)$ at point f along an arbitrary unit-norm function $d(\widehat{u}) \in L^2(\mathbb{S}^2)$, where the derivative is denoted by $(D_d G)(f)$, and then solve $(D_d G)(f) = 0$ for all possible $d(\widehat{u})$. We first note that for an arbitrary functional of the form $I_\mathcal{B}(f) = \langle \mathcal{B}f, f \rangle$, the derivative along d, denoted by $(D_d I_\mathcal{B})(f)$, is given by

$$
\begin{aligned}
(D_d I_\mathcal{B})(f) &\triangleq \lim_{\epsilon \to 0} \frac{I_\mathcal{B}(f + \epsilon d) - I_\mathcal{B}(f)}{\epsilon} \\
&= \lim_{\epsilon \to 0} \frac{\langle \mathcal{B}(f + \epsilon d), f + \epsilon d \rangle - \langle \mathcal{B}f, f \rangle}{\epsilon} \\
&= \langle \mathcal{B}f, d \rangle + \langle \mathcal{B}d, f \rangle.
\end{aligned}
\tag{8.67}
$$

If the operator is self-adjoint, then the derivative simplifies to

$$
\boxed{(D_d I_\mathcal{B})(f) = \langle \mathcal{B}f, d \rangle + \langle d, \mathcal{B}f \rangle = 2\mathfrak{Re}\big(\langle \mathcal{B}f, d \rangle\big).}
\tag{8.68}
$$

Since the two operators $\mathcal{B}_h$ and $\mathcal{B}_s$ that we used in (8.57) and (8.62) are indeed chosen to be self-adjoint, we can write the derivate of $G(f)$ along d as

$$
\boxed{(D_d G)(f) = 2\mathfrak{Re}\big(\langle \mu_h \mathcal{B}_h f + \mu_s \mathcal{B}_s f + f, d \rangle\big).}
\tag{8.69}
$$

Setting $(D_d G)(f) = 0$ and noting that $d(\widehat{u})$ is an arbitrary function in $L^2(\mathbb{S}^2)$, we conclude that the function that maximizes $G(f)$ must satisfy

$$
\boxed{\mu_h \mathcal{B}_h f + \mu_s \mathcal{B}_s f + f = o,}
\tag{8.70}
$$

where o is the zero element of $L^2(\mathbb{S}^2)$.

8.12.3   Elaborating on the solution

Rearranging (8.70) we obtain

$$
-\mu_s \mathcal{B}_s f - \mu_h \mathcal{B}_h f = f.
\tag{8.71}
$$

From this abstract equation we seek to write it in more recognizable forms. The first form is a standard integral equation and will be revealed to be a spatial formulation, that is, f will be directly in the form $f(\widehat{u}) \in L^2(\mathbb{S}^2)$. The second form is a standard infinite-dimensional matrix equation and will be revealed as a spectral formulation, that is, f will be expressed in terms of its vector spectral representation $\mathbf{f}$. In the time-frequency case the analogs developed in (Franks, 1969, Ch. 6) correspond to a time domain integral equation and a frequency domain integral equation.

Spatial formulation

We first intend to write (8.71) as an integral equation based on the action of operators in the spatial domain. Note that $(\mathcal{B}_h f)(\widehat{\boldsymbol{u}})$ is spatially defined as $h(\widehat{\boldsymbol{u}})f(\widehat{\boldsymbol{u}})$. In recasting $\mathcal{B}_s$ in the spatial domain we define the symmetric kernel

$$B_s(\widehat{\boldsymbol{u}},\widehat{\boldsymbol{v}}) = \sum_{\ell,m}\sum_{p,q} s_{\ell,p}^{m,q} Y_\ell^m(\widehat{\boldsymbol{u}})\overline{Y_p^q(\widehat{\boldsymbol{v}})}, \quad s_{\ell,p}^{m,q} = \overline{s_{p,\ell}^{q,m}} \tag{8.72}$$

and then

$$(\mathcal{B}_s f)(\widehat{\boldsymbol{u}}) = \int_{\mathbb{S}^2} B_s(\widehat{\boldsymbol{u}},\widehat{\boldsymbol{v}})f(\widehat{\boldsymbol{v}})\,ds(\widehat{\boldsymbol{v}}).$$

And therefore (8.71) is equivalent to

$$-\mu_s \int_{\mathbb{S}^2} B_s(\widehat{\boldsymbol{u}},\widehat{\boldsymbol{v}})f(\widehat{\boldsymbol{v}})\,ds(\widehat{\boldsymbol{v}}) - \mu_h h(\widehat{\boldsymbol{u}})f(\widehat{\boldsymbol{u}}) = f(\widehat{\boldsymbol{u}}), \tag{8.73}$$

which is a Fredholm integral equation of the second kind; see (Debnath and Mikusiński, 1999, pp. 223–224).

Spectral formulation

We can also represent (8.71) in terms of spherical harmonic coefficients of f and those of the output signals $(\mathcal{B}_s f)(\widehat{\boldsymbol{u}})$ and $(\mathcal{B}_h f)(\widehat{\boldsymbol{u}})$. We take the inner product of both sides of (8.71) with the spherical harmonic Y_ℓ^m to obtain

$$-\mu_s \langle \mathcal{B}_s f, Y_\ell^m \rangle - \mu_h \langle \mathcal{B}_h f, Y_\ell^m \rangle = \langle f, Y_\ell^m \rangle, \tag{8.74}$$

which in expanded form is

$$-\mu_s \sum_{p,q} s_{\ell,p}^{m,q}(f)_p^q - \mu_h \sum_{p,q} (f)_p^q \int_{\mathbb{S}^2} h(\widehat{\boldsymbol{u}}) Y_p^q(\widehat{\boldsymbol{u}})\overline{Y_\ell^m(\widehat{\boldsymbol{u}})}\,ds(\widehat{\boldsymbol{u}}) = (f)_\ell^m, \tag{8.75}$$

where we have used (8.61) and the results in Problem 8.2.

To put (8.75) into a purely spectral form, we utilize (8.65) to obtain

$$-\mu_s \sum_{p,q} s_{\ell,p}^{m,q}(f)_p^q - \mu_h \sum_{p,q} h_{\ell,p}^{m,q}(f)_p^q = (f)_\ell^m,$$

which of course is quite general and its existence could have been inferred directly from (8.71).

In matrix form, that is, gathering the matrix elements into matrices and gathering the components $(f)_\ell^m$ into a vector spectral representation of f, we obtain

$$-\mu_s \mathbf{S} f - \mu_h \mathbf{H} f = \mathbf{f}, \tag{8.76}$$

where $\mathbf{S}$ is the operator matrix associated with operator $\mathcal{B}_s$ given in (8.60), and $\mathbf{H}$ is the operator matrix associated with operator $\mathcal{B}_h$ given in (8.59).

Remark 8.9. Many problems of interest are formulated with f being band-limited and then (8.76) becomes a finite-dimensional linear algebra problem. Other problems may be infinite-dimensional but well approximated as a finite-dimensional one. In this way the spectral formulation is often preferred computationally and conceptually.   □

Spatial vs Spectral

Spatial integral equation (8.73) and spectral matrix equation (8.76) are two equivalent ways of viewing the same quadratic variational problem. As in the time-frequency analog some problems are simpler in one domain than the other and standard problems emerge as special cases best suited to one domain or the other (Franks, 1969, Ch. 6).

In the following problems we see how the earlier problems of Section 8.4 and Section 8.5 fit into this framework.

Problems ——

8.5 (Spatial concentration revisited). Suppose that we wish to optimally concentrate the energy of a bandlimited function f with the maximum degree L to a spatial region $R \subset \mathbb{S}^2$ as formulated in Section 8.4, p. 261.

What is the choice of the spatial quadratic functional (determined by operator $\mathcal{B}_h$) and the spectral quadratic functional (determined by operator $\mathcal{B}_s$) in the 2-sphere Franks framework that recovers the eigenvector problem (8.29)?

8.6 (Spectral concentration revisited). Suppose that we wish to optimally concentrate the energy of an already spacelimited function f in spatial region $R \subset \mathbb{S}^2$ to maximum spectral degree L as formulated in Section 8.5, p. 265.

What is the choice of the spatial quadratic functional (determined by operator $\mathcal{B}_h$) and the spectral quadratic functional (determined by operator $\mathcal{B}_s$) in the 2-sphere Franks framework that recovers the integral equation eigenfunction problem (8.42)?

8.13   Spatio-spectral analysis on 2-sphere

8.13.1   Introduction and motivation

In time-frequency analysis, short-time Fourier transform (STFT) (Cohen, 1989; Durak and Arikan, 2003) has been used in many applications to analyze *localized* spectral contents of signals. It has been been widely applied to perform localized frequency analysis and spectral estimation of stationary signals. For non-stationary signals, it is used for analyzing speech signals and radar signals for inverse synthetic aperture radar applications, as it reveals the time dependence and evolution of the signal spectrum.

It is true that any signal on 2-sphere can be expanded in terms of spherical harmonics. However, these functions are not spatially concentrated and therefore not well suited for joint spatio-spectral analysis. That is, the spherical harmonic coefficient of a signal f, denoted by $\langle f, Y_\ell^m \rangle$, reveals the *global* contribution of Y_ℓ^m in the signal on the entire 2-sphere and not in any particular region R.

For localized spectral analysis on 2-sphere, most of the existing methods employ spherical wavelets (Audet, 2011; Wiaux et al., 2008; McEwen et al., 2007; Antoine and Vandergheynst, 1999; Narcowich and Ward, 1996; Starck et al., 2006), which enable *space-scale* decomposition of a signal on the sphere. With a suitable choice of the mother wavelet, they can reveal directional information (Audet, 2011; Wiaux et al., 2008; McEwen et al., 2007).

An alternative to the wavelet (that is, space-scale) approach is a "space-spectral" approach that we consider next. This is inspired by an early work (Simons et al., 1997) and closely follows the derivations in (Khalid et al., 2012a, 2013b).

Our goal here is to develop a tool to obtain *spatially localized spectral information* about a signal defined on 2-sphere as an analog of STFT. This will enable us to answer questions like, "In a certain localized region on 2-sphere, what is the contribution of a specific spherical harmonic Y_ℓ^m?" Development of spatio-spectral analysis techniques can be of interest in various fields of science and engineering, such as analysis of planetary gravity and topography data in geophysics (Wieczorek and Simons, 2005; Simons et al., 1997; Audet, 2011).

8.13.2   Procedure and SLSHT definition

The idea is very simple and given as follows. A *localized* azimuthally symmetric window function

$$h \in \mathcal{H}^0(\mathbb{S}^2)$$

(which has the property $h(\theta, \phi) = h(\theta)$) is rotated as desired by (φ, ϑ) around the zy-axes to mask the signal $f \in \mathbb{S}^2$ under study and then the spherical harmonic transform (7.36) is applied to this masked version of the signal. By localized we mean that most of the energy of the window $h(\theta)$ in spatial domain should be concentrated in a small region around the north pole — the north pole acts as an origin for the definition of the window. We will later discuss choices of appropriate window functions and will make use of what we learnt in previous sections and in particular Section 8.9 and Section 8.10. We call this procedure spatially localized spherical harmonics transform (SLSHT). Figure 8.3 shows the transform. Mathematically, we can write

$$g(\vartheta, \varphi; \ell, m) \triangleq \int_{\mathbb{S}^2} \big(\mathcal{D}(\varphi, \vartheta, 0)h\big)(\widehat{\boldsymbol{u}}) f(\widehat{\boldsymbol{u}}) \overline{Y_\ell^m(\widehat{\boldsymbol{u}})}\, ds(\widehat{\boldsymbol{u}}), \qquad (8.77)$$

where $\mathcal{D}(\varphi, \vartheta, 0)$ is the rotation operator defined in Section 7.11, p. 226, with the first rotation angle around the z-axis being zero. Note that the output of such transform $g(\vartheta, \varphi; \ell, m)$ resides in both spatial and spherical domains. It is a function of applied rotation φ and ϑ and the degree and order of the spherical harmonic transform. The SLSHT represents the contribution of the spherical harmonic Y_ℓ^m within a rotationally symmetric spatial domain centered on co-latitude $\theta = \vartheta$ and longitude $\phi = \varphi$ for the signal f. We can think of (ϑ, φ) parameterizing a point such as $\widehat{\boldsymbol{v}}$ on 2-sphere. However, as much as possible, we will deliberately use (ϑ, φ) in $g(\vartheta, \varphi; \ell, m)$ to explicitly remind ourselves about the rotation operator involved in the SLSHT.

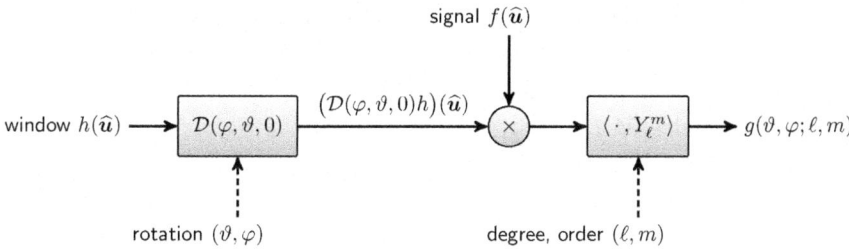

Figure 8.3 Spatially localized spherical harmonic transform (SLSHT) operation. The signal f is *localized* using azimuthally symmetric window function $h(\theta)$ rotated by (φ, ϑ) under rotation operator $\mathcal{D}(\varphi, \vartheta, 0)$ and the spherical harmonic transform is applied to the masked signal.

8.13.3   SLSHT expansion

It is interesting to see how the transformed signal $g(\vartheta, \varphi; \ell, m)$ is related to the spherical harmonic coefficients of the original signal f, denoted by $(f)_p^q$, and those of the azimuthally symmetric window function h, denoted by $(h)_s^0$. From (7.161), p. 244, we know that

$$\langle \mathcal{D}(\varphi, \vartheta, 0)h, Y_s^t \rangle = \sqrt{\frac{4\pi}{2s+1}} \, \overline{Y_s^t(\vartheta, \varphi)}(h)_s^0.$$

And therefore

$$\left(\mathcal{D}(\varphi, \vartheta, 0)h\right)(\widehat{\boldsymbol{u}}) = \sum_{s,t} \sqrt{\frac{4\pi}{2s+1}} \, \overline{Y_s^t(\vartheta, \varphi)}(h)_s^0 Y_s^t(\widehat{\boldsymbol{u}}).$$

Note that on the right-hand side of the above equation $\overline{Y_s^t(\vartheta, \varphi)}$ is fixed for a given (ϑ, φ) and acts as a constant coefficient. It is $Y_s^t(\widehat{\boldsymbol{u}})$ which is a function on the 2-sphere. Replacing this along with $f = \sum_{p,q}(f)_p^q Y_p^q(\widehat{\boldsymbol{u}})$ into (8.77) results in

$$g(\vartheta, \varphi; \ell, m) = \int_{\mathbb{S}^2} \sum_{s,t} \sqrt{\frac{4\pi}{2s+1}} \, \overline{Y_s^t(\vartheta, \varphi)}(h)_s^0 Y_s^t(\widehat{\boldsymbol{u}}) \sum_{p,q}(f)_p^q Y_p^q(\widehat{\boldsymbol{u}}) \overline{Y_\ell^m(\widehat{\boldsymbol{u}})} \, ds(\widehat{\boldsymbol{u}})$$

$$= \sum_{s,t} \sqrt{\frac{4\pi}{2s+1}} \, \overline{Y_s^t(\vartheta, \varphi)}(h)_s^0 \sum_{p,q}(f)_p^q \int_{\mathbb{S}^2} Y_s^t(\widehat{\boldsymbol{u}}) Y_p^q(\widehat{\boldsymbol{u}}) \overline{Y_\ell^m(\widehat{\boldsymbol{u}})} \, ds(\widehat{\boldsymbol{u}}).$$

$$(8.78)$$

And using the shorthand notion for the integral involving three spherical harmonics, defined in (7.104), p. 222, we can simplify (8.78) to become

$$\boxed{g(\vartheta, \varphi; \ell, m) = \sum_{s,t} \sqrt{\frac{4\pi}{2s+1}} \, \overline{Y_s^t(\vartheta, \varphi)}(h)_s^0 \sum_{p,q}(f)_p^q y(s, t; p, q; \ell, m).}$$

$$(8.79)$$

8.13.4   SLSHT distribution and matrix representation

Let us assume that the signal of interest f is bandlimited with a finite degree $L_f < \infty$ and the window (which is under our control) is also bandlimited with a finite degree $L_h < \infty$.

Using the single-index notations $n = \ell(\ell+1) + m$, $r = p(p+1) + q$ and $u = s(s+1) + t$,[4] we can write (8.79) as

$$g(\vartheta, \varphi; n) = \sum_{u=0}^{N_h} \sqrt{\frac{4\pi}{2\lfloor\sqrt{u}\rfloor + 1}} \overline{Y_u(\vartheta, \varphi)}(h)_{\lfloor\sqrt{u}\rfloor}^0 \sum_{r=0}^{N_f} (f)_u y(u; r; n), \qquad (8.80)$$

where $N_h = L_h^2 + 2L_h$ and $N_f = L_f^2 + 2L_f$. Similar to the discussion following (7.110), the maximum possible degree for a SLSHT component is $L_g = L_f + L_h$, which results in the maximum possible n being $N_g = L_g^2 + 2L_g$.

For a given $\{\vartheta, \varphi\}$, we call all possible $N_g + 1$ SLSHT spectral components the "SLSHT distribution" and represent it in vector form as

$$\mathbf{g}(\vartheta, \varphi) \triangleq \big(g(\vartheta, \varphi; 0), g(\vartheta, \varphi; 1), \ldots, g(\vartheta, \varphi; N_g)\big)'. \qquad (8.81)$$

From (8.80), we can express the SLSHT distribution in matrix form

$$\boxed{\mathbf{g}(\vartheta, \varphi) = \mathbf{\Psi}(\vartheta, \varphi)\mathbf{f},} \qquad (8.82)$$

where $\mathbf{f}$ is the vector spectral representation of f of size $(N_f + 1) \times 1$ and $\mathbf{\Psi}(\vartheta, \varphi)$ is the transformation matrix of size $(N_g + 1) \times (N_f + 1)$ given by

$$\mathbf{\Psi}(\vartheta, \varphi) \triangleq \begin{pmatrix} \psi_{0,0} & \psi_{0,1} & \cdots & \psi_{0,N_f} \\ \psi_{1,0} & \psi_{1,1} & \cdots & \psi_{1,N_f} \\ \vdots & \vdots & \ddots & \vdots \\ \psi_{N_g,0} & \psi_{N_g,1} & \cdots & \psi_{N_g,N_f} \end{pmatrix}, \qquad (8.83)$$

with entries

$$\psi_{n,r} \equiv \psi_{n,r}(\vartheta, \varphi) = \sum_{u=0}^{N_h} \sqrt{\frac{4\pi}{2\lfloor\sqrt{u}\rfloor + 1}} \overline{Y_u(\vartheta, \varphi)}(h)_{\lfloor\sqrt{u}\rfloor}^0 y(u; r; n). \qquad (8.84)$$

Remark 8.10. The matrix form in (8.82) projects the spectral representation of the signal f to the joint spatio-spectral domain. The size of the transformation matrix is dependent on the maximum spectral degree of the input signal f and the window function h. The value of matrix elements is dependent on the applied rotation $\{\vartheta, \varphi\}$ and the window function h, the choice of which will be discussed shortly.   □

8.13.5   Signal inversion

The aim is recovery of all spherical coefficients of the original signal from its SLSHT distribution $\mathbf{g}(\vartheta, \varphi)$. This is possible and the procedure, which averages the SLSHT distribution over all possible rotations, is proven in the following theorem.

---
[4] Note that $s = \lfloor\sqrt{u}\rfloor$.

Theorem 8.1 (Signal inversion from SLSHT distribution). *Let $\mathbf{g}(\vartheta, \varphi)$ represent the SLSHT distribution of the signal f using azimuthally symmetric window function h, where the distribution is of the form (8.81) and each SLSHT component given by (8.80). Then the vector spectral representation of f denoted by $\mathbf{f}$ can be recovered from $\mathbf{g}(\vartheta, \varphi)$, up to a multiplicative factor, by averaging the SLSHT distribution over all possible rotations in the spatial domain. More specifically,*

$$\hat{\mathbf{f}} = \int_0^{2\pi}\!\!\int_0^{\pi} \mathbf{g}(\vartheta, \varphi) \, \sin\vartheta \, d\vartheta \, d\varphi$$
$$= \sqrt{4\pi}(h)_0^0 \times \left((f)_0, (f)_1, \ldots, (f)_{N_f}, 0, \ldots, 0\right)', \tag{8.85}$$

where $\hat{\mathbf{f}}$ is a column vector of length $N_g + 1$ where only the first $N_f + 1$ elements can be non-zero. □

Proof. Integrating (8.81) over the spatial domain and using (8.82) gives

$$\int_0^{2\pi}\!\!\int_0^{\pi} \mathbf{g}(\vartheta, \varphi) \, \sin\vartheta \, d\vartheta \, d\varphi = \left(\int_{\mathbb{S}^2} \mathbf{\Psi}(\hat{\boldsymbol{v}}) \, ds(\hat{\boldsymbol{v}})\right)\mathbf{f}, \tag{8.86}$$

where we used simplifying notation $\hat{\boldsymbol{v}} \equiv \{\vartheta, \varphi\}$ and $ds(\hat{\boldsymbol{v}}) = \sin\vartheta \, d\varphi \, d\vartheta$. The integral of kernel matrix $\mathbf{\Psi}$ is obtained by integrating each matrix element $\psi_{n,r}(\hat{\boldsymbol{v}})$ in (8.84), that is,

$$\int_{\mathbb{S}^2} \psi_{n,r}(\hat{\boldsymbol{v}}) \, ds(\hat{\boldsymbol{v}}) = \sum_{u=0}^{N_h} \sqrt{\frac{4\pi}{2\lfloor\sqrt{u}\rfloor + 1}}(h)_{\lfloor\sqrt{u}\rfloor}^0 y(u; r; n) \int_{\mathbb{S}^2} \overline{Y_u(\hat{\boldsymbol{v}})} \, ds(\hat{\boldsymbol{v}}). \tag{8.87}$$

According to the definition of spherical harmonic (7.19), the integral of spherical harmonics over the 2-sphere is zero except for the zeroth-order DC harmonic. That is, the integral in the above equation is non-zero only for $u = 0$ and we obtain

$$\int_{\mathbb{S}^2} \psi_{n,r}(\hat{\boldsymbol{v}}) \, ds(\hat{\boldsymbol{v}}) = 4\pi(h)_0^0 y(0; r, n). \tag{8.88}$$

From (7.104) and using the orthonormality of spherical harmonics $y(0; r; n)$ reduces to

$$y(0; r; n) = \int_{\mathbb{S}^2} Y_0(\hat{\boldsymbol{u}})Y_r(\hat{\boldsymbol{u}})\overline{Y_n(\hat{\boldsymbol{u}})} \, ds(\hat{\boldsymbol{u}})$$
$$= \sqrt{\frac{1}{4\pi}} \int_{\mathbb{S}^2} Y_r(\hat{\boldsymbol{u}})\overline{Y_n(\hat{\boldsymbol{u}})} \, ds(\hat{\boldsymbol{u}})$$
$$= \sqrt{\frac{1}{4\pi}}\delta_{r,n}.$$

Incorporating these results in (8.87), we get

$$\int_{\mathbb{S}^2} \psi_{n,r}(\hat{\boldsymbol{v}}) \, ds(\hat{\boldsymbol{v}}) = \sqrt{4\pi}(h)_0^0\delta_{r,n}, \tag{8.89}$$

which upon substitution in (8.86) completes the proof. □

> **Remark 8.11.** From the result in Theorem 8.1, we can see that we only need to know the DC component of the window function $(h)_0^0$ in order to recover the signal exactly from its SLSHT distribution.   □

8.14   Optimal spatio-spectral concentration of window function

The effectiveness of SLSHT distribution depends on the chosen window function. For a higher resolution in one domain, the window should be narrower in that domain. However, from the spherical uncertainty principle (8.55), there is a fundamental limit on simultaneous concentration of a signal in both spatial and spectral domains (Narcowich and Ward, 1996). If a window is chosen to obtain the desired resolution in one domain, it is said to be optimal if it is also optimally localized in the other domain (Cohen, 1989).

We compare several normalized energy azimuthally symmetric window functions jointly in spatial and spectral domains from the perspective of the uncertainty principle on 2-sphere. The windows are rectangular window, triangular window, cosine window, Hamming window, Hanning window, the window used in (Simons et al., 1997), Gaussian window and the eigenfunction window (defined below), all of which are parameterized by θ_0 denoting the window co-latitude truncation width. The first five windows are defined in (Harris, 1978). The window in (Simons et al., 1997) is obtained by truncating the rectangular window in the spectral domain within the main spectral lobe. The Gaussian window is a unit energy function that decays exponentially with the truncation width and its variance is chosen such that 99% of the energy lies within the truncation width.

Eigenfunction window

To obtain the eigenfunction window $h(\theta)$, we first set $N^{(0)}$ in (8.52), p. 270, to one and find $L_h = (2\pi/\theta_0) - 1$. That is we would like to get as the solution to (8.51) a single most concentrated eigenfunction in the polar cap region R_{θ_0}. In this case $\mathbf{D}^{(0)}$ is an $(L_h + 1) \times (L_h + 1)$ symmetric matrix where the entries are given by

$$D_{\ell,p}^{0,0} = \int_{R_{\theta_0}} Y_\ell^0(\widehat{\boldsymbol{u}}) Y_p^0(\widehat{\boldsymbol{u}}) \, ds(\widehat{\boldsymbol{u}}).$$

Window comparison results

Figure 8.4 plots (a) variance σ_S^2 in spatial domain (8.53), (b) variance σ_L^2 in spectral domain (8.54) and (c) uncertainty product in (8.55) for the eight different azimuthally symmetric window functions under consideration. The parameter θ_0 varies from $\pi/18 = 10°$ to $\pi/3 = 60°$. Figure 8.4a and Figure 8.4b confirm our expectation that generally the variance in spatial domain increases with the truncation width and the variance in spectral domain decreases with the truncation width. The rectangular window has the worst spectral localization, because the discontinuity at spatial truncation points increases its variance in the spectral domain. The window used in (Simons et al., 1997) performs very well in the spectral domain, but poorly in the spatial domain. The figures also show that

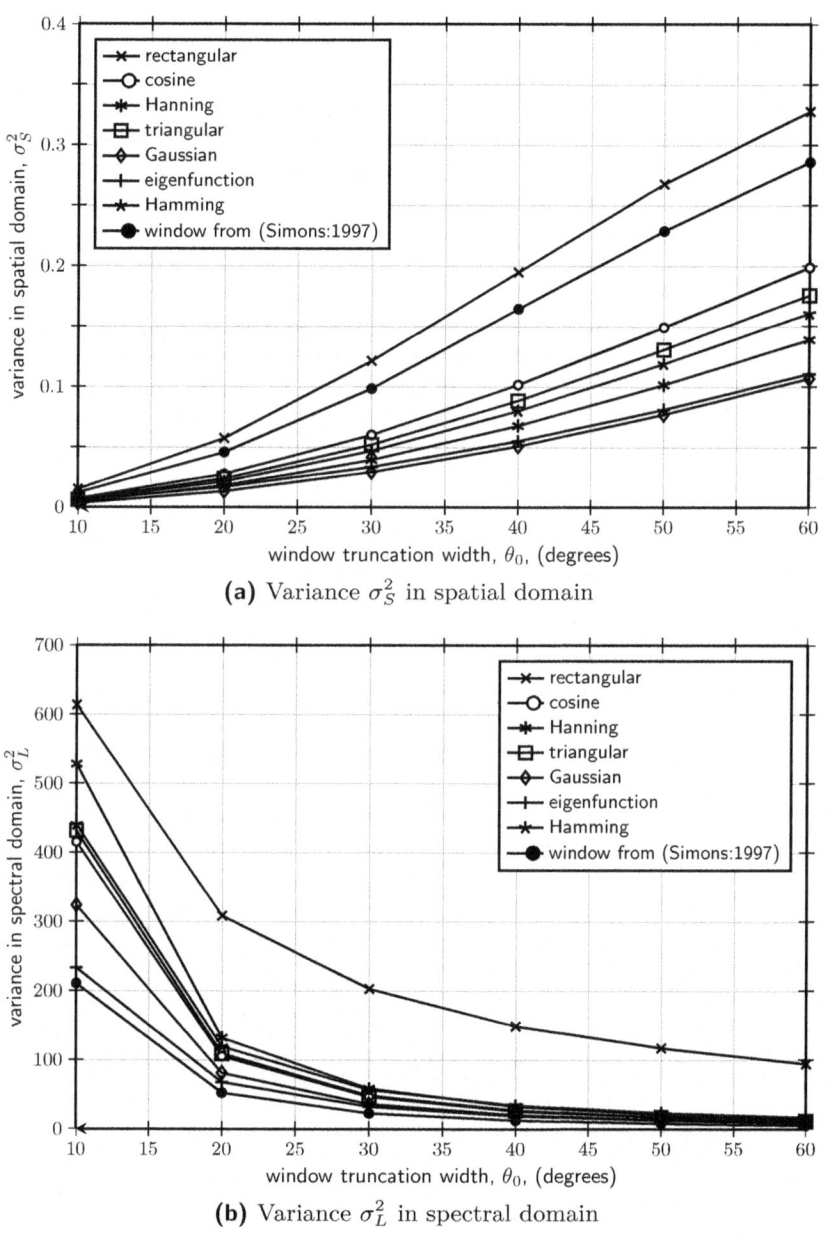

(a) Variance σ_S^2 in spatial domain

(b) Variance σ_L^2 in spectral domain

Figure 8.4 (a) Variance σ_S^2 in spatial domain and (b) variance σ_L^2 in spectral domain for different types of azimuthally symmetric window functions.

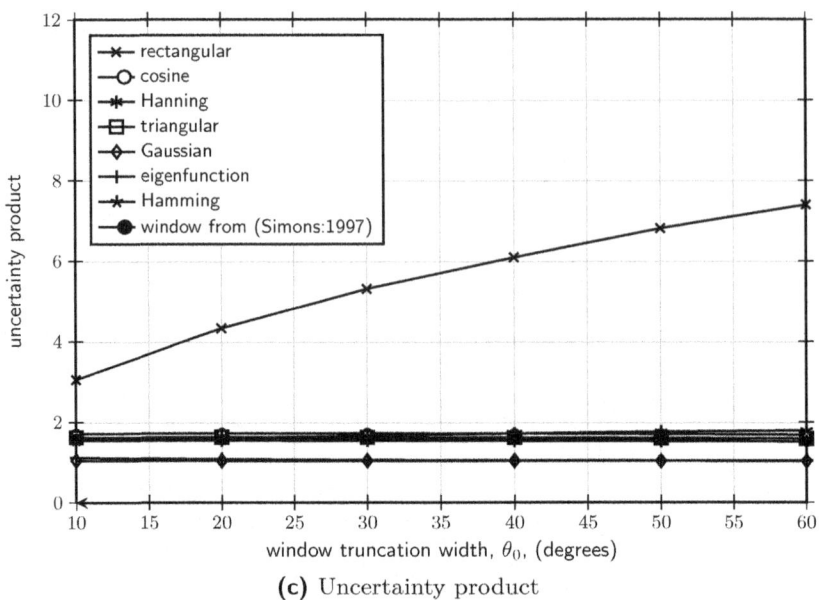

(c) Uncertainty product

Figure 8.4 (Continued) (c) Uncertainty product in (8.55) for different types of azimuthally symmetric window functions.

the Gaussian window and an eigenfunction window exhibit better overall localization behavior. Comparatively, these two windows have the lowest variances
in both domains. Figure 8.4c confirms that both the eigenfunction window and
the Gaussian window get very close to the lower bound of 1 for the uncertainty
product of (8.55).

The optimal truncation width depends on the required resolution in spatial
and spectral domains. According to Figure 8.4, the spatial variance σ_S^2 of eigenfunction window is very close to that of Gaussian window. However, its spectral
variance σ_L^2 is lower than that of Gaussian window, especially at lower truncation widths. The Gaussian window and eigenfunction window are plotted for
truncation width $\theta_0 = \pi/8 = 22.5°$ in both spatial and spectral domains in
Figure 8.5. Both windows are normalized to unit energy and chosen such that
99% of their energy lies within the truncation width. It is observed that the
eigenfunction window has smaller bandwidth and its energy is more uniformly
distributed relative to Gaussian window. Thus, the eigenfunction window can
be a good choice for window function in the SLSHT distribution.

8.14.1   SLSHT on Mars topographic data

We apply the spatially localized spherical harmonics transform (SLSHT) using
the eigenfunction window to study an example from Mars. We use a high-
resolution Mars topographic map (size = 800 × 800) and illustrate that the
SLSHT distribution reveals localized spectral contributions of spherical harmonics in the spatial domain, for example, in a well-defined mountainous area.

To calculate the spherical harmonic components and triple product, we use

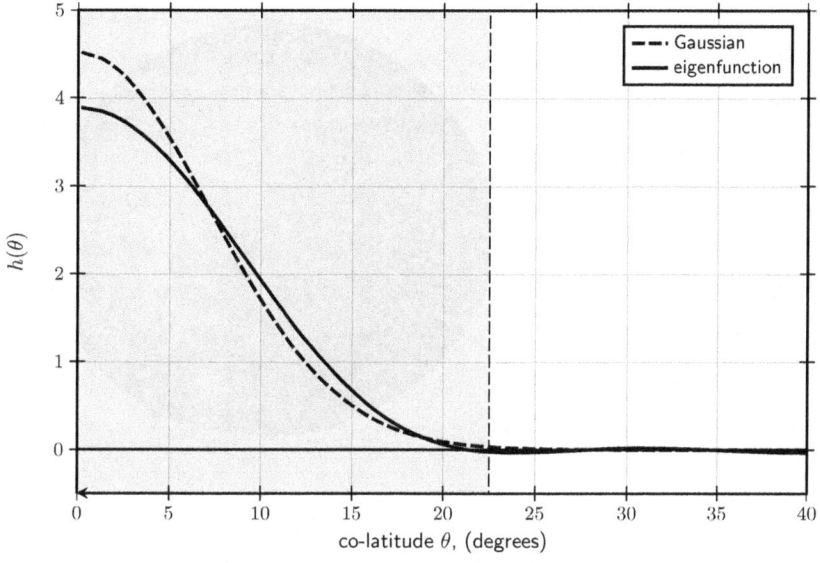

(a) Window spatial shape versus co-latitude θ.

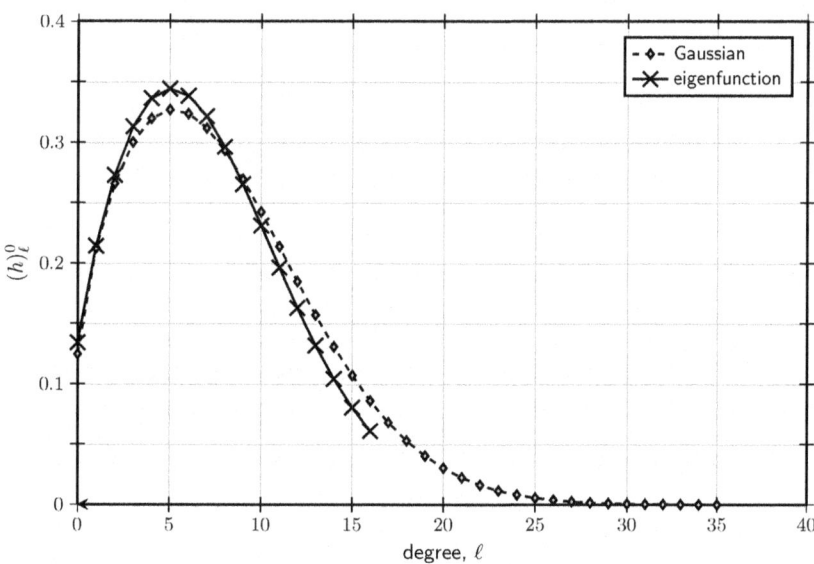

(b) Window spectral shape versus degree ℓ.

Figure 8.5 Comparison of Gaussian window and eigenfunction window functions in (a) spatial domain and (b) spectral domain, for truncation width $\theta_0 = \pi/8$. The eigenfunction window has a smaller spectral bandwidth relative to the Gaussian window.

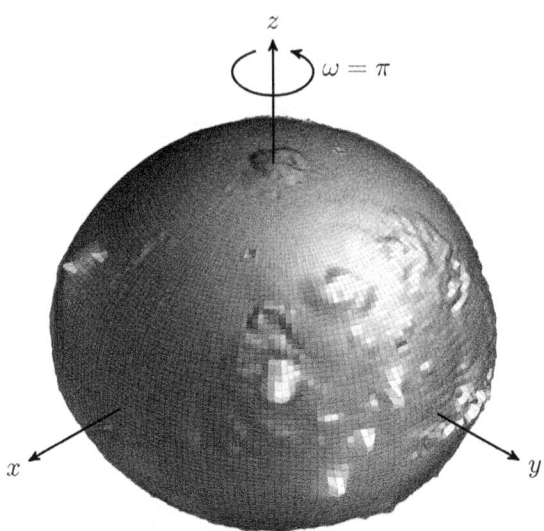

Figure 8.6 Mars topographic map $f(\theta, \phi)$, obtained using a bandlimited spectral model of Mars with maximum spectral degree $L_f = 150$. The standard International Astronomical Union map has been rotated through $\omega = \pi$ about the z-axis to bring more interesting features to the viewpoint. The height map is exaggerated.

equiangular sampling with $N \times N$ samples on the 2-sphere as $\theta_j = \pi j / N$, $\phi_k = 2\pi k / N$ for $j, k \in \{0, 1, \ldots, N{-}1\}$ as described in Section 7.6, p. 214, and (Driscoll and Healy, 1994). For computing the SLSHT distribution using an eigenfunction window with bandwidth L_h, we use a very high-resolution $N = 800 \gg 2(L_h + 1)$ to obtain smooth plots. For inverse spherical harmonic transform of a function with maximum spherical harmonics degree L_f, we use the minimum resolution $N = 2(L_f + 1)$ (Driscoll and Healy, 1994).

Figure 8.6 shows the Mars topographic map f in the spatial domain, which is obtained using its spherical harmonics topographic model up to degree $L_f = 150$ (Wieczorek, 2007). We use the unit energy normalized Mars signal with DC component eliminated. An area of interest is around Olympus Mons, the highest mountain on Mars and one of the highest known mountains in the solar system. The spherical coordinates of Olympus Mons are $\theta = 71.6°$ and $\phi = 224°$. In order to bring this to the front view in Figure 8.6, a rotation of $\omega = \pi$ is applied to the Mars map and hence the location of Olympus Mons in Figure 8.6 distinctively appears at $\phi = 46°$.

Figure 8.7a shows the energy per degree E_ℓ, for $\ell \in \{0, 1, \ldots, 45\}$, using (7.46) in the spectrum of Mars. Almost 90% of the energy lies in degrees less than 10. The higher-degree spherical harmonics contain very little energy and as such the corresponding spherical harmonic coefficients do not reveal information about the region of their contributions in the signal f. However, we wish to calculate the energy contribution by spherical harmonics in the region around the large volcanoes. For this purpose, we define the *energy localization ratio per degree* $E'_\ell(R)$ as a measure of energy contribution by all order spherical harmonics for

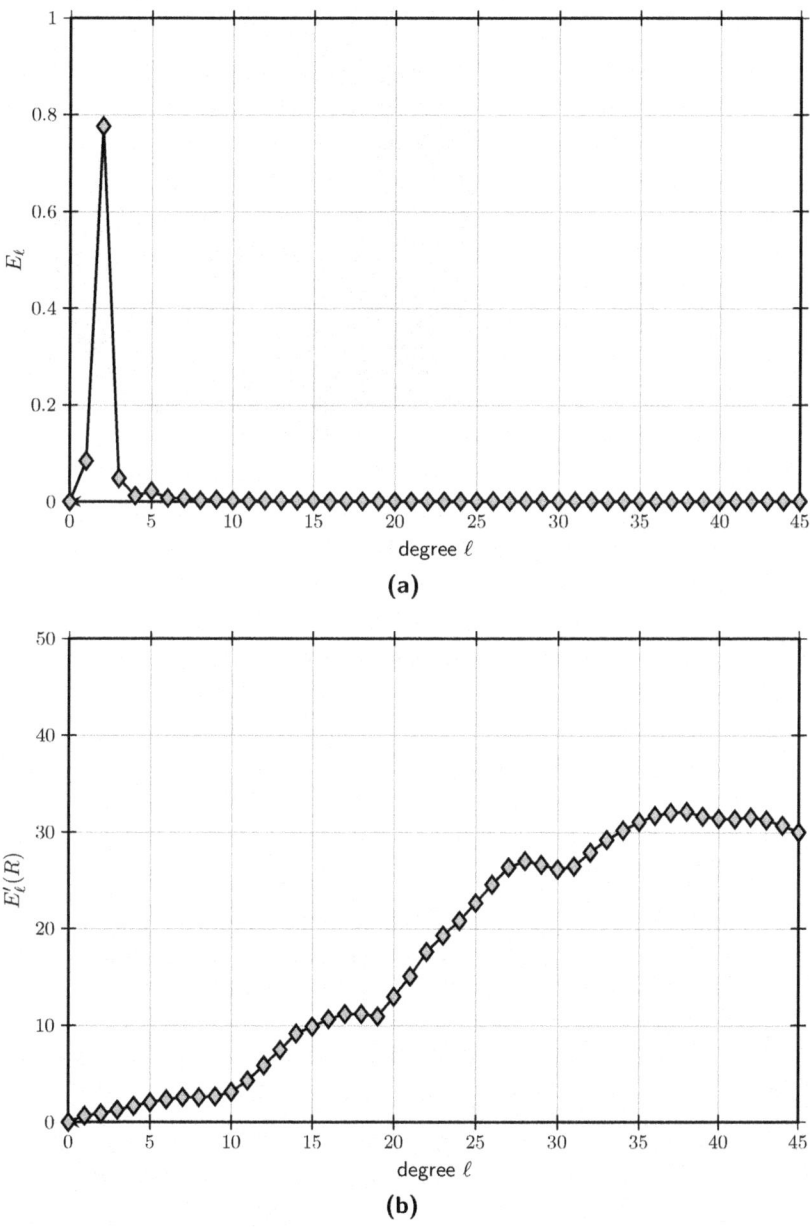

(a)

(b)

Figure 8.7 (a) Energy per spherical harmonic degree E_ℓ using (7.46), and (b) energy
localization ratio per degree $E'_\ell(R)$ using (8.90) for region R defined in the text
and for components of SLSHT distribution $\mathbf{g}(\vartheta, \varphi)$ in (8.79) of Mars topographic
map for degree $0 \leq \ell \leq 45$.

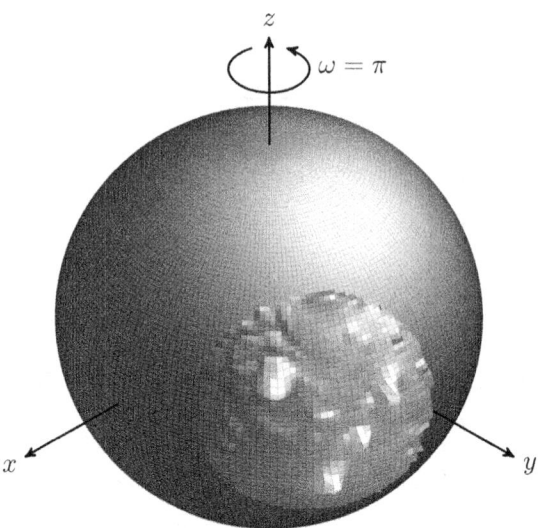

Figure 8.8 Mars topographic map masked to R, the region around the large volcanoes, as discussed in the text. It is used in the computation of $E'_\ell(R)$ in (8.90) with results in Figure 8.7b.

a particular degree in the localized region R as

$$E'_\ell(R) \triangleq \sum_{m=-\ell}^{\ell} \frac{\displaystyle\int_R \big|g(\widehat{\boldsymbol{v}};\ell,m)\big|^2 ds(\widehat{\boldsymbol{v}})}{\displaystyle\int_{\mathbb{S}^2} \big|g(\widehat{\boldsymbol{v}};\ell,m)\big|^2 ds(\widehat{\boldsymbol{v}})}, \tag{8.90}$$

where R denotes the spherical cap region of width $\theta_0 = 22.5°$ centered at $\theta = 90°$ and $\phi = 240°$; that is, $16°$ to the east and $18.4°$ to the south of Olympus Mons (at $\phi = 60°$ from the viewpoint of Figure 8.6). The region R is shown clearly in Figure 8.8. The region R only covers 3.81% of the area of the whole sphere and captures the magnitude of the SLSHT distribution component around the volcanoes.

We have determined all SLSHT components of the form $g(\vartheta, \varphi; \ell, m)$ for $\ell \in \{0, 1, \ldots, 45\}$ and all orders $m \in \{-\ell, -\ell+1, \ldots, \ell\}$ using an eigenfunction window with truncation width $\theta_0 = 22.5°$. Figure 8.7b shows the energy localization ratio per degree $E'_\ell(R)$, which indicates that the higher-degree spherical harmonics, despite their low energy content in the overall spectrum, have their energy localized in region R.

Figure 8.9 and Figure 8.10 show the magnitude of first-order ($m = 1$) distribution components of the form $g(\vartheta, \varphi; \ell, 1)$ for degrees in the range of $\ell \in \{1, 2, \ldots, 12\}$ and $\ell \in \{34, 35, \ldots, 45\}$, respectively. These distribution components indicate that the contribution of higher-order spherical harmonics is mainly around the region where volcanoes are located. As expected from Figure 8.7b, the higher-order SLSHT components exhibit more distinctive localized concentration in the regions of interest.

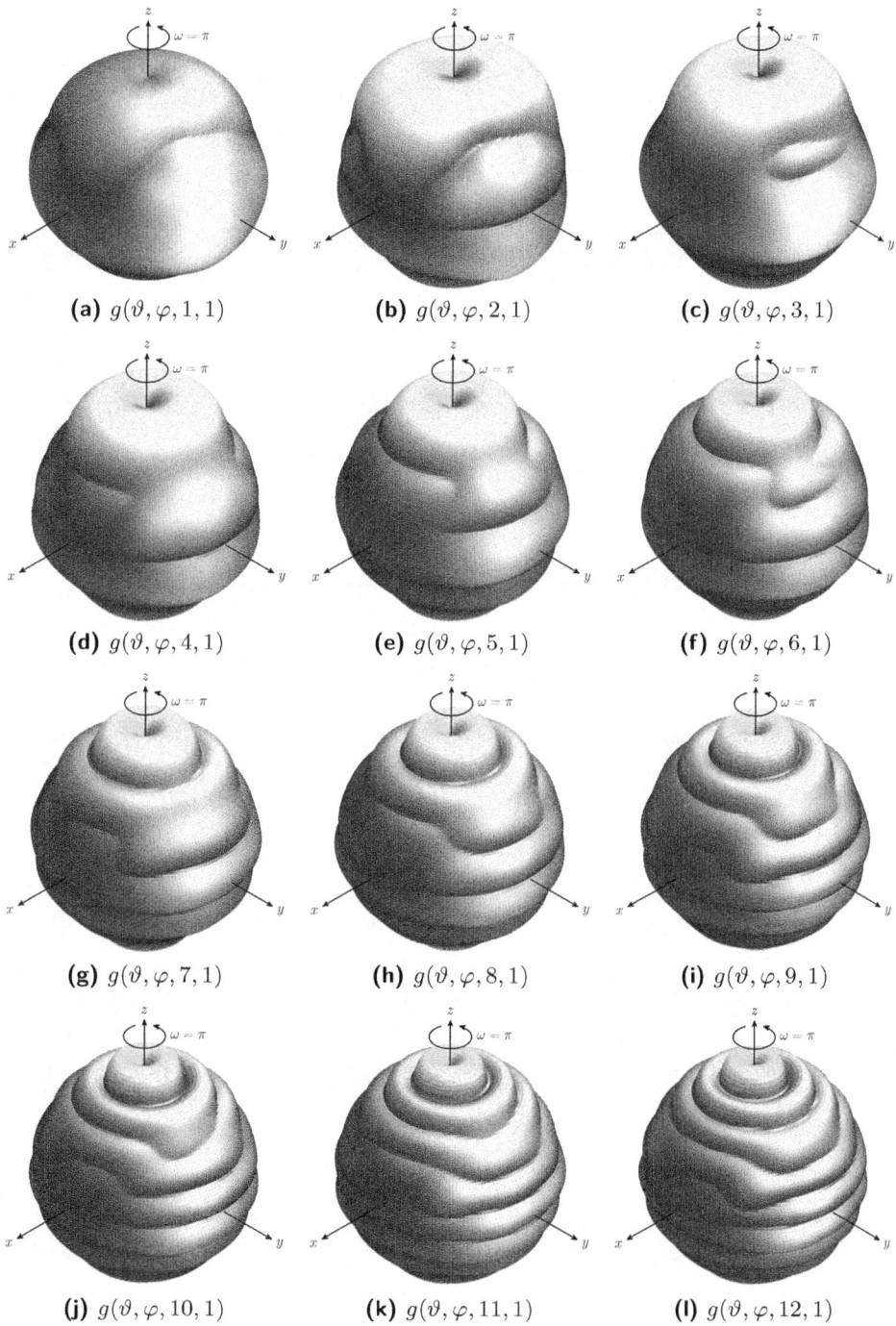

(a) $g(\vartheta, \varphi, 1, 1)$ **(b)** $g(\vartheta, \varphi, 2, 1)$ **(c)** $g(\vartheta, \varphi, 3, 1)$

(d) $g(\vartheta, \varphi, 4, 1)$ **(e)** $g(\vartheta, \varphi, 5, 1)$ **(f)** $g(\vartheta, \varphi, 6, 1)$

(g) $g(\vartheta, \varphi, 7, 1)$ **(h)** $g(\vartheta, \varphi, 8, 1)$ **(i)** $g(\vartheta, \varphi, 9, 1)$

(j) $g(\vartheta, \varphi, 10, 1)$ **(k)** $g(\vartheta, \varphi, 11, 1)$ **(l)** $g(\vartheta, \varphi, 12, 1)$

Figure 8.9 SLSHT distribution components $g(\vartheta, \varphi; 1, 1)$ to $g(\vartheta, \varphi; 12, 1)$ for degrees $\ell = 1$ to $\ell = 12$ and order $m = 1$ for the Mars topography data are shown from left to right and top to bottom.

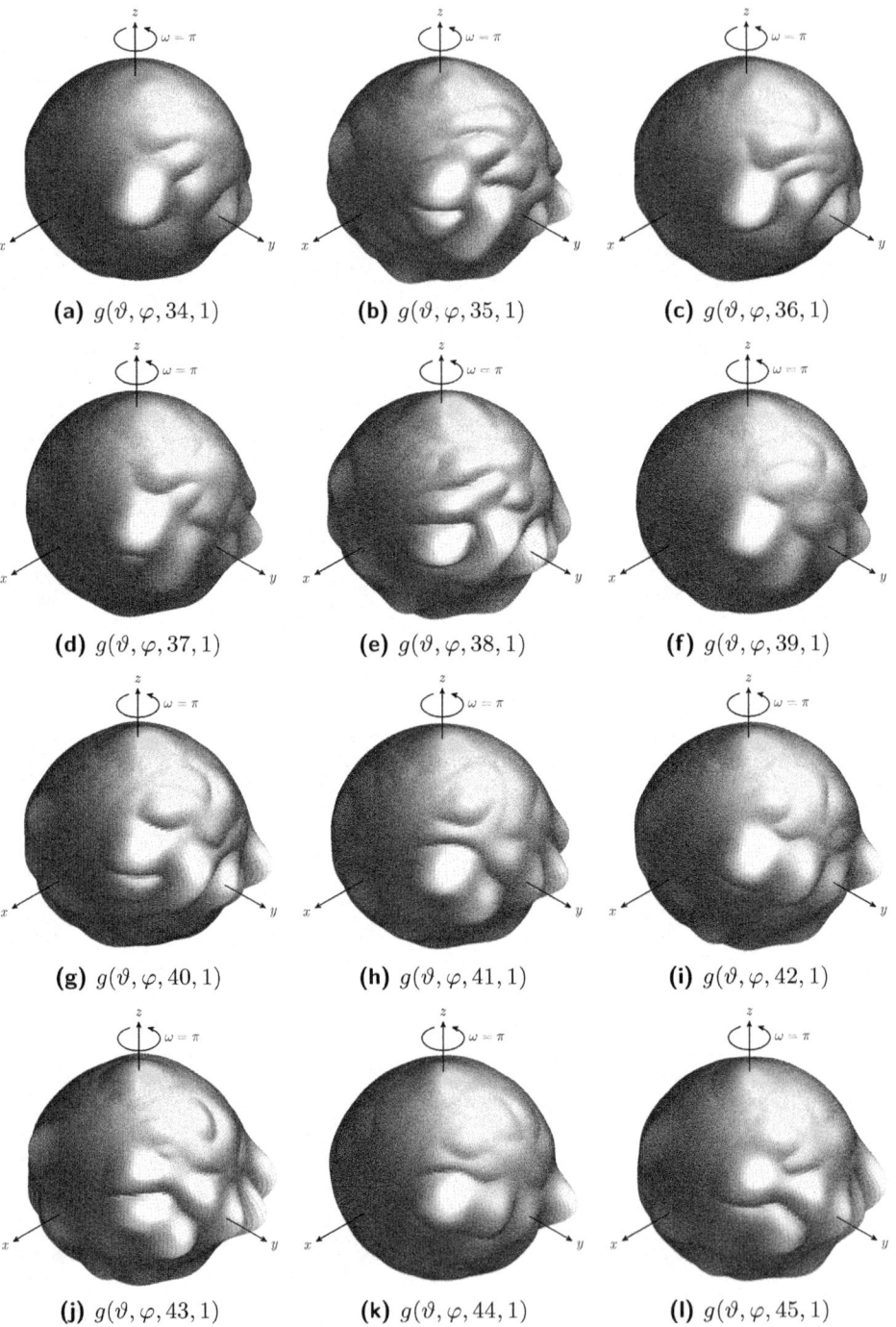

(a) $g(\vartheta, \varphi, 34, 1)$ **(b)** $g(\vartheta, \varphi, 35, 1)$ **(c)** $g(\vartheta, \varphi, 36, 1)$

(d) $g(\vartheta, \varphi, 37, 1)$ **(e)** $g(\vartheta, \varphi, 38, 1)$ **(f)** $g(\vartheta, \varphi, 39, 1)$

(g) $g(\vartheta, \varphi, 40, 1)$ **(h)** $g(\vartheta, \varphi, 41, 1)$ **(i)** $g(\vartheta, \varphi, 42, 1)$

(j) $g(\vartheta, \varphi, 43, 1)$ **(k)** $g(\vartheta, \varphi, 44, 1)$ **(l)** $g(\vartheta, \varphi, 45, 1)$

Figure 8.10 SLSHT distribution components $g(\vartheta, \varphi; 34, 1)$ to $g(\vartheta, \varphi; 45, 1)$ for degrees $\ell = 34$ to $\ell = 45$ and order $m = 1$ for the Mars topography data are shown from left to right and top to bottom.

9 Convolution on 2-sphere

9.1 Introduction

One of the most basic operations on signals in any domain, including 2-sphere, is linear filtering or convolution. Yet, unlike conventional time-domain signals, for signals on the 2-sphere this is not consistently well defined and there exist competing definitions.

So the first aim of this chapter is to study existing definitions of convolution on 2-sphere in the literature and identify their properties, advantages and shortcomings. We determine the relationship between various definitions and show that two seemingly different definitions are essentially the same (which stems from the azimuthally symmetric or isotropic convolution property inherent in those definitions) (Kennedy et al., 2011). Our framework reveals that none of the existing formulations are natural extensions of convolution in time domain. For example, they are not commutative. Changing the role of signal and filter would change the outcome of convolution. Moreover, in one definition, the domain of convolution output does not remain on the 2-sphere.

Recognizing that a well-posed general definition for convolution on 2-sphere which is anisotropic and commutative is more difficult to formulate and that a true parallel with Euclidean convolution may not exist, in the second part of this chapter we use the power of abstract techniques and the tools we have studied to show that a commutative anisotropic convolution on the 2-sphere can indeed be simply constructed. We discuss additional properties of the proposed convolution, especially in the spectral domain, and present some clarifying examples.

We begin with the familiar case of convolution in time domain or convolution on the real line, which shall guide our subsequent developments for convolution on the 2-sphere.

9.2 Convolution on real line revisited

On the real line, convolution for two functions, $f(t)$ and $h(t)$, is defined as follows

$$\big(h \star f\big)(s) \triangleq \int_{-\infty}^{\infty} h(s-t)f(t)\,dt.$$

When written this way, we tend to regard $f(t)$ as the input and $h(t)$ as the filter. But, of course, by elementary techniques (change of variable) we find

$$\big(h \star f\big)(s) = \int_{-\infty}^{\infty} h(s-t)f(t)\,dt = \int_{-\infty}^{\infty} h(t)f(s-t)\,dt = \big(f \star h\big)(s),$$

which shows convolution is commutative. It is preferable that we did not rely on specific properties of the above integral and, rather, abstract the problem to explore the underlying source of the commutativity so that we can apply it in a more general context later in this chapter. To do this, we revisit the above in an operator framework and initially make a simple problem a bit less straightforward, but more useful for generalizing.

Define $\mathcal{T}_s$ as a *translation operator* parameterized by s

$$\big(\mathcal{T}_s f\big)(t) \triangleq f(t-s),$$

which shifts the function horizontally by s, and $\mathcal{R}$ as the *time reversal (about the origin)* operator

$$\big(\mathcal{R}f\big)(t) \triangleq f(-t).$$

Then convolution can be written using a composition of these operators

$$\big(h \star f\big)(s) = \int_{-\infty}^{\infty} \big(\mathcal{T}_s \circ \mathcal{R}h\big)(t) f(t)\, dt.$$

We can observe that $\mathcal{T}_s$ and $\mathcal{R}$ do not commute and their inverses are $\mathcal{T}_{-s}$ and $\mathcal{R}$, respectively.

Example 9.1. We have

$$\big(\mathcal{T}_s \circ \mathcal{R}g\big)(t) = \big(\mathcal{T}_s g\big)(-t) = g(s-t),$$

which differs from

$$\big(\mathcal{R} \circ \mathcal{T}_s g\big)(t) = \big(\mathcal{R}g\big)(t-s) = g(-t-s).$$

This shows that the operators $\mathcal{T}_s$ and $\mathcal{R}$ do not commute. Similarly,

$$\big(\mathcal{T}_{-s} \circ \mathcal{R}g\big)(t) = \big(\mathcal{T}_{-s}g\big)(-t) = g(-s-t) = g(-t-s)$$

differs from

$$\big(\mathcal{R} \circ \mathcal{T}_{-s}g\big)(t) = \big(\mathcal{R}g\big)(t+s) = g(-t+s) = g(s-t).$$

Furthermore, two time reverses recover the original function

$$\big(\mathcal{R} \circ \mathcal{R}g\big)(t) = \big(\mathcal{R}g\big)(-t) = g\big(-(-t)\big) = g(t),$$

and so indeed $\mathcal{R}$ is its own inverse. Finally, we see also that

$$\big(\mathcal{T}_{-s} \circ \mathcal{T}_s g\big)(t) = \big(\mathcal{T}_{-s}g\big)(t-s) = g(t-s+s) = g(t),$$

which shows that $\mathcal{T}_s$ and $\mathcal{T}_{-s}$ are inverses.   □

As operators, $\mathcal{T}_s$ and $\mathcal{R}$, have an additional property of being *isometries* and thereby the Lebesgue measure is invariant under their actions.[1] Stripped of this gobbledygook, if we integrate the function $f(t)$ over the entire real line then we get the same answer as when we integrate $f(-t)$ or $f(t+s)$ or $f(s-t)$, etc. So $\mathcal{T}_s \circ \mathcal{R}$ is an isometry and so is its inverse $\mathcal{R} \circ \mathcal{T}_{-s}$. This means we can apply $\mathcal{R} \circ \mathcal{T}_{-s}$, or any other isometry, to the functions in the *integrand* and leave the

---

[1] The Lebesgue measure is invariant under rotations, translations and reflections, that is, invariant under all isometries of $\mathbb{R}^N$.

integral unchanged. It amounts to an isometric affine change of variables and we have

$$\int_{-\infty}^{\infty} (\mathcal{T}_s \circ \mathcal{R}h)(t) f(t)\, dt = \int_{-\infty}^{\infty} (\mathcal{R} \circ \mathcal{T}_{-s} \circ \mathcal{T}_s \circ \mathcal{R}h)(t)(\mathcal{R} \circ \mathcal{T}_{-s}f)(t)\, dt$$

$$= \int_{-\infty}^{\infty} h(t)(\mathcal{R} \circ \mathcal{T}_{-s}f)(t)\, dt.$$

However, we note that, from Example 9.1,

$$(\mathcal{R} \circ \mathcal{T}_{-s}g)(t) = g(s-t) = (\mathcal{T}_s \circ \mathcal{R}g)(t), \quad \forall g.$$

That is,

$$\boxed{\int_{-\infty}^{\infty} (\mathcal{T}_s \circ \mathcal{R}h)(t) f(t)\, dt = \int_{-\infty}^{\infty} (\mathcal{T}_s \circ \mathcal{R}f)(t) h(t)\, dt.}$$

This is why the convolution commutes because the expression is invariant to the exchange of f and h.

Involution

Abstracting slightly, we have

$$\int_{-\infty}^{\infty} (\mathcal{U}h)(t) f(t)\, dt = \int_{-\infty}^{\infty} (\mathcal{U}f)(t) h(t)\, dt,$$

whenever operator $\mathcal{U}$ and its adjoint satisfy

$$\mathcal{U} = \mathcal{U}^{-1} \quad \text{and} \quad \mathcal{U}^\star \circ \mathcal{U} = \mathcal{U} \circ \mathcal{U}^\star = \mathcal{I},$$

and is an isometry with respect to the real-line Lebesgue measure. This is the "involutory" property and it is stronger than the unitary property (see (4.40), p. 135, $\mathcal{U}^{-1} = \mathcal{U}^\star$), which it also implies.

> **Remark 9.1 (Commutativity from involution).** In short, an operator which is an involution is an operator which is its own inverse and this plays an essential role in securing commutativity. This is the property we need to impose in the operator kernel when later we seek a definition of convolution on the 2-sphere which is commutative. □

The road forward

This material is useful background to consider how convolution of the 2-sphere should be defined. Next we cover notions of convolution on the 2-sphere from the literature, before picking up some of the abstract observations above to guide a new definition of convolution on the 2-sphere which is commutative.

9.1. Show that the adjoint of operator $\mathcal{T}_s$ is

$$\mathcal{T}_s^\star = \mathcal{T}_{-s}$$

and that operator $\mathcal{R}$ is self-adjoint.

9.2. One might believe that the negative sign in the definition for convolution is merely convention and not crucial. That is, the operator equation

$$b(s) = \int_{-\infty}^{\infty} h(s+t) f(t) \, dt$$

has the right stuff to be an alternative definition of convolution. Show that this definition does not give the commutativity property.

9.3. An important variant on the previous problem is the notion of cross-correlation in a deterministic setting

$$c(s) = \int_{-\infty}^{\infty} h(s+t) \overline{f(t)} \, dt,$$

which is defined for complex-valued signals. Analyze this equation in the operator setting to infer its properties. (Obviously, for real-valued signals, it is no different from the form in the previous problem, which tells you something.)

9.3   Spherical convolution of type 1

Here we study a notion of convolution on the 2-sphere essentially equivalent (but not identical) to that developed in (Driscoll and Healy, 1994). It is motivated by the observation that convolution in $\mathbb{R}^2$ is given by

$$(h \star f)(\boldsymbol{u}) \triangleq \int_{\mathbb{R}^2} h(\boldsymbol{u} - \boldsymbol{v}) f(\boldsymbol{o} + \boldsymbol{v}) \, d\boldsymbol{v}, \quad \boldsymbol{u} \in \mathbb{R}^2, \tag{9.1}$$

where $\boldsymbol{o}$ denotes the origin in $\mathbb{R}^2$.

Type 1 convolution tries to emulate the convolution in Euclidean spaces by replacing translations with full rotations in SO(3) and integrating over all possible rotation angles characterizing SO(3). Let $f \in L^2(\mathbb{S}^2)$ be the desired signal with spherical harmonic coefficients $(f)_\ell^m \triangleq \langle f, Y_\ell^m \rangle$ and $h \in L^2(\mathbb{S}^2)$ be the filter with spherical harmonic coefficients $(h)_\ell^m \triangleq \langle h, Y_\ell^m \rangle$. Type 1 convolution between filter h and signal f evaluated at a point $\widehat{\boldsymbol{u}} \in \mathbb{S}^2$ is defined as follows:

$$\boxed{(h \odot f)(\widehat{\boldsymbol{u}}) \triangleq \frac{1}{2\pi} \int_0^{2\pi} \int_0^\pi \int_0^{2\pi} h(\boldsymbol{R}^{-1}\widehat{\boldsymbol{u}}) f(\boldsymbol{R}\widehat{\boldsymbol{\eta}}) \, d\omega \, \sin\vartheta \, d\vartheta \, d\varphi,} \tag{9.2}$$

where $\widehat{\boldsymbol{\eta}} = (1, 0, 0)' \in \mathbb{S}^2$ is the north pole. $\boldsymbol{R}$ is the rotation matrix, $\boldsymbol{R} \equiv \boldsymbol{R}_{\varphi\vartheta\omega}^{(zyz)}$, as defined in (7.10), p. 185, and is parameterized by three Euler rotation angles $(\varphi, \vartheta, \omega)$ around zyz-axes with $\varphi \in [0, 2\pi)$, $\vartheta \in [0, \pi]$, $\omega \in [0, 2\pi)$.

This definition is similar to that in (Driscoll and Healy, 1994). Note, however, that the authors there had defined "left convolution" as

$$\left(h \,\widetilde{\odot}\, f\right)(\widehat{\boldsymbol{u}}) \triangleq \int_0^{2\pi}\!\!\int_0^\pi\!\!\int_0^{2\pi} h(\boldsymbol{R}\widehat{\boldsymbol{\eta}}) f(\boldsymbol{R}^{-1}\widehat{\boldsymbol{u}}) \, d\omega \, \sin\vartheta \, d\vartheta \, d\varphi, \quad \widehat{\boldsymbol{u}} \in \mathbb{S}^2. \qquad (9.3)$$

For the purpose of comparing with the type 2 convolution later, we prefer the definition (9.2).

Let us examine the convolution (9.2) a bit further. First the signal f is evaluated at point $\boldsymbol{R}\widehat{\boldsymbol{\eta}}$, which is the rotation of the north pole by $\boldsymbol{R}$. Comparing with (9.1), this is somewhat analogous to translating the origin $\boldsymbol{o}$ to $\boldsymbol{v}$ for evaluating the signal f in $\mathbb{R}^2$. Then the filter is rotated by $(\varphi, \vartheta, \omega)$ and its value at point $\widehat{\boldsymbol{u}}$ is used to multiply with $f(\boldsymbol{R}\widehat{\boldsymbol{\eta}})$ and finally averaging over all possible rotations is done. Based on (7.113), p. 227, we can rewrite (9.2) in rotation operator form as

$$\left(h \odot f\right)(\widehat{\boldsymbol{u}}) = \frac{1}{2\pi} \int_0^{2\pi}\!\!\int_0^\pi\!\!\int_0^{2\pi} \left(\mathcal{D}(\varphi,\vartheta,\omega)h\right)(\widehat{\boldsymbol{u}}) f(\boldsymbol{R}_{\varphi\vartheta\omega}^{(zyz)}\widehat{\boldsymbol{\eta}}) \, d\omega \, \sin\vartheta \, d\vartheta \, d\varphi. \qquad (9.4)$$

The important observation is that in $\boldsymbol{R}\widehat{\boldsymbol{\eta}}$, the first rotation of the north pole by ω around the z-axis is ineffectual since $\boldsymbol{R}_\omega^{(z)}\widehat{\boldsymbol{\eta}} = \widehat{\boldsymbol{\eta}}$. Consequently,

$$\boxed{f(\boldsymbol{R}_{\varphi\vartheta\omega}^{(zyz)}\widehat{\boldsymbol{\eta}}) = f(\boldsymbol{R}_{\varphi\vartheta}^{(zy)}\widehat{\boldsymbol{\eta}}) = f(\vartheta,\varphi) = f(\widehat{\boldsymbol{v}}).}$$

For a point such as $\widehat{\boldsymbol{v}} \equiv \widehat{\boldsymbol{v}}(\vartheta,\varphi) = \boldsymbol{R}_{\varphi\vartheta}^{(zy)}\widehat{\boldsymbol{\eta}} = (\sin\vartheta\cos\varphi, \sin\vartheta\sin\varphi, \cos\vartheta)'$ on 2-sphere. As a result, the convolution is simplified to

$$\left(h \odot f\right)(\widehat{\boldsymbol{u}}) = \frac{1}{2\pi} \int_0^{2\pi}\!\!\int_0^\pi f(\vartheta,\varphi) \int_0^{2\pi} \left(\mathcal{D}(\varphi,\vartheta,\omega)h\right)(\widehat{\boldsymbol{u}}) \, d\omega \, \sin\vartheta \, d\vartheta \, d\varphi$$

$$= \frac{1}{2\pi} \int_0^{2\pi}\!\!\int_0^\pi f(\vartheta,\varphi) \int_0^{2\pi} \left(\mathcal{D}_z(\varphi) \circ \mathcal{D}_y(\vartheta) \circ \mathcal{D}_z(\omega)h\right)(\widehat{\boldsymbol{u}}) \, d\omega \, \sin\vartheta \, d\vartheta \, d\varphi, \qquad (9.5)$$

where we have deliberately broken the rotation operator into its three constituent parts. Noting that the rotation operator is linear and hence the order of rotation and integrals can be switched, we define

$$\boxed{h_0(\widehat{\boldsymbol{u}}) \triangleq \frac{1}{2\pi} \int_0^{2\pi} \left(\mathcal{D}_z(\omega)h\right)(\widehat{\boldsymbol{u}}) \, d\omega} \qquad (9.6)$$

to arrive at

$$\left(h \odot f\right)(\widehat{\boldsymbol{u}}) = \int_0^{2\pi}\!\!\int_0^\pi \left(\mathcal{D}_z(\varphi) \circ \mathcal{D}_y(\vartheta)h_0\right)(\widehat{\boldsymbol{u}}) f(\vartheta,\varphi) \, \sin\vartheta \, d\vartheta \, d\varphi.$$

Now recall the discussion in Section 7.12, p. 242. In particular, we proved that projection into the subspace of azimuthally symmetric signals $\mathcal{H}^0(\mathbb{S}^2)$ can be, in fact, achieved by averaging the desired signal over all rotations about the z-axis. Referring to (9.6), we see that filter h is essentially *transformed into an azimuthally symmetric filter*, denoted by h_0, before being applied as the convolution filter. We summarize the results in the following theorem.

Theorem 9.1 (Spherical convolution of type 1). *Type 1 convolution between filter h and signal f*

$$(h \odot f)(\widehat{\boldsymbol{u}}) = \frac{1}{2\pi} \int_0^{2\pi} \int_0^{\pi} \int_0^{2\pi} h(\boldsymbol{R}^{-1}\widehat{\boldsymbol{u}}) f(\boldsymbol{R}\widehat{\boldsymbol{\eta}}) \, d\omega \, \sin\vartheta \, d\vartheta \, d\varphi$$

$$= \frac{1}{2\pi} \int_0^{2\pi} \int_0^{\pi} \int_0^{2\pi} (\mathcal{D}(\varphi,\vartheta,\omega)h)(\widehat{\boldsymbol{u}}) f(\boldsymbol{R}_{\varphi\vartheta\omega}^{(zyz)}\widehat{\boldsymbol{\eta}}) \, d\omega \, \sin\vartheta \, d\vartheta \, d\varphi$$

is identical to convolution over only two angles (ϑ, φ) with the azimuthally symmetric version of h, denoted by h_0. That is,

$$(h \odot f)(\widehat{\boldsymbol{u}}) = \int_0^{2\pi} \int_0^{\pi} (\mathcal{D}_z(\varphi) \circ \mathcal{D}_y(\vartheta)h_0)(\widehat{\boldsymbol{u}}) f(\vartheta,\varphi) \, \sin\vartheta \, d\vartheta \, d\varphi, \qquad (9.7)$$

where

$$h_0(\widehat{\boldsymbol{u}}) = \frac{1}{2\pi} \int_0^{2\pi} (\mathcal{D}_z(\omega)h)(\widehat{\boldsymbol{u}}) \, d\omega = \sum_{\ell=0}^{\infty} (h)_\ell^0 Y_\ell^0(\widehat{\boldsymbol{u}}),$$

which only depends on the zero-order spherical harmonic coefficients of h.   □

It is also worthwhile to examine type 1 convolution in spectral domain. That is, we are interested in finding $\langle f \odot h, Y_\ell^m \rangle$, which is given by the following theorem.

Theorem 9.2. *The spherical harmonic coefficients of type 1 convolution are given by*

$$\langle h \odot f, Y_\ell^m \rangle = \sqrt{\frac{4\pi}{(2\ell+1)}} (h)_\ell^0 (f)_\ell^m. \qquad (9.8)$$

   □

This confirms that in fact only the azimuthally symmetric harmonics of the filter appear in the output. Equation (9.8) is the "frequency domain" version of (9.7) and exhibits a multiplication property. Due to the $m = 0$ action on the second argument h in (9.8), this convolution definition is not commutative. That is,

$$f \odot h \neq h \odot f.$$

Furthermore, there is the scaling factor which depends on degree ℓ. A proof is given in (up to a scaling) (Driscoll and Healy, 1994, Theorem 1). Here, we provide an alternative proof.

Proof. In our proof, we do not assume that the filter is already transformed into an azimuthally symmetric filter. Rather, we start by determining the spherical harmonic coefficients of (9.5):

$$\langle h \odot f, Y_\ell^m \rangle = \frac{1}{2\pi} \int_0^{2\pi} \int_0^{\pi} \int_0^{2\pi} \langle \mathcal{D}(\varphi,\vartheta,\omega)h, Y_\ell^m \rangle f(\vartheta,\varphi) \, d\omega \, \sin\vartheta \, d\vartheta \, d\varphi.$$

We note that $\langle \mathcal{D}(\varphi, \vartheta, \omega)h, Y_\ell^m \rangle$ is the spherical harmonic coefficient of the rotated filter and using (7.114), p. 227, we can write

$$
\langle h \circledcirc f, Y_\ell^m \rangle = \frac{1}{2\pi} \sum_{m'=-\ell}^{\ell} (h)_\ell^{m'}
$$

$$
\times \int_0^{2\pi} \int_0^{\pi} \int_0^{2\pi} f(\vartheta, \varphi) D_{m,m'}^{\ell}(\varphi, \vartheta, \omega) \, d\omega \, \sin\vartheta \, d\vartheta \, d\varphi. \tag{9.9}
$$

Now we use the spherical harmonic expansion $f(\vartheta, \varphi) = \sum_{s,t}(f)_s^t Y_s^t(\vartheta, \varphi)$ and note that according to (7.129), p. 233,

$$
Y_s^t(\vartheta, \varphi) = \sqrt{\frac{2s+1}{4\pi}} \overline{D_{t,0}^s(\varphi, \vartheta, 0)}
$$

$$
= \sqrt{\frac{2s+1}{4\pi}} \overline{D_{t,0}^s(\varphi, \vartheta, \omega)},
$$

where the second equality follows from the fact that $e^{i0\omega} = e^{i0} = 1$ in the definition of the Wigner D-matrix $D_{t,0}^s(\varphi, \vartheta, \omega)$ in (7.115), p. 227. Therefore, the triple integral over SO(3) rotation angles in (9.9) can be regarded as the inner product of two Wigner D-matrices (weighted by $(f)_s^t$) and (9.9) can be written as

$$
\langle h \circledcirc f, Y_\ell^m \rangle = \frac{1}{2\pi} \sum_{m'=-\ell}^{\ell} (h)_\ell^{m'} \sum_{s,t} \sqrt{\frac{2s+1}{4\pi}} (f)_s^t \langle D_{m,m'}^{\ell}, D_{t,0}^s \rangle_{\mathrm{SO}(3)}
$$

$$
= \frac{1}{2\pi} \frac{8\pi^2}{2\ell+1} \sum_{m'=-\ell}^{\ell} (h)_\ell^{m'} \sum_{s,t} \sqrt{\frac{2s+1}{4\pi}} (f)_s^t \, \delta_{\ell,s} \, \delta_{m,t} \, \delta_{m',0},
$$

where the last step is taken using the orthogonality of Wigner D-matrices over SO(3) rotation angles in (7.127), p. 231. We finally obtain

$$
\langle h \circledcirc f, Y_\ell^m \rangle = \sqrt{\frac{4\pi}{(2\ell+1)}} (h)_\ell^0 \, (f)_\ell^m. \tag{9.10}
$$

$\square$

9.3.1   Type 1 convolution operator matrix and kernel

Upon substituting

$$
f(\vartheta, \varphi) = Y_p^q(\vartheta, \varphi) = \sum_{s,t} \delta_{p,s} \delta_{q,t} Y_s^t(\vartheta, \varphi)
$$

into (9.10), we conclude that the operator matrix elements for type 1 convolution are given by

$$
\boxed{b_{\ell,p}^{m,q} = \langle h \circledcirc Y_p^q, Y_\ell^m \rangle = (h)_\ell^0 \delta_{\ell,p} \delta_{m,q}.} \tag{9.11}
$$

And therefore the operator matrix has a diagonal structure where the diagonal elements of a given degree $\ell = p$ and different orders $m = q$ are all $(h)^0_\ell$.

$$
\mathbf{B} = \begin{pmatrix}
(h)^0_0 & 0 & 0 & 0 & 0 & 0 & 0 & 0 & 0 & \cdots \\
0 & (h)^0_1 & 0 & 0 & 0 & 0 & 0 & 0 & 0 & \cdots \\
0 & 0 & (h)^0_1 & 0 & 0 & 0 & 0 & 0 & 0 & \cdots \\
0 & 0 & 0 & (h)^0_1 & 0 & 0 & 0 & 0 & 0 & \cdots \\
0 & 0 & 0 & 0 & (h)^0_2 & 0 & 0 & 0 & 0 & \cdots \\
0 & 0 & 0 & 0 & 0 & (h)^0_2 & 0 & 0 & 0 & \cdots \\
0 & 0 & 0 & 0 & 0 & 0 & (h)^0_2 & 0 & 0 & \cdots \\
0 & 0 & 0 & 0 & 0 & 0 & 0 & (h)^0_2 & 0 & \cdots \\
0 & 0 & 0 & 0 & 0 & 0 & 0 & 0 & (h)^0_2 & \cdots \\
\vdots & \vdots & \vdots & \vdots & \vdots & \vdots & \vdots & \vdots & \vdots & \ddots
\end{pmatrix}, \tag{9.12}
$$

Using (7.89), p. 219, the operator kernel is given by

$$
\begin{aligned}
B(\widehat{\boldsymbol{u}}, \widehat{\boldsymbol{v}}) &= \sum_{\ell,m} \sum_{p,q} b^{m,q}_{\ell,p} Y^m_\ell(\widehat{\boldsymbol{u}}) \overline{Y^q_p(\widehat{\boldsymbol{v}})} \\
&= \sum_{\ell=0}^\infty (h)^0_\ell \sum_{m=-\ell}^\ell Y^m_\ell(\widehat{\boldsymbol{u}}) \, \overline{Y^m_\ell(\widehat{\boldsymbol{v}})} \\
&= \sum_{\ell=0}^\infty (h)^0_\ell \frac{(2\ell + 1)}{4\pi} P_\ell(\widehat{\boldsymbol{u}} \cdot \widehat{\boldsymbol{v}}),
\end{aligned}
$$

where the last equality follows from addition theorem (7.30), p. 193.

Reflecting on rotations

The argument that rotations in SO(3) are the analog of translations in $\mathbb{R}^2$ is strictly not correct. The problem with choosing "all SO(3)" rotations is that the conventional Euclidean convolution is not formulated (and should not be formulated) with all (proper) isometries in $\mathbb{R}^2$. This has been explicitly taken into account in defining convolution on $\mathbb{R}^2$ in (9.1). In $\mathbb{R}^2$, whilst the two translations are isometric mappings and involved in the definition for convolution, the isometry corresponding to a rotation in $\mathbb{R}^2$ is not involved; nor is the reflection isometry (or associated improper rotations) involved. In contrast, under the actions of all elements of SO(3) a function in $L^2(\mathbb{S}^2)$ can reposition to all possible orientations on the 2-sphere in a proper sense. The analogy in the 2D would have the set of all 2D translations plus a rotation — a playing card flicked on the floor can also take an arbitrary orientation. However, $\mathbb{R}^2$ convolution is defined without the rotation and the orientation is fixed.

We have already elaborated how inclusion of this additional rotation for defining type 1 convolution results in azimuthal averaging of h. As a result, the only relevant filter spherical harmonic coefficients are $(h)^0_\ell$, as can be seen from (9.8). Were rotation included in the $\mathbb{R}^2$ convolution (triple integral instead of double integral) then we would also observe rotation averaging as we see in type 1 spherical convolution.

Problem

9.4. On domain $\mathbb{R}^2$, define the operator $\mathcal{D}_\theta$ for $\theta \in [0, 2\pi)$ to be the rotation in the x-y plane of the function $h(\boldsymbol{u})$, $\boldsymbol{u} \in \mathbb{R}^2$,

$$(\mathcal{D}_\theta h)(\boldsymbol{u}) = h(\boldsymbol{R}_\theta^{-1}\boldsymbol{u}),$$

where $\boldsymbol{R}_\theta$ is the 2D rotation matrix

$$\boldsymbol{R}_\theta \triangleq \begin{pmatrix} \cos\theta & -\sin\theta \\ \sin\theta & \cos\theta \end{pmatrix}.$$

Show that if the convolution in $\mathbb{R}^2$ were defined as

$$\big(h \odot f\big)(\boldsymbol{u}) \triangleq \frac{1}{2\pi} \int_{\mathbb{R}^2} \int_0^{2\pi} \big(\mathcal{T}_{\boldsymbol{u}} \circ \mathcal{R} \circ \mathcal{D}_\theta h\big)(\boldsymbol{v}) f(\boldsymbol{o} + \boldsymbol{v}) \, d\theta \, d\boldsymbol{v}, \quad \boldsymbol{u} \in \mathbb{R}^2, \quad (9.13)$$

where $\mathcal{T}_{\boldsymbol{u}}$ and $\mathcal{R}$ are the translation by $\boldsymbol{u}$ and space reversal operators, respectively, and $\boldsymbol{o}$ denotes the origin in $\mathbb{R}^2$, then this convolution would amount to a conventional convolution in $\mathbb{R}^2$ defined in (9.1)

$$\big(h_0 \star f\big)(\boldsymbol{u}) = \int_{\mathbb{R}^2} h_0(\boldsymbol{u} - \boldsymbol{v}) f(\boldsymbol{o} + \boldsymbol{v}) \, d\boldsymbol{v}, \quad \boldsymbol{u} \in \mathbb{R}^2,$$

but with an "averaged" filter kernel as

$$h_0(\boldsymbol{v}) \triangleq \frac{1}{2\pi} \int_0^{2\pi} \big(\mathcal{D}_\theta h\big)(\boldsymbol{v}) \, d\theta.$$

Due to the averaging action on the first argument h in this new convolution, the convolution is not commutative and $h * f \neq f * h$.

Remark 9.2. If you believe (9.13) is insane then reflect on the fact it is the $\mathbb{R}^2$ analogy of spherical convolution of type 1, Theorem 9.1, p. 298, defined on $\mathbb{S}^2$; as pointed out in (Kennedy et al., 2011). This explains use choice of operator symbol $\odot$ for both $\mathbb{S}^2$ and $\mathbb{R}^2$. $\square$

9.4   Spherical convolution of type 2

Using another binary operand symbol $\circledast$, the definition for type 2 convolution (or isotropic convolution) comes from (Tegmark et al., 1996; Cui and Freeden, 1997; Seon, 2006) and takes the form

$$\big(h_0 \circledast f\big)(\widehat{\boldsymbol{u}}) \triangleq \sqrt{\frac{1}{2\pi}} \int_{\mathbb{S}^2} h_0(\widehat{\boldsymbol{u}} \cdot \widehat{\boldsymbol{v}}) f(\widehat{\boldsymbol{v}}) \, ds(\widehat{\boldsymbol{v}}), \quad \widehat{\boldsymbol{u}} \in \mathbb{S}^2. \qquad (9.14)$$

This definition can be viewed as a linear operator on $L^2(\mathbb{S}^2)$, mapping $f(\cdot) \in L^2(\mathbb{S}^2)$ to $\big(h_0 \circledast f\big)(\cdot) \in L^2(\mathbb{S}^2)$ parameterized by the univariate function $h_0(\cdot) \in$

$L^2(-1,+1)$. Here, $h_0(\cdot)$ is the kernel and can be regarded as defining a filter response. This filter response only depends on the angular distance between the two points $\hat{u}$ and $\hat{v}$ (see (7.2), p. 176) and not their relative orientation. To clarify further, for a given point of interest $\hat{u}$, all points satisfying a fixed $\zeta = \hat{u} \cdot \hat{v} = \cos \Delta$ (Δ being the angular distance between $\hat{u}$ and $\hat{v}$) form a circularly symmetric region around $\hat{u}$ and invoke the same filter response $h_0(\zeta)$. Hence, the convolution is *isotropic*.

On the surface, the two definitions of convolution (9.2) and (9.14) do not bear much similarity. Type 2 convolution, (9.14), convolves an input function $f \in L^2(\mathbb{S}^2)$ with a univariate filter kernel $h_0 \in L^2(-1,+1)$. This differs from type 1 convolution, (9.2), which takes a filter in $L^2(\mathbb{S}^2)$. In this context, we pose the following questions:

Q1: What is the relationship between type 1 and 2 convolutions (9.2) and (9.14) for signals defined on $\mathbb{S}^2$?

Q2: Can these two definitions be understood in a common framework?

Q3: Do these notions represent the most general form of convolution for signals defined on $\mathbb{S}^2$?

9.4.1   Characterization of type 2 convolution

Since $\zeta = \hat{u} \cdot \hat{v} = \cos \Delta \in [-1,+1]$ and $h_0(\zeta) \in L^2(-1,+1)$, we can expand $h_0(\zeta)$ based on normalized Legendre polynomials (2.32), which are complete in $L^2(-1,+1)$, as

$$h_0(\cos \Delta) = \sum_{\ell=0}^{\infty} (h)_\ell \sqrt{\frac{2\ell+1}{2}} P_\ell(\cos \Delta), \tag{9.15}$$

where $(h)_\ell$ are the Legendre polynomial coefficients of h_0. In fact, we can think of $\sqrt{1/(2\pi)}h_0(\cos \Delta)$ as an azimuthally symmetric function in $L^2(\mathbb{S}^2)$, which using (7.29), p. 193, and (7.76), p. 213, is expressed as

$$\sqrt{\frac{1}{2\pi}}h_0(\cos \Delta) = \sum_{\ell=0}^{\infty} (h)_\ell^0 Y_\ell^0(\Delta) \tag{9.16}$$

$$= \sum_{\ell=0}^{\infty} (h)_\ell^0 \sqrt{\frac{2\ell+1}{4\pi}} P_\ell(\cos \Delta),$$

where $(h)_\ell^0 \equiv (h)_\ell$ is the zero-order spherical harmonic coefficient of degree ℓ. Now, using the addition theorem (7.30), p. 193, we replace the Legendre polynomial of degree ℓ with

$$P_\ell(\cos \Delta) = P_\ell(\hat{u} \cdot \hat{v}) = \frac{4\pi}{2\ell+1} \sum_{m=-\ell}^{\ell} Y_\ell^m(\hat{u})\overline{Y_\ell^m(\hat{v})}. \tag{9.17}$$

And hence the filter kernel in (9.16) becomes

$$\sqrt{\frac{1}{2\pi}}h_0(\hat{u} \cdot \hat{v}) = \sum_{\ell=0}^{\infty} \sqrt{\frac{4\pi}{2\ell+1}}(h)_\ell^0 \sum_{m=-\ell}^{\ell} Y_\ell^m(\hat{u})\overline{Y_\ell^m(\hat{v})}. \tag{9.18}$$

Using (9.18) in (9.14) and the spherical harmonic expansion

$$f(\widehat{\boldsymbol{v}}) = \sum_{p,q}(f)_p^q Y_p^q(\widehat{\boldsymbol{v}}),$$

we arrive at

$$\big(h_0 \circledast f\big)(\widehat{\boldsymbol{u}}) = \sum_{\ell=0}^{\infty} \sqrt{\frac{4\pi}{2\ell+1}}(h)_\ell^0 \sum_{m=-\ell}^{\ell} Y_\ell^m(\widehat{\boldsymbol{u}}) \sum_{p,q}(f)_p^q \int_{\mathbb{S}^2} \overline{Y_\ell^m(\widehat{\boldsymbol{v}})} Y_p^q(\widehat{\boldsymbol{v}})\, ds(\widehat{\boldsymbol{v}})$$

$$= \sum_{\ell,m} \sqrt{\frac{4\pi}{2\ell+1}}(h)_\ell^0 (f)_\ell^m Y_\ell^m(\widehat{\boldsymbol{u}}),$$

from which we glean that the spherical harmonic coefficients of $\big(h_0 \circledast f\big)(\widehat{\boldsymbol{u}})$ are given by

$$\langle h_0 \circledast f, Y_\ell^m \rangle = \sqrt{\frac{4\pi}{2\ell+1}}(h)_\ell^0 (f)_\ell^m. \tag{9.19}$$

9.4.2   Equivalence between type 1 and 2 convolutions

It should have become apparent by now that despite their initial different appearances, type 1 and 2 convolutions are essentially the same isotropic operators. In type 1 convolution, (9.2), even if we start with a non-azimuthally symmetric filter h in $L^2(\mathbb{S}^2)$, the extra initial rotation by all possible ω around the z-axis will turn the filter into an azimuthally symmetric one, $h_0 \in \mathcal{H}^0(\mathbb{S}^2)$. Any component of the filter function h orthogonal to $\mathcal{H}^0(\mathbb{S}^2) \subset L^2(\mathbb{S}^2)$ plays no role in the final result. So we can comfortably start with the projection of h into the space of azimuthally symmetric signals $\mathcal{H}^0(\mathbb{S}^2)$ and use type 2 convolution. In the end, only the zero-order spherical harmonic coefficients of the filter are used to define the operator action in spectral domain and its kernel. We summarize our findings in the following theorem.

Theorem 9.3 (Convolution equivalence). *Type 1 convolution between signal* $f \in L^2(\mathbb{S}^2)$ *and filter* $h \in L^2(\mathbb{S}^2)$ *with spherical harmonic coefficients* $(h)_\ell^m$

$$\big(h \odot f\big)(\widehat{\boldsymbol{u}}) \triangleq \frac{1}{2\pi} \int_0^{2\pi}\!\!\int_0^{\pi}\!\!\int_0^{2\pi} h(\boldsymbol{R}^{-1}\widehat{\boldsymbol{u}}) f(\boldsymbol{R}\widehat{\boldsymbol{\eta}})\, d\omega\, \sin\vartheta\, d\vartheta\, d\varphi$$

is equivalent to the following type 2 convolution

$$\big(h_0 \circledast f\big)(\widehat{\boldsymbol{u}}) \triangleq \int_{\mathbb{S}^2} h_0(\widehat{\boldsymbol{u}} \cdot \widehat{\boldsymbol{v}}) f(\widehat{\boldsymbol{v}})\, ds(\widehat{\boldsymbol{v}}),$$

where the convolution kernel h_0 *belongs to* $\mathcal{H}^0(\mathbb{S}^2) \subset L^2(\mathbb{S}^2)$ *and is induced by projection of* $h \in L^2(\mathbb{S}^2)$ *into the subspace of azimuthally symmetric signals* $\mathcal{H}^0(\mathbb{S}^2)$ *through operator* $\mathcal{B}^0$ *as follows*

$$h_0(\widehat{\boldsymbol{u}} \cdot \widehat{\boldsymbol{v}}) = h_0(\cos\Delta) = \big(\mathcal{B}^0 h\big)(\cos\Delta), \quad 0 \le \Delta \le \pi$$

$$= \sum_{\ell=0}^{\infty}(h)_\ell^0 Y_\ell^0(\Delta). \tag{9.20}$$

In spectral domain, we have

$$\langle h \circledcirc f, Y_\ell^m \rangle = \langle h_0 \circledast f, Y_\ell^m \rangle = \sqrt{\frac{4\pi}{2\ell+1}} (h)_\ell^0 (f)_\ell^m,$$

where $(f)_\ell^m = \langle f, Y_\ell^m \rangle$.

The convolution operator matrix is fully diagonal with elements

$$b_{\ell,p}^{m,q} = (h)_\ell^0 \delta_{\ell,p} \delta_{m,q} \qquad (9.21)$$

and the operator kernel is

$$B(\widehat{\boldsymbol{u}}, \widehat{\boldsymbol{v}}) = \sum_{\ell=0}^{\infty} (h)_\ell^0 \frac{(2\ell+1)}{4\pi} P_\ell(\widehat{\boldsymbol{u}} \cdot \widehat{\boldsymbol{v}}).$$

$\square$

The convolution operator is isotropic *if and only if* its corresponding infinite matrix is of the special diagonal form (9.21) as shown in (9.12).

Note that compared to (9.14) in the definition of type 2 convolution in Theorem 9.3, we have absorbed the normalization factor $\sqrt{1/2\pi}$ into h_0.

Example 9.2. As an example, in (Bülow, 2004) a parametric form of smoothing (low-pass filtering) based on spherical diffusion was introduced, which acts on a given signal defined on $\mathbb{S}^2$ and corresponds to

$$(h)_\ell \triangleq \exp(-\ell(\ell+1)\varsigma) \qquad (9.22)$$

in (9.21), where parameter $\varsigma \geq 0$, called "kt" in (Bülow, 2004), controls the spherical harmonic "bandwidth" of the filter. Of course, any well-formed sequence, $(h)_\ell$, that monotonically decreases to zero with ℓ (sufficiently quickly), in the spirit of (9.22), can act as a low-pass filter. $\square$

Example 9.3. A standard example of isotropic convolution, from (Seon, 2006), uses the von Mises-Fisher distribution for the 2-sphere which depends on the "concentration parameter" $\kappa \geq 0$,

$$h_0(\cos \Delta) = \frac{\kappa \exp(\kappa \cos \Delta)}{4\pi \sinh \kappa}. \qquad (9.23)$$

Thereby the function on the 2-sphere, h, in type 1 convolution, would have projection $\mathcal{B}^0 h$, see (7.157), proportional to the well-known von Mises-Fisher distribution. When $\kappa \gg 1$ the $(h)_\ell$ corresponding to (9.23) can be shown to be approximated by (9.22) with $\varsigma = (2\kappa)^{-1}$, (Seon, 2006). $\square$

Remark 9.3. Theorem 9.3 takes a given filter $h(\cdot)$ in $L^2(\mathbb{S}^2)$ for type 1 convolution and synthesizes a kernel $h_0(\cdot)$ in $\mathcal{H}^0(\mathbb{S}^2)$ to be used in type 2 isotropic convolution. Conversely, we can ask for a given isotropic filter kernel $h_0(\cdot)$ in $\mathcal{H}^0(\mathbb{S}^2)$ what the corresponding admissible $h(\cdot)$ in $L^2(\mathbb{S}^2)$ are. The answer, by the orthogonality of the Legendre polynomials, is any $h \in L^2(\mathbb{S}^2)$ satisfying

$$(h)_\ell^0 = \langle h, Y_\ell^0 \rangle = \sqrt{\frac{2\ell+1}{4\pi}} \int_{-1}^{+1} h_0(\zeta) P_\ell(\zeta) \, d\zeta.$$

$\square$

Synopsis

The developed framework in this and the last section indicated that SO(3) methods (Driscoll and Healy, 1994; Bülow, 2004) are not a natural extension of conventional convolution defined on $\mathbb{R}^2$, given in (9.1). The simpler type 2 convolution formula, (9.14), has the equivalent isotropic action. So we can simply refer to both type 1 and type 2 convolutions as isotropic convolution. These convolutions have the advantage that their evaluation in the spherical harmonic domain is particularly simple and thereby attractive for applications that involve an isotropic or azimuthally symmetric filter.

It also became clear from the earlier discussion that type 1 convolution integrates over three Euler angles, in contrast to $\mathbb{R}^2$ convolution (9.1), which integrates over 2D translations in the plane. This extra integration is the source of the difficulties. That is, the integration over the first rotation $\omega \in [0, 2\pi)$ through matrix $\boldsymbol{R}_\omega^{(z)}$ about the north pole $\widehat{\boldsymbol{\eta}} \in \mathbb{S}^2$ zeroes the contribution from filter spherical harmonic coefficients with $m \neq 0$. As a result, isotropic convolution is simply not commutative.

9.5   Spherical convolution of type 3

In this section, we briefly review another definition of convolution in the literature for signals in $L^2(\mathbb{S}^2)$, which is non-isotropic or anisotropic; see (Yeo et al., 2008). Hence, it is expected to be more general than isotropic convolution discussed above. However, it will soon become clear that it has its own shortcomings. Specifically, type 3 convolution is defined as (Yeo et al., 2008)

$$h \, \square \, f = g(\vartheta, \varphi, \omega) \triangleq \int_{\mathbb{S}^2} \big(\mathcal{D}(\varphi, \vartheta, \omega)h\big)(\widehat{\boldsymbol{u}}) f(\widehat{\boldsymbol{u}}) \, ds(\widehat{\boldsymbol{u}}), \qquad (9.24)$$

in which the filter kernel h is rotated by operator $\mathcal{D}(\varphi, \vartheta, \omega)$, multiplied by the desired signal $f(\widehat{\boldsymbol{u}})$ and then integration over all possible points $\widehat{\boldsymbol{u}}$ on 2-sphere is performed.

The advantage of (9.24) compared to isotropic (type 1 and 2) convolution is that the kernel filter can be, in general, non-azimuthally symmetric and the convolution operator has some directional bias. It is anisotropic inasmuch that it is not isotropic because the operator does not transform h into an azimuthally symmetric one. However, referring to (9.24), it becomes immediately clear that this construction takes two functions in $L^2(\mathbb{S}^2)$ and produces an output $g(\varphi, \vartheta, \omega)$ whose domain does not belong to $\mathbb{S}^2$ (it is a function of three Euler angles) and this is not desired for most practical use. Another shortcoming is that such convolution is not commutative in general and $h \, \square \, f \neq f \, \square \, h$.[2]

9.5.1   Alternative characterization of type 3 convolution

Before concluding this section, we present an alternative form of type 3 convolution which is expressed in terms of the spherical harmonic coefficients of $h(\widehat{\boldsymbol{u}})$, denoted by $(h)_\ell^m$, and those of $f(\widehat{\boldsymbol{u}})$, denoted by $(f)_\ell^m$. This presentation will

---

[2] This should become more clear in Section 9.6, where we show that $\mathcal{D}(\varphi, \vartheta, \omega)^{-1} = \mathcal{D}(-\omega, -\vartheta, -\varphi) \neq \mathcal{D}(\varphi, \vartheta, \omega)$ in general, unless a very specific condition is met.

be useful in later sections when we develop a commutative anisotropic convolution with output in $L^2(\mathbb{S}^2)$. The key identity we need is to specify the spherical harmonic coefficients of $\mathcal{D}(\varphi, \vartheta, \omega)h$ borrowed from Section 7.11, p. 226,

$$
\left\langle \mathcal{D}(\varphi, \vartheta, \omega)h, Y_s^t \right\rangle = \sum_{t'=-s}^{s} D_{t,t'}^s(\varphi, \vartheta, \omega)(h)_s^{t'}, \tag{9.25}
$$

where $D_{t,t'}^s(\varphi, \vartheta, \omega)$ is the Wigner D-matrix. Corresponding to the coefficients (9.25) we have the spatial function

$$
\big(\mathcal{D}(\varphi, \vartheta, \omega)h\big)(\widehat{\boldsymbol{u}}) = \sum_{s,t} Y_s^t(\widehat{\boldsymbol{u}}) \sum_{t'=-s}^{s} D_{t,t'}^s(\varphi, \vartheta, \omega)(h)_s^{t'}. \tag{9.26}
$$

And we manipulate an expansion for $f(\widehat{\boldsymbol{u}})$, for reasons that will become apparent, as follows

$$
\begin{aligned}
f(\widehat{\boldsymbol{u}}) &= \sum_{p,q}(f)_p^q Y_p^q(\widehat{\boldsymbol{u}}) \\
&= \sum_{p,q}(-1)^q(f)_p^q \overline{Y_p^{-q}(\widehat{\boldsymbol{u}})} \\
&= \sum_{p,q}(-1)^q(f)_p^{-q} \overline{Y_p^q(\widehat{\boldsymbol{u}})},
\end{aligned} \tag{9.27}
$$

where we have used the conjugate property of the spherical harmonics (7.26), p. 192. Then (9.26) and (9.27) can be substituted into (9.24) and manipulated

$$
\begin{aligned}
g(\vartheta, \varphi, \omega) &= \int_{\mathbb{S}^2} \big(\mathcal{D}(\varphi, \vartheta, \omega)h\big)(\widehat{\boldsymbol{u}}) f(\widehat{\boldsymbol{u}})\, ds(\widehat{\boldsymbol{u}}) \\
&= \int_{\mathbb{S}^2} \sum_{s,t} Y_s^t(\widehat{\boldsymbol{u}}) \sum_{t'=-s}^{s} D_{t,t'}^s(\varphi, \vartheta, \omega)(h)_s^{t'} \sum_{p,q}(-1)^q(f)_p^{-q} \overline{Y_p^q(\widehat{\boldsymbol{u}})}\, ds(\widehat{\boldsymbol{u}}) \\
&= \sum_{s,t} \sum_{p,q}(-1)^q(f)_p^{-q} \delta_{s,p}\delta_{t,q} \sum_{t'=-s}^{s} D_{t,t'}^s(\varphi, \vartheta, \omega)(h)_s^{t'},
\end{aligned}
$$

where we have used the orthonormality of the spherical harmonics — this is why we used an expansion in the conjugate of the spherical harmonics in (9.27). This yields the sought result

$$
h \,\square\, f = \sum_{s,t}(-1)^t(f)_s^{-t} \sum_{t'=-s}^{s} D_{t,t'}^s(\varphi, \vartheta, \omega)(h)_s^{t'}. \tag{9.28}
$$

9.5. Given the generalized Parseval relation, (2.57), p. 72,

$$\langle h, f \rangle = \int_{\mathbb{S}^2} h(\widehat{\boldsymbol{u}}) \overline{f(\widehat{\boldsymbol{u}})} \, ds(\widehat{\boldsymbol{u}}) = \sum_{\ell, m} (h)_\ell^m \overline{(f)_\ell^m} \tag{9.29}$$

and the conjugation property, (7.26), p. 192, of the spherical harmonics, or otherwise, show

$$\int_{\mathbb{S}^2} h(\widehat{\boldsymbol{u}}) f(\widehat{\boldsymbol{u}}) \, ds(\widehat{\boldsymbol{u}}) = \sum_{\ell, m} (-1)^m (h)_\ell^m (f)_\ell^{-m}. \tag{9.30}$$

9.6   Commutative anisotropic convolution

9.6.1   Requirements for convolution on 2-sphere

The developed framework so far pointed out the fact that a proper definition for anisotropic convolution, which respects some important and expected properties of normal convolution in $\mathbb{R}^2$, such as being commutative, is missing in the past literature.

Towards developing a commutative anisotropic convolution, we lay down the following requirements (Sadeghi et al., 2012):

R1: it should take two functions in $L^2(\mathbb{S}^2)$ and generate a function in $L^2(\mathbb{S}^2)$;

R2: it should, at its heart, be a double integral as the 2-sphere is two-dimensional;

R3: it should be commutative and in the integrand it should involve the use of an involutory operator; that is, an operator which is its own inverse. And it also needs to be an isometry with respect to the measure on the 2-sphere, $ds(\widehat{\boldsymbol{u}})$.

Here, we formulate and study the properties of a commutative anisotropic (type 4) convolution on 2-sphere.

9.6.2   A starting point

The convolution on the real line in Section 9.2 forms an implicit prescription for constructing a suitable notion of convolution on the 2-sphere. We believe that not all of isometries expressible in SO(3) should be involved (nor the improper rotations not captured by SO(3)). Only two degrees of freedom available in SO(3) should participate, in the same way that by analogy only two of the isometries in the $\mathbb{R}^2$ case participate in convolution. Our initial candidate is

$$g_\omega(\vartheta, \varphi) = \int_{\mathbb{S}^2} \big(\mathcal{D}(\varphi, \vartheta, \omega) h\big)(\widehat{\boldsymbol{u}}) f(\widehat{\boldsymbol{u}}) \, ds(\widehat{\boldsymbol{u}}), \tag{9.31}$$

where $\mathcal{D}(\varphi, \vartheta, \omega)$ is the standard rotation operator, defined in Section 7.11, p. 226, parameterized in the Euler angles in the zyz convention

$$\mathcal{D}(\varphi, \vartheta, \omega) = \mathcal{D}_z(\varphi) \circ \mathcal{D}_y(\vartheta) \circ \mathcal{D}_z(\omega), \tag{9.32}$$

but the treatment of angle ω in (9.31) is unspecified at this point. The angle ω might be a constant or a function of ϑ and φ.

The definition in (9.31) differs in philosophy from that in (9.24) given by (Yeo et al., 2008). Here we can make the identification that the output is a function of two arguments ϑ, φ whenever $\omega = \omega(\vartheta, \varphi)$, or $\omega = \omega_0$ (some constant), or ω is regarded as a parameter.

Against the requirements laid down in Section 9.6.1, (9.31) has the right beginnings, but is too general. In all respects this equation is very close in structure to the 1D and 2D Euclidean cases. It satisfies requirements R1 and R2. In general it does not satisfy requirement R3 (although it does satisfy the isometry property) and this is where we focus next.

9.6.3   Establishing commutativity

We recall our observation that involution is the source of commutativity for convolution, Remark 9.1, p. 295, which guides our development here. The isometry property of $\mathcal{D}(\varphi, \vartheta, \omega)$ under measure $ds(\widehat{\boldsymbol{u}})$, and indeed its inverse $\mathcal{D}(\varphi, \vartheta, \omega)^{-1}$, in (9.31), means that it can be written

$$\begin{aligned}
g_\omega(\vartheta, \varphi) &= \int_{\mathbb{S}^2} \big(\mathcal{D}(\varphi, \vartheta, \omega)h\big)(\widehat{\boldsymbol{u}}) f(\widehat{\boldsymbol{u}})\, ds(\widehat{\boldsymbol{u}}) \\
&= \int_{\mathbb{S}^2} h(\widehat{\boldsymbol{u}}) \big(\mathcal{D}(\varphi, \vartheta, \omega)^{-1} f\big)(\widehat{\boldsymbol{u}})\, ds(\widehat{\boldsymbol{u}}) \\
&= \int_{\mathbb{S}^2} h(\widehat{\boldsymbol{u}}) \big(\mathcal{D}(-\omega, -\vartheta, -\varphi) f\big)(\widehat{\boldsymbol{u}})\, ds(\widehat{\boldsymbol{u}}),
\end{aligned}$$

where, by elementary geometric considerations (see also (7.122), p. 229),

$$\boxed{\mathcal{D}(\varphi, \vartheta, \omega)^{-1} = \mathcal{D}(-\omega, -\vartheta, -\varphi).} \tag{9.33}$$

Our objective is to select ω to have (9.33) equal $\mathcal{D}(\varphi, \vartheta, \omega)$, so that it becomes an involution.

As we saw in Section 7.11, p. 226, a negative rotation about the y-axis by $-\vartheta$ is outside the range of admissible ϑ values ($[0, \pi]$), but algebraically and geometrically it is perfectly fine. The key identity was (7.123), p. 229, repeated here,

$$\boxed{\mathcal{D}(\varphi, \vartheta, \omega) = \mathcal{D}(\pi + \varphi, -\vartheta, \pi + \omega),} \tag{9.34}$$

where the z-axis rotations have $(\mathrm{mod}\, 2\pi)$ omitted to avoid clutter.

We can equate (9.33) and (9.34) with the identification

$$\pi + \omega \equiv -\varphi \pmod{2\pi}. \tag{9.35}$$

That is, we have the involution[3]

$$
\begin{aligned}
\mathcal{D}(\varphi, \vartheta, \pi - \varphi)^{-1} &= \mathcal{D}(-\pi + \varphi, -\vartheta, -\varphi) \\
&= \mathcal{D}(\varphi, \vartheta, \pi - \varphi),
\end{aligned}
\tag{9.36}
$$

where the initial z-axis rotation is through angle

$$
\boxed{\omega \equiv -\pi - \varphi \equiv \pi - \varphi \pmod{2\pi}.}
$$

In summary, we began with (9.31) with ω unspecified and we have shown that it must be chosen to be a function[4] of φ according to (9.35) for the overall rotation operation to be an involution. This yields the desired commutativity. Using a new binary operand symbol $\odot$ for commutative anisotropic convolution, we establish that

$$
\begin{aligned}
\big(h \odot f\big)(\vartheta, \varphi) &= \int_{\mathbb{S}^2} \big(\mathcal{D}(\varphi, \vartheta, \pi - \varphi)h\big)(\widehat{\boldsymbol{u}}) f(\widehat{\boldsymbol{u}}) \, ds(\widehat{\boldsymbol{u}}) \tag{9.37} \\
&= \int_{\mathbb{S}^2} h(\widehat{\boldsymbol{u}}) \big(\mathcal{D}(\varphi, \vartheta, \pi - \varphi)f\big)(\widehat{\boldsymbol{u}}) \, ds(\widehat{\boldsymbol{u}}) \\
&= \int_{\mathbb{S}^2} \big(\mathcal{D}(\varphi, \vartheta, \pi - \varphi)f\big)(\widehat{\boldsymbol{u}}) h(\widehat{\boldsymbol{u}}) \, ds(\widehat{\boldsymbol{u}}) \\
&= \big(f \odot h\big)(\vartheta, \varphi),
\end{aligned}
$$

where we used (9.36). One can also show that $\omega = \pi - \varphi$ is a sufficient condition for the convolution to be commutative. This can be done by inserting $\omega = \pi - \varphi$ in $\mathcal{D}(\varphi, \vartheta, \omega)$ and verifying using (9.34) that $\mathcal{D}(\varphi, \vartheta, \pi - \varphi)^{-1} = \mathcal{D}(-\pi + \varphi, -\vartheta, -\varphi) = \mathcal{D}(\varphi, \vartheta, \pi - \varphi)$.

In summary, we have derived a commutative anisotropic convolution on 2-sphere

$$
\boxed{\big(h \odot f\big)(\vartheta, \varphi) = \int_{\mathbb{S}^2} \big(\mathcal{D}(\varphi, \vartheta, \pi - \varphi)h\big)(\widehat{\boldsymbol{u}}) f(\widehat{\boldsymbol{u}}) \, ds(\widehat{\boldsymbol{u}}).}
\tag{9.38}
$$

Remark 9.4. The proposed convolution can be interpreted as a mapping of anisotropic convolution defined on SO(3) in (9.24) to $\mathbb{S}^2$ with the constraint that ω varies with the longitude φ as $\omega = \pi - \varphi \pmod{2\pi}$. Since, ω must be chosen as a function of φ, it cannot be freely controlled at each spatial position. $\square$

We now explore this definition through geometric interpretations and some spectral analysis.

---

[3] Again $\pi - \varphi \pmod{2\pi}$ might be better in this equation.
[4] Initially, we permitted ω to be a function of ϑ and φ, but have shown that it is independent of ϑ.

9.6. By considering the component rotations of the operator $\mathcal{D}(\varphi, \vartheta, \pi - \varphi)$, given by

$$\mathcal{D}(\varphi, \vartheta, \pi - \varphi) = \mathcal{D}_z(\varphi) \circ \mathcal{D}_y(\vartheta) \circ \mathcal{D}_z(\pi - \varphi),$$

show that it is an involution. That is, it satisfies

$$\mathcal{D}(\varphi, \vartheta, \pi - \varphi) \circ \mathcal{D}(\varphi, \vartheta, \pi - \varphi) = \mathcal{I},$$

where $\mathcal{I}$ is the identity or trivial operator.

9.7. The previous problem can be revisited, and the result proven, not through operators, but by the equivalent rotation of the coordinate axes (see (7.113), p. 227)

$$\left(\mathcal{D}_z(\varphi) \circ \mathcal{D}_y(\vartheta) \circ \mathcal{D}_z(\pi - \varphi)f\right)(\widehat{\boldsymbol{u}}) = f\left(\boldsymbol{R}^{(z)}_{-\pi+\varphi} \boldsymbol{R}^{(y)}_{-\vartheta} \boldsymbol{R}^{(z)}_{-\varphi} \widehat{\boldsymbol{u}}\right).$$

Show that $\boldsymbol{R}^{(z)}_{-\pi+\varphi} \boldsymbol{R}^{(y)}_{-\vartheta} \boldsymbol{R}^{(z)}_{-\varphi}$ is an involuntary 3×3 matrix for all ϑ and φ by proving:

$$\left(\boldsymbol{R}^{(z)}_{-\pi+\varphi} \boldsymbol{R}^{(y)}_{-\vartheta} \boldsymbol{R}^{(z)}_{-\varphi}\right)\left(\boldsymbol{R}^{(z)}_{-\pi+\varphi} \boldsymbol{R}^{(y)}_{-\vartheta} \boldsymbol{R}^{(z)}_{-\varphi}\right) = \mathbf{I}_3, \quad \forall \vartheta, \ \varphi,$$

where $\mathbf{I}_3$ is the 3×3 identity matrix.

9.8. In real-line convolution we factored the key operator into two portions; one part doing a time reversal and the other part doing translation. Show that there is an analogous situation for the 2-sphere commutative convolution. That is, how can the operator be factored to yield a similar interpretation?

9.6.4   Graphical depiction

In Figure 9.1 we present a sequence of images to depict the commutative convolution in action. Figure 9.1a depicts a simplified filter indicated by an asymmetric region on the 2-sphere. The filter, of course, in general has support on the whole 2-sphere. An ω rotation (by π) is shown in Figure 9.1b and may be associated with flipping or "time reversing" the filter. Figure 9.1c is the first ω rotation (by $\pi - \varphi$); Figure 9.1d shows the second ϑ rotation and, finally, Figure 9.1e shows the filter after the third φ rotation.

Furthermore, Figure 9.2 shows that an intrinsic single rotation along axis

$$\widehat{\boldsymbol{w}}(\varphi) \triangleq \left(\cos(\pi + \varphi), \ \sin(\pi + \varphi), \ 0\right)' \tag{9.39}$$

by a single rotation ϑ, with rotation matrix $\boldsymbol{R}^{(\widehat{\boldsymbol{w}}(\varphi))}_{\vartheta}$, realizes as a single rotation the same result as the three rotations that take Figure 9.1c to Figure 9.1e.

Finally, Figure 9.3 graphically corroborates that the convolution (9.37) is indeed commutative, through two examples. In it, the key point is that Figure 9.3b and Figure 9.3c are isometries and when integrated yield the same result; similarly for Figure 9.3d and Figure 9.3e.

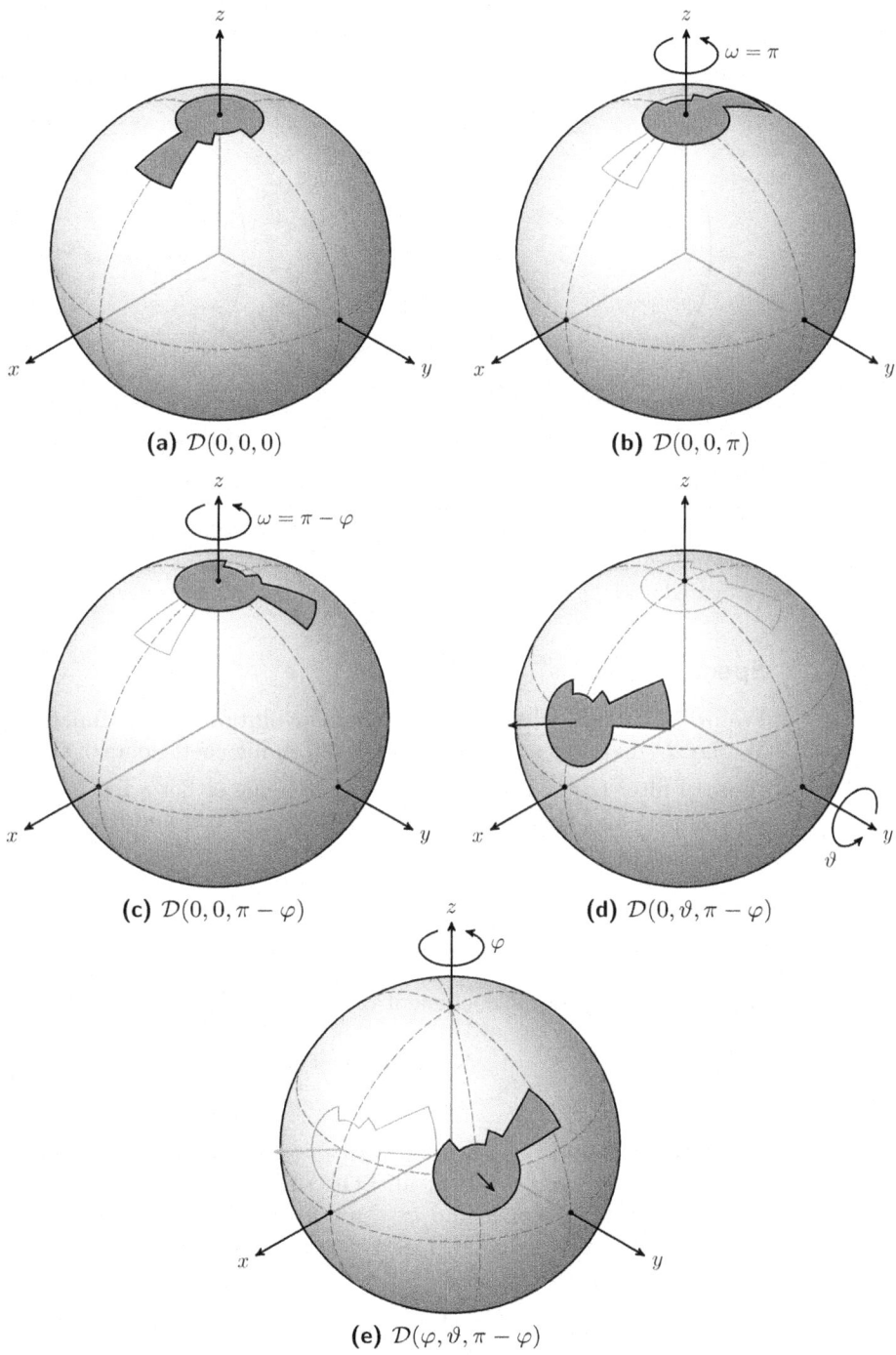

(a) $\mathcal{D}(0,0,0)$

(b) $\mathcal{D}(0,0,\pi)$

(c) $\mathcal{D}(0,0,\pi-\varphi)$

(d) $\mathcal{D}(0,\vartheta,\pi-\varphi)$

(e) $\mathcal{D}(\varphi,\vartheta,\pi-\varphi)$

Figure 9.1 Demonstration of the action of the commutative convolution kernel. A nominal asymmetrical support region for the kernel $h(\widehat{\boldsymbol{u}})$ is transformed under the action of operator $\mathcal{D}(\varphi,\vartheta,\pi-\varphi)$ according to the component rotations indicated.

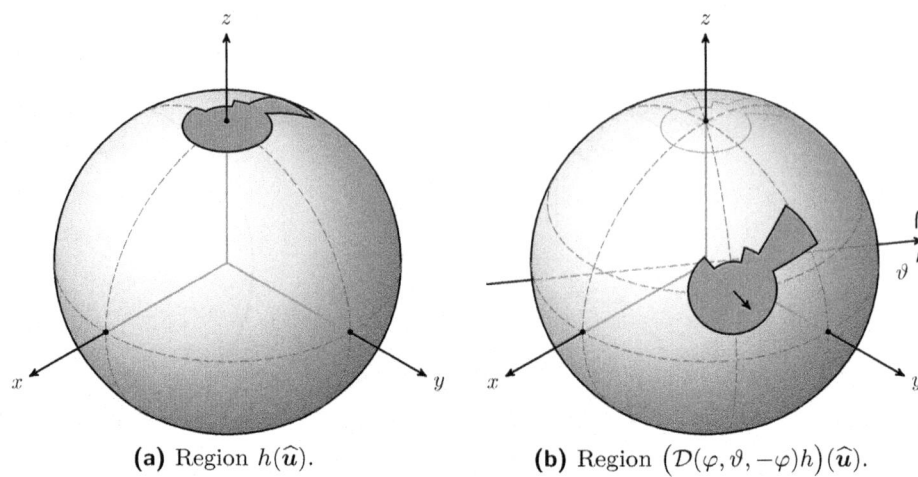

(a) Region $h(\widehat{\boldsymbol{u}})$.   **(b)** Region $\big(\mathcal{D}(\varphi,\vartheta,-\varphi)h\big)(\widehat{\boldsymbol{u}})$.

Figure 9.2 Single intrinsic rotation $\boldsymbol{R}_{\vartheta}^{(\widehat{\boldsymbol{w}(\varphi)})}$ version of $\mathcal{D}(\varphi,\vartheta,-\varphi)$ where $\widehat{\boldsymbol{w}}(\varphi)$ is given by (9.39). (a) Nominal support region of some filter kernel $h(\widehat{\boldsymbol{u}})$. (b) Transformed support region $\big(\mathcal{D}(\varphi,\vartheta,-\varphi)h\big)(\widehat{\boldsymbol{u}})$ under equivalent intrinsic rotation about indicated axis.

9.6.5   Spectral analysis

We utilize (9.28), which specifies the convolution output function g in terms of Wigner D-matrices and the spherical harmonic coefficients of the signal $(f)_s^t$ and those of filter $(h)_s^{t'}$. We first start with ω being either a fixed value or a function of ϑ and φ and then specialize to the case of commutative convolution where $\omega = \pi - \varphi$. According to (9.28), repeated here with $g_\omega(\vartheta,\varphi)$ instead of general $g(\vartheta,\varphi,\omega)$

$$g_\omega(\vartheta,\varphi) = \sum_{s,t}(-1)^t(f)_s^{-t}\sum_{t'=-s}^{s} D_{t,t'}^{s}(\varphi,\vartheta,\omega)(h)_s^{t'}, \qquad (9.40)$$

the spherical harmonic coefficient of g is given by

$$\langle g_\omega, Y_\ell^m\rangle = \sum_{s,t}(-1)^t(f)_s^{-t}\sum_{t'=-s}^{s}(h)_s^{t'}\big\langle D_{t,t'}^{s}(\omega), Y_\ell^m\big\rangle_{\mathbb{S}^2}, \qquad (9.41)$$

where the notation $\big\langle D_{t,t'}^{s}(\omega), Y_\ell^m\big\rangle_{\mathbb{S}^2}$ is used to emphasize that the inner product is taken over $\mathbb{S}^2$ and not over all $\mathrm{SO}(3)$ rotation angles and ω is either a fixed value or a function of ϑ and φ. But for brevity, we drop the dependence on ω and $\mathbb{S}^2$ in the remainder of derivations. Let us examine $\big\langle D_{t,t'}^{s}, Y_\ell^m\big\rangle$ in more detail:

$$\big\langle D_{t,t'}^{s}, Y_\ell^m\big\rangle = \int_0^{2\pi}\int_0^{\pi} D_{t,t'}^{s}(\varphi,\vartheta,\omega)\overline{Y_\ell^m(\vartheta,\varphi)}\sin\vartheta\,d\vartheta\,d\varphi. \qquad (9.42)$$

Recall that according to (7.129), p. 233,

$$\overline{Y_\ell^m(\vartheta,\varphi)} = \sqrt{\frac{2\ell+1}{4\pi}}\,D_{m,0}^{\ell}(\varphi,\vartheta,0), \qquad (9.43)$$

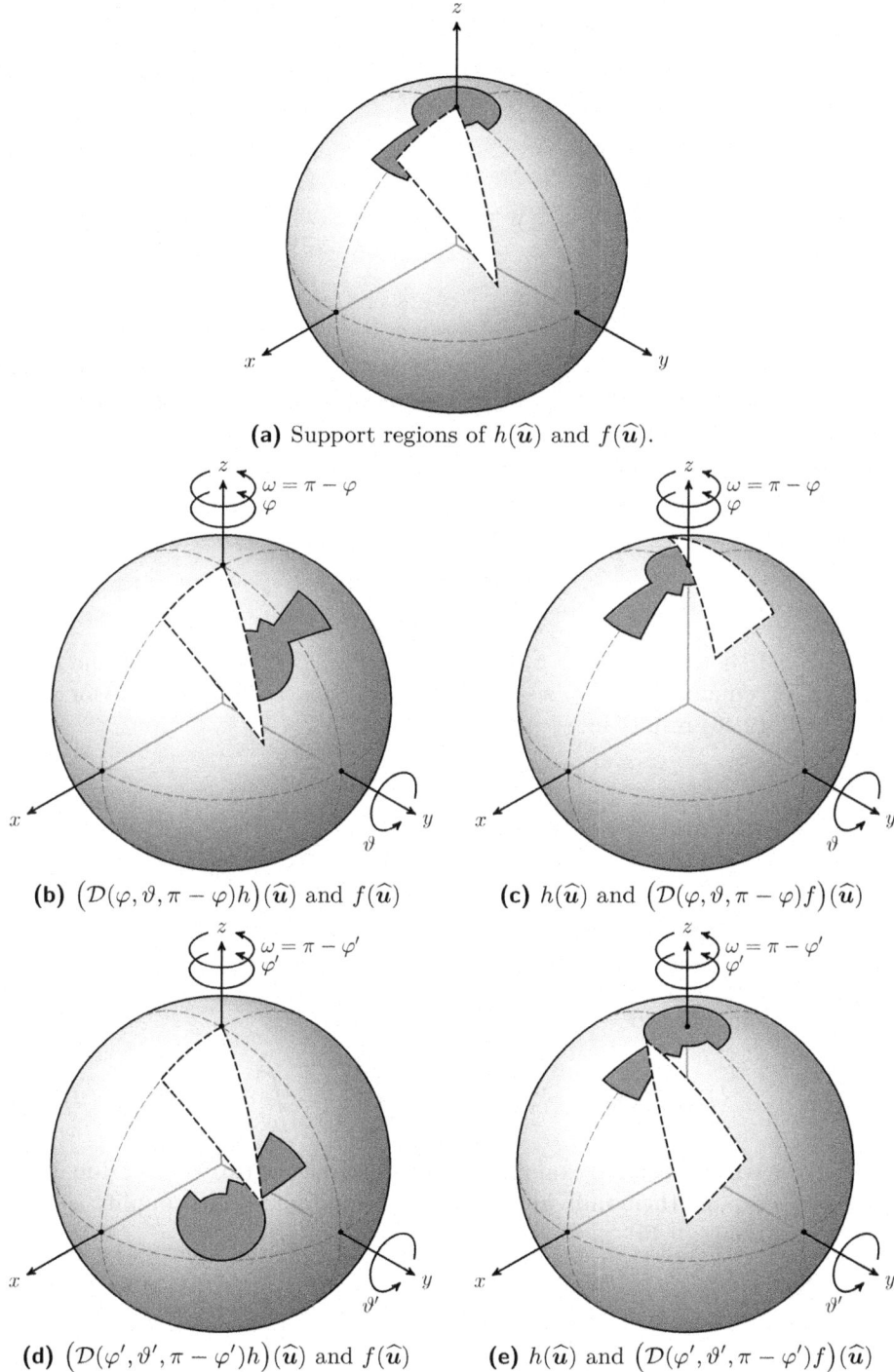

(a) Support regions of $h(\widehat{\boldsymbol{u}})$ and $f(\widehat{\boldsymbol{u}})$.

(b) $\big(\mathcal{D}(\varphi,\vartheta,\pi-\varphi)h\big)(\widehat{\boldsymbol{u}})$ and $f(\widehat{\boldsymbol{u}})$

(c) $h(\widehat{\boldsymbol{u}})$ and $\big(\mathcal{D}(\varphi,\vartheta,\pi-\varphi)f\big)(\widehat{\boldsymbol{u}})$

(d) $\big(\mathcal{D}(\varphi',\vartheta',\pi-\varphi')h\big)(\widehat{\boldsymbol{u}})$ and $f(\widehat{\boldsymbol{u}})$

(e) $h(\widehat{\boldsymbol{u}})$ and $\big(\mathcal{D}(\varphi',\vartheta',\pi-\varphi')f\big)(\widehat{\boldsymbol{u}})$

Figure 9.3 Commutative convolution on the support regions of two functions, dark keyhole $h(\widehat{\boldsymbol{u}})$ and light triangular $f(\widehat{\boldsymbol{u}})$. Figures (b)/(d) and (c)/(e) are identical up to a rotation corroborating that convolution $h \odot f$ is the same as $f \odot h$.

where the Wigner D-matrix is defined as

$$D^\ell_{m,m'}(\varphi,\vartheta,\omega) \triangleq e^{-im\varphi}d^\ell_{m,m'}(\vartheta)e^{-im'\omega}, \tag{9.44}$$

and

$$d^\ell_{m,m'}(\vartheta) = \sum_n (-1)^{n-m'+m} \frac{\sqrt{(\ell+m')!\,(\ell-m')!(\ell+m)!\,(\ell-m)!}}{(\ell+m'-n)!\,(n)!\,(\ell-n-m)!\,(n-m'+m)!}$$

$$\times \cos\left(\frac{\vartheta}{2}\right)^{2\ell-2n+m'-m} \sin\left(\frac{\vartheta}{2}\right)^{2n-m'+m}, \tag{9.45}$$

where the sum is over all n such that denominator terms do not become negative. That is, $\max(0, m'-m) \le n \le \min(\ell+m', \ell-m)$. Plugging (9.43)–(9.45) into (9.42) results in

$$\langle D^s_{t,t'}, Y^m_\ell \rangle = \sqrt{\frac{2\ell+1}{4\pi}}$$

$$\times \int_0^{2\pi}\int_0^\pi e^{-it\varphi}d^s_{t,t'}(\vartheta)e^{-it'\omega}e^{-im\varphi}d^\ell_{m,0}(\vartheta)\,\sin\vartheta\,d\vartheta\,d\varphi. \tag{9.46}$$

It is time to make a decision about ω. We saw that for the convolution to be commutative, $\omega = \pi - \varphi$ should be chosen. Using this value for ω and rearranging (9.46) results in

$$\langle D^s_{t,t'}, Y^m_\ell \rangle = \sqrt{\frac{2\ell+1}{4\pi}}$$

$$\times \int_0^\pi d^s_{t,t'}(\vartheta)d^\ell_{m,0}(\vartheta)\int_0^{2\pi} e^{-i(t-t'+m)\varphi}e^{-it'\pi}\,d\varphi\,\sin\vartheta\,d\vartheta, \tag{9.47}$$

where, upon orthogonality of exponentials over $[0, 2\pi]$, the inner integral is non-zero only when $t - t' + m = 0$ or only when $t' = t + m$ and the inner product simplifies to

$$\langle D^s_{t,t+m}, Y^m_\ell \rangle = 2\pi(-1)^{(m+t)}\sqrt{\frac{2\ell+1}{4\pi}}\int_0^\pi d^s_{t,t+m}(\vartheta)d^\ell_{m,0}(\vartheta)\,\sin\vartheta\,d\vartheta. \tag{9.48}$$

We now need to examine the inner product of two Wigner d-matrices. Let us first expand them individually for the given indices and simplify as much as possible. Using (9.45)

$$d^s_{t,t+m}(\vartheta) = \sum_n (-1)^{n-m}c_1(s,t,m,n)\cos\left(\frac{\vartheta}{2}\right)^{2s-2n+m}\sin\left(\frac{\vartheta}{2}\right)^{2n-m}, \tag{9.49}$$

where $c_1(s,t,m,n)$ is defined by

$$c_1(s,t,m,n) \triangleq \frac{\sqrt{(s+t+m)!\,(s-t-m)!\,(s+t)!\,(s-t)!}}{(s+t+m-n)!\,(n)!\,(s-n-t)!\,(n-m)!}$$

and

$$d^\ell_{m,0}(\vartheta) = \sum_{n'}(-1)^{n'+m}c_2(\ell,m,n')\cos\left(\frac{\vartheta}{2}\right)^{2\ell-2n'-m}\sin\left(\frac{\vartheta}{2}\right)^{2n'+m}, \qquad (9.50)$$

where $c_2(\ell,m,n')$ is defined by

$$c_2(\ell,m,n') \triangleq \frac{\sqrt{(\ell)!\,(\ell)!\,(\ell+m)!\,(\ell-m)!}}{(\ell-n')!\,(n')!\,(\ell-n'-m)!\,(n'+m)!}.$$

Therefore, the integral in (9.48) is expanded as

$$\int_0^\pi d^s_{t,t+m}(\vartheta)d^\ell_{m,0}(\vartheta)\sin\vartheta\,d\vartheta = \sum_n\sum_{n'}(-1)^{(n+n')}c_1(s,t,m,n)\,c_2(\ell,m,n')$$

$$\times \int_0^\pi \cos\left(\frac{\vartheta}{2}\right)^{2s+2\ell-2n-2n'}\sin\left(\frac{\vartheta}{2}\right)^{2n+2n'}\sin\vartheta\,d\vartheta. \quad (9.51)$$

Upon a change of variable $\cos(\vartheta/2)=x$, the integral becomes

$$\int_0^\pi \cos\left(\frac{\vartheta}{2}\right)^{2s+2\ell-2n-2n'}\sin\left(\frac{\vartheta}{2}\right)^{2n+2n'}\sin\vartheta\,d\vartheta \qquad (9.52)$$

$$= 4\int_0^1 x^{(2s+2\ell-2n-2n'+1)}(1-x^2)^{(n+n')}dx$$

$$= 2\frac{(s+\ell-n-n')!\,(n+n')!}{(s+\ell+1)!} \triangleq c_3(s,\ell,n,n').$$

Substituting all the results back in (9.48), we obtain

$$\langle D^s_{t,t+m},Y^m_\ell\rangle = 2\pi(-1)^{(m+t)}\sqrt{\frac{2\ell+1}{4\pi}}\sum_n\sum_{n'}(-1)^{(n+n')}$$

$$\times c_1(s,t,m,n)\,c_2(\ell,m,n')\,c_3(s,\ell,n,n'). \quad (9.53)$$

Finally, the spherical harmonic coefficient of commutative convolution output is presented in the following theorem.

Theorem 9.4 (Commutative anisotropic convolution). *The convolution between filter h and signal f, defined as*

$$(h\odot f)(\vartheta,\varphi) \triangleq \int_{\mathbb{S}^2}(\mathcal{D}(\varphi,\vartheta,\pi-\varphi)h)(\widehat{\boldsymbol{u}})f(\widehat{\boldsymbol{u}})\,ds(\widehat{\boldsymbol{u}}),$$

has the following spherical harmonic coefficient,

$$\langle h\odot f,Y^m_\ell\rangle = \sqrt{\pi(2\ell+1)}(-1)^m\sum_{s,t}(f)^{-t}_s(h)^{t+m}_s\sum_n\sum_{n'}(-1)^{(n+n')}$$

$$\times c_1(s,t,m,n)\,c_2(\ell,m,n')\,c_3(s,\ell,n,n'),$$

where $(f)^{-t}_s = \langle f,Y^{-t}_s\rangle$ and $(h)^{t+m}_s = \langle h,Y^{t+m}_s\rangle$. Note that $(h)^{t+m}_s$ is only non-zero when $-s\le t+m\le s$ and $(f)^{-t}_s$ is only non-zero when $-s\le -t\le s$

. *Therefore, the range of valid t is* $\max(-s, -s - m) \le t \le \min(s, s - m)$. *Moreover the constants c_1 to c_3 are given by*

$$c_1(s, t, m, n) \triangleq \frac{\sqrt{(s+t+m)!\,(s-t-m)!\,(s+t)!\,(s-t)!}}{(s+t+m-n)!\,(n)!\,(s-n-t)!\,(n-m)!}$$

$$c_2(\ell, m, n') \triangleq \frac{\sqrt{(\ell)!\,(\ell)!\,(\ell+m)!\,(\ell-m)!}}{(\ell-n')!\,(n')!\,(\ell-n'-m)!\,(n'+m)!}$$

$$c_3(s, \ell, n, n') \triangleq 2\frac{(s+\ell-n-n')!\,(n+n')!}{(s+\ell+1)!}.$$

The range of summations over n and n' is given by $\max(0, m) \le n \le \min(s - t, s+t+m)$ and $\max(0, -m) \le n' \le \min(\ell, \ell-m)$, respectively, with the further condition that $n + n' \le s + \ell$.   □

Remark 9.5. By some algebraic manipulations and using additional properties of Wigner d-matrices such as

$$d^\ell_{m,m'}(\vartheta) = (-1)^{m'-m} d^\ell_{m',m}(\vartheta) = d^\ell_{-m',-m}(\vartheta),$$

one can verify from the spectral analysis of the convolution that it is indeed commutative. Of course, the proof of commutativity in spatial domain presented in Section 9.6.3 is much more clear and conceptually valuable.   □

Remark 9.6. We provide some remarks on the computational complexity of commutative convolution. Let L_f and L_h denote the maximum spectral degree of f and h, respectively. According to (9.40), the maximum spectral degree of the convolution output is $L_g = \min(L_f, L_h)$. Direct computation of (9.38) on an equiangular grid along the co-latitude and longitude requires sampling of the filter and signal on the same grid and its computational complexity will be $O(N^4)$ where $N \ge \max(L_f, L_h)$. On the other hand, if we try to adapt FFT-based efficient techniques in (Wandelt and Górski, 2001) originally proposed for type 3 (anisotropic) convolution for computing commutative convolution, the complexity will be $O(L^4 + N^3 \log N)$. In this approach, the convolution output is first (unnecessarily) sampled over all Euler angles φ, ϑ, and ω and is then specialized to $\omega = \pi - \varphi$. Therefore, there is redundancy in this type of computation. A faster approach is proposed in (Khalid et al., 2012b), where $\omega = \pi - \varphi$ is used in (7.153), p. 240 for the decoupling of terms in (9.40) which can reduce the computational complexity to $O(LN^2 \log N)$. This reduction is not possible in the general type 3 convolution,.   □

9.6.6   Operator matrix elements

It is worthwhile to identify the operator matrix elements corresponding to the commutative convolution. For this purpose, we let $f(\hat{\boldsymbol{u}}) = Y^q_p(\hat{\boldsymbol{u}})$ with $(f)^{-t}_s =$

$\langle Y_p^q, Y_s^{-t} \rangle = \delta_{s,p} \delta_{t,-q}$ in the result of Theorem 9.4 to obtain

$$b_{\ell,p}^{m,q} = \langle h \odot Y_p^q, Y_\ell^m \rangle = \sqrt{\pi(2\ell+1)}(-1)^m (h)_p^{-q+m} \sum_n \sum_{n'} (-1)^{(n+n')}$$

$$\times c_1(p,-q,m,n)\, c_2(\ell,m,n')\, c_3(p,\ell,n,n'). \quad (9.54)$$

The operator kernel is accordingly found using (7.89), p. 219.

Example 9.4. It is interesting to examine the structure of the operator matrix and understand how it may mix different degrees and orders. For this purpose, let us assume that filter h is strictly bandlimited to degree $L_h = 2$ and is identical to 1 inside the band. That is,

$$(h)_\ell^m = \begin{cases} 1 & \ell \in \{0,1,\ldots,L_h\},\ m \in \{-\ell,\ldots,\ell\}, \\ 0 & \text{otherwise.} \end{cases}$$

According to (9.54), this poses an immediate restriction on input degrees $p \leq L_h$. Also, $-p \leq -q + m \leq p$ or $-p + q \leq m \leq p + q$. That is, only very limited output orders m can be present for any output degree ℓ. The result of evaluating the first 16×9 elements of the operator matrix is presented below, where the elements with absolute value lower than 0.0001 are set to zero:

$$\mathbf{B} = \begin{pmatrix}
3.5449 & 1.7725 & 0 & 1.7725 & 1.1816 & -0.5908 & 0 & -0.5908 & 1.1816 & \cdots \\
0 & 2.0467 & 2.0467 & 0 & 1.4472 & 0 & 0 & 1.4472 & 0 & \cdots \\
0 & 1.0233 & 2.0467 & 1.0233 & 1.0233 & 1.0233 & 0 & 1.0233 & 1.0233 & \cdots \\
0 & 0 & 2.0467 & 2.0467 & 0 & 1.4472 & 0 & 0 & 1.4472 & \cdots \\
0 & 1.6180 & 0 & 0 & 1.5853 & 0.9708 & 1.5853 & 0 & 0 & \cdots \\
0 & 0 & 0 & 0 & 0.6472 & 1.5853 & 1.5853 & 0.6472 & 0 & \cdots \\
0 & 0 & 0 & 0 & 0.2642 & 1.0569 & 1.5853 & 1.0569 & 0.2642 & \cdots \\
0 & 0 & 0 & 0 & 0 & 0.6472 & 1.5853 & 1.5853 & 0.6472 & \cdots \\
0 & 0 & 0 & 1.6180 & 0 & 0 & 1.5853 & 0.9708 & 1.5853 & \cdots \\
0 & 0 & 0 & 0 & 1.3981 & 1.3981 & 0 & 0 & 0 & \cdots \\
0 & -0.8562 & 0 & 0 & 0 & 0.8562 & 0 & 0 & 0 & \cdots \\
0 & 0 & 0 & 0 & 0 & 0 & 0 & 0 & 0 & \cdots \\
0 & 0 & 0 & 0 & 0 & 0 & 0 & 0 & 0 & \cdots \\
0 & 0 & 0 & 0 & 0 & 0 & 0 & 0 & 0 & \cdots \\
0 & 0 & 0 & -0.8562 & 0 & 0 & 0 & 0.8562 & 0 & \cdots \\
0 & 0 & 0 & 0 & 0 & 0 & 0 & 1.3981 & 1.3981 & \cdots \\
\vdots & \vdots & \vdots & \vdots & \vdots & \vdots & \vdots & \vdots & \vdots & \ddots
\end{pmatrix}.$$

This operator matrix is reasonably sparse, but it clearly mixes different degrees. Columns to the right of what is shown (where $p > L_h$) are all identical to zero. However, rows below what is shown can be in general non-zero subject to $-p + q \leq m \leq p + q$. $\qquad\square$

9.6.7   Special case — one function is azimuthally symmetric

Let $h(\widehat{\boldsymbol{u}})$ be an azimuthally symmetric function, which implies $(h)_s^t = \langle h, Y_s^t \rangle = 0$ for $t \neq 0$. For this special case, we can write (9.28) with $\omega = \pi - \varphi$ as

$$\left(h \odot f\right)(\vartheta, \varphi) = g_{\pi - \varphi}(\vartheta, \varphi) = \sum_{s,t}(-1)^t d_{t,0}^s(\vartheta) e^{-it\varphi}(f)_s^{-t}(h)_s^0, \qquad (9.55)$$

which can be expressed using relation (9.43) and (9.44) as

$$\left(h \odot f\right)(\vartheta, \varphi) = \sum_{s,t}(-1)^t \sqrt{\frac{4\pi}{2s+1}}\, \overline{Y_s^t(\vartheta, \varphi)}(f)_s^{-t}(h)_s^0. \qquad (9.56)$$

Using conjugate symmetry and orthonormal property of spherical harmonics, we obtain

$$\langle h \odot f, Y_\ell^m \rangle = \sqrt{\frac{4\pi}{2\ell+1}}(h)_\ell^0(f)_\ell^m, \qquad (9.57)$$

which is also equal to $\langle f \odot h, Y_\ell^m \rangle$. By commutativity, a simplified form of multiplication in spherical harmonic domain results if either the signal (nominally f) or filter (nominally h) is azimuthally symmetric.

In comparison, type 1 convolution proposed in (Driscoll and Healy, 1994) is not commutative and from (9.8) we observe

$$\langle h \circledcirc f, Y_\ell^m \rangle = \sqrt{\frac{4\pi}{2\ell+1}}(h)_\ell^0(f)_\ell^m,$$

$$\langle f \circledcirc h, Y_\ell^m \rangle = \sqrt{\frac{4\pi}{2\ell+1}}(f)_\ell^0(h)_\ell^0 \delta_{m,0},$$

where $h(\widehat{\boldsymbol{u}})$ is an azimuthally symmetric function.

9.7   Alt-azimuth anisotropic convolution on 2-sphere

9.7.1   Background

We consider another convolution on the 2-sphere, which can be relevant for some applications. In Figure 9.4 we show a radio telescope mounted on an alt-azimuth mount. We could also call this a tilt-and-pan mechanical system.

In alt-azimuth anisotropic convolution, the initial rotation of the filter by ω around the z-axis is set to zero. Therefore, the filter goes through altitude change by ϑ and then azimuth change by φ and then is convolved with the signal. This is shown in Figure 9.5. Mathematically, the output of alt-azimuth or tilt-and-pan convolution is

$$g_0(\vartheta, \varphi) = \int_{\mathbb{S}^2}\left(\mathcal{D}(\varphi, \vartheta, 0)h\right)(\widehat{\boldsymbol{u}})f(\widehat{\boldsymbol{u}})\,ds(\widehat{\boldsymbol{u}}).$$

Its output in spherical harmonic domain can be found in a similar manner to Section 9.6.5. In going from (9.46) to (9.47), we should set $\omega = 0$ and as a result $t = -m$ should be chosen, but t' remains an unconstrained variable and can take any value between $-t$ and t. See Problem 9.9 below.

Figure 9.4 The rotations in an alt-azimuth or tilt-and-pan convolution correspond to the extrinsic rotations through two axes of an alt-azimuth mount here displayed with one antenna of the Very Large Array (VLA) radio telescope. The movement mathematically corresponds to a first rotation through co-latitude ϑ ("tilt"), rotation about the y-axis, followed by a second rotation through longitude φ ("pan"), rotation about the z-axis. Because of lack of commutativity this rotation differs from the more widely known "pan-tilt" naming convention used for similar mechanical systems, which actually refers to a set of intrinsic rotations.

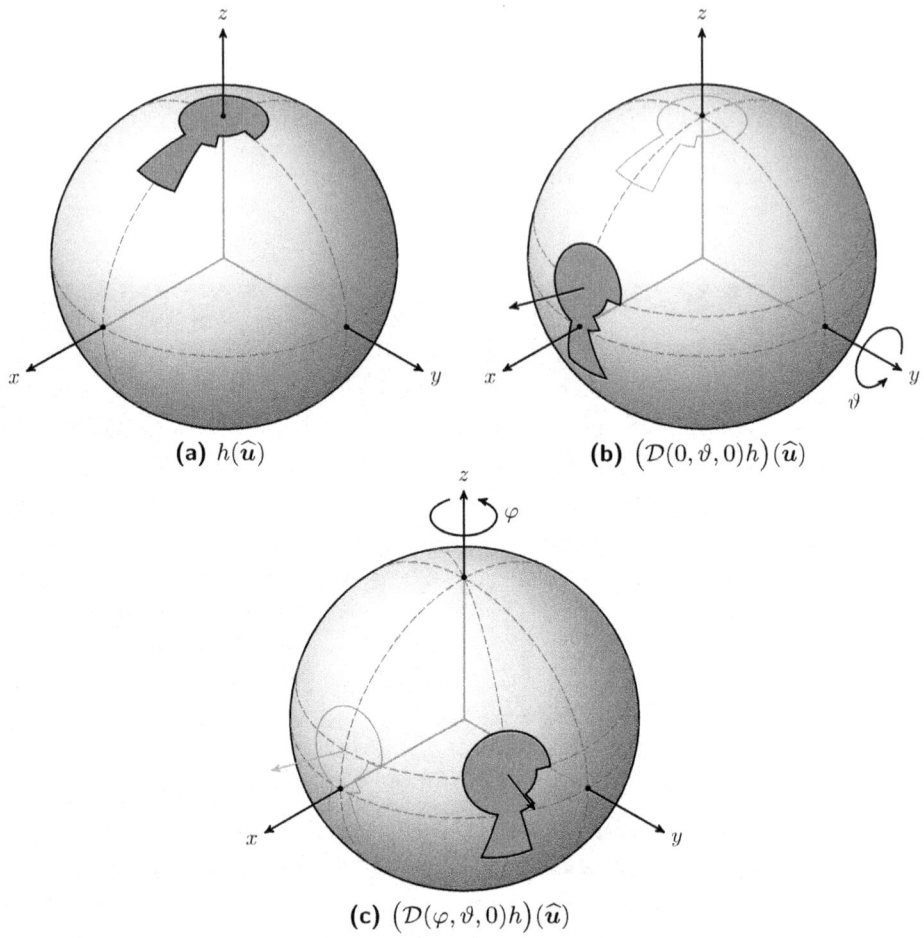

(a) $h(\widehat{\boldsymbol{u}})$ **(b)** $\big(\mathcal{D}(0,\vartheta,0)h\big)(\widehat{\boldsymbol{u}})$

(c) $\big(\mathcal{D}(\varphi,\vartheta,0)h\big)(\widehat{\boldsymbol{u}})$

Figure 9.5 A schematic of alt-azimuth or tilt-and-pan convolution.

Problem

9.9. Determine the spherical harmonic coefficient, $\langle g_0, Y_\ell^m \rangle$, for the alt-azimuth convolution in terms of the spherical harmonic coefficients of signal $f(\widehat{\boldsymbol{u}})$ and filter $h(\widehat{\boldsymbol{u}})$, and simplify as much as possible.

10 Reproducing kernel Hilbert spaces

10.1 Background to RKHS

Reproducing kernel Hilbert spaces (RKHS) are a class of separable Hilbert spaces formulated by Polish mathematician Nachman Aronszajn (1907–1980) (Aronszajn, 1950), but also attributed to fellow countryman Stefan Bergman (1895–1977). They are *function* spaces, that is, the vectors are functions whose domain is the bounded closed interval $\Omega = [a, b]$.

From one perspective all separable complex Hilbert spaces of Hilbert dimension $\aleph_0$ are structurally the same according to the Hilbert space isomorphism; see Theorem 2.17 (p. 99 and surrounding material).[1] This raises some questions:

Q1: Given that a reproducing kernel Hilbert space is a separable function space on domain Ω, how can it be significantly different from $L^2(\Omega)$ or, for that matter, significantly different from ℓ^2?

Q2: What is the relationship between a reproducing kernel Hilbert space on domain Ω and the space of square integrable functions on domain Ω, $L^2(\Omega)$?

A partial answer is that generally one has *applications* in mind and the choice of having a function space versus a sequence (infinite indexed sequence space such as ℓ^2) comes down to *tuning the representation of the elements* of the Hilbert space to match some intended application. As indicated above, in this case we want the representation of the elements to be functions. So we shall focus on the relationship between the functions in $L^2(\Omega)$ and the functions in the reproducing kernel Hilbert space. However, we are not going to jump into the definition of reproducing kernel Hilbert spaces straight away. Instead, we are going to systematically construct a new Hilbert space representation built on a Fourier characterization of $L^2(\Omega)$. This leads to a class of Hilbert spaces more general than reproducing kernel Hilbert spaces, which we call Fourier weighted Hilbert spaces, but sharing a lot of the algebraic structure, as well as the inherited Hilbert geometric structure.

In the first part of this chapter we use the terminology "Fourier weighted Hilbert space" as a surrogate for reproducing kernel Hilbert space until we nail down some additional mandatory conditions which are more technical in nature.

---

[1] In what follows, when we say Hilbert space we mean complex separable Hilbert space. Real separable Hilbert spaces can also be considered in an analogous way.

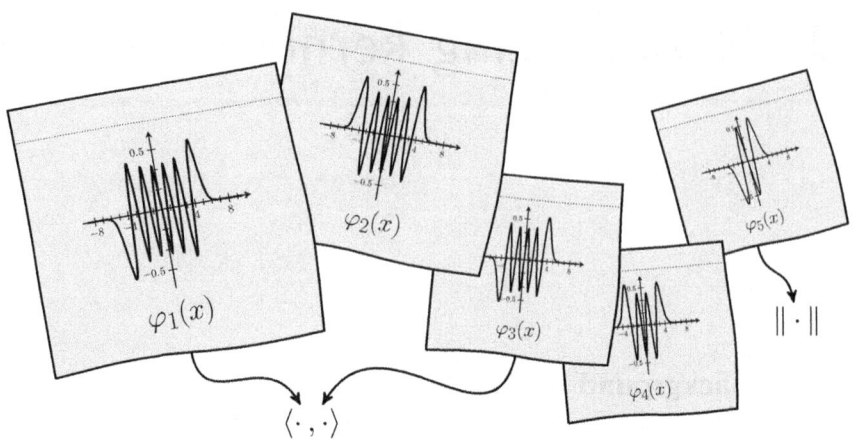

Figure 10.1 Sticky-note interpretation of vectors in a Hilbert space and the use of functions only as labels for abstract Hilbert space vectors.

10.1.1   Functions as sticky-note labels

One abstract way to think of functions is as the contents of sticky-note labels attached to each vector in the separable complex Hilbert space (of Hilbert dimension $\aleph_0$). This is depicted in Figure 10.1. In the figure a sticky-note adorned with a representation of the vector is attached to each abstract vector in the Hilbert space. In this case they are adorned with functions. Functions can be useful for computing the norm and inner product through the appropriate integral expressions, but the values for these norms and inner products really only depend on the abstract vector and not the function label. That is, if the sticky-notes were unmarked then the norms and inner products would still exist. Any two labeled vectors can be combined to compute a complex number through an inner product and each labeled vector has a norm. Esoteric spaces such as reproducing kernel Hilbert spaces are just another labeling system as far as abstract Hilbert spaces are concerned.

If we were only concerned with the Hilbert space then using functions is simply an aid to help our limited brains make sense of what is going on.[2] Of course, because functions can be used as labels then as engineers and physicists we get excited[3] because we can imagine the functions as representing signals or physical waveforms, etc.

The term "reproducing kernel Hilbert space" conjures up something exotic and deep, a terminology suitable for impressing guests at a cocktail party. However, armed with the isomorphism concept, a connection between $L^2(\Omega)$ and a Fourier weighted Hilbert space, which we denote by $\mathcal{H}_\lambda$ (subsequently with additional conditions this will be the RKHS and denoted $\mathcal{H}_K$), must exist. This is shown in Figure 10.2, which is just a re-casting of Figure 2.12, p. 100. The reason for

---

[2]   Our brains are even more profoundly limited than we realize because converting inner product computations to integration does not make things easier.

[3]   It doesn't take much to get engineers and physicists excited, which is a sad reflection on their dull lives.

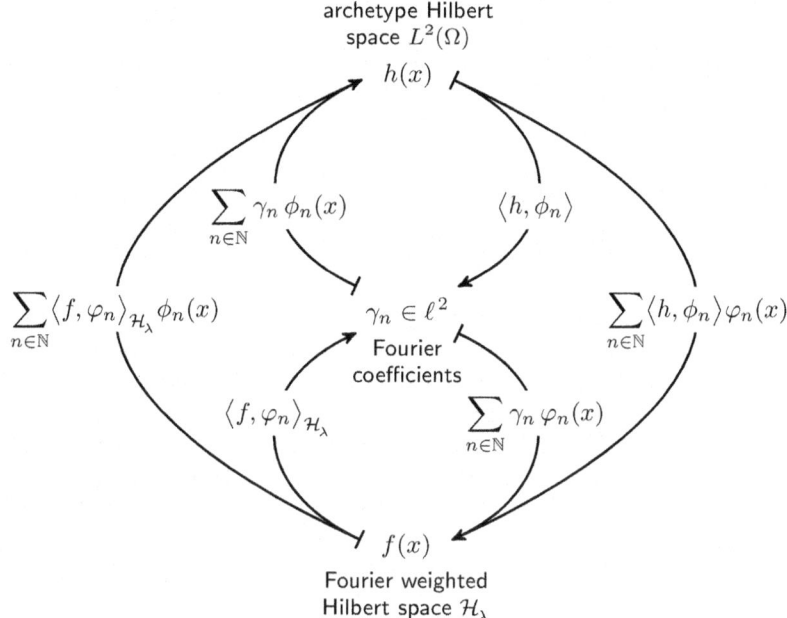

Figure 10.2 Isomorphic separable Hilbert spaces $L^2(\Omega)$, ℓ^2 and $\mathcal{H}_\lambda$. In establishing the isomorphism we have shown a set of Fourier coefficients, γ_n, which are shared between the two Hilbert spaces. In essence $\gamma_n = \langle h, \phi_n \rangle = \langle f, \varphi_n \rangle_{\mathcal{H}_\lambda}$ where $\langle \cdot, \cdot \rangle$ is the usual inner product in $L^2(\Omega)$, and $\langle \cdot, \cdot \rangle_{\mathcal{H}_\lambda}$ is an inner product for the Fourier weighted Hilbert space defined later in Section 10.3.

seeking an isomorphism with $L^2(\Omega)$ is that we are targeting function spaces and $L^2(\Omega)$ is the standard or archetypal function space. The isomorphism property is generally proven by use of an intermediate canonical ℓ^2 space and this space is almost always useful. That is, all three spaces prove useful. In Figure 10.2, $\mathcal{H}_\lambda$ represents the Fourier weighted Hilbert space, $L^2(\Omega)$ is the archetypal function space, and ℓ^2 is the space of Fourier coefficients. There are two sets of orthonormal functions indicated in the diagram, which alludes to the property that there exists a complete orthonormal sequence of functions, shown as $\{\varphi_n\}_{n \in \mathbb{N}}$ for $\mathcal{H}_\lambda$. For $L^2(\Omega)$ we have the complete orthonormal functions denoted $\{\phi_n\}_{n \in \mathbb{N}}$.[4] The isomorphism, as described, simply changes the label of the Hilbert space vectors and not the vectors themselves.

Remark 10.1 (Fourier weighted Hilbert space as function labels). To this point we have only assumed the Fourier weighted Hilbert space (subsequently RKHS), $\mathcal{H}_\lambda$, is separable, but already some of the mystery is reduced. If $L^2(\Omega)$ is a fancy function labeling system for the vectors in the Hilbert space, then also $\mathcal{H}_\lambda$ is another function labeling system.   □

---

[4] Previously in the book, primarily in Chapter 6, p. 160, we used the symbol φ to denote general complete orthonormal sequences and ϕ to denote complete orthonormal eigenfunction sequences (in a well-defined sense the best specific choice of complete orthonormal sequence). Our treatment here will prove to be consistent.

In Part I of this book, we have seen how to construct a variety of "different" function spaces when we looked at various classical orthogonal functions, which were generally defined on different domains. If *we stick to one specific domain*, Ω, then we have seen two ways to create new Hilbert space representations. These are:

1: forming *subspaces* of $L^2(\Omega)$, Section 2.6, p. 63;

2: tinkering with the inner product by introducing *weighting or other related modifications*, Section 2.4.7, p. 50.

Fourier weighted Hilbert spaces (subsequently RKHS) broadly fit in the latter category and, generally do not fit in the former subspace category.

We shall take a *constructive* approach in describing a Fourier weighted Hilbert space. Somewhat implicit in the above, we want to have a Hilbert space with some *additional external structure to enrich it* to make it more interesting for applications. We want the vectors in the Hilbert space be constrained in some way to be functions with additional desirable properties. Given the intention is to find a new function space, then we need to clarify what it is about $L^2(\Omega)$ that we do not like because otherwise, for most applications, it generally seems to suffice.

Problem

10.1. A description was given describing functions as the contents of sticky-note labels on a separable Hilbert space. In general it would be unpleasant to have to place a sticky-note on every function. How would Cantor and Hilbert propose to reduce the amount of work here?

10.1.2   What is wrong with $L^2(\Omega)$?

The space $L^2(\Omega)$ has been studied earlier in the book and is often presented as the archetypal Hilbert space where the vectors are functions. As a function space it is extremely useful because it deals with *finite energy functions or signals*. But $L^2(\Omega)$ is a weird space — weirder than what is really intended. In a technical sense the vectors in $L^2(\Omega)$ are not individual functions, but equivalence classes where equivalence means functions are equal almost everywhere (a.e.); see (2.24), p. 49 and surrounding material. Even when we start with a complete orthonormal sequence of functions which are themselves "infinitely smooth" (such as the complex exponentials and other standard orthogonal functions; see Section 2.5, p. 52), infinite series formed from them can be discontinuous. So the space contains the functions we expect and want to work with plus additional weird functions we generally have little interest in (and may have difficulty conceptualizing). It seems this is the price to be paid if we want to have completeness in a function space. But, somewhat surprisingly, it need not be the case, as we should explore.

Summarizing briefly the problems, $L^2(\Omega)$ ends up having weird vectors for two reasons:

1: the norm cannot discriminate between two pointwise different functions if they are equal almost everywhere (a.e.), which means the elements of $L^2(\Omega)$ are not actually functions, but equivalence classes of functions;

2: insisting on Cauchy sequences converging implies we must include discontinuous functions into the picture even if we start with a complete orthonormal sequence of "infinitely smooth" functions.

10.2   Constructing Hilbert spaces from continuous functions

Let us set about trying to construct an infinite-dimensional separable Hilbert space where all the elements are *functions* with some degree of smoothness and are at least continuous. We need a definition to be more precise.

Definition 10.1 (Class C^k and smooth functions). *A function f is said to be of class C^k if all the derivatives $f', f'', \ldots, f^{(k)}$ exist and are continuous (the continuity is automatic for all the derivatives except for $f^{(k)}$). The function f is said to be of class C^∞, or smooth, if it has derivatives of all orders. For continuous functions C^0 can be written as C.* □

So here we are initially expecting the functions to be in at least $C(\Omega)$ (at least continuous on domain Ω), a property that is not present for most "functions" in $L^2(\Omega)$. We should also understand that what we are seeking is a more specific (less general) and more structurally constrained type of Hilbert space. One motivation is a belief that in the real world we do not expect to see signals with unnatural abrupt changes including what we see in the step function and Dirac delta function (impulse).

Remark 10.2 (Why infinite dimensional?). The reason for seeking an *infinite-dimensional* Hilbert space rather than a finite-dimensional one when the objective is smoothness should be clear. When forming complete orthonormal sequences, the constituent elements are not just continuous, but often smooth, that is, in C^∞. An example is the set of complex exponentials in classical Fourier series. If there are only a finite number of dimensions, the superposition of a finite number of C^∞ functions is in C^∞. Therefore, the problem is only non-trivial when we consider infinite-dimensional Hilbert spaces. □

10.2.1   Completing continuous functions

First Attempt

The first attempt to construct a suitable Hilbert space of functions is to start with the space of continuous functions $C(\Omega)$. This can be made a Banach space by picking any norm and completing the space; Section 2.3.6, p. 35. The complete space has the property that all Cauchy sequences converge and the limits generally include functions introduced in the completion process. If we choose the sup-norm

$$\|f\|_{\mathrm{sup}} = \sup_{x\in\Omega}\bigl|f(x)\bigr|,$$

then this can be shown to yield a complete space without the completion step being necessary. That is, all Cauchy limits using the sup-norm are continuous functions. However, the sup-norm does not satisfy the parallelogram law (Theorem 2.2, p. 39) and so continuous functions cannot be extended to a Hilbert space with this norm. To be more clear, this means there is no inner product that induces the sup-norm. This fails our construction objective to get a Hilbert space, but at least as a Banach space all the vectors are continuous functions.

Second attempt

What about other norms? Start again with the space of continuous functions $C(\Omega)$ and consider the standard $L^2(\Omega)$-norm this time.[5] When we complete this space, this yields the standard $L^2(\Omega)$ Banach space, but it includes non-continuous functions (or equivalence classes of a.e.-equal non-continuous functions) and, thus, not all functions are in $C(\Omega)$. This norm does satisfy the parallelogram law and so we can construct an inner product from the norm, which is the usual inner product. Therefore, completing continuous functions with a standard finite energy norm yields the Hilbert space we are trying to avoid, that is, $L^2(\Omega)$.

Synopsis

So we need to be a bit more clever in our construction (or possibly what we are seeking is not possible). In a sense $L^2(\Omega)$ seems to be the only option (because we need a norm to satisfy the parallelogram law) and what we should really target is a limited subclass (but not subspace) of the finite energy functions that constitute $L^2(\Omega)$, so as to meet our smoothness/continuity objective.

Problem
_____

10.2. Show that for the sup-norm, continuous functions cannot be extended to a Hilbert space with this norm.

_____

10.3   Fourier weighted Hilbert spaces

Here we subject $L^2(\Omega)$ to surgery and tinker with its inner product workings, Frankenstein-style, to create a modified Hilbert space. We start with an arbitrary complete orthonormal sequence. We note that the rate of convergence of a Fourier series when using this complete orthonormal sequence in some way encodes the degree of smoothness of the function it represents. Functions with slower decaying Fourier series tend to be more discontinuous, and functions with faster decaying Fourier series tend to be smoother. The $L^2(\Omega)$ inner product is then modified to penalize functions whose Fourier series converges too slowly such that the induced norm under the modified inner product is infinite — this penalty implies they are excluded from the modified Hilbert space. The goal is to find an inner product modification that excludes all discontinuous functions.

_____

[5] We know this is a folly.

10.3.1   Pass the scalpel, nurse

Recall in $L^2(\Omega)$ the inner product satisfies the generalized Parseval relation, from Definition 2.22, p. 70,

$$\langle f, g \rangle = \int_\Omega f(x)\overline{g(x)}\,dx = \sum_{n \in \mathbb{N}} (f)_n \overline{(g)_n}, \tag{10.1}$$

where

$$(f)_n \triangleq \langle f, \phi_n \rangle = \int_\Omega f(x)\overline{\phi_n(x)}\,dx, \tag{10.2}$$

$$(g)_n \triangleq \langle g, \phi_n \rangle = \int_\Omega g(x)\overline{\phi_n(x)}\,dx \tag{10.3}$$

are the Fourier coefficients with respect to some arbitrary choice of complete orthonormal sequence in $L^2(\Omega)$:

$$\{\phi_n\}_{n \in \mathbb{N}}, \quad \text{where } \langle \phi_p, \phi_q \rangle = \delta_{p,q}, \quad \forall p, q \in \mathbb{N},$$

over the domain Ω. (The specific choice of complete orthonormal sequence is not relevant at this stage and any choice can be taken as suitable. Once chosen, it is important to regard it as fixed for the subsequent analysis.) Then for a function f to be in $L^2(\Omega)$ we have

$$\boxed{f \in L^2(\Omega) \iff \|f\|^2 = \langle f, f \rangle = \sum_{n \in \mathbb{N}} |(f)_n|^2 < \infty.} \tag{10.4}$$

Before we operate, we need to review the nature of the Fourier series of functions in $L^2(\Omega)$. We can look at an example of a discontinuous function in $L^2(\Omega)$ to see at what rate its Fourier series converges.

Example 10.1 (A discontinuous function). Let $\Omega = [-\pi, \pi]$ and consider $L^2(\Omega)$ with inner product (10.1). Choose the standard complex exponential Fourier basis:

$$\phi_k(x) \triangleq \frac{e^{ikx}}{\sqrt{2\pi}}, \quad k \in \mathbb{Z}.$$

These form a complete orthonormal sequence for functions in $L^2(\Omega)$. Here we have indexed with doubly infinite, but still countable $k \in \mathbb{Z}$.

Consider

$$f(x) \triangleq \mathbf{1}_{[-\pi/2, \pi/2]}(x), \quad -\pi \le x \le \pi,$$

which is shown in Figure 10.3. Function f has Fourier coefficients, (10.2), given by

$$(f)_k = \langle f, \phi_k \rangle$$

$$= \int_{-\pi}^{\pi} \mathbf{1}_{[-\pi/2, \pi/2]}(x)\frac{e^{-ikx}}{\sqrt{2\pi}}\,dx$$

$$= 2\int_0^{\pi/2} \frac{\cos kx}{\sqrt{2\pi}}\,dx$$

$$= \sqrt{\frac{2}{\pi}}\frac{\sin(k\pi/2)}{k},$$

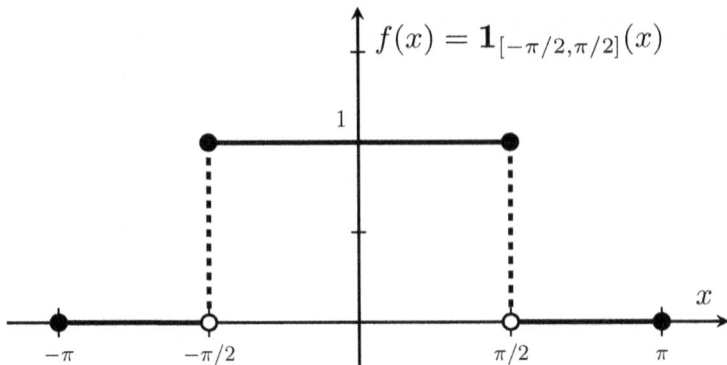

Figure 10.3 Discontinuous function in Example 10.1.

and it can be synthesized, in the sense of convergence in the mean, as

$$f(x) = \sum_{k=-\infty}^{\infty} (f)_k \phi_k(x)$$

$$= \sum_{k=-\infty}^{\infty} \frac{\sin(k\,\pi/2)}{k\pi} e^{ikx}.$$

Considering the $\Omega = [-\pi, \pi]$ domain, the energy of $f(x)$ is given by

$$\|f\|^2 = \pi,$$

since $f(x) = 1$ in $|x| \leq \pi/2$ and is zero elsewhere. In the Fourier coefficient domain, the sequence

$$\{(f)_k\}_{k\in\mathbb{Z}} = \sqrt{\frac{2}{\pi}} \times \Big\{ \dots, 0, \frac{1}{5}, 0, -\frac{1}{3}, 0, 1, \underbrace{\frac{\pi}{2}}_{k=0 \text{ term}}, 1, 0, -\frac{1}{3}, 0, \frac{1}{5}, 0, \dots \Big\}$$

has energy, in the ℓ^2-norm, expressed as

$$\|\{(f)_k\}_{k\in\mathbb{Z}}\|^2 = \sum_{k=-\infty}^{\infty} |(f)_k|^2 = \frac{\pi}{2} + \frac{4}{\pi} \underbrace{\Big(\frac{1}{1^2} + \frac{1}{3^2} + \frac{1}{5^2} + \cdots \Big)}_{\pi/8} = \pi,$$

which is equal to the previously computed energy Ω-domain energy $\|f\|^2$ by the Parseval relation.   □

Of interest here, for later consideration, is the nature of the decay of the Fourier coefficients and the terms in the series for $\|f\|^2$ as $|k| \to \infty$. The former are like $O(1/k)$ and the latter are like $O(1/k^2)$ as $k \to \infty$. We claim that the discontinuity of the function has an impact on the convergence rate of the Fourier series. That is, the discontinuity gives a low decay rate of the Fourier coefficients. In Table 10.1 we tabulate some characteristics of functions and associated Fourier series decay rate that draw from classical Fourier series analysis. The insight here is that we can penalize and thereby exclude less smooth functions by insisting the asymptotic rate of decay to zero is faster than some threshold rate.

Table 10.1 Function characteristics and the decay rates of complex exponential Fourier series coefficients, $(f)_k$. (Here $k \in \mathbb{Z}$ following Example 10.1, but more generally we would prefer to use $n \in \mathbb{N}$ and $(f)_n$.) Also note that a function being analytic implies it is in class C^∞, but not vice versa.

Characteristic	$(f)_k$ decay rate	Class
impulses (Dirac)	1	not continuous
step functions	$1/k$	not continuous
ramp functions	$1/k^2$	class C
piecewise quadratic	$1/k^3$	class C^2
piecewise cubic (spline)	$1/k^4$	class C^3
analytic	ρ^k for $|\rho| < 1$	within class C^∞

10.3.2   Forming a new inner product

For what follows we will make comparison with the structure of the inner product, from (10.1), written in the following form:

$$\langle f, g \rangle = \sum_{n \in \mathbb{N}} (f)_n \overline{(g)}_n. \tag{10.5}$$

For functions f, g on domain Ω with Fourier coefficients (10.2) and (10.3), consider a new candidate inner product

$$\langle f, g \rangle_{\mathcal{H}_\lambda} \triangleq \sum_{n \in \mathbb{N}} \frac{(f)_n \overline{(g)}_n}{\lambda_n}, \tag{10.6}$$

which introduces a real bounded weighting sequence

$$\lambda \triangleq \{\lambda_n\}_{n \in \mathbb{N}}, \quad \text{where} \quad \lambda_n \in \mathbb{R} \text{ and } |\lambda_n| \le B < \infty. \tag{10.7}$$

In the notation for the inner product, $\mathcal{H}_\lambda$ denotes the candidate new Hilbert space. We shall review in a moment what restrictions might apply to this bounded weighting sequence.[6]

Remark 10.3. In (10.6) with $\lambda_n = 1$, $\forall n \in \mathbb{N}$, then we just have the $L^2(\Omega)$ inner product, (10.5). $\qquad\square$

If we look at (10.6) more closely we note that it is a little odd. This is because the Fourier coefficients on the right-hand side are the ones computed with respect to the $L^2(\Omega)$ inner product and not with respect to the inner product $\langle \cdot, \cdot \rangle_{\mathcal{H}_\lambda}$ on

---

[6]   Later we will be interested in the case when it is a sequence of positive values and belongs to ℓ^2.

the left-hand side. Making this more explicit, (10.6) can be written,

$$
\begin{aligned}
\langle f, g \rangle_{\mathcal{H}_\lambda} &= \sum_{n \in \mathbb{N}} \frac{\langle f, \phi_n \rangle \overline{\langle g, \phi_n \rangle}}{\lambda_n} \\
&= \sum_{n \in \mathbb{N}} \frac{\langle f, \phi_n \rangle \langle \phi_n, g \rangle}{\lambda_n},
\end{aligned}
\tag{10.8}
$$

where the inner products on the right-hand side, without subscript, are understood to be the $L^2(\Omega)$ inner products. In the formulation we require that f and g have well-defined and finite Fourier coefficients (with respect to the complete orthonormal sequence $\{\phi_n\}_{n \in \mathbb{N}}$ in $L^2(\Omega)$). Recall our objective is to have a space that includes continuous functions and excludes discontinuous (and possibly other) functions.

> **Remark 10.4.** In the earlier development we attempted to complete the space of continuous functions with various norms to get a Hilbert space of continuous functions. We failed. Now we make a leap of insight and realize that it was a mistake to think our goal Hilbert space should include *all* continuous functions. Not only should we perhaps reject discontinuous functions, but also continuous functions which contribute to synthesizing non-continuous functions under the completion process.   □

10.3.3   Inner product conditions

For (10.6) to be a valid inner product certain conditions need to be satisfied — adapted from Definition 2.11, p. 37:

$$
\langle f, g \rangle_{\mathcal{H}_\lambda} = \overline{\langle g, f \rangle_{\mathcal{H}_\lambda}}
\tag{10.9a}
$$

$$
\langle \alpha_1 f_1 + \alpha_2 f_2, g \rangle_{\mathcal{H}_\lambda} = \alpha_1 \langle f_1, g \rangle_{\mathcal{H}_\lambda} + \alpha_2 \langle f_2, g \rangle_{\mathcal{H}_\lambda},
\tag{10.9b}
$$

$$
\langle f, f \rangle_{\mathcal{H}_\lambda} \geq 0, \text{ and } \langle f, f \rangle_{\mathcal{H}_\lambda} = 0 \implies f = o,
\tag{10.9c}
$$

where $\alpha_1, \alpha_2 \in \mathbb{C}$.

We focus on the third condition, (10.9c), as this gives requirements on the weighting sequence, $\{\lambda_n\}_{n \in \mathbb{N}}$. The induced norm of any non-zero function f needs to be positive

$$
\|f\|_{\mathcal{H}_\lambda}^2 \triangleq \langle f, f \rangle_{\mathcal{H}_\lambda} = \sum_{n \in \mathbb{N}} \frac{|(f)_n|^2}{\lambda_n} > 0, \quad \forall f \neq o,
\tag{10.10}
$$

and this implies

$$
\boxed{\lambda_n > 0, \quad \forall n \in \mathbb{N}.}
\tag{10.11}
$$

Furthermore, it is clear, once we have (10.11),

$$
\langle f, f \rangle_{\mathcal{H}_\lambda} = 0 \iff (f)_n = 0, \quad \forall n \in \mathbb{N}.
$$

The right-hand side is the same as $f = o$, which can be written $f(x) = 0$ a.e. Collectively, these results give (10.9c). The validity of (10.9a) and (10.9b) is left for Problem 10.4 and Problem 10.5.

In summary, inner product candidate (10.6) does form a valid inner product provided $\lambda_n > 0,\ \forall n \in \mathbb{N}$.

10.3.4   Finite norm condition

We raise some questions:

Q1: If a function is in our Fourier weighted Hilbert space then what is the defining condition and how do we test that?

Q2: What are the general properties we can expect such a function to satisfy, and in particular, what is its relationship to $L^2(\Omega)$?

In analogy with the condition for $f \in L^2(\Omega)$, (10.4), we have the condition

$$
f \in \mathcal{H}_\lambda \iff \|f\|_{\mathcal{H}_\lambda}^2 = \langle f, f \rangle_{\mathcal{H}_\lambda} = \sum_{n \in \mathbb{N}} \frac{|(f)_n|^2}{\lambda_n} < \infty. \tag{10.12}
$$

By (10.7), if $f \in \mathcal{H}_\lambda$ then $f \in L^2(\Omega)$, but not conversely in general. So $\mathcal{H}_\lambda$ is picky about which finite $L^2(\Omega)$-energy functions are in the space and which are not in the space. This is implicit because the inner product in $\mathcal{H}_\lambda$ requires the $L^2(\Omega)$ Fourier coefficients to exist. However, explicitly this is relatively easy to show; see Problem 10.3.

Some properties

We are yet to see that $\mathcal{H}_\lambda$, with (10.11), is a Hilbert space. But we can see immediately some structure and properties:

P1: $f \in \mathcal{H}_\lambda$ means f has a finite $\mathcal{H}_\lambda$ induced norm.

P2: The inner product is a "weighted inner product" — expressed in terms of the Fourier coefficients with respect to the orthonormal sequence $\{\phi_n\}_{n \in \mathbb{N}}$ in $L^2(\Omega)$; the weights $\{\lambda_n\}_{n \in \mathbb{N}}$ are positive and bounded, (10.7).

P3: The inner product, (10.6), induces a weighted Fourier coefficient ℓ^2-norm, (10.10).

P4: Because the norm (10.10) is generally different from the $L^2(\Omega)$-norm, then $\mathcal{H}_\lambda$ in general will *not be a subspace* of $L^2(\Omega)$.

10.3. From (10.7) and (10.11) the real elements of the weighting sequence satisfy

$$0 < \lambda_n \leq B, \quad \forall n \in \mathbb{N},$$

where $B < \infty$. Show that

$$f \in \mathcal{H}_\lambda \implies f \in L^2(\Omega),$$

where the former condition is (10.12) and the latter is (10.4).

10.4. Show that the inner product given in (10.6), with the additional condition $\lambda_n > 0$, $\forall n \in \mathbb{N}$, satisfies the first requirement for an inner product (10.9a):

$$\langle f, g \rangle_{\mathcal{H}_\lambda} = \overline{\langle g, f \rangle_{\mathcal{H}_\lambda}}$$

10.5. Show that the inner product given in (10.6), with the additional condition $\lambda_n > 0$, $\forall n \in \mathbb{N}$, satisfies the second requirement for an inner product (10.9b):

$$\langle \alpha_1 f_1 + \alpha_2 f_2, g \rangle_{\mathcal{H}_\lambda} = \alpha_1 \langle f_1, g \rangle_{\mathcal{H}_\lambda} + \alpha_2 \langle f_2, g \rangle_{\mathcal{H}_\lambda},$$

where $\alpha_1, \alpha_2 \in \mathbb{C}$.

10.6. Consider the non-negative alternative to (10.11), given by

$$\lambda_n \geq 0, \ \forall n \in \mathbb{N}.$$

Suppose $\lambda_n = 0$ for some n. What modifications, if any, should be necessary in this case and what does this imply for the completeness of $\{\phi_n\}_{n \in \mathbb{N}}$? If $\lambda_n = 0$, for $n \in \{N+1, N+2, \ldots\}$, for some finite $N < \infty$, what happens in this case?

10.3.5   Setting up an isomorphism

The following is a definition of a positive operator. All operators we consider in this section are of this type.

Definition 10.2 (Positive operator). *An operator $\mathcal{L}$ on $\mathcal{H}$ is called positive if it is self-adjoint and*

$$\langle \mathcal{L}f, f \rangle \geq 0, \quad \forall f \in \mathcal{H}.$$

To indicate a positive operator we can use the notation $\mathcal{L} \geq 0$.   □

Definition 10.3 (Strictly positive operator). *An operator $\mathcal{L}$ on $\mathcal{H}$ is called strictly positive if it is self-adjoint and*

$$\langle \mathcal{L}f, f \rangle > 0, \quad \forall f \in L^2(\Omega) \quad \text{satisfying} \quad f \neq o.$$

To indicate a strictly positive operator we can use the notation $\mathcal{L} > 0$.   □

Remark 10.5. We have largely focussed on bounded operators in this book. However, in one direction the isomorphism, as an operator, can be unbounded. This is certainly true if, for example, we have a compact operator for the isomorphism in one direction because then the inverse, representing the isomorphism in the reverse direction, must be unbounded.   □

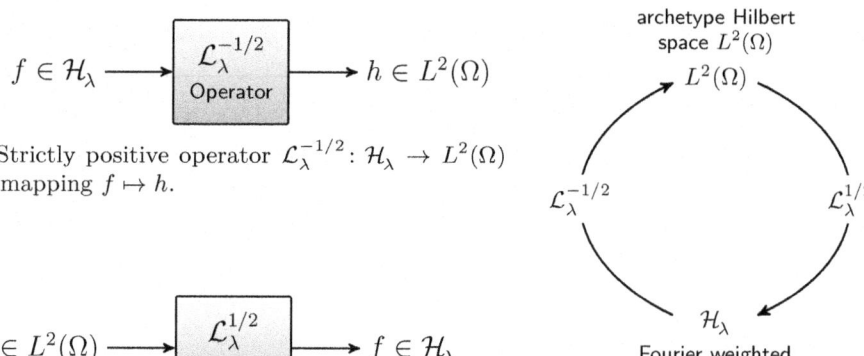

(a) Strictly positive operator $\mathcal{L}_\lambda^{-1/2} : \mathcal{H}_\lambda \to L^2(\Omega)$ mapping $f \mapsto h$.

(b) Strictly positive operator $\mathcal{L}_\lambda^{-1/2} : L^2(\Omega) \to \mathcal{H}_\lambda$ mapping $h \mapsto f$.

(c) Isomorphism diagram

Figure 10.4 Two strictly positive operators, $\mathcal{L}_\lambda^{-1/2}$ and $\mathcal{L}_\lambda^{1/2}$, and the isomorphism they define between $\mathcal{H}_\lambda$ and $L^2(\Omega)$.

Isomorphism operators

The functions that are finite in the $\mathcal{H}_\lambda$-norm, (10.10), will be shown to be related to certain finite energy functions, using the $L^2(\Omega)$-norm, through the weights $\{\lambda_n\}_{n \in \mathbb{N}}$. It is evident from (10.8) that we can distribute the positive real weight $\lambda_n > 0$ into two parts to rewrite the inner product as:

$$\langle f, g \rangle_{\mathcal{H}_\lambda} = \sum_{n \in \mathbb{N}} \frac{\langle f, \phi_n \rangle}{\sqrt{\lambda_n}} \frac{\langle \phi_n, g \rangle}{\sqrt{\lambda_n}}, \tag{10.13}$$

where we have elected to take the positive square root of $\lambda_n > 0$. The right-hand side of (10.13) involves only the $L^2(\Omega)$-norm and can be simplified with a simple trick by further distributing each $1/\sqrt{\lambda_n}$ term to the respective components of functions f and g in the ϕ_n direction. That is, we want to map $f \in \mathcal{H}_\lambda$ in the following way[7]

$$\sum_{n \in \mathbb{N}} (f)_n \phi_n(x) \longmapsto \sum_{n \in \mathbb{N}} \frac{(f)_n \phi_n(x)}{\sqrt{\lambda_n}}. \tag{10.14}$$

This motivates us to define an operator,[8]

$$\begin{aligned} \mathcal{L}_\lambda^{-1/2} : \mathcal{H}_\lambda &\longrightarrow L^2(\Omega), \\ f &\longmapsto h, \end{aligned} \tag{10.15}$$

---

[7]   On the input side of the operator we can well use $L^2(\Omega)$ (remember that $f \in \mathcal{H}_\lambda$ implies $f \in L^2(\Omega)$). But unless $f \in \mathcal{H}_\lambda$, the output can be unbounded and not in $L^2(\Omega)$. The operator is primarily being set up to be an isomorphism to map a function in $\mathcal{H}_\lambda$ to a function in $L^2(\Omega)$.

[8]   The notation for the operator should not concern us, but is helpful later. The $-1/2$ can be taken as the power of the λ_n used in a multiplicative way. To avoid cumbersome notation, subscript λ in $\mathcal{L}_\lambda{}^{-1/2}$ is used to denote the dependence on the pair $\{\phi_n\}_{n \in \mathbb{N}}$ and $\{\lambda_n\}_{n \in \mathbb{N}}$.

shown in Figure 10.4a, where $h = \mathcal{L}_\lambda^{-1/2} f$, and is explicitly given by

$$
\left(\mathcal{L}_\lambda^{-1/2} f \right)(x) \triangleq \sum_{n \in \mathbb{N}} \frac{\langle f, \phi_n \rangle \phi_n(x)}{\sqrt{\lambda_n}} = \sum_{n \in \mathbb{N}} \frac{(f)_n \phi_n(x)}{\sqrt{\lambda_n}}. \tag{10.16}
$$

Using this, by simple algebra, (10.13) becomes

$$
\langle f, g \rangle_{\mathcal{H}_\lambda} = \sum_{n \in \mathbb{N}} \langle \mathcal{L}_\lambda^{-1/2} f, \phi_n \rangle \langle \phi_n, \mathcal{L}_\lambda^{-1/2} g \rangle.
$$

Then, because orthonormal sequence $\{\phi_n\}_{n \in \mathbb{N}}$ is complete in $L^2(\Omega)$, we can invoke the generalized Parseval relation, (2.57), p. 72, to obtain an expression that links the two inner products

$$
\langle f, g \rangle_{\mathcal{H}_\lambda} = \langle \mathcal{L}_\lambda^{-1/2} f, \mathcal{L}_\lambda^{-1/2} g \rangle. \tag{10.17}
$$

Furthermore, by looking at the induced norms, or substituting $g = f$ if you like, we have

$$
\| f \|_{\mathcal{H}_\lambda} = \left\| \mathcal{L}_\lambda^{-1/2} f \right\|,
$$

which can be used to relate the conditions under which functions belong to the two spaces, (10.4) and (10.12).

Operator $\mathcal{L}_\lambda^{-1/2}$ is invertible and we denote the inverse by $\mathcal{L}_\lambda^{1/2} : L^2(\Omega) \to \mathcal{H}_\lambda$. We have

$$
\mathcal{L}_\lambda^{1/2} \circ \mathcal{L}_\lambda^{-1/2} = \mathcal{I},
$$

where $\mathcal{I}$ is the identity operator, and the inverse mapping is explicitly given by

$$
\left(\mathcal{L}_\lambda^{1/2} f \right)(x) \triangleq \sum_{n \in \mathbb{N}} \sqrt{\lambda_n} \langle f, \phi_n \rangle \phi_n(x) = \sum_{n \in \mathbb{N}} \sqrt{\lambda_n} (f)_n \phi_n(x), \tag{10.18}
$$

which is shown in Figure 10.4b. The isomorphism which uses both (10.16) and (10.18) is shown in Figure 10.4c. A more detailed picture of the isomorphism is given later in Section 10.3.9.

Problems

10.7. Show that operator $\mathcal{L}_\lambda^{-1/2}$, regarded as an operator on $L^2(\Omega)$, satisfies

$$
\langle \mathcal{L}_\lambda^{-1/2} f, f \rangle \geq 0, \quad \forall f \in L^2(\Omega).
$$

10.8. Show that that operator $\mathcal{L}_\lambda^{1/2}$ is strictly positive, $\mathcal{L}_\lambda^{1/2} > 0$. That is, show

$$
\langle \mathcal{L}_\lambda^{1/2} f, f \rangle > 0, \quad \forall f \in L^2(\Omega) \quad \text{satisfying} \quad f \neq o.
$$

10.3.6   Function test condition

If we have a function f we can test if it is in $\mathcal{H}_\lambda$ by building a related function $g = \mathcal{L}_\lambda^{-1/2} f$ and testing if g is in $L^2(\Omega)$; that is, by checking if it has finite energy or if $\|g\|^2 < \infty$. We write this test as:

$$\boxed{\mathcal{L}_\lambda^{-1/2} f \in L^2(\Omega) \implies f \in \mathcal{H}_\lambda.} \tag{10.19}$$

Furthermore, because $\lambda_n > 0$, $\forall n \in \mathbb{N}$ then this condition is invertible:

$$f \in \mathcal{H}_\lambda \implies \mathcal{L}_\lambda^{-1/2} f \in L^2(\Omega). \tag{10.20}$$

This is more of theoretical interest because testing whether a function is in $L^2(\Omega)$ is really no easier than testing if a function is in $\mathcal{H}_\lambda$. The test just means we can relate a function in $\mathcal{H}_\lambda$ to something we regard as more familiar (testing if a function has finite $L^2(\Omega)$ energy).

Problem _____

10.9. Find test conditions similar to (10.19)–(10.20), which can be written as

$$\mathcal{L}_\lambda^{-1/2} f \in L^2(\Omega) \iff f \in \mathcal{H}_\lambda,$$

relating $L^2(\Omega)$ and $\mathcal{H}_\lambda$, but expressed in terms of the $\mathcal{L}_\lambda^{1/2}$ operator.

_____

10.3.7   Fundamental operator

The isomorphism operators, (10.16) and (10.18), can be obtained from an underlying more fundamental operator, $\mathcal{L}_\lambda h \colon L^2(\Omega) \to L^2(\Omega)$,[9] shown in Figure 10.5a and defined here:

$$\boxed{(\mathcal{L}_\lambda h)(x) \triangleq \sum_{n \in \mathbb{N}} \lambda_n \langle h, \phi_n \rangle \phi_n(x) = \sum_{n \in \mathbb{N}} \lambda_n (h)_n \phi_n(x).} \tag{10.21}$$

Here the $\{\lambda_n\}_{n \in \mathbb{N}}$ appear without the $\pm 1/2$ powers, and it is straightforward to see that $\mathcal{L}_\lambda > 0$, that is, it is strictly positive; see Problem 10.8 for the style of argument. A general power $s \in \mathbb{R}$ can also be used and the corresponding operators $\mathcal{L}_\lambda^s$ are well-defined given $\mathcal{L}_\lambda$ (where implicitly $s = 1$) is strictly positive. For $s < 1$, including the case we have considered when $s = -1/2$, (10.15), one needs to be slightly more precise about the input space; see (Cucker and Smale, 2002, p. 27).

Square root operator

The operator $\mathcal{L}_\lambda^s$ with $s = 1/2$, which is given in (10.18) can be thought of as the square root operator of $\mathcal{L}_\lambda$, as shown in Figure 10.5b,

$$\mathcal{L}_\lambda = \mathcal{L}_\lambda^{1/2} \circ \mathcal{L}_\lambda^{1/2}.$$

The proof of this is left to Problem 10.12.

_____

[9] Clearly, the output, $(\mathcal{L}_\lambda h)(x)$, is not going to be onto $L^2(\Omega)$; that is, not a surjection.

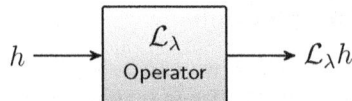

(a) Strictly positive operator $\mathcal{L}_\lambda$.

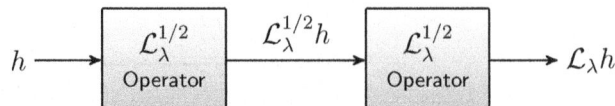

(b) Decomposition of strictly positive operator $\mathcal{L}_\lambda = \mathcal{L}_\lambda^{1/2} \circ \mathcal{L}_\lambda^{1/2}$.

Figure 10.5 Two representations of strictly positive operator $\mathcal{L}_\lambda \colon L^2(\Omega) \to L^2(\Omega)$ mapping $h \mapsto \mathcal{L}_\lambda h$, and defined in (10.21).

Synopsis

In summary, operator $\mathcal{L}_\lambda > 0$ furnishes directly the key components $\{\lambda_n\}_{n\in\mathbb{N}}$ and $\{\phi_n\}_{n\in\mathbb{N}}$ used in defining $\mathcal{H}_\lambda$. Its square root operator $\mathcal{L}_\lambda^{1/2} > 0$, and the square root inverse, define the isomorphism between the new space $\mathcal{H}_\lambda$ and archetype Hilbert space $L^2(\Omega)$. So with the inner product (10.17), repeated here:

$$\langle f, g \rangle_{\mathcal{H}_\lambda} = \langle \mathcal{L}_\lambda^{-1/2} f, \mathcal{L}_\lambda^{-1/2} g \rangle,$$

then $\mathcal{H}_\lambda$ is a Hilbert space. For a slight generalization see (Cucker and Smale, 2002, p. 29).[10]

Problems

10.10. Show that with λ_n real and positive, for all $n \in \mathbb{N}$, then the operator

$$(\mathcal{L}_\lambda f)(x) \triangleq \sum_{n\in\mathbb{N}} \lambda_n \langle f, \phi_n \rangle \phi_n(x)$$

is self-adjoint.

10.11. Show that the operator $\mathcal{L}_\lambda$ in (10.21) is a strictly positive operator.

10.12. Show that the operator $\mathcal{L}_\lambda$ has the following operator decomposition

$$\mathcal{L}_\lambda = \mathcal{L}_\lambda^{1/2} \circ \mathcal{L}_\lambda^{1/2}.$$

10.3.8   Weighting sequence considerations

We make a simple observation:

---

[10] On p. 29, the indicated isomorphism should read $A^s \colon \mathcal{L}_\nu^2(X) \to \mathbb{E}$ (in their notation).

Remark 10.6 (Special case). The sequence $\lambda_n = 1$, $\forall n \in \mathbb{N}$ means that $\mathcal{H}_\lambda \equiv L^2(\Omega)$, which is perfectly well-defined despite the fact that λ has infinite ℓ^2-norm. Even if λ_n converges to zero as $n \to 0$, this is not strong enough to be able to say anything significant because the converge rate is the important issue.   □

Of course with a trivial weighting sequence, as in Remark 10.6, this does not yield anything interesting or new. There are many behaviors in the λ_n that might be considered, which similarly are not of much interest. For example, having the bound on the λ_n, given by B, different from 1 adds little. That is, we can safely take $0 < \lambda_n \leq 1$, $\forall n \in \mathbb{N}$.

Case $\lambda_n \to 0$

Now consider $\lambda_n \to 0$ as $n \to \infty$, and where $\lambda_n > 0$, $\forall n$. Then the $(f)_n$ needs to decay *more rapidly* to zero for $\|\mathcal{L}_\lambda^{-1/2} f\|$ to be finite (so that $f \in \mathcal{H}_\lambda$) in comparison to the decay to zero required for $\|f\|$ to be finite (so that $f \in L^2(\Omega)$). So promisingly, a function f, with Fourier coefficients $(f)_n$, can be in $\mathcal{H}_\lambda$ only if the *Fourier coefficients go to zero sufficiently fast*. There are two aspects to this decay requirement:

(a) there is a decay requirement to counteract the growth of $1/\sqrt{\lambda_n} \to \infty$ as $n \to \infty$ (since $\lambda_n \to 0$). That is, $(f)_n$ needs to decay such that

$$\frac{(f)_n}{\sqrt{\lambda_n}} \to 0;$$

(b) there is an additional decay requirement overall to be in $L^2(\Omega)$ as $n \to \infty$.

Problem 10.13 and Problem 10.14 flesh out this point a bit more by considering both effects. But Example 10.2, below, is a useful starting point.

Remark 10.7 (Low-pass functions). In familiar engineering terms, a function in $\mathcal{H}_\lambda$, say $f \in \mathcal{H}_\lambda = \mathcal{L}_\lambda^{1/2} h$, is a smoothed or low-pass filtered version of function h in $L^2(\Omega)$. As such, $\mathcal{L}_\lambda^{1/2}$ can be regarded as a type of *low-pass filter*, and $\mathcal{L}_\lambda^{-1/2}$ is its inverse (emphasizes higher-order Fourier coefficients, such as the "unsharp mask" used in image processing).[11]   □

Next we give an example that at least one implicit discontinuous function cannot belong to the space $\mathcal{H}_\lambda$ for an explicit λ satisfying $\lambda_n \to 0$.

Example 10.2. Let us define our space $\mathcal{H}_\lambda$ in terms of a complete orthonormal sequence $\{\phi_n\}_{n \in \mathbb{N}}$ in $L^2(\Omega)$ (which we do not need to know for this exercise) and the weights

$$\lambda_n = \frac{1}{n}, \quad \forall n \in \mathbb{N}.$$

Suppose we have a candidate function, $f \in L^2(\Omega)$, with Fourier coefficients

$$(f)_n = \frac{1}{n}, \quad \forall n \in \mathbb{N}. \tag{10.22}$$

---

[11] In the sense that high frequencies are increasingly attenuated. Also we tend to think in terms of the orthonormal sequence being the normalized complex exponentials, but the idea is the same for the general case.

We note that these converge sufficiently fast to zero, as $n \to \infty$, to satisfy

$$\sum_{n \in \mathbb{N}} (f)_n \phi_n \in L^2(\Omega),$$

because the coefficients (10.22) are square summable. Example 10.1 is an example of a function that has this rate of decay. It is relatively standard to expect that such a function f, defined by (10.22), would have step-like discontinuities; see Table 10.1 and (Strang, 2007, pp. 317–327).

We can apply our test, given in Section 10.3.6, to see if f belongs to $\mathcal{H}_\lambda$. That is, we build a new function h and see if $h \in L^2(\Omega)$. We have

$$h(x) = \left(\mathcal{L}_\lambda^{-1/2} f\right)(x) = \sum_{n \in \mathbb{N}} \frac{(f)_n \phi_n(x)}{\sqrt{\lambda_n}} = \sum_{n \in \mathbb{N}} \frac{1}{\sqrt{n}} \phi_n(x),$$

which does not converge, because

$$\sum_{n \in \mathbb{N}} \left(\frac{1}{\sqrt{n}}\right)^2 = \sum_{n \in \mathbb{N}} \frac{1}{n}$$

is the harmonic series which diverges. Therefore, since $h = \mathcal{L}_\lambda^{-1/2} f \notin L^2(\Omega)$ then $f \notin \mathcal{H}_\lambda$. □

To complete our description of the isomorphism we should like to say something about orthonormal sequences in $\mathcal{H}_\lambda$. We had mentioned in Example 10.2 that $\{\phi_n\}_{n \in \mathbb{N}}$ is orthonormal in $L^2(\Omega)$, but it is not orthonormal in $\mathcal{H}_\lambda$. In the next section we clarify this point and find the natural orthonormal sequence in $\mathcal{H}_\lambda$ labeled in Figure 10.2, p. 323, as $\{\varphi_n\}_{n \in \mathbb{N}}$.

Problems

10.13. The Riemann zeta function of a complex variable s is given by

$$\zeta(s) = \sum_{n=1}^{\infty} \frac{1}{n^s},$$

and is known to converge for $\mathfrak{Re}(s) > 1$. Suppose

$$\lambda_n = \frac{1}{n^{1/2+\epsilon}}, \qquad (10.23)$$

where $\epsilon > 0$. Show $\{\lambda_n\}_{n \in \mathbb{N}} \in \ell^2$. Discuss the result when ϵ is very small.

10.14. Suppose $\{\lambda_n\}_{n \in \mathbb{N}}$ is given by (10.23). Further suppose that a function $f \in L^2(\Omega)$ has Fourier coefficients (assuming ν is large enough), satisfying

$$(f)_n = \frac{1}{n^\nu}.$$

What range of ν values results in $f \in \mathcal{H}_\lambda$ so that it satisfies (10.12), p. 331? Discuss the result.

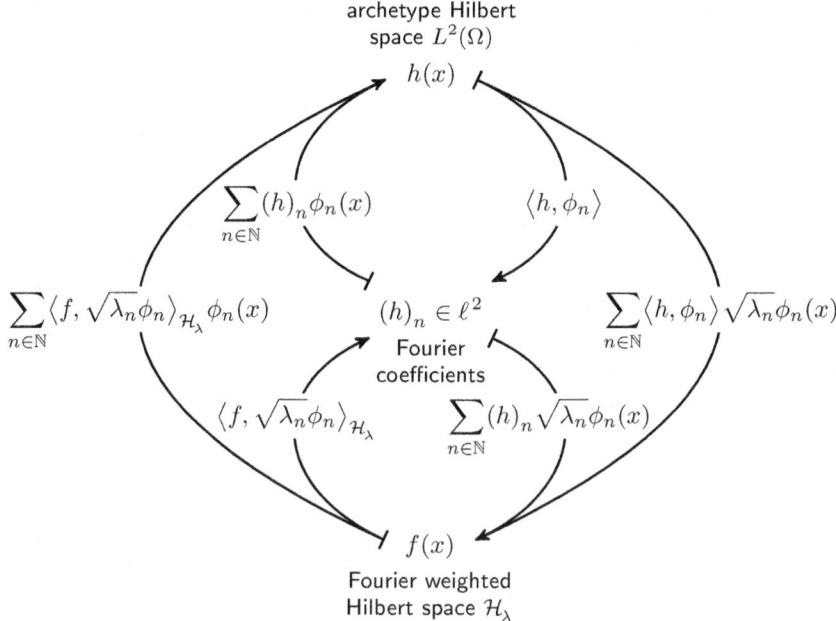

Figure 10.6 Isomorphism between $\mathcal{H}_\lambda$ and $L^2(\Omega)$. Here $h = \mathcal{L}_\lambda^{-1/2} f$ where $\mathcal{L}_\lambda^{-1/2}$ is given in (10.16) and $\mathcal{L}_\lambda^{1/2}$ is given in (10.18).

10.3.9   Orthonormal sequences

Here we develop expressions for the natural orthonormal sequence in $\mathcal{H}_\lambda$ using two approaches — an isomorphism approach which uses previously developed isomorphism diagrams and an algebraic approach which characteristically tends to be devoid of insight. In a sense these expressions are implicit in the work we have done to this point, but seeing the result from different perspectives should help to clarify things.

Isomorphism approach

We can collect the key identities from the previous sections to form an isomorphism diagram shown in Figure 10.6. It has an identical structure to the generic diagram Figure 10.2, p. 323. To interpret the expression for the map $f \mapsto h$ we derive the following simple identity, based on the definition for the inner product (10.6),

$$
\begin{aligned}
\langle f, \sqrt{\lambda_n} \phi_n \rangle_{\mathcal{H}_\lambda} &= \sum_{m \in \mathbb{N}} \frac{\langle f, \phi_m \rangle \langle \phi_m, \sqrt{\lambda_n} \phi_n \rangle}{\lambda_m} \\
&= \sum_{m \in \mathbb{N}} \frac{1}{\lambda_m} \langle f, \phi_m \rangle \delta_{m,n} \sqrt{\lambda_n} \\
&= \frac{(f)_n}{\sqrt{\lambda_n}}.
\end{aligned}
$$

With this identity, operator $\mathcal{L}_\lambda^{-1/2}$ in (10.16) can be expressed in terms of the $\mathcal{H}_\lambda$ inner product such that the map $f \mapsto h$ becomes

$$
\begin{aligned}
h(x) &= \left(\mathcal{L}_\lambda^{-1/2} f\right)(x) \\
&= \sum_{n \in \mathbb{N}} \frac{(f)_n}{\sqrt{\lambda_n}} \phi_n(x) \\
&= \sum_{n \in \mathbb{N}} \langle f, \sqrt{\lambda_n} \phi_n \rangle_{\mathcal{H}_\lambda} \phi_n(x),
\end{aligned}
$$

as shown on the leftmost upwards branch in Figure 10.6.

Comparing Figure 10.6 with Figure 10.2 we can infer

$$
\varphi_n(x) \equiv \sqrt{\lambda_n} \phi_n(x), \quad \forall n \in \mathbb{N}.
$$

That is, the orthonormal functions in $\mathcal{H}_\lambda$ must be

$$
\{\varphi_n\}_{n \in \mathbb{N}} \equiv \{\sqrt{\lambda_n} \phi_n\}_{n \in \mathbb{N}}. \tag{10.24}
$$

To confirm this is indeed an orthonormal sequence in $\mathcal{H}_\lambda$ we use an algebraic approach next.

Algebraic approach

By definition, $\{\phi_n\}_{n \in \mathbb{N}}$ is a complete orthonormal sequence for $L^2(\Omega)$. That is,

$$
\langle \phi_p, \phi_q \rangle = \delta_{p,q}, \quad \forall p, q \in \mathbb{N},
$$

and

$$
f = \sum_{n \in \mathbb{N}} \langle f, \phi_n \rangle \phi_n, \quad \forall f \in L^2(\Omega),
$$

where equality is in the sense of convergence in the mean. Given the close link between $L^2(\Omega)$ and $\mathcal{H}_\lambda$ above, we can ask if there is a related complete orthonormal sequence for $\mathcal{H}_\lambda$ with respect to the inner product (10.6).

Consider

$$
\begin{aligned}
\langle \sqrt{\lambda_p} \phi_p, \sqrt{\lambda_q} \phi_q \rangle_{\mathcal{H}_\lambda} &= \sum_{n \in \mathbb{N}} \frac{\langle \sqrt{\lambda_p} \phi_p, \phi_n \rangle}{\sqrt{\lambda_n}} \frac{\langle \phi_n, \sqrt{\lambda_q} \phi_q \rangle}{\sqrt{\lambda_n}} \\
&= \sum_{n \in \mathbb{N}} \frac{\sqrt{\lambda_p}}{\sqrt{\lambda_n}} \delta_{pn} \frac{\sqrt{\lambda_q}}{\sqrt{\lambda_n}} \delta_{nq} \\
&= \delta_{p,q},
\end{aligned}
$$

where we have used completeness of $\{\phi_n\}_{n \in \mathbb{N}}$ in $L^2(\Omega)$. Therefore,

$$
\{\sqrt{\lambda_n} \phi_n\}_{n \in \mathbb{N}}
$$

is an orthonormal sequence for $\mathcal{H}_\lambda$. In fact, it is complete in $\mathcal{H}_\lambda$ because $\{\phi_n\}_{n \in \mathbb{N}}$ is complete in $L^2(\Omega)$.

Function expansion

The natural expansion for function f in $\mathcal{H}_\lambda$ is then

$$f(x) = \sum_{n\in\mathbb{N}} \zeta_n \underbrace{\sqrt{\lambda_n}\phi_n(x)}_{\text{orthonormal in } \mathcal{H}_\lambda}$$

$$= \sum_{n\in\mathbb{N}} \zeta_n \varphi_n(x),$$

for $\mathcal{H}_\lambda$-Fourier coefficients

$$\zeta_n \triangleq \langle f, \sqrt{\lambda_n}\phi_n\rangle_{\mathcal{H}_\lambda}$$

$$= \langle f, \varphi_n\rangle_{\mathcal{H}_\lambda}$$

and requires $\{\zeta_n\}_{n\in\mathbb{N}} \in \ell^2$.

> **Remark 10.8.** The same function $f(x)$ shown above can be expanded in the $L^2(\Omega)$-orthonormal sequence $\{\phi_n\}_{n\in\mathbb{N}}$, given by
>
> $$f(x) = \sum_{n\in\mathbb{N}} (f)_n \phi_n(x),$$
>
> with $L^2(\Omega)$-Fourier coefficients
>
> $$(f)_n \triangleq \langle f, \phi_n\rangle.$$
>
> But remember if $f \in L^2(\Omega)$, and therefore $\{(f)_n\}_{n\in\mathbb{N}} \in \ell^2$, then it is possible that the same function $f \notin \mathcal{H}_\lambda$ because $\{\zeta_n\}_{n\in\mathbb{N}} \notin \ell^2$. In the $\mathcal{H}_\lambda$ expansion above, the orthonormal functions are being deflated and the Fourier coefficients are being inflated inversely as $n \to \infty$ to leave f invariant. □

10.3.10 Isomorphism equations

In summary, an isomorphism between $\mathcal{H}_\lambda$ and $L^2(\Omega)$ has been established and captured through:

$$f \in \mathcal{H}_\lambda \iff \mathcal{L}_\lambda^{-1/2} f \in L^2(\Omega), \tag{10.25a}$$

$$\mathcal{L}_\lambda^{1/2} h \in \mathcal{H}_\lambda \iff h \in L^2(\Omega), \tag{10.25b}$$

$$\langle f, g\rangle_{\mathcal{H}_\lambda} = \langle \mathcal{L}_\lambda^{-1/2} f, \mathcal{L}_\lambda^{-1/2} g\rangle, \tag{10.25c}$$

$$\langle \mathcal{L}_\lambda^{1/2} f, \mathcal{L}_\lambda^{1/2} g\rangle_{\mathcal{H}_\lambda} = \langle f, g\rangle, \tag{10.25d}$$

$$\|f\|_{\mathcal{H}_\lambda} = \|\mathcal{L}_\lambda^{-1/2} f\|, \tag{10.25e}$$

$$\|\mathcal{L}_\lambda^{1/2} f\|_{\mathcal{H}_\lambda} = \|f\|, \tag{10.25f}$$

$$\begin{array}{c}\{\sqrt{\lambda_n}\phi_n\}_{n\in\mathbb{N}} \text{ is orthonormal} \\ \text{and complete in } \mathcal{H}_\lambda\end{array} \iff \begin{array}{c}\{\phi_n\}_{n\in\mathbb{N}} \text{ is orthonormal} \\ \text{and complete in } L^2(\Omega)\end{array}, \tag{10.25g}$$

where, combining (10.16) and (10.18), we have

$$\left(\mathcal{L}_\lambda^{\pm1/2} f\right)(x) \triangleq \sum_{n\in\mathbb{N}} \lambda_n^{\pm1/2}\langle f, \phi_n\rangle\phi_n(x) = \sum_{n\in\mathbb{N}} \lambda_n^{\pm1/2}(f)_n\phi_n(x).$$

A pictorial form of these isomorphism equations which omits the ℓ^2 space mappings is shown in Figure 10.7.

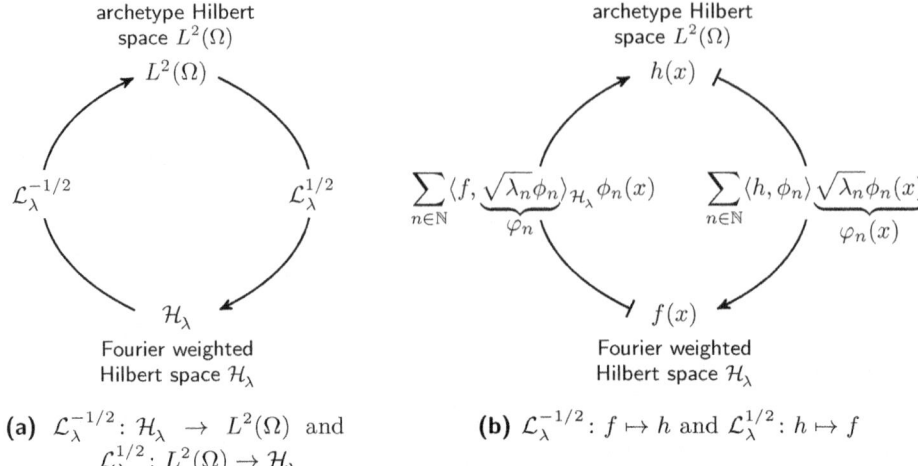

(a) $\mathcal{L}_\lambda^{-1/2} : \mathcal{H}_\lambda \rightarrow L^2(\Omega)$ and $\mathcal{L}_\lambda^{1/2} : L^2(\Omega) \rightarrow \mathcal{H}_\lambda$

(b) $\mathcal{L}_\lambda^{-1/2} : f \mapsto h$ and $\mathcal{L}_\lambda^{1/2} : h \mapsto f$

Figure 10.7 Isomorphism between $\mathcal{H}_\lambda$ and $L^2(\Omega)$. Here $h = \mathcal{L}_\lambda^{-1/2} f$ where $\mathcal{L}_\lambda^{-1/2}$ is given in (10.16) and $\mathcal{L}_\lambda^{1/2}$ is given in (10.18).

The road forward

We have crept marginally towards our goal of having only continuous functions in our space. The result also makes sense because the functions in $L^2(\Omega)$ that are admissible as vectors in $\mathcal{H}_\lambda$ have to be "low-pass filtered" versions of finite energy functions. One cannot obtain a step-like discontinuous signal as the output of a low-pass operator for a finite energy input when the λ_n weights decay fast enough.

What we now are interested in are the conditions on $\lambda_n \rightarrow 0$ as $n \rightarrow \infty$ to ensure only continuous functions are in the corresponding space $\mathcal{H}_\lambda$. To find such conditions we can cast our development in a form that involves a kernel of an operator implicit in the preceding development.

In Example 10.2, p. 337, the orthonormal sequence in $L^2(\Omega)$, $\{\phi_n\}_{n\in\mathbb{N}}$, was not explicitly given and changing it does not change the conclusion reached. However, the orthonormal sequence does have a role and strictly speaking our notation for operators $\mathcal{L}_\lambda^{-1/2}$ and the like should have reflected its role. This will be rectified by using a kernel which as a bivariate function holds all the information about the weighting sequence $\{\lambda_n\}_{n\in\mathbb{N}}$, and the sequence $\{\phi_n\}_{n\in\mathbb{N}}$.

10.4 O Kernel, Kernel, wherefore art thou Kernel?

10.4.1 Kernel of an integral operator

To this point we have seen that it is relatively straightforward to define a new Hilbert space which uses a modified inner product that weighs the Fourier coefficients with respect to some reference $L^2(\Omega)$ space and choice of complete orthonormal sequence. We shall now see that it is possible to characterize some of these Fourier weighted Hilbert spaces using a kernel of an integral operator on $L^2(\Omega)$. None of this should be surprising given we have an orthonormal sequence,

$\{\phi_n\}_{n\in\mathbb{N}}$, and a sequence of weights, $\{\lambda_n\}_{n\in\mathbb{N}}$, and this seems awfully familiar to a having a set of eigenfunctions and eigenvalues (as the notation suggests). The notation was deliberate and shall be shown to be consistent with earlier.

Hilbert-Schmidt integral operator

We can distill what we found in Chapter 6, p. 160, and summarized in Figure 6.1, p. 170. The integral operator, the Hilbert-Schmidt integral operator, on $L^2(\Omega)$:

$$(\mathcal{L}_K f)(x) = \int_\Omega K(x,y) f(y)\, dy, \qquad (10.26)$$

with kernel $K(x,y)$ satisfying:

C1: $\displaystyle\iint_{\Omega\times\Omega} |K(x,y)|^2 dx\, dy < \infty$ (Hilbert-Schmidt),

C2: $K(x,y) = \overline{K(y,x)}$ (Hermitian symmetry),

is compact and self-adjoint. The inner product, which characterizes the norm and orthogonality, is the usual one for $L^2(\Omega)$,

$$\langle f, g \rangle = \int_\Omega f(x)\overline{g(x)}\, dx,$$

which was given in Section 2.4.6, p. 47.

By the spectral theory of compact self-adjoint operators it has an associated set of orthogonal eigenfunctions, which can be taken as orthonormal, $\{\phi_n(x)\}_{n\in\mathbb{N}}$, and real eigenvalues, $\{\lambda_n\}_{n\in\mathbb{N}}$ tending to zero,

$$(\mathcal{L}_K \phi_n)(x) = \lambda_n \phi_n(x), \quad n \in \mathbb{N},$$

which enables us to represent the kernel in the following expansion (see Theorem 6.2, p. 164), first given by Schmidt in his 1905 thesis:[12]

$$K(x,y) = \sum_{n\in\mathbb{N}} \lambda_n \phi_n(x)\overline{\phi_n(y)} \qquad \text{(in the mean).} \qquad (10.27)$$

This kernel representation is interpreted in the sense of convergence in the mean. Mercer, in his work, was motivated to correct this "defect" (Mercer, 1909, p. 439), striving for a stronger sense of equality, that being uniform convergence, for the kernel expansion,[13] but this is only possible to guarantee if a number of stronger conditions are imposed.

> **Remark 10.9.**   Even if we impose a condition of continuity on the kernel $K(x,y)$, meaning as a function it is continuous in both x and y on $\Omega \times \Omega$, the expansion (10.27) is still not necessarily uniformly convergent. The next objective is to find sufficient conditions on the kernel which are stronger than continuity, so as to guarantee uniform convergence.   □

---

[12] Although Schmidt attributed the result to Hilbert according to the comments given in (Mercer, 1909, p. 439).

[13] The kernel expansion (10.27) is *not* "Mercer's theorem," but is often confused with it. We shall clarify what Mercer's theorem is in the next subsection.

Remark 10.10. The spectral theory of compact self-adjoint operators requires $\lambda_n \to 0$ as $n \to \infty$. This condition is strengthened by the Hilbert-Schmidt condition to give $\{\lambda_n\}_{n \in \mathbb{N}} \in \ell^2$, which is the case we want to work with. □

10.4.2 Mercer's theorem

The Schmidt expansion for the kernel (10.27) implies λ_n are real and, with the Hilbert-Schmidt condition on the kernel, the sequence $\{\lambda_n\}_{n \in \mathbb{N}}$ is in ℓ^2. This last condition is evident from Theorem 6.3, p. 167:

$$\|\mathcal{L}_K\|_{\mathrm{op}} \triangleq \sup_{\|f\|=1} \|\mathcal{L}_K f\| \leq \|K\| = \Big(\sum_{n \in \mathbb{N}} \lambda_n^2\Big)^{1/2},$$

given $\|K\| < \infty$ (meaning $K \in L^2(\Omega \times \Omega)$).

In the formulation of Mercer's theorem we must impose a few additional conditions, the most important of which is the non-zero eigenvalues need to satisfy $\lambda_n > 0$ — real and positive non-zero eigenvalues. In contrast, Theorem 6.2, p. 164, has $\lambda_n \in \mathbb{R}$ only, meaning that it can have real and possibly mixed negative, zero, and positive eigenvalues. It could have infinitely many positive and infinitely many negative eigenvalues in general.

If we identify the eigenvalues $\{\lambda_n\}_{n \in \mathbb{N}}$ from the integral equation using the kernel $K(x, y)$ with our Fourier weighting then having $\lambda_n > 0$ is perfectly matched to the requirements for Fourier weighted Hilbert spaces — recall $\lambda_n > 0$, from (10.11), is required to have a valid inner product and norm (to be an inner product space).

Theorem 10.1 (Mercer's theorem). *Consider the compact self-adjoint integral operator on $L^2(\Omega)$*

$$(\mathcal{L}_K f)(x) = \int_\Omega K(x, y) f(y)\, dy,$$

satisfying conditions:

C1: $\displaystyle\iint_{\Omega \times \Omega} |K(x, y)|^2 dx\, dy < \infty$ (Hilbert-Schmidt),

C2: $K(x, y) = \overline{K(y, x)}$ (Hermitian symmetry),

C3: $\Omega = [a, b]$ (bounded domain),

C4: $K(x, y)$ *continuous in x and y*   (continuity of kernel).

If the eigenvalues $\{\lambda_n\}_{n \in \mathbb{N}}$ corresponding to the eigenfunctions $\{\phi_n(x)\}_{n \in \mathbb{N}}$ of operator $\mathcal{L}_K$ satisfy condition:

C5: $\lambda_n > 0, \quad n \in \mathbb{N}$ (positivity),

then the kernel expansion

$$\boxed{K(x, y) = \sum_{n \in \mathbb{N}} \lambda_n \phi_n(x)\overline{\phi_n(y)} \quad \text{(uniformly convergent)}} \tag{10.28}$$

is uniformly convergent. □

Proof. Riesz has given a short elegant proof in (Riesz and Sz.-Nagy, 1990, pp. 245–246). The original proof is in (Mercer, 1909).   □

Of course, few people might care that the kernel expansion is uniformly convergent rather than just convergent in the mean (equivalently strongly convergent). The importance of the result is its implications, and the (additional) conditions on the kernel are well matched to specific applications.

Remarks on Mercer's theorem

R1: There are many references in the literature referring to expansion (10.28) being Mercer's theorem. As we have pointed out the expansion was given by Schmidt and attributed to Hilbert. The real Mercer's theorem is a statement about imposing stronger conditions that strengthen the expansion convergence from convergence in the mean, (10.27), to uniform convergence, (10.28).

R2: The *continuity* of the kernel $K(x, y)$ is enough to guarantee that $\mathcal{L}_K f$ is a continuous function even if $f \in L^2(\Omega)$. In short,

$$f \in L^2(\Omega) \implies \mathcal{L}_K f \in C(\Omega),$$

when $K(x, y)$ is continuous; see (Cucker and Smale, 2002, p. 33).

R3: Given the preceding item, if we take $f = \phi_n$ (an eigenfunction) then continuity implies

$$\mathcal{L}_K \phi_n = \lambda_n \phi_n \in C(\Omega). \qquad (10.29)$$

That is, $\phi_n \in C(\Omega)$ — such eigenfunctions are continuous functions.

R4: Again with *continuity* of the kernel $K(x, y)$, it is easy to broaden Mercer's theorem in the following way (Mercer, 1909; Riesz and Sz.-Nagy, 1990). Rather than having no negative eigenvalues we can have a finite number. These negative-eigenvalue eigenfunctions are also continuous. Therefore, a linear combination of a finite number of negative-eigenvalue continuous functions is still continuous and does not damage the theorem claims. Similarly, the result stays unchanged with an infinite number of negative eigenvalues and a finite number (or zero) positive eigenvalues. The only case that potentially breaks the uniform convergence is when there are an infinite number of positive and infinite number of negative eigenvalues.

R5: We understand, from Definition 2.27, p. 80, that uniform convergence implies pointwise convergence and so the kernel expansion also converges pointwise. But uniform convergence is a stronger result.

R6: If we have an operator $\mathcal{L}_K$ with Hilbert-Schmidt kernel $K(x, y)$ that does *not* satisfy the eigenvalue condition of Mercer's theorem, then the iterate $\mathcal{L}_K^2 \triangleq \mathcal{L}_K \circ \mathcal{L}_K$ surely does. That is, iterating an operator implies its equivalent kernel

$$K_2(x, y) \triangleq \int_\Omega K(x, z) K(z, y) \, dz$$

has an expansion

$$K_2(x, y) = \sum_{n \in \mathbb{N}} \lambda_n^2 \phi_n(x) \overline{\phi_n(y)},$$

which converges uniformly, where eigenfunctions $\{\phi_n\}_{n \in \mathbb{N}}$ corresponding to non-zero eigenvalues are common to $\mathcal{L}_K$ and $\mathcal{L}_K^2$.

R7: We note that relative to the weaker Hilbert-Schmidt kernel expansion conditions, for the Mercer case there is a strong condition that Ω is bounded. There are technical headaches when trying to move to unbounded domains.

R8: We have skirted around the issue of when any eigenvalues are zero. This does not lead to any great problems as the operator $\mathcal{L}_K$ annihilates these eigenfunctions. In any case, our case of interest is when $\lambda_n > 0$, $\forall n \in \mathbb{N}$.

10.4.3 Square summable weighting

Schmidt synthesis

Before moving on, we have seen there is much structure when the real weights satisfy

$$\{\lambda_n\}_{n \in \mathbb{N}} \in \ell^2$$

and when we have a complete orthonormal sequence $\{\phi_n\}_{n \in \mathbb{N}}$ in $L^2(\Omega)$. In that case we can synthesize a Hilbert-Schmidt kernel in the sense of convergence in the mean,

$$K(x, y) = \sum_{n \in \mathbb{N}} \lambda_n \phi_n(x) \overline{\phi_n(y)},$$

and thereby construct a compact self-adjoint integral equation operator $\mathcal{L}_K$ which has $\{\lambda_n\}_{n \in \mathbb{N}} \in \ell^2$ as its real eigenvalues and $\{\phi_n\}_{n \in \mathbb{N}}$ as its eigenfunctions.[14] However, this is insufficient to yield a valid inner product under our earlier construction — (10.11) requires the eigenvalues to be positive in addition to being real.

Mercer synthesis

With the additional positivity of the real eigenvalues,

$$\lambda_n > 0, \quad \forall n \in \mathbb{N},$$

$\Omega = [a, b]$ being closed and bounded, and continuity of the kernel, then Mercer's theorem tells us the kernel expansion in terms of the eigenfunctions and eigenvalues not only converges in the mean, but converges uniformly. Furthermore, each of the eigenfunctions must be continuous and the operator $\mathcal{L}_K$ is also strictly positive in the sense

$$\langle \mathcal{L}_K f, f \rangle > 0, \quad \forall f \in L^2(\Omega) \quad \text{satisfying} \quad f \neq o.$$

---

[14] If any eigenvalues are zero this is easily handled and for the moment we can assume $\lambda_n \neq 0$, $\forall n \in \mathbb{N}$.

Recall, that $\lambda_n > 0$ is matched to the requirement to yield a valid inner product (10.11). It is an essential requirement to construct our RKHS considered next.

> **Remark 10.11 (New notation).** To this point we have used $\mathcal{H}_\lambda$ to represent the class of Fourier weighted Hilbert spaces. With the additional Mercer conditions, we use the new notation $\mathcal{H}_K$ to make it clear that we are dealing with a special subclass of $\mathcal{H}_\lambda$. Furthermore, all results given for $\mathcal{H}_\lambda$ with operator $\mathcal{L}_\lambda$ regarding the carry over to the space $\mathcal{H}_K$ with strictly positive operator $\mathcal{L}_K > 0$ — results such as the isomorphism equations (10.25).   □

10.5   Reproducing kernel Hilbert spaces

10.5.1   Complete orthonormal functions

In the specific Mercer-type kernel context the inner product and induced norm will be denoted

$$\langle f, g \rangle_{\mathcal{H}_K} \quad \text{and} \quad \|f\|_{\mathcal{H}_K} = \langle f, f \rangle_{\mathcal{H}_K}^{1/2}$$

with subscript $\mathcal{H}_K$ used in place of the previous subscript $\mathcal{H}_\lambda$.

The Mercer-type kernel expansion is in terms of the (continuous) orthonormal[15] eigenfunctions, $\{\phi_n\}_{n \in \mathbb{N}}$, of the integral equation operator on $L^2(\Omega)$. In $\mathcal{H}_K$, the related orthonormal sequence of functions was found to be, repeated here from (10.24),

$$\varphi_n(x) = \sqrt{\lambda_n} \phi_n(x), \quad \forall n \in \mathbb{N}. \tag{10.30}$$

That is, they are orthonormal with respect to the inner product in $\mathcal{H}_K$, (10.6):

$$\langle \varphi_n, \varphi_m \rangle_{\mathcal{H}_K} = \delta_{n,m}, \quad \forall n, m \in \mathbb{N},$$

and complete:

$$f(x) = \sum_{n \in \mathbb{N}} \langle f, \varphi_n \rangle_{\mathcal{H}_K} \varphi_n(x), \quad \forall f \in \mathcal{H}_K,$$

which can be understood in the sense of convergence in the mean or uniform convergence. Table 10.2 updates the earlier findings to reflect the restriction on $\mathcal{H}_K$ under the conditions in Mercer's theorem. So it is of interest to express the kernel expansion, (10.28), in terms of this orthonormal sequence of functions in $\mathcal{H}_K$, since it is the space in which we are interested.

10.5.2   Completeness relation and Dirac delta functions

The Dirac delta function is also known as the impulse function. But this association can be misleading because it is really a description relevant for $L^2(\Omega)$. For other spaces, functions which act like the Dirac delta need not be impulses. In fact the Dirac delta function for $L^2(\Omega)$ is a bit of a strange beast because as a function it is not in $L^2(\Omega)$. For other nicer spaces like the one we are examining, this need not be the case.

---

[15] These $\{\phi_n\}_{n \in \mathbb{N}}$ are normalized in $L^2(\Omega)$ and orthogonal but not normalized in $\mathcal{H}_K$.

Table 10.2 RKHS $\mathcal{H}_K$ versus $L^2(\Omega)$

Hilbert space	Orthonormal sequence	Inner product		
		Notation	Expression	Alt expression
$L^2(\Omega)$	$\left\{\phi_n(x)\right\}_{n\in\mathbb{N}}$	$\langle f,g\rangle$	$\displaystyle\int_\Omega f(x)\overline{g(x)}\,dx$	$\displaystyle\sum_{n\in\mathbb{N}}(f)_n\overline{(g)}_n$
$\mathcal{H}_K$	$\left\{\sqrt{\lambda_n}\phi_n(x)\right\}_{n\in\mathbb{N}}$	$\langle f,g\rangle_{\mathcal{H}_K}$	$\displaystyle\sum_{n\in\mathbb{N}}\frac{\langle f,\phi_n\rangle\langle\phi_n,g\rangle}{\lambda_n}$	$\displaystyle\sum_{n\in\mathbb{N}}\frac{(f)_n\overline{(g)}_n}{\lambda_n}$

The $\mathcal{L}_K$ operator kernel has an alternative form, directly from (10.30),

$$K(x,y) = \sum_{n\in\mathbb{N}} \lambda_n \phi_n(x)\overline{\phi_n(y)}$$

$$= \sum_{n\in\mathbb{N}} \varphi_n(x)\overline{\varphi_n(y)},$$

which is clearly reminiscent of a "completeness relation" (compare this with Theorem 2.13, p. 93). That is, define $\delta_K(\cdot)$ through

$$\boxed{K(x,y) = \delta_K(x-y) \triangleq \sum_{n\in\mathbb{N}} \varphi_n(x)\overline{\varphi_n(y)},} \qquad (10.31)$$

where $\delta_K(\cdot)$ could be defined as a "Dirac delta function for $\mathcal{H}_K$." Later we establish properties to justify this analogy and notation.

Next we revisit the $L^2(\Omega)$ delta function properties to gain some inspiration and clear the fog a little.[16]

$L^2(\Omega)$ Dirac delta function

The familiar engineering/physics notion of the Dirac delta function of being an *impulse* in $L^2(\Omega)$ does not lend itself to ready generalization. Rather, its "sifting property" defined in $L^2(\Omega)$, (2.78), p. 93, provides a path to generalization:

$$f(x) = \int_\Omega f(y)\delta(x-y)\,dy,$$
$$= \int_\Omega R(x,y)f(y)\,dy, \quad \forall f \in L^2(\Omega), \qquad (10.32)$$

where the kernel is defined as $R(x,y) \triangleq \delta(x-y)$. Kernel $R(x,y)$ is clearly not Hilbert-Schmidt. Nonetheless the sifting property can be viewed as an integral operator equation. The kernel in this case can be regarded as a *univariate* function because it is a function only of the difference $x - y$.

---

[16] ...over the great Grimpen Mire there hung a dense, white fog...

Since the delta function is real and symmetrical, we can define

$$\delta_y(x) \triangleq \delta(x - y) \equiv \overline{\delta(y - x)},$$

where $y \in \Omega$ is a parameter. (We note that $\delta_y(\cdot) \notin L^2(\Omega)$.) This means we can recast the sifting property as an inner product:

$$\langle f(\cdot), \delta_y(\cdot) \rangle = \int_\Omega f(x)\overline{\delta_y(x)}\, dx = f(y), \quad \forall y \in \Omega \ \ \forall f \in L^2(\Omega). \tag{10.33}$$

This is a "reproducing property" of the Dirac delta function in $L^2(\Omega)$. That is, putting a function $f \in L^2(\Omega)$ into an integral equation given by (10.32) or (10.33) with the special Dirac delta function kernel yields the same function $f \in L^2(\Omega)$ out when we evaluate for all possible $y \in \Omega$. The function $\delta_y(\cdot)$ is also called the "point-evaluation function at the point $y \in \Omega$" but, in this case, it does not belong to the space $L^2(\Omega)$.

10.5.3   Reproducing kernel property

Army types are hard enough to live with, but to think they are having children is pushing the boundary of acceptable. So what is this nonsense about reproducing colonels?

The sifting property in $L^2(\Omega)$, above, can be viewed in terms of the integral equation or an inner product. To find the analogous result for $\mathcal{H}_K$ it is far simpler to work from the inner product viewpoint (as we shall see). Following the $L^2(\Omega)$ manipulations, we write the kernel $K(x, y)$ in (10.31) as a univariate function, by treating y as a parameter,

$$K_y(x) \triangleq K(x, y), \quad \forall y \in \Omega. \tag{10.34}$$

Then the kernel function can be written in many useful ways:

$$K_y(x) = \sum_{n \in \mathbb{N}} \lambda_n \phi_n(x)\overline{\phi_n(y)},$$

$$= \sum_{n \in \mathbb{N}} \left(\lambda_n \overline{\phi_n(y)}\right) \phi_n(x),$$

$$= \sum_{n \in \mathbb{N}} (K_y)_n \phi_n(x), \quad \forall y \in \Omega.$$

This means, from the last two lines, that as a function in $L^2(\Omega)$, K_y has Fourier coefficients

$$(K_y)_n = \langle K_y, \phi_n \rangle = \lambda_n \overline{\phi_n(y)}, \quad \forall y \in \Omega. \tag{10.35}$$

However, this time $K_y(\cdot) \in \mathcal{H}_K$; see Problem 10.15.

Recalling the definition of an inner product in $\mathcal{H}_K$, (10.6), we evaluate the following inner product:

$$\langle f(\cdot), K_y(\cdot)\rangle_{\mathcal{H}_K} = \sum_{n\in\mathbb{N}} \frac{(f)_n \overline{(K_y)_n}}{\lambda_n},$$

$$= \sum_{n\in\mathbb{N}} \frac{(f)_n \overline{\lambda_n \phi_n(y)}}{\lambda_n},$$

$$= \sum_{n\in\mathbb{N}} (f)_n \phi_n(y) = \sum_{n\in\mathbb{N}} \langle f, \phi_n\rangle \phi_n(y).$$

That is,

$$\langle f(\cdot), K_y(\cdot)\rangle_{\mathcal{H}_K} = f(y), \quad \forall y \in \Omega, \ \forall f \in \mathcal{H}_K, \tag{10.36}$$

which is the "reproducing property" of the kernel K. This can also be written more briefly as $f(y) = \langle f, K_y\rangle_{\mathcal{H}_K}$, $\forall f \in \mathcal{H}_K$. This reproducing property can be written directly in terms of the kernel without the intermediate definition given in (10.34):

$$\langle f(\cdot), K(\cdot, y)\rangle_{\mathcal{H}_K} = f(y), \quad \forall y \in \Omega, \ \forall f \in \mathcal{H}_K.$$

In summary, we have found our RKHS and we note that the point-evaluation function at the point $y \in \Omega$ is in $\mathcal{H}_K$, that is, $K_y(\cdot) \in \mathcal{H}_K$. In the following theorem K_y is generated from the Mercer kernel, as given in (10.34), and to each such Mercer kernel K we can associated a unique RKHS.

Theorem 10.2 (Reproducing kernel Hilbert space). *There exists a unique Hilbert space $\mathcal{H}_K$ of functions on Ω satisfying the following conditions:*

C1: $K_y \in \mathcal{H}_K$, $\forall y \in \Omega$;

C2: *the span of the set $\{K_y : y \in \Omega\}$ is dense in $\mathcal{H}_K$;*

C3: $f(y) = \langle f, K_y\rangle_{\mathcal{H}_K}$, $\forall f \in \mathcal{H}_K$.

Moreover, $\mathcal{H}_K$ consists of continuous functions.   ☐

Proof. C3 was proven above, and C1 is covered in Problem 10.15. For the remainder see (Cucker and Smale, 2002).   ☐

10.5.4   Feature map and kernel trick

As we have seen, the kernel in $\mathcal{H}_K$ can be interpreted in many ways. Here is yet another way

$$K(x, y) = \sum_{n\in\mathbb{N}} \varphi_n(x)\overline{\varphi_n(y)}$$

$$= \langle \{\varphi_n(x)\}_{n\in\mathbb{N}}, \{\varphi_n(y)\}_{n\in\mathbb{N}}\rangle_{\ell^2}, \tag{10.37}$$

which is an apparently trite manipulation. Against instinct, we formalize this by defining a mapping, called the feature map,

$$\Phi \colon \Omega \longrightarrow \ell^2,$$
$$x \longmapsto \{\varphi_n(x)\}_{n\in\mathbb{N}} \equiv \{\sqrt{\lambda_n}\phi_n(x)\}_{n\in\mathbb{N}}. \tag{10.38}$$

That is, for each point $x \in \Omega \subset \mathbb{R}$ we generate a sequence in ℓ^2. Then the kernel in (10.37) can be written

$$\boxed{K(x,y) = \langle \Phi(x), \Phi(y)\rangle_{\ell^2}.} \tag{10.39}$$

This is simple manipulation but initially without clear motivation.

The ℓ^2 space here is called the feature space and the inner product of two feature vectors in the infinite-dimensional space is simply given by the evaluation of the kernel as a function. This is the central observation of the "kernel trick" — to compute an inner product one does not need to do an infinite-dimensional ℓ^2-inner product, but just use evaluations of the kernel as a function (Aizerman et al., 1964). Normally one chooses a simple continuous Hermitian positive definite kernel function $K(x,y)$ at the start rather than designing for particular eigenfunctions $\{\phi_n\}_{n\in\mathbb{N}}$ and positive eigenvalues $\{\lambda_n\}_{n\in\mathbb{N}}$ of the associated integral equation. These eigenfunctions and eigenvalues are not needed to work in the feature space. The right-hand side of (10.37) is indeed complicated, as would be more general inner product expressions, but the left-hand side of (10.37) involves only a simple kernel evaluation.

> **Remark 10.12.** The ability to compute inner products in an infinite dimensional space by just evaluating the kernel as in (10.39) can be exploited in other ways in the RKHS $\mathcal{H}_K$. To a point $y_0 \in \Omega$ we associate a point spread function $\delta_K(x - y_0)$ (or interpolating function) centered at $y_0 \in \Omega$, just as defined in (10.31). Therefore, we can think of synthesizing functions in our RKHS from a weighted sum of point spread functions based on "samples" at a finite number of points. This idea is explored in Problem 10.16.   □

Problems

10.15 (Kernel in RKHS). Prove that $K_y(\cdot) = K(\cdot, y) \in \mathcal{H}_K$, $\forall y \in \Omega$, where $K(x,y)$ is the Mercer kernel of the RKHS $\mathcal{H}_K$.

10.16 (RKHS inner product evaluation). Consider two sets, $\{x_1, x_2, \ldots, x_N\}$ and $\{y_1, y_2, \ldots, y_M\}$, of points belonging to the interval Ω. Then, by Problem 10.15,

$$K_{x_n}(\cdot) = K(\cdot, x_n) \in \mathcal{H}_K, \quad n = 1, 2, \ldots, N,$$

and

$$K_{y_m}(\cdot) = K(\cdot, y_m) \in \mathcal{H}_K, \quad m = 1, 2, \ldots, M$$

where $K(x,y)$ is the Mercer kernel of the RKHS $\mathcal{H}_K$.

To samples $\{x_n\}_{n=1}^N$ we can associate complex valued samples $\{\beta_n\}_{n=1}^N$, and to samples $\{y_m\}_{m=1}^M$ we can associate complex valued samples $\{\alpha_m\}_{m=1}^M$. Then with synthesized functions

$$f(\cdot) \triangleq \sum_{m=1}^M \alpha_m K_{y_m}(\cdot) \in \mathcal{H}_K \quad \text{and} \quad g(\cdot) \triangleq \sum_{n=1}^N \beta_n K_{x_n}(\cdot) \in \mathcal{H}_K,$$

show that the kernel reproducing property, (10.36), implies

$$\langle f, g \rangle_{\mathcal{H}_K} = \langle f(\cdot), g(\cdot) \rangle_{\mathcal{H}_K} = \sum_{m=1}^M \sum_{n=1}^N \alpha_m \overline{\beta_n} K(x_n, y_m).$$

10.6 RKHS on 2-sphere

10.6.1 RKHS construction

Here we bring together two tracks in this book, these being the emphasis on the generality of Hilbert space methods, presented in Part I and Part II, and the particular focus on the processing of signals on the 2-sphere, presented in Chapters 7, 8, and 9 of Part III. It is natural to ask what form an RKHS takes on for functions/signals defined on the domain $\mathbb{S}^2$, the 2-sphere, rather than a finite closed interval on the real line Ω. See Section 7.3, p. 189 onwards, for the definitions of $L^2(\mathbb{S}^2)$ and the spherical harmonic $Y_\ell^m(\widehat{x})$ of degree $\ell \in \{0, 1, 2, \ldots\}$ and order $m \in \{-\ell, -\ell+1, \ldots, \ell\}$, where $\widehat{x} \in \mathbb{S}^2$ is a point in the domain. In addition, $(f)_\ell^m = \langle f, Y_\ell^m \rangle$ denotes the spherical harmonic transform of $f \in L^2(\mathbb{S}^2)$, see (7.36), p. 195, which is expressed in terms of the inner product (7.17), p. 189. Finally, we use the shorthand $\sum_{\ell,m} \triangleq \sum_{\ell=0}^\infty \sum_{m=-\ell}^\ell$, see (2.71), p. 88.

For a construction of a class of RKHSs on the 2-sphere based on the spherical harmonics, which are orthonormal in $L^2(\mathbb{S}^2)$, we present a theorem which recasts the earlier RKHS results.

Theorem 10.3. *Define a square summable sequence of positive real weights*

$$\{\lambda_\ell^m : \lambda_\ell^m \in \mathbb{R}, \ \lambda_\ell^m > 0\}_{\ell,m} \in \ell^2, \tag{10.40}$$

and corresponding sequence of spherical harmonics

$$\{Y_\ell^m(\widehat{x})\}_{\ell,m} \tag{10.41}$$

with matching degrees $\ell \in \{0, 1, 2, \ldots\}$, and orders $m \in \{-\ell, -\ell+1, \ldots, \ell\}$.

Then the set of functions on the 2-sphere satisfying the following induced norm condition forms a Hilbert space, $\mathcal{H}_K$,

$$f \in \mathcal{H}_K \iff \|f\|_{\mathcal{H}_K} \triangleq \langle f, f \rangle_{\mathcal{H}_K}^{1/2} < \infty \tag{10.42}$$

with inner product

$$\langle f, g \rangle_{\mathcal{H}_K} \triangleq \sum_{\ell, m} \frac{\langle f, Y_\ell^m \rangle \langle Y_\ell^m, g \rangle}{\lambda_\ell^m}. \tag{10.43}$$

where f, g are two functions defined on domain $\mathbb{S}^2$. Furthermore, if we define the Hermitian symmetric[17] kernel

$$K(\widehat{\boldsymbol{x}}, \widehat{\boldsymbol{y}}) \triangleq \sum_{\ell, m} \lambda_\ell^m Y_\ell^m(\widehat{\boldsymbol{x}}) \overline{Y_\ell^m(\widehat{\boldsymbol{y}})}, \tag{10.44}$$

then we have the properties:

P1: *reproducing kernel property:*

$$\langle f(\cdot), K(\cdot, \widehat{\boldsymbol{y}}) \rangle_{\mathcal{H}_K} = f(\widehat{\boldsymbol{y}}), \quad \forall f \in \mathcal{H}_K, \tag{10.45}$$

P2: *kernel inclusion property:*

$$K(\cdot, \widehat{\boldsymbol{y}}) \in \mathcal{H}_K, \quad \forall \widehat{\boldsymbol{y}} \in \mathbb{S}^2, \tag{10.46}$$

P3: *eigenfunction property:*

$$(\mathcal{L}_K Y_\ell^m)(\widehat{\boldsymbol{x}}) = \lambda_\ell^m Y_\ell^m(\widehat{\boldsymbol{x}}), \tag{10.47}$$

for all $\ell \in \{0, 1, 2, \dots\}$, $m \in \{-\ell, -\ell + 1, \dots, \ell\}$, where

$$(\mathcal{L}_K f)(\widehat{\boldsymbol{x}}) \triangleq \int_{\mathbb{S}^2} K(\widehat{\boldsymbol{x}}, \widehat{\boldsymbol{y}}) f(\widehat{\boldsymbol{y}}) \, ds(\widehat{\boldsymbol{y}}), \quad f \in L^2(\mathbb{S}^2), \tag{10.48}$$

is a compact operator on $L^2(\mathbb{S}^2)$,

P4: *strictly positive operator property:*

$$\langle \mathcal{L}_K f, f \rangle > 0, \quad \forall f \in L^2(\mathbb{S}^2) \text{ satisfying } f \neq o, \tag{10.49}$$

where $o \in L^2(\mathbb{S}^2)$ is the zero vector.

Moreover, $\mathcal{H}_K$ consists only of continuous functions. □

Remark 10.13. Not every positive definite kernel $K(\widehat{\boldsymbol{x}}, \widehat{\boldsymbol{y}})$ can be put in this form (10.44) because we have restricted attention to using spherical harmonics. However, the theorem can be generalized by using a suitable orthonormal sequence $\{\varphi_n(\widehat{\boldsymbol{x}})\}_{n \in \mathbb{Z}}$ which is complete in $L^2(\mathbb{S}^2)$, in place of the sequence of spherical harmonics given in (10.41); and positive real $\{\lambda_n\}_{n \in \mathbb{Z}} \in \ell^2$ replacing the eigenvalues (10.40). This generalization is complete because the operator $\mathcal{L}_K$ in (10.48) is compact and self-adjoint and by the spectral theorem has a countable number of eigenfunctions that are orthogonal and complete in $L^2(\mathbb{S}^2)$ which can be taken as $\{\varphi_n(\widehat{\boldsymbol{x}})\}_{n \in \mathbb{Z}}$. □

---

[17] Obviously $K(\widehat{\boldsymbol{x}}, \widehat{\boldsymbol{y}}) = \overline{K(\widehat{\boldsymbol{y}}, \widehat{\boldsymbol{x}})}$, given the $\{\lambda_\ell^m\}_{\ell, m}$ are real.

10.6.2   Isomorphism

It is important to emphasize that the only difference between a conventional Hilbert space of functions such as $L^2(\mathbb{S}^2)$ and an RKHS of functions defined on the same domain such as are given in $\mathcal{H}_K$ is in the representation of the vectors and the definition of the inner product. All Hilbert spaces of the same dimension over the same scalar field are isomorphic, see Definition 2.29 and Theorem 2.17, p. 99. The vectors in one Hilbert space can be mapped to vectors in the other Hilbert space in a way that evaluations of inner products (and norms) are preserved, and the Fourier coefficients are equal. Here we show how to set up the isomorphism between the $L^2(\mathbb{S}^2)$ and $\mathcal{H}_K$. The isomorphism is given by

$$\mathcal{L}_K^{-1/2} : \mathcal{H}_K \longrightarrow L^2(\mathbb{S}^2) \qquad (10.50)$$
$$f \longmapsto h$$

and

$$\mathcal{L}_K^{1/2} : L^2(\mathbb{S}^2) \longrightarrow \mathcal{H}_K \qquad (10.51)$$
$$h \longmapsto f$$

where $\mathcal{L}_K^{1/2}$ is the strictly positive square root of the operator $\mathcal{L}_K$ in (10.48) satisfying $\mathcal{L}_K = \mathcal{L}_K^{1/2} \circ \mathcal{L}_K^{1/2}$, and $\mathcal{L}_K^{-1/2}$ is its inverse. These mappings are straightforward once one recognizes that

$$(\mathcal{L}_K^p f)(\widehat{\boldsymbol{x}}) = \sum_{\ell,m} (\lambda_\ell^m)^p \langle f, Y_\ell^m \rangle Y_\ell^m,$$

for any real power $p \in \mathbb{R}$, noting that $\lambda_\ell^m > 0$ (positive real). Note that $\mathcal{L}_K$ is compact, but $\mathcal{L}_K^{1/2}$ need not be. We can expect that $\mathcal{L}_K^{-1/2}$ is not bounded.

10.6.3   Two representation of vectors

Expansion using orthonormal sequence

Now we are interested in how to represent the vectors in the RKHS $\mathcal{H}_K$ (10.42). The spherical harmonics can be used in a Fourier synthesis expansion to represent vectors in $L^2(\mathbb{S}^2)$ since they are complete and orthonormal. They have many ideal properties that have made them a compelling representation and have dominated applications and theory as we have seen in previous chapters. With $\mathcal{H}_K$ it might be regarded as a death-nell for this new Hilbert space if we have to abandon spherical harmonics for something less workable. Fortunately that is not the case.

As an instance of (10.24), for $\mathcal{H}_K$ defined in (10.42) in Theorem 10.3 the sequence of functions

$$\{\varphi_\ell^m\}_{\ell,m}, \qquad (10.52)$$

where

$$\boxed{\varphi_\ell^m(\widehat{\boldsymbol{x}}) \triangleq \sqrt{\lambda_\ell^m} Y_\ell^m(\widehat{\boldsymbol{x}}),}$$

is orthonormal under inner product (10.43) and is complete. Therefore we have

$$f(\widehat{\boldsymbol{x}}) = \sum_{\ell,m} \langle f, \varphi_\ell^m \rangle_{\mathcal{H}_K} \varphi_\ell^m(\widehat{\boldsymbol{x}})$$

and

$$f \in \mathcal{H}_K \iff \left\{ \langle f, \varphi_\ell^m \rangle_{\mathcal{H}_K} \right\}_{\ell,m} \in \ell^2.$$

Hence, we see that the spherical harmonics are preserved in shape, are orthogonal but are *not normalized* in $\mathcal{H}_K$. It is because the spherical harmonics are attenuated for higher degrees ℓ (noting that condition (10.40) implies $\lambda_\ell^m \to 0$ as $\ell \to \infty$) relative to their mapped isomorphic counterparts in $L^2(\mathbb{S}^2)$ (10.50), that the functions in $\mathcal{H}_K$ are smoother, low-pass and indeed continuous.

Inner products (10.43), in $\mathcal{H}_K$, can be alternatively evaluated in the RKHS Fourier domain using the orthonormal sequence in $\mathcal{H}_K$ (10.52),

$$\langle f, g \rangle_{\mathcal{H}_K} \triangleq \sum_{\ell,m} \langle f, \varphi_\ell^m \rangle_{\mathcal{H}_K} \overline{\langle g, \varphi_\ell^m \rangle}_{\mathcal{H}_K}$$

$$= \sum_{\ell,m} \langle f, \varphi_\ell^m \rangle_{\mathcal{H}_K} \langle \varphi_\ell^m, g \rangle_{\mathcal{H}_K}$$

which is the generalized Parseval relation (2.57), p. 72. Note that, in comparison, in (10.43) the inner product in $\mathcal{H}_K$ is defined in terms of the $L^2(\mathbb{S}^2)$ Fourier domain.

Expansion using kernel

As revealed in the analogous case of Problem 10.16, an RKHS with kernel $K(\widehat{\boldsymbol{x}}, \widehat{\boldsymbol{y}})$ offers a second way to represent vectors because of the two properties (10.45) and (10.46). First we rewrite (10.46) as follows:

$$\begin{aligned} K_{\widehat{\boldsymbol{y}}}(\widehat{\boldsymbol{x}}) &\triangleq K(\widehat{\boldsymbol{x}}, \widehat{\boldsymbol{y}}), \quad \forall \widehat{\boldsymbol{x}}, \widehat{\boldsymbol{y}} \in \mathbb{S}^2, \\ K_{\widehat{\boldsymbol{y}}} &\in \mathcal{H}_K, \quad \forall \widehat{\boldsymbol{y}} \in \mathbb{S}^2. \end{aligned} \qquad (10.53)$$

Now suppose we have a finite number P of points on the 2-sphere, that is, $\widehat{\boldsymbol{y}}_p \in \mathbb{S}^2$, and corresponding samples $\alpha_p \in \mathbb{C}$, for $p \in \{1, 2, \ldots, P\}$, such that we can define

$$f(\cdot) \triangleq \sum_{p=1}^{P} \alpha_p K_{\widehat{\boldsymbol{y}}_p}(\cdot) = \sum_{p=1}^{P} \alpha_p K(\cdot, \widehat{\boldsymbol{y}}_p) \in \mathcal{H}_K.$$

Similarly, suppose we have Q samples $\beta_q \in \mathbb{C}$ located at $\widehat{\boldsymbol{x}}_q \in \mathbb{S}^2$, for $q \in \{1, 2, \ldots, Q\}$, such that we can define

$$g(\cdot) \triangleq \sum_{q=1}^{Q} \beta_q K_{\widehat{\boldsymbol{x}}_q}(\cdot) = \sum_{q=1}^{Q} \beta_q K(\cdot, \widehat{\boldsymbol{x}}_q) \in \mathcal{H}_K.$$

Then the reproducing property (10.45) leads to

$$\boxed{ \langle f(\cdot), g(\cdot) \rangle_{\mathcal{H}_K} = \sum_{p=1}^{P} \sum_{q=1}^{Q} \alpha_p \overline{\beta_q} K(\widehat{\boldsymbol{x}}_q, \widehat{\boldsymbol{y}}_p). }$$

10.6.4   Closed-form isotropic reproducing kernels

The reproducing kernel property enables simple computation of inner products of functions when expanded in functions derived from the same kernel, that is, using the "kernel inclusion property" (10.46). When contemplating inner product evaluation therefore it is of interest to find *closed-form* kernel functions rather than the general kernel expansion (10.44), which would involve an infinite complexity or at least possibly a high degree of computation. These closed-form kernel functions need to have eigenvalues λ_ℓ^m which are real, positive, and decay to zero sufficiently fast to satisfy (10.40) of Theorem 10.3.

Isotropic property

Some restriction is necessary to obtain a class of closed-form reproducing kernels which are isotropic. Here we use the term "isotropic" to refer to the property that the kernel $K(\widehat{\boldsymbol{x}}, \widehat{\boldsymbol{y}})$ only depends on the angle $\widehat{\boldsymbol{x}} \cdot \widehat{\boldsymbol{y}}$ between the two arguments and not on the individual directions $\widehat{\boldsymbol{x}}$ and $\widehat{\boldsymbol{y}}$.

We consider the case where the eigenvalues satisfy

$$\boxed{\lambda_\ell^m = \lambda_\ell,} \tag{10.54}$$

for all degrees $\ell \in \{0, 1, 2, \ldots\}$, and for all orders $m \in \{-\ell, -\ell+1, \ldots, \ell\}$. Then (10.44) becomes

$$K(\widehat{\boldsymbol{x}}, \widehat{\boldsymbol{y}}) = \sum_{\ell=0}^{\infty} \lambda_\ell \sum_{m=-\ell}^{\ell} Y_\ell^m(\widehat{\boldsymbol{x}})\overline{Y_\ell^m(\widehat{\boldsymbol{y}})}$$

$$= k(\widehat{\boldsymbol{x}} \cdot \widehat{\boldsymbol{y}}), \quad \widehat{\boldsymbol{x}}, \widehat{\boldsymbol{y}} \in \mathbb{S}^2, \tag{10.55}$$

where we have defined the real-valued[18] univariate kernel function

$$\boxed{k(z) \triangleq \frac{1}{4\pi} \sum_{\ell=0}^{\infty} (2\ell + 1)\lambda_\ell P_\ell(z), \quad -1 \leq z \leq +1} \tag{10.56}$$

where $P_\ell(\cdot)$ is the Legendre polynomial of degree ℓ, see Section 2.5.1, p. 53, and we have used the addition theorem of spherical harmonics of (7.30), p. 193, repeated here

$$\sum_{m=-\ell}^{\ell} Y_\ell^m(\widehat{\boldsymbol{x}})\overline{Y_\ell^m(\widehat{\boldsymbol{y}})} = \frac{(2\ell+1)}{4\pi} P_\ell(\widehat{\boldsymbol{x}} \cdot \widehat{\boldsymbol{y}}).$$

In essence, the term $(2\ell + 1)$ captures the multiplicity of the λ_ℓ eigenvalue. Also, as given in (Kennedy et al., 2011, p. 663),[19]

$$\boxed{\lambda_\ell = 2\pi \int_{-1}^{+1} k(z)P_\ell(z)\,dz, \quad \ell \in \{0, 1, 2, \ldots\},} \tag{10.57}$$

---

[18] Note that because the Legendre polynomials are real and the eigenvalues are real then the univariate kernel function must be real-valued.

[19] In (Seon, 2006), a similar development is given but is marred by the omission on the leading 2π term.

which means we can recover the eigenvalues from the univariate kernel function $k(\cdot)$ in (10.56). In principle, one could start with a candidate univariate kernel function $k(\cdot)$ and use (10.57) to test that all eigenvalues are positive as required in Theorem 10.3. In fact, it is preferable to start with $k(\cdot)$ because, by the kernel trick, it is all that is needed to evaluate inner products. Knowledge of the values of the eigenvalues beyond knowing they are positive and decay sufficiently quickly is not needed.

Hilbert-Schmidt property

We can specify the requirement (10.40) in Theorem 10.3 to incorporate the properties that the eigenvalues are with multiplicity, order-invariant, real-valued and positive

$$\{\lambda_\ell^m = \lambda_\ell : \lambda_\ell^m \in \mathbb{R}, \ \lambda_\ell^m > 0\}_{\ell,m} \in \ell^2$$
$$\Longleftrightarrow \{\sqrt{2\ell+1}\lambda_\ell : \lambda_\ell \in \mathbb{R}, \ \lambda_\ell > 0\}_{\ell=0}^\infty \in \ell^2. \tag{10.58}$$

This condition can be written more plainly as

$$\sum_{\ell=0}^{\infty} (2\ell+1)\lambda_\ell^2 < \infty, \quad \lambda_\ell \in \mathbb{R}, \ \lambda_\ell > 0. \tag{10.59}$$

Now because the Parseval relation for Legendre polynomials can be written (see Section 2.5.1, p. 53)

$$\int_{-1}^{+1} k^2(z)\,dz = \frac{1}{8\pi^2}\sum_{\ell=0}^{\infty}(2\ell+1)\lambda_\ell^2,$$

we see that Hilbert-Schmidt property (10.59) is also equivalent to the simple test

$$\int_{-1}^{+1} k^2(z)\,dz < \infty \tag{10.60}$$

on the univariate kernel function.

Closed-form kernel construction

Any univariate kernel function $k(\cdot)$ defined on domain $[-1,+1]$ can be expanded in terms of Legendre polynomials, see (2.33) in Section 2.5.1, p. 53. To have a Mercer kernel, we also require the Fourier coefficients[20] to be positive and the corresponding eigenvalues, $\{\lambda_\ell\}_{\ell=0}^\infty$, to decay sufficiently quickly as prescribed in (10.59) or equivalently in(10.60). We can summarize this construction in the following theorem.

---

[20] The actual Legendre Fourier coefficients are easily related to the eigenvalues and positivity of one implies the positivity of the other.

Theorem 10.4 (2-sphere RKHS from a univariate kernel function). *Suppose a real-valued function $k(\cdot)$ defined on domain $[-1, +1]$ satisfies conditions:*

C1: *Mercer condition:* $\lambda_\ell = 2\pi \displaystyle\int_{-1}^{+1} k(z) P_\ell(z)\,dz > 0, \quad \forall \ell \in \{0, 1, 2, \ldots\},$

C2: *Hilbert-Schmidt condition:* $\displaystyle\int_{-1}^{+1} k^2(z)\,dz < \infty.$

Then there exists an RKHS on the 2-sphere, $\mathcal{H}_K$, with (isotropic) kernel

$$K(\widehat{\boldsymbol{x}}, \widehat{\boldsymbol{y}}) = k(\widehat{\boldsymbol{x}} \cdot \widehat{\boldsymbol{y}})$$
$$= \sum_{\ell, m} \lambda_\ell Y_\ell^m(\widehat{\boldsymbol{x}}) \overline{Y_\ell^m(\widehat{\boldsymbol{y}})}$$

where

$$k(z) = \frac{1}{4\pi} \sum_{\ell=0}^{\infty} (2\ell + 1) \lambda_\ell P_\ell(z).$$

Furthermore, the inner product is given by

$$\langle f, g \rangle_{\mathcal{H}_K} = \sum_{\ell=0}^{\infty} \frac{1}{\lambda_\ell} \sum_{m=-\ell}^{\ell} (f)_\ell^m \overline{(g)_\ell^m},$$

where $(f)_\ell^m = \langle f, Y_\ell^m \rangle$, $(g)_\ell^m = \langle g, Y_\ell^m \rangle$ are the spherical harmonic transforms of $f, g \in L^2(\mathbb{S}^2)$, respectively. □

Generally it may be difficult or tedious to confirm that all the eigenvalues for a candidate closed-form real-valued univariate kernel function $k(\cdot)$ are positive. Another strategy is to find functions defined on domain $[-1, +1]$ with known Legendre series expansions with Fourier coefficients which are non-negative and use these as the basis for construction. We now illustrate this approach.

Example 10.3 (Cui and Freeden). An identity equivalent to

$$\frac{1}{4\pi} \sum_{\ell=1}^{\infty} \frac{P_\ell(z)}{\ell(\ell + 1)} = \frac{1}{4\pi} - \frac{1}{2\pi} \log\left(1 + \sqrt{\frac{1 - z}{2}}\right), \quad -1 \le z \le +1 \quad (10.61)$$

has been used to construct an RKHS kernel on the 2-sphere (Cui and Freeden, 1997). Comparing (10.61) with (10.56) we infer the eigenvalue expressions are

$$\lambda_\ell = \begin{cases} \lambda_0 > 0 & \ell = 0, \\ \dfrac{1}{\ell(\ell + 1)(2\ell + 1)} & \ell \in \{1, 2, \ldots\}, \end{cases}$$

and do satisfy (10.58). Therefore,

$$\boxed{K(\widehat{\boldsymbol{x}}, \widehat{\boldsymbol{y}}) = \frac{(1 + \lambda_0)}{4\pi} - \frac{1}{2\pi} \log\left(1 + \sqrt{\frac{1 - (\widehat{\boldsymbol{x}} \cdot \widehat{\boldsymbol{y}})}{2}}\right), \quad \lambda_0 > 0} \quad (10.62)$$

is a reproducing kernel. Each $Y_\ell^m(\widehat{\boldsymbol{x}})$ is an eigenfunction of the integral equation with the above kernel with eigenvalue $\lambda_\ell > 0$. The choice $\lambda_0 = 1$ is standard (Cui and Freeden, 1997). □

Problems

10.17 (Lebedev kernel RKHS). The following identity is an adaption of the Legendre polynomial expansion given in (Lebedev, 1972, pp. 55-60):

$$\frac{1}{4\pi} \sum_{\ell=1}^{\infty} \frac{P_\ell(z)}{(2\ell-1)(2\ell+3)} = \frac{1}{12\pi} - \frac{1}{8\pi}\sqrt{\frac{1-z}{2}}, \quad -1 \le z \le +1. \quad (10.63)$$

Use this identity to show

$$K(\hat{\boldsymbol{x}}, \hat{\boldsymbol{y}}) = \frac{(1+3\lambda_0)}{12\pi} - \frac{1}{8\pi}\sqrt{\frac{1-(\hat{\boldsymbol{x}}\cdot\hat{\boldsymbol{y}})}{2}}, \quad \lambda_0 > 0$$

is a reproducing kernel, and compare with the Cui and Freden kernel (10.62). Give the expression for the eigenvalues λ_ℓ and show they are positive.

10.18 (Legendre generating function kernel RKHS). The generating function for the Legendre polynomials is given by

$$\sum_{\ell=0}^{\infty} \rho^\ell P_\ell(z) = \frac{1}{\sqrt{1-2z\rho+\rho^2}}, \quad -1 \le z \le +1. \quad (10.64)$$

where ρ is the indeterminate. Use this identity to show

$$K(\hat{\boldsymbol{x}}, \hat{\boldsymbol{y}}) = \frac{1}{\sqrt{1-2(\hat{\boldsymbol{x}}\cdot\hat{\boldsymbol{y}})\rho+\rho^2}}$$

is a reproducing kernel for $0 < \rho < 1$. Give the expression for the eigenvalues λ_ℓ and show they are positive and decay exponentially for $0 < \rho < 1$.

10.19 (von Mises-Fisher distribution kernel RKHS). The von Mises-Fisher distribution for the 2-sphere (Seon, 2006) can be used for inspiration to define perhaps the most natural 2-sphere Mercer kernel. Show that

$$K(\hat{\boldsymbol{x}}, \hat{\boldsymbol{y}}) = \frac{\kappa \exp\big(\kappa\,(\hat{\boldsymbol{x}}\cdot\hat{\boldsymbol{y}})\big)}{4\pi \sinh \kappa}$$

is a reproducing kernel for any choice of the real-valued "concentration parameter" $\kappa \ge 0$. The identity given in (Seon, 2006, Equation (17)) may be helpful to prove this result.

10.7 RKHS synopsis

This chapter has looked at the theory of reproducing kernel Hilbert spaces (RKHSs) using the techniques from earlier in the book. RKHSs have been presented as a special representation of complex separable Hilbert spaces where the vectors are functions defined on a bounded closed interval $\Omega = [a, b]$ (or as functions defined on the 2-sphere $\mathbb{S}^2$). We summarize the key findings and point to further reading:

R1: (RKHS Characterization). For every continuous, positive definite Mercer kernel $K(x, y)$ we can associate a unique RKHS and for all RKHSs there exists a unique kernel; see Theorem 10.2, p. 350. In this way we can fully characterize the class of all RKHSs; see (Aronszajn, 1950; Cucker and Smale, 2002).

R2: (Kernel and associated eigenstructure). Associated with each continuous, positive definite RKHS kernel $K(x, y)$ we have a countable set of eigenvalues, $\{\lambda_n\}_{n \in \mathbb{N}} \in \ell^2$, such that the λ_n are real and positive, and corresponding eigenfunctions, $\{\phi_n\}_{n \in \mathbb{N}}$ are continuous functions (and can be chosen orthonormal); see (10.29), p. 345. These countable sets of eigenfunctions and eigenvalues can be obtained from the integral equation which has $K(x, y)$ as its kernel; see (10.26), p. 343. This means the kernel has all the information that is held by a countable sets of eigenfunctions and eigenvalues; see (10.27), p. 343, and (10.28), p. 344.

R3: (RKHS vectors are low pass functions). The inner product in the RKHS is a modification of the standard $L^2(\Omega)$ inner product expressed in terms of the eigenfunctions and eigenvalues; see (10.8), p. 330. In particular the eigenvalues appear as an inverse weight in the inner product. This means only those functions whose Fourier coefficients decay significantly fast to zero (relative to the weaker requirement to be in $L^2(\Omega)$) belong to the RKHS; see (10.12), p. 331.

R4: (RKHS vectors are continuous functions). Each RKHS is a true function space where every vector is a continuous function. This is not strictly true of $L^2(\Omega)$ where each vector is an equivalence class (or set) of functions; see Theorem 10.2, p. 350. Of course, $L^2(\Omega)$ contains equivalence classes of discontinuous functions and equivalence classes of some continuous functions that are not in the RKHS. Because the Fourier coefficients of functions in the RKHS decay so rapidly then this is why they are continuous, as must be the eigenfunctions; see Theorem 10.2, p. 350, and (10.29), p. 345.

R5: (RKHS isomorphism with $L^2(\Omega)$). Despite their differing vectors we can set up an isomorphism between the vectors in $L^2(\Omega)$ and the vectors in the RKHS; see (10.25), p. 341. The existence of such an isomorphism follows from separability, equal cardinality and both being complex Hilbert spaces. The isomorphism can be expressed in terms of the square root of the strictly positive definite integral operator $\mathcal{L}_K > 0$ with kernel K and its inverse; that is, with $\mathcal{L}_K$ replacing $\mathcal{L}_\lambda$ in the isomorphism equations (10.25), p. 341.

R6: (RKHS kernel reproducing property). The RKHS kernel has a reproducing property that is analogous to the sifting property of the Dirac delay function for $L^2(\Omega)$; see (10.36), p. 350. However, for the RKHS this kernel also (in a precise sense) belongs to the RKHS; see Problem 10.15. This kernel and its reproducing property are central to representing the RKHS and computing efficiently the inner product using kernel evaluations; see Theorem 10.2 and Problem 10.16.

R7: (RKHS reference). A rigorous treatment of RKHS in the context of learning theory is given in the 49-page research report (Cucker and Smale, 2002). The paper goes beyond just RKHS and looks at various function and operator norm bounds. The more relevant parts are:

- spaces associated to a kernel, p. 11;
- Theorem 2, the spectral theorem and the treatment of powers τ of a strictly positive operator, p. 27;
- the isomorphism and inclusion mapping associated with a compact strictly positive operator, p. 29; and
- most of Chapter III, which includes Mercer's theorem, feature mapping and the equivalence of the operator kernel viewpoint and the weighting the Fourier coefficients viewpoint, pp. 32–37.

An RKHS, $\mathcal{H}_K$, is seen to be an inclusion into the space of continuous functions, $C(\Omega)$. It can be thought of as a subset and not a subspace, as we have described. An important observation is that much of the space of continuous functions is not in $\mathcal{H}_K$.

R8: (RKHS continuous evaluation functional approach). There is a chapter on RKHS in the book (Máté, 1989, Chapter 3, pp. 122–166). Some relevant parts are:

- it defines the RKHS through the continuous evaluation functional approach;
- there is a small section devoted to the question of separability of an RKHS, pp. 126–127;
- there are a number of good examples, but some of the spaces considered include Hardy space H^2 and Sobolev spaces; and
- there is a short proof that for every strictly positive symmetric kernel there exists a unique RKHS, pp. 125–126.

R9: (Mercer's theorem). There is generally a misunderstanding with Mercer's theorem being confused with the general kernel expansion developed by Schmidt and Hilbert given in (10.27), p. 343. Mercer's theorem provides sufficient conditions that strengthens the sense of convergence of the kernel expansion — from convergence in the mean, (10.27), p. 343, to uniform convergence, (10.28), p. 344; see Theorem 10.1, p. 344. This underpins the continuity of the functions in the RKHS. Furthermore, the conditions on the kernel match well with applications and constructions such as in the theory of stochastic processes. Finally, we recommend some reading material related to Mercer's theorem:

- the original work in English by Mercer (Mercer, 1909);
- the text by Riesz and Sz.-Nagy which has a short proof of the theorem (Riesz and Sz.-Nagy, 1990, pp. 245–246); and
- the mathematical foundations of learning research report which covers the theorem (Cucker and Smale, 2002, Chapter III).

Answers to problems in Part I

Answers to problems in Chapter 1

Answer to Problem 1.1. (page 7): Vectors, sequences and functions are the most useful cases. Linear mappings called operators can form a vector space as can matrices. In addition, vectors of functions and functions of vectors can form a vector space. □

Answer to Problem 1.2. (page 16): Move guest n to room $2n$ for all $n \in \mathbb{N}$. This frees up a countable number of rooms, those with odd room numbers. Then move new guest m into room $2m - 1$ for all $m \in \mathbb{N}$. □

Answer to Problem 1.3. (page 16): Now Hilbert is losing his patience. He gets the buses to line up and assigns a natural number $b \in \mathbb{N}$ to each bus. Within each bus the passengers are assigned a natural number $p \in \mathbb{N}$ and subsequently a "height" $b + p$.

Now the problem looks simpler than the argument we used for the rationals $\mathbb{Q}$. There are only a finite number of guests of a given height and these can be ordered. For example, passenger 1 on bus 3 has the same height as passenger 2 on bus 2, and there are only 3 passengers of such height 4. So in $(b, p) \equiv$ (bus, passenger) notation; the guest rooms assignments are:

(b,p)	(1,1)	(1,2)	(2,1)	(1,3)	(2,2)	(3,1)	(1,4)	(2,3)	(3,2)	(4,1)	(1,5)	$\cdots$
height	2	3	3	4	4	4	5	5	5	5	6	$\cdots$
Room	1	2	3	4	5	6	7	8	9	10	11	$\cdots$

Of course, this is one of many possible solutions. □

Answer to Problem 1.4. (page 17): Suppose a transcendental number is rational, say p/q. This is the root of the polynomial $f(x) = qx - p$ and so must be algebraic. This contradicts the definition of a transcendental number. □

362

Answer to Problem 1.5. (page 20): For a finite set of size N, the set of all its subsets contains 2^N elements. Since real numbers, written in binary format, represent all subsets of $\aleph_0$, it is justifiable to generalize and write $\mathfrak{c} = 2^{\aleph_0}$. $\square$

Answer to Problem 1.6. (page 20): Write the real numbers in decimal format, which can be mapped into their corresponding binary format. So if by the argument in the answer to Problem 1.5, we could write $\mathfrak{c} = 2^{\aleph_0}$, then we can also argue $\mathfrak{c} = 10^{\aleph_0}$. Clearly, this can be generalized by taking any base (other than 2 or 10) for the representation of real numbers. $\square$

Answer to Problem 1.7. (page 20): We have the following table

Operation	Symbolic notation	Example
Addition	$\mathfrak{c} + \aleph_0 = \mathfrak{c}$	$\mathbb{R}_0 \cup \mathbb{N}$
	$\mathfrak{c} + \mathfrak{c} = \mathfrak{c}$	$\mathbb{R}_0 \cup \mathbb{R}_1$
Multiplication	$\mathfrak{c} \cdot \aleph_0 = \mathfrak{c}$	$\mathbb{R} = \cup_{n \in \mathbb{Z}} \mathbb{R}_n$
	$\mathfrak{c} \cdot \mathfrak{c} = \mathfrak{c}$	$\mathbb{R}^2 = \{(\alpha, \beta) : \alpha, \beta \in \mathbb{R}\}$
Exponentiation	$\mathfrak{c}^{\aleph_0} = \mathfrak{c}$	Set of all single-valued functions from $\mathbb{Q}$ to $\mathbb{R}$
	$\mathfrak{c}^{\mathfrak{c}} > \mathfrak{c}$	Set of all single-valued functions from $\mathbb{R}$ to $\mathbb{R}$
	$2^{\mathfrak{c}} > \mathfrak{c}$	Set of indicator functions over all subsets of $\mathbb{R}$

where

$$\mathbb{R}_n \triangleq \{x : x \in \mathbb{R}, \ n \le x < n+1\}$$

is taken as our "reference" set. $\square$

Answer to Problem 1.8. (page 20): Any continuous function $f_1 : \mathbb{R} \to \mathbb{R}$ is completely characterized by its values over the set of rationals. That is, if $f_2 : \mathbb{R} \to \mathbb{R}$ is also continuous and $f_1(x) = f_2(x)$ for every x in $\mathbb{Q}$, then $f_1 = f_2$. Because of this, there can be at most as many continuous functions from $\mathbb{R}$ to $\mathbb{R}$ as there are functions of any kind from $\mathbb{Q}$ to $\mathbb{R}$. The cardinality of the set of all possible functions from $\mathbb{Q}$ to $\mathbb{R}$ is $|\mathbb{R}|^{|\mathbb{Q}|} = (2^{\aleph_0})^{\aleph_0} = 2^{(\aleph_0 \cdot \aleph_0)} = 2^{\aleph_0} = \mathfrak{c}$. Therefore, $\mathfrak{c}$ is an upper bound on the cardinality of the set of all continuous functions.

On the other hand, consider a subset of all continuous functions such as constant functions. The cardinality of all constant functions is simply $|\mathbb{R}| = \mathfrak{c}$. Therefore, $\mathfrak{c}$ is also a lower bound on the cardinality of the set of all continuous functions. And by combining the two arguments, we see that the cardinality of the set of all continuous functions has to be equal to $\mathfrak{c}$. $\square$

Answer to Problem 1.9. (page 20): The set of all possible single-valued functions $f : \mathbb{R} \to \mathbb{R}$ has a cardinality $|\mathbb{R}|^{|\mathbb{R}|} = \mathfrak{c}^{\mathfrak{c}}$ (for any point on the real line, we can freely assign a function value on the real line). This involves exponentiation of $\mathfrak{c}$ and therefore the size of the set must be greater than $\mathfrak{c}$. $\square$

Answer to Problem 1.10. (page 22): The union of the umbrella intervals defines a function that dominates the rational numbers (in the sense that the integral is greater than or equal to). If none of the umbrellas were to overlap then the total width of all umbrellas is $\epsilon + \epsilon/2 + \epsilon/4 + \cdots = 2\epsilon$. So at most 2ϵ, and at least ϵ, of the real line stays dry. The remainder of the real line gets wet. In the limit as $\epsilon \to 0$ the dry portion goes to zero, since the dominating umbrella function integral goes to zero as $\epsilon \to 0$. □

Answer to Problem 1.11. (page 22): Having established that algebraic numbers are countable, we can construct an umbrella series with finite sum as in the construction of the rational number case. So the integral of such a function will be zero. □

Answers to problems in Chapter 2

Answer to Problem 2.1. (page 26): The numbers 1 and $i = \sqrt{-1}$ are linearly independent over the real scalar field and linearly dependent over the complex scalar field. So in general, linear independence depends on the scalar field assumed. □

Answer to Problem 2.2. (page 28): Since the triangle inequality, (2.4), p. 26, $\|f + g\| \le \|f\| + \|g\|$ is valid for any $f, g \in \mathcal{H}$, choose $g = -f$. Then, we obtain $\|f - f\| = \|o\| = 0 \le \|f\| + \|-f\| = 2\|f\|$. Since according to Definition 2.4 $\|f\| = 0$ if and only if $f = o$, we conclude that in all other cases $\|f\|$ must be strictly greater than zero. □

Answer to Problem 2.3. (page 28): A normed space is a special case of a metric space by: (1) letting $\mathcal{M}$ be a (complex) vector space, and (2) defining the distance as

$$d(f, g) = \|f - g\|.$$

The properties of a norm then guarantee the properties for the metric are satisfied. Certainly metric spaces are more general, but do not have enough structure that you could claim they subsume normed spaces. □

Answer to Problem 2.4. (page 29): For every ϵ desired in the Cauchy proof, choose $\epsilon_0 = \epsilon/2$ for the convergent sequence to find N and then use the triangle inequality (2.4), p. 26. □

Answer to Problem 2.5. (page 29): The Cauchy sequence $f_n = 1/n$ takes values in the interval $(0, 1]$ for all $n \in \mathbb{N}$, but does not converge to a point in this set because 0 is excluded. There are more sophisticated examples later in the book, beginning in Section 2.3.4, p. 32. □

Answer to Problem 2.6. (page 38): Let us write $\langle f, g \rangle = |\langle f, g \rangle| e^{i\theta}$ and $\langle g, f \rangle = \overline{\langle f, g \rangle} = |\langle f, g \rangle| e^{-i\theta}$.

$$\|f + g\| = \langle f + g, f + g \rangle^{1/2} = \left(\langle f, f \rangle + \langle g, g \rangle + \langle f, g \rangle + \langle g, f \rangle \right)^{1/2}$$

$$= \left(\|f\|^2 + \|g\|^2 + \langle f, g \rangle + \overline{\langle f, g \rangle} \right)^{1/2}$$

$$= \left(\|f\|^2 + \|g\|^2 + 2|\langle f, g \rangle| \cos \theta \right)^{1/2}$$

$$\leq \left(\|f\|^2 + \|g\|^2 + 2\|f\| \, \|g\| \cos \theta \right)^{1/2} \tag{A.1}$$

$$\leq \left(\|f\|^2 + \|g\|^2 + 2\|f\| \, \|g\| \right)^{1/2} = \|f\| + \|g\|,$$

where (A.1), p. 365, is by the Cauchy-Schwarz inequality. □

Answer to Problem 2.7. (page 42): Direct computation yields

$$\|a + b\|^2 - \|a - b\|^2 = \sum_n \left((\alpha_n + \beta_n)^2 - (\alpha_n - \beta_n)^2 \right)$$

$$= \sum_n (\alpha_n^2 + 2\alpha_n \beta_n + \beta_n^2) - \sum_n (\alpha_n^2 - 2\alpha_n \beta_n + \beta_n^2) = 4 \sum_n \alpha_n \beta_n \equiv 4 \langle a, b \rangle,$$

which yields the desired result. □

Answer to Problem 2.8. (page 42): Somewhat similar to Problem 2.7. Firstly,

$$\|a + b\|^2 - \|a - b\|^2 = \sum_n \left(|\alpha_n + \beta_n|^2 - |\alpha_n - \beta_n|^2 \right)$$

$$= \sum_n \left((\alpha_n + \beta_n)\overline{(\alpha_n + \beta_n)} - (\alpha_n - \beta_n)\overline{(\alpha_n - \beta_n)} \right)$$

$$= \sum_n \left(|\alpha_n|^2 + \alpha_n \overline{\beta_n} + \overline{\alpha_n} \beta_n + |\beta_n|^2 \right) - \sum_n \left(|\alpha_n|^2 - \alpha_n \overline{\beta_n} - \overline{\alpha_n} \beta_n + |\beta_n|^2 \right)$$

$$= 2 \sum_n \left(\alpha_n \overline{\beta_n} + \overline{\alpha_n} \beta_n \right),$$

and, secondly,

$$i\|a + ib\|^2 - i\|a - ib\|^2 = i \sum_n \left(|\alpha_n + i\beta_n|^2 - |\alpha_n - i\beta_n|^2 \right)$$

$$= i \sum_n \left((\alpha_n + i\beta_n)\overline{(\alpha_n + i\beta_n)} - (\alpha_n - i\beta_n)\overline{(\alpha_n - i\beta_n)} \right)$$

$$= i \sum_n \left(|\alpha_n|^2 - i\alpha_n \overline{\beta_n} + i\overline{\alpha_n} \beta_n + |\beta_n|^2 \right) - i \sum_n \left(|\alpha_n|^2 + i\alpha_n \overline{\beta_n} + i\overline{\alpha_n} \beta_n + |\beta_n|^2 \right)$$

$$= 2 \sum_n \left(\alpha_n \overline{\beta_n} - \overline{\alpha_n} \beta_n \right).$$

Then (2.13), p. 41, follows. □

Answer to Problem 2.9. (page 42): In (2.12), p. 39, we evaluate the two expressions

$$\|f + g\|^2 + \|f - g\|^2 = (3/2)^2 = (1/2)^2 = 5/2,$$

$$2 \left(\|f\|^2 + \|g\|^2 \right) = 2(1^2 + (1/2)^2) = 5/2,$$

and so the parallelogram law, (2.12), p. 39, is satisfied. But of course, it needs to hold for all choices of functions, not just these particular f and g — it is just lucky here. In Problem 2.10 we see that there exist other functions such that the parallelogram law fails. Therefore, we cannot construct an inner product for this space. $\qquad\square$

Answer to Problem 2.10. (page 42): The calculations are summarized in the following table, all functions are defined on the closed interval $[0,1]$:

Expression	$\|\cdot\|_1$	$\|\cdot\|_2$	$\|\cdot\|_{\max}$
$h(x) = 1 - x$	$1/2$	$1/\sqrt{3}$	1
$g(x) = x$	$1/2$	$1/\sqrt{3}$	1
$h(x) + g(x) = 1$	1	1	1
$h(x) - g(x) = 1 - 2x$	$1/2$	$1/3$	1
$\|h + g\|^2 + \|h - g\|^2$	$3/2$	$4/3$	2
	$\neq$	$\|$	$\neq$
$2\big(\|h\|^2 + \|g\|^2\big)$	$1/2$	$4/3$	4

The parallelogram law holds only for $\|\cdot\|_2$, which, by observation, can be induced from the inner product

$$\langle f, g \rangle = \int_0^1 f(x)\overline{g(x)}\, dx.$$

$\qquad\square$

Answer to Problem 2.11. (page 45): Suppose that $f_1, f_2, \ldots, f_n$ are n mutually orthogonal elements of $\mathcal{H}$ and assume that the linear combination $\sum_{k=1}^n \alpha_k f_k = o$. We need to show that $\alpha_1 = \alpha_2 = \cdots = \alpha_n = 0$. For any $1 \le j \le n$, we can write

$$0 = \langle o, f_j \rangle = \Big\langle \sum_{k=1}^n \alpha_k f_k, f_j \Big\rangle = \sum_{k=1}^n \alpha_k \langle f_k, f_j \rangle = \alpha_j \|f_j\|^2,$$

where the last equality follows from the assumption of orthogonality ($\langle f_k, f_j \rangle = \delta_{j,k}$). Since f_j is a non-zero vector, it must be that $\alpha_j = 0$. Therefore, orthogonal vectors in $\mathcal{H}$ are also linearly independent. $\qquad\square$

Answer to Problem 2.12. (page 45): We have

$$\|f - g\|^2 = \langle f - g, f - g \rangle = \langle f, f \rangle - \langle f, g \rangle - \langle g, f \rangle + \langle g, g \rangle$$
$$= \|f\|^2 - \langle f, g \rangle - \langle g, f \rangle + \|g\|^2.$$

Given f and g are orthonormal; we have $\|f\| = \|g\| = 1$ and $\langle f, g \rangle = \langle g, f \rangle = 0$. Then $\|f - g\|^2 = 2$. $\qquad\square$

Answer to Problem 2.13. (page 50): In $L^2(0,1)$ the vectors are not functions, but equivalence classes of functions. So every function that equals the zero function a.e. belongs to the class that constitutes the zero vector. For example, the third function considered in Figure 1.5, p. 21, $f_3(x) = \mathbf{1}_{\mathbb{Q}}(x)$, satisfies $\|f_3\| =$

0 and is as worthy as $o(x)$ to be a candidate zero vector. Of course, the zero function is a sane choice as a representative of the equivalence class and any other choice, such as $f_3(x)$ which differs only on a set of measure zero from $o(x)$, is just being annoying. ☐

Answer to Problem 2.14. (page 50): *Reflexivity:* With $g = f$ then $f - g$ is pointwise zero everywhere and certainly its norm is zero. *Symmetry:* for the norm we have $\|f - g\| = \|-(f - g)\| = \|g - f\|$, so $\|f - g\| = 0$ implies $\|g - f\| = 0$, or $f \sim g$ implies $g \sim f$. *Transitivity:* starting with $\|f - g\| = 0$ and $\|g - h\| = 0$ then, by (2.24), p. 49, $f = g$ a.e. and $g = h$ a.e. and, therefore, $f = h$ a.e. which implies $\|f - h\| = 0$. ☐

Answer to Problem 2.15. (page 50): Consider the set $\pi\mathbb{Q}\cap(0, 1]$, which includes points such as $\pi/4$, and $134\pi/1487$, and excludes zero and points outside the domain. Since π is irrational (indeed transcendental) then $\pi\mathbb{Q}\setminus\{0\}$ contains no rational numbers and is countable. ☐

Answer to Problem 2.16. (page 56): See the table below which specifies the Legendre polynomials $P_\ell(x)$ coefficients, in the indicated ℓ column, for $\ell \in \{0, 1, \dots, 15\}$.

ℓ	0	1	2	3	4	5	6	7	8	9	10	11	12	13	14	15
×	1	1	$\frac{1}{2}$	$\frac{1}{2}$	$\frac{1}{8}$	$\frac{1}{8}$	$\frac{1}{16}$	$\frac{1}{16}$	$\frac{1}{128}$	$\frac{1}{128}$	$\frac{1}{256}$	$\frac{1}{256}$	$\frac{1}{1024}$	$\frac{1}{1024}$	$\frac{1}{2048}$	$\frac{1}{2048}$
1	1	0	−1	0	3	0	−5	0	35	0	−63	0	231	0	−429	0
x	0	1	0	−3	0	15	0	−35	0	315	0	−693	0	3003	0	−6435
x^2	0	0	3	0	−30	0	105	0	−1260	0	3465	0	−18018	0	45045	0
x^3	0	0	0	5	0	−70	0	315	0	−4620	0	15015	0	−90090	0	255255
x^4	0	0	0	0	35	0	−315	0	6930	0	−30030	0	225225	0	−765765	0
x^5	0	0	0	0	0	63	0	−693	0	18018	0	−90090	0	765765	0	−2909907
x^6	0	0	0	0	0	0	231	0	−12012	0	90090	0	−1021020	0	4849845	0
x^7	0	0	0	0	0	0	0	429	0	−25740	0	218790	0	−2771340	0	14549535
x^8	0	0	0	0	0	0	0	0	6435	0	−109395	0	2078505	0	−14549535	0
x^9	0	0	0	0	0	0	0	0	0	12155	0	−230945	0	4849845	0	−37182145
x^{10}	0	0	0	0	0	0	0	0	0	0	46189	0	−1939938	0	22309287	0
x^{11}	0	0	0	0	0	0	0	0	0	0	0	88179	0	−4056234	0	50702925
x^{12}	0	0	0	0	0	0	0	0	0	0	0	0	676039	0	−16900975	0
x^{13}	0	0	0	0	0	0	0	0	0	0	0	0	0	1300075	0	−35102025
x^{14}	0	0	0	0	0	0	0	0	0	0	0	0	0	0	5014575	0
x^{15}	0	0	0	0	0	0	0	0	0	0	0	0	0	0	0	9694845

The second row shows the leading multiplicative fraction of the polynomial which needs to multiply all the indicated integer-valued polynomial coefficients. ☐

Answer to Problem 2.17. (page 56): See (Lebedev, 1972, pp. 50–51). ☐

Answer to Problem 2.18. (page 56): Take the inner product of both sides of Legendre series (2.33), p. 55, with $P_\ell(x)$ and use orthogonality. Otherwise, if needed, see (Lebedev, 1972, p. 53). ☐

Answer to Problem 2.19. (page 56): We note that even/odd degree Legendre polynomials are even/odd functions. By throwing out all the odd degree Legendre polynomials, we can only represent all piecewise smooth even functions. "All" because the remaining odd degree Legendre polynomials are orthogonal to even functions and would have zero coefficient and Theorem 2.5, p. 55, can be invoked. Therefore, piecewise smooth odd functions or piecewise smooth functions that are neither even or odd cannot be represented by even degree Legendre polynomials.   □

Answer to Problem 2.20. (page 59): Take the weighted inner product of both sides of Hermite series (2.40), p. 58, with $H_m(x)$ and use orthogonality. Otherwise, if needed, see (Lebedev, 1972, p. 68).   □

Answer to Problem 2.21. (page 59): Hermite functions can be written

$$\varrho_n(x) = \sqrt{w(x)}\zeta_n(x) = e^{-x^2/2}\zeta_n(x) = (2^n n!\sqrt{\pi})^{-1/2}e^{-x^2/2}H_n(x),$$
$$n \in \{0, 1, \ldots\}.$$

And thereby, the inner product, (2.39), p. 58, can be written

$$\langle \zeta_m, \zeta_n \rangle_w = \int_{-\infty}^{\infty} e^{-x^2/2}\zeta_m(x)e^{-x^2/2}\zeta_n(x)\, dx = \langle \varrho_m, \varrho_n \rangle = \delta_{n,m},$$
$$m, n \in \{0, 1, \ldots\},$$

where $\langle \cdot, \cdot \rangle$ is the un-weighted inner product on the real line $\mathbb{R}$. This shows $\varrho_n(x)$ are orthonormal on $L^2(\mathbb{R})$ with weighting $w(x) = 1$.   □

Answer to Problem 2.22. (page 65): $\mathfrak{M}_k$ is a manifold because for $a = \alpha e_k$ and $b = \beta e_k$, $a + b = (\alpha + \beta)e_k \in \mathfrak{M}_k$. Proving that $\mathfrak{M}_k$ is closed is done similarly to Example 2.11:

Assume that c is an accumulation point of $\mathfrak{M}_k$. Then for any $\epsilon > 0$, there exists an $a \in \mathfrak{M}_k$ for which $\|a - c\| = \|\alpha e_k - c\| < \epsilon$. Since ℓ^2 is closed, $c \in \ell^2$ and is of the form $c = \sum_{n=1}^{\infty} \gamma_n e_n$. From this we conclude that

$$\|\alpha e_k - c\|^2 = (\alpha - \gamma_k)^2 + \sum_{n=1, n\neq k}^{\infty} \gamma_n^2 < \epsilon.$$

Hence, $\gamma_n = 0$ for all $n \neq k$, which means that c is of the form $c = \gamma e_k$ and belongs to $\mathfrak{M}_k$.   □

Answer to Problem 2.23. (page 66): Pick $a = \alpha e_k \in \mathfrak{M}_k$ and $b = \beta e_n \in \mathfrak{M}_n$. Then,

$$\langle a, b \rangle = \langle \alpha e_k, \beta e_n \rangle = \alpha\overline{\beta}\langle e_k, e_n \rangle = \alpha\overline{\beta}\delta_{k,n}.$$

□

Answer to Problem 2.24. (page 72): The generalized Parseval relation can be shown by expanding f as a Fourier series expansion (2.54) in Definition 2.23, p. 71,

$$\langle f, g \rangle = \left\langle \sum_{n=1}^{\infty} \langle f, \varphi_n \rangle \varphi_n, g \right\rangle = \sum_{n=1}^{\infty} \langle f, \varphi_n \rangle \langle \varphi_n, g \rangle = \sum_{n=1}^{\infty} \langle f, \varphi_n \rangle \overline{\langle g, \varphi_n \rangle}.$$

□

Answer to Problem 2.25. (page 73): This is simple logic. If any orthonormal sequence were complete and countable then the space would be separable by definition.   □

Answer to Problem 2.26. (page 78): Weak convergence of f_n to f means

$$\langle f_n, g \rangle \to \langle f, g \rangle, \quad \forall g \in \mathcal{H}.$$

Now we choose $g = f$ to get

$$\langle f_n, f \rangle \to \langle f, f \rangle = \|f\|^2$$

and

$$\langle f, f_n \rangle = \overline{\langle f_n, f \rangle} \to \overline{\langle f, f \rangle} = \|f\|^2.$$

Now using the above results and the assumption $\lim_{n\to\infty} \|f_n\| = \|f\|$ expand $\|f - f_n\|^2$ as

$$\|f - f_n\|^2 = \langle f - f_n, f - f_n \rangle = \|f\|^2 + \|f_n\|^2 - \langle f, f_n \rangle - \langle f_n, f \rangle$$
$$\to \|f\|^2 + \|f_n\|^2 - \|f\|^2 - \|f\|^2 \to 0.$$

Therefore, strong convergence is established. $\qquad\square$

Answer to Problem 2.27. (page 80): Indeed it does not tend to a point in the space. But the sequence given is not Cauchy in the first place and so does not converge. So this does not cause any problems for the completeness of $L^2(0, 2\pi)$.
$\square$

Answer to Problem 2.28. (page 81): Assume that the sequence f_n converges uniformly to f. That is, for any $\epsilon > 0$, there exists N such that $|f_n(x) - f(x)| < \epsilon$ for all $n > N$. Now

$$|f_n(x) - f(x)| < \epsilon, \quad \forall x \in \Omega \iff \sup_{x \in \Omega} |f_n(x) - f(x)| < \epsilon.$$

Therefore, for any $\epsilon > 0$, there exists N such that $\sup_{x \in \Omega} |f_n(x) - f(x)| < \epsilon$ for all $n > N$. The converse is also true by reversing the order of arguments. $\qquad\square$

Answer to Problem 2.29. (page 95): This is just a repeated application of the generalized Parseval relation. $\qquad\square$

Answer to Problem 2.30. (page 95): This is only trivially different from the calculation leading up to (2.79), p. 94. $\qquad\square$

Answer to Problem 2.31. (page 95): The weighting is $\ell + 1/2$ and follows directly by using (2.84), p. 95. $\qquad\square$

Answer to Problem 2.32. (page 96): The qualification that applies for the weighting of the Legendre polynomials, Problem 2.31, p. 95, also applies here. From (2.36), p. 57, we see the relevant expression for the weighting term is

$$w(x) = e^{-x^2/2},$$

but we also need the Hermite polynomials to be normalized as in (2.38), p. 58. Then the result follows. $\qquad\square$

Answer to Problem 2.33. (page 96): The generalized Laguerre polynomials in (2.85), p. 96, are not normalized but we can rearrange this equation to the form

$$\int_0^\infty \frac{n!}{\Gamma(n + \alpha + 1)} x^\alpha e^{-x} L_n^{(\alpha)}(x) L_m^{(\alpha)}(x) \, dx = \delta_{n,m},$$

with which we can identify the index-dependent general weighting term

$$w_n(x) = \frac{n!}{\Gamma(n + \alpha + 1)} x^\alpha e^{-x},$$

which is in the form of the completeness relation index-dependent weighting (2.84), p. 95.   □

Answer to Problem 2.34. (page 96): Without loss of generality, we can simplify (2.86), p. 96, to imply that we need to prove

$$\delta(\theta) = \sin\theta\,\delta(\cos\theta).$$

Consider change of variables $u = \cos\theta$ in the following:

$$\int_0^\pi \delta(\cos\theta)f(\cos\theta)\left|\frac{d\cos\theta}{d\theta}\right|d\theta = \int_{+1}^{-1} \delta(u)f(u)\,du,$$

from which $\delta(\theta) = \sin\theta\,\delta(\cos\theta)$ is implied.   □

Answer to Problem 2.35. (page 102):

$$\langle \widetilde{f}, \widetilde{g} \rangle = \left\langle \sum_{n=1}^{\infty}\langle f, \varphi_n\rangle\widetilde{\varphi}_n, \sum_{m=1}^{\infty}\langle g, \varphi_m\rangle\widetilde{\varphi}_m \right\rangle$$

$$= \sum_{n=1}^{\infty}\sum_{m=1}^{\infty}\langle f, \varphi_n\rangle\overline{\langle g, \varphi_m\rangle}\langle\widetilde{\varphi}_n, \widetilde{\varphi}_m\rangle = \sum_{n=1}^{\infty}\langle f, \varphi_n\rangle\overline{\langle g, \varphi_n\rangle} = \langle f, g\rangle,$$

where we used the generalized Parseval relation (2.57), p. 72.   □

Answers to problems in Part II

Answers to problems in Chapter 3

This introductory chapter had no problems.

Answers to problems in Chapter 4

Answer to Problem 4.1. (page 113): Define a modified right shift operator $\mathcal{S}_R^U$ which takes any element $a = \{\alpha_n\}_{n=1}^{\infty} \in \ell^2$ to $\mathcal{S}_R^U a = \{(n-1)\alpha_{n-1}\}_{n=1}^{\infty}$ and $\alpha_0 = 0$. Therefore, $\mathcal{S}_R^U e_k = k e_{k+1}$ and $\lim_{k\to\infty}\|\mathcal{S}_R^U e_k\| = \infty$. Therefore, the operator is unbounded and the output space of $\mathcal{S}_R^U$ is not ℓ^2. $\qquad\square$

Answer to Problem 4.2. (page 115): Assume that the linear operator $\mathcal{B}$ is continuous at $f_0 = o = g - g$ for all $g \in \mathcal{H}$. Defining $h \triangleq f + g$, (4.4), p. 114, becomes

$$\|f - o\| = \|f + g - g\| = \|h - g\| < \delta \implies$$
$$\left\|\mathcal{B}f - \mathcal{B}(g - g)\right\| = \|\mathcal{B}f + \mathcal{B}g - \mathcal{B}g\| = \|\mathcal{B}h - \mathcal{B}g\| < \epsilon, \quad \forall f, g, h \in \mathcal{H},$$

where the last two equalities follow from the linearity of operator. Therefore, the operator is continuous around every $g \in \mathcal{H}$ and hence everywhere continuous. $\quad\square$

Answer to Problem 4.3. (page 115): A bounded linear operator is continuous everywhere. It follows that a nowhere continuous operator is unbounded. $\qquad\square$

Answer to Problem 4.4. (page 116):

$$\|\mathcal{B}_n f - \mathcal{B}f\|^2 = \langle \mathcal{B}_n f - \mathcal{B}f, \mathcal{B}_n f - \mathcal{B}f\rangle$$
$$= \|\mathcal{B}_n f\|^2 + \|\mathcal{B}f\|^2 - \langle \mathcal{B}_n f, \mathcal{B}f\rangle - \langle \mathcal{B}f, \mathcal{B}_n f\rangle.$$

Now use $g = \mathcal{B}f$ in (4.5), p. 115, to conclude that $\langle \mathcal{B}_n f, \mathcal{B}f\rangle \to \|\mathcal{B}f\|^2$ and $\langle \mathcal{B}f, \mathcal{B}_n f\rangle \to \|\mathcal{B}f\|^2$. Therefore,

$$\|\mathcal{B}_n f - \mathcal{B}f\|^2 = \|\mathcal{B}_n f\|^2 - \|\mathcal{B}f\|^2.$$

Now, invoking $\|\mathcal{B}_n f\| \to \|\mathcal{B}f\|$ results in strong convergence of the operators. $\square$

Answer to Problem 4.5. (page 116): For the first part we write

$$\|\mathcal{B}_1 + \mathcal{B}_2\|_{\mathrm{op}} = \sup_{\|f\|=1} \|\mathcal{B}_1 f + \mathcal{B}_2 f\|$$

$$\leq \sup_{\|f\|=1} \|\mathcal{B}_1 f\| + \sup_{\|f\|=1} \|\mathcal{B}_2 f\| = \|\mathcal{B}_1\|_{\mathrm{op}} + \|\mathcal{B}_2\|_{\mathrm{op}}.$$

For the second part we write

$$\|\alpha \mathcal{B}_1\|_{\mathrm{op}} = \sup_{\|f\|=1} \|\alpha \mathcal{B}_1 f\|$$

$$= |\alpha| \sup_{\|f\|=1} \|\mathcal{B}_1 f\| = |\alpha| \|\mathcal{B}_1\|_{\mathrm{op}}.$$

For the third part we write

$$\|\mathcal{B}_2 \circ \mathcal{B}_1\|_{\mathrm{op}} = \sup_{\|f\|=1} \|(\mathcal{B}_2 \circ \mathcal{B}_1) f\|$$

$$= \sup_{\|f\|=1} \|\mathcal{B}_2(\mathcal{B}_1 f)\|$$

$$\leq \|\mathcal{B}_2\|_{\mathrm{op}} \sup_{\|f\|=1} \|\mathcal{B}_1 f\| = \|\mathcal{B}_2\|_{\mathrm{op}} \|\mathcal{B}_1\|_{\mathrm{op}}.$$

$\square$

Answer to Problem 4.6. (page 120): For any element $f = \sum_{m=1}^{\infty} \langle f, \varphi_m \rangle \varphi_m \in \mathcal{H}$ and using Parseval relation and Cauchy-Schwarz inequality, we can bound the norm of $\mathcal{B}f$ as

$$\|\mathcal{B}f\|^2 = \left\| \sum_{n=1}^{\infty} \langle \mathcal{B}f, \varphi_n \rangle \varphi_n \right\|^2 = \sum_{n=1}^{\infty} \sum_{m=1}^{\infty} \left| \langle f, \varphi_m \rangle b_{n,m}^{(\varphi)} \right|^2$$

$$\leq \sum_{n=1}^{\infty} \sum_{m=1}^{\infty} |\langle f, \varphi_m \rangle|^2 \sum_{m=1}^{\infty} \left| b_{n,m}^{(\varphi)} \right|^2 = \|f\| B^2.$$

Therefore, the norm of $\mathcal{B}$ is upper bounded as

$$\|\mathcal{B}\|_{\mathrm{op}} \leq B = \left(\sum_{n=1}^{\infty} \sum_{m=1}^{\infty} \left| b_{n,m}^{(\varphi)} \right|^2 \right)^{1/2}.$$

$\square$

Answer to Problem 4.7. (page 120): This can be easily shown by noting that $f = \sum_{m=1}^{N} \langle f, \varphi_m \rangle \varphi_m$ for a set of N basis functions $\{\varphi_m\}_{m=1}^{N}$ and in this case the operator matrix becomes an $N \times N$ matrix with elements $b_{n,m}^{(\varphi)} = \langle \mathcal{B}\varphi_m, \varphi_n \rangle$ for $n, m = 1, 2, \ldots, N$. $\square$

Answer to Problem 4.8. (page 120): Writing input f in terms of $\{\varphi_m\}_{m=1}^{\infty}$ and then applying the operator we have

$$\mathcal{B}f = \mathcal{B}\left(\sum_{m=1}^{\infty} \langle f, \varphi_m \rangle \varphi_m \right) = \sum_{m=1}^{\infty} \langle f, \varphi_m \rangle \mathcal{B}\varphi_m.$$

From this we can find the projection of $\mathcal{B}f$ along ζ_n in the output space as

$$\langle \mathcal{B}f, \zeta_n \rangle = \sum_{m=1}^{\infty} \langle f, \varphi_m \rangle \langle \mathcal{B}\varphi_m, \zeta_n \rangle,$$

from which we conclude that the elements of operator matrix are $b_{n,m}^{(\zeta,\varphi)} = \langle \mathcal{B}\varphi_m, \zeta_n \rangle$. $\square$

Answer to Problem 4.9. (page 134):

$$\langle \mathcal{L}_K f, g \rangle = \int_\Omega \int_\Omega K(x,y) f(y) \, dy \, \overline{g(x)} \, dx$$

$$= \int_\Omega f(y) \int_\Omega K(x,y) \overline{g(x)} \, dx \, dy = \int_\Omega f(y) \int_\Omega \overline{\overline{K(x,y)} g(x)} \, dx \, dy.$$

Noting that

$$K(x,y) = \sum_{n=1}^\infty \sum_{m=1}^\infty k_{n,m}^{(\varphi)} \overline{\varphi_m(y)} \varphi_n(x),$$

we can write the kernel for the adjoint operator as

$$\overline{K(x,y)} = \sum_{n=1}^\infty \sum_{m=1}^\infty \overline{k_{n,m}^{(\varphi)}} \, \overline{\varphi_n(x)} \varphi_m(y).$$

Now defining $(k_{m,n}^{(\varphi)})^\star = \overline{k_{n,m}^{(\varphi)}}$, the adjoint kernel is given by

$$K^\star(y,x) = \sum_{n=1}^\infty \sum_{m=1}^\infty (k_{m,n}^{(\varphi)})^\star \overline{\varphi_n(x)} \varphi_m(y),$$

from which we conclude that

$$\langle \mathcal{L}_K f, g \rangle = \int_\Omega f(y) \int_\Omega \overline{K^\star(y,x) g(x)} \, dx \, dy = \langle f, \mathcal{L}_K^\star g \rangle.$$

$\square$

Answer to Problem 4.10. (page 134): For the first relation we write

$$\langle \alpha \mathcal{B} f, g \rangle = \alpha \langle \mathcal{B} f, g \rangle = \alpha \langle f, \mathcal{B}^\star g \rangle = \langle f, \overline{\alpha} \mathcal{B}^\star g \rangle,$$

from which we conclude that $(\alpha \mathcal{B})^\star = \overline{\alpha} \mathcal{B}^\star$.

For the second relation we write

$$\langle (\mathcal{B}_2 \circ \mathcal{B}_1) f, g \rangle = \langle \mathcal{B}_2 (\mathcal{B}_1 f), g \rangle = \langle \mathcal{B}_1 f, \mathcal{B}_2^\star g \rangle$$

$$= \langle f, \mathcal{B}_1^\star (\mathcal{B}_2^\star g) \rangle = \langle f, (\mathcal{B}_1^\star \circ \mathcal{B}_2^\star) g \rangle,$$

from which we conclude that $(\mathcal{B}_2 \circ \mathcal{B}_1)^\star = \mathcal{B}_1^\star \circ \mathcal{B}_2^\star$.

For the third relation we write

$$\langle (\mathcal{B}_1 + \mathcal{B}_2) f, g \rangle = \langle \mathcal{B}_1 f + \mathcal{B}_2 f, g \rangle = \langle \mathcal{B}_1 f, g \rangle + \langle \mathcal{B}_2 f, g \rangle$$

$$= \langle f, \mathcal{B}_1^\star g \rangle + \langle f, \mathcal{B}_2^\star g \rangle = \langle f, (\mathcal{B}_1^\star + \mathcal{B}_2^\star) g \rangle,$$

from which we conclude that $(\mathcal{B}_1 + \mathcal{B}_2)^\star = \mathcal{B}_1^\star + \mathcal{B}_2^\star$.

$\square$

Answer to Problem 4.11. (page 134): For all $f, g \in \mathcal{H}$ and for all bounded operators $\mathcal{B}$ and $\mathcal{B}^\star$ we can write

$$\langle (\mathcal{B}^\star)^\star f, g \rangle = \langle f, \mathcal{B}^\star g \rangle = \langle \mathcal{B} f, g \rangle.$$

Choosing $g = \varphi_n$ for all $n \in \mathbb{N}$ where $\{\varphi_n\}_{n=1}^\infty$ is a complete orthonormal sequence in $\mathcal{H}$, results in

$$\alpha_n = \langle (\mathcal{B}^\star)^\star f, \varphi_n \rangle = \langle \mathcal{B} f, \varphi_n \rangle.$$

And therefore

$$(\mathcal{B}^\star)^\star f = \mathcal{B} f = \sum_{n=1}^\infty \alpha_n \varphi_n,$$

from which we conclude that $(\mathcal{B}^\star)^\star = \mathcal{B}$.

$\square$

Answer to Problem 4.12. (page 134):

$$\langle \mathcal{B}f, g \rangle = \langle f, \mathcal{B}^\star g \rangle = \langle \mathcal{I}f, \mathcal{B}^\star g \rangle = \langle (\mathcal{B}^{-1} \circ \mathcal{B})f, \mathcal{B}^\star g \rangle$$
$$= \langle f, (\mathcal{B}^{-1} \circ \mathcal{B})(\mathcal{B}^\star g)^\star \rangle = \langle f, (\mathcal{B}^\star \circ (\mathcal{B}^{-1})^\star)(\mathcal{B}^\star g) \rangle$$
$$= \langle f, (\mathcal{B}^\star \circ (\mathcal{B}^{-1})^\star \circ \mathcal{B}^\star)g \rangle,$$

where the second-last equality follows from Problem 4.10. From the above we conclude that $(\mathcal{B}^{-1})^\star \circ \mathcal{B}^\star$ must be the identity operator and hence $(\mathcal{B}^{-1})^\star = (\mathcal{B}^\star)^{-1}$.   □

Answer to Problem 4.13. (page 134): We prove the result by contradiction. First assume that $\|\mathcal{B}^\star\|_{\mathrm{op}} < \|\mathcal{B}\|_{\mathrm{op}}$. Using the definition of operator norm we can write

$$\|\mathcal{B}\|_{\mathrm{op}}^2 = \sup_{\|f\|=1} \|\mathcal{B}f\|^2 = \sup_{\|f\|=1} |\langle \mathcal{B}f, \mathcal{B}f \rangle| = \sup_{\|f\|=1} |\langle (\mathcal{B}^\star \circ \mathcal{B})f, f \rangle|$$
$$\leq \sup_{\|f\|=1} \|(\mathcal{B}^\star \circ \mathcal{B})f\| \|f\| \leq \sup_{\|f\|=1} \|\mathcal{B}^\star \circ \mathcal{B}\|_{\mathrm{op}} \|f\|^2 = \|\mathcal{B}^\star \circ \mathcal{B}\|_{\mathrm{op}}$$
$$\leq \|\mathcal{B}^\star\|_{\mathrm{op}} \|\mathcal{B}\|_{\mathrm{op}} < \|\mathcal{B}\|_{\mathrm{op}}^2,$$

where in the first line we have also used the definition of operator adjoint. The first inequality is the Cauchy-Schwarz inequality, the second inequality is a consequence of the definition of operator norm, the third inequality is a result from Problem 4.5, p. 116, and finally the last inequality is from our assumption, which results in the contradiction $\|\mathcal{B}\|_{\mathrm{op}} < \|\mathcal{B}\|_{\mathrm{op}}$.

Similarly, let us assume that $\|\mathcal{B}^\star\|_{\mathrm{op}} > \|\mathcal{B}\|_{\mathrm{op}}$ and write

$$\|\mathcal{B}^\star\|_{\mathrm{op}}^2 = \sup_{\|f\|=1} \|\mathcal{B}^\star f\|^2 = \sup_{\|f\|=1} |\langle \mathcal{B}^\star f, \mathcal{B}^\star f \rangle| = \sup_{\|f\|=1} |\langle (\mathcal{B} \circ \mathcal{B}^\star)f, f \rangle|$$
$$\leq \sup_{\|f\|=1} \|(\mathcal{B} \circ \mathcal{B}^\star)f\| \|f\| \leq \sup_{\|f\|=1} \|\mathcal{B} \circ \mathcal{B}^\star\|_{\mathrm{op}} \|f\|^2 = \|\mathcal{B} \circ \mathcal{B}^\star\|_{\mathrm{op}}$$
$$\leq \|\mathcal{B}\|_{\mathrm{op}} \|\mathcal{B}^\star\|_{\mathrm{op}} < \|\mathcal{B}^\star\|_{\mathrm{op}}^2,$$

where in the first line we have also used the definition of operator adjoint. Again our assumption results in a contradiction. Therefore, we conclude that $\|\mathcal{B}^\star\|_{\mathrm{op}} = \|\mathcal{B}\|_{\mathrm{op}}$.   □

Answer to Problem 4.14. (page 134): We need a notion of convergence in a Banach space without relying on an inner product that reduces to the weak convergence notion we have for Hilbert spaces. The key observation is all bounded linear functionals in Hilbert space look like $\langle \cdot, g \rangle$ for some $g \in \mathcal{H}$. Every bounded linear functional can be written this way, by the Riesz representation theorem; see Section 4.9, p. 131. The Hilbert space notion of weak convergence therefore is like saying we have convergence for every bounded linear functional.

Definition A.1 (Weak convergence in Banach space). *A sequence of vectors* $\{f_n\}$ *in a Banach space* $\mathcal{H}$ *is called weakly convergent to a vector* $f \in \mathcal{H}$ *if* $g(f_n) \to g(f)$ *as* $n \to \infty$, *for every bounded linear functional* $g(\cdot)$ *on* $\mathcal{H}$.   □

The functional output is in $\mathbb{C}$, $g(\cdot) \colon \mathcal{H} \mapsto \mathbb{C}$ similar to $\langle \cdot, g \rangle \colon \mathcal{H} \mapsto \mathbb{C}$, for $g \in \mathcal{H}$ (here the symbol g for a functional in Banach space should not be confused with the symbol g for a vector in Hilbert space).   □

Answer to Problem 4.15. (page 135):

$$\langle \mathcal{B}a, b \rangle = \sum_{n=1}^{\infty} \frac{\alpha_n}{n} \beta_n = \sum_{n=1}^{\infty} \alpha_n \frac{\beta_n}{n} = \langle a, \mathcal{B}b \rangle.$$

To find the operator norm, we write

$$\|\mathcal{B}a\| = \sum_{n=1}^{\infty} \left(\frac{\alpha_n}{n}\right)^2 \le \sum_{n=1}^{\infty} \alpha_n^2 = \|a\|.$$

Now, $e_1 = \{1, 0, 0, \dots\}$ achieves the equality and hence $\|\mathcal{B}\|_{\mathrm{op}} = 1$. $\square$

Answer to Problem 4.16. (page 135): Suppose $\mathcal{B}_1$ and $\mathcal{B}_2$ are each self-adjoint and commute: $\mathcal{B}_2 \circ \mathcal{B}_1 = \mathcal{B}_1 \circ \mathcal{B}_2$. Now

$$\langle (\mathcal{B}_2 \circ \mathcal{B}_1)f, g \rangle = \langle f, (\mathcal{B}_2 \circ \mathcal{B}_1)^{\star} g \rangle = \langle f, (\mathcal{B}_1^{\star} \circ \mathcal{B}_2^{\star})g \rangle$$
$$= \langle f, (\mathcal{B}_1 \circ \mathcal{B}_2)g \rangle = \langle f, (\mathcal{B}_2 \circ \mathcal{B}_1)g \rangle,$$

where we have used the results in Problem 4.10, p. 134, for the second equality. From the above we conclude that $\mathcal{B}_2 \circ \mathcal{B}_1 = (\mathcal{B}_2 \circ \mathcal{B}_1)^{\star}$ and hence $\mathcal{B}_2 \circ \mathcal{B}_1$ is self-adjoint.

Now assume that $\mathcal{B}_1$ and $\mathcal{B}_2$ are each self-adjoint and $(\mathcal{B}_2 \circ \mathcal{B}_1)$ is also self-adjoint. Hence, $(\mathcal{B}_2 \circ \mathcal{B}_1)^{\star} = \mathcal{B}_2 \circ \mathcal{B}_1$. But since $(\mathcal{B}_2 \circ \mathcal{B}_1)^{\star} = \mathcal{B}_1^{\star} \circ \mathcal{B}_2^{\star} = \mathcal{B}_1 \circ \mathcal{B}_2$, we conclude that $\mathcal{B}_1$ and $\mathcal{B}_2$ must commute. $\square$

Answer to Problem 4.17. (page 135): Using the results in Problem 4.10, p. 134, we can write $(\mathcal{B}_1 \circ \mathcal{B}_2 \circ \mathcal{B}_1)^{\star} = \mathcal{B}_1^{\star} \circ \mathcal{B}_2^{\star} \circ \mathcal{B}_1^{\star} = \mathcal{B}_1 \circ \mathcal{B}_2 \circ \mathcal{B}_1$ and is hence self-adjoint. Similarly we have $(\mathcal{B}_2 \circ \mathcal{B}_1 \circ \mathcal{B}_2)^{\star} = \mathcal{B}_2^{\star} \circ \mathcal{B}_1^{\star} \circ \mathcal{B}_2^{\star} = \mathcal{B}_2 \circ \mathcal{B}_1 \circ \mathcal{B}_2$. $\square$

Answer to Problem 4.18. (page 137): Applying another operator $\mathcal{P}_{\mathfrak{M}}$ to

$$\mathcal{P}_{\mathfrak{M}} f = \sum_{n \in M} \langle f, \varphi_n \rangle \varphi_n,$$

we obtain

$$\mathcal{P}_{\mathfrak{M}}(\mathcal{P}_{\mathfrak{M}} f) = \sum_{n \in M} \langle \mathcal{P}_{\mathfrak{M}} f, \varphi_n \rangle \varphi_n = \sum_{n \in M} \langle \sum_{m \in M} \langle f, \varphi_m \rangle \varphi_m, \varphi_n \rangle \varphi_n$$
$$= \sum_{n \in M} \sum_{m \in M} \langle f, \varphi_m \rangle \langle \varphi_m, \varphi_n \rangle \varphi_n = \sum_{n \in M} \langle f, \varphi_n \rangle \varphi_n = \mathcal{P}_{\mathfrak{M}} f,$$

where we have used the orthonormality of $\langle \varphi_m, \varphi_n \rangle = \delta_{m,n}$.

To show that $\mathcal{P}_{\mathfrak{M}}$ is self-adjoint, we write

$$\langle \mathcal{P}_{\mathfrak{M}} f, g \rangle = \langle \sum_{n \in M} \langle f, \varphi_n \rangle \varphi_n, g \rangle = \sum_{n \in M} \langle f, \varphi_n \rangle \langle \varphi_n, g \rangle$$
$$= \sum_{n \in M} \langle f, \varphi_n \rangle \langle \overline{g}, \overline{\varphi}_n \rangle = \sum_{n \in M} \langle \overline{\varphi}_n, \overline{f} \rangle \langle \overline{g}, \overline{\varphi}_n \rangle = \langle \sum_{n \in M} \langle \overline{g}, \overline{\varphi}_n \rangle \overline{\varphi}_n, \overline{f} \rangle$$
$$= \langle f, \sum_{n \in M} \overline{\langle \overline{g}, \overline{\varphi}_n \rangle} \varphi_n \rangle = \langle f, \sum_{n \in M} \langle g, \varphi_n \rangle \varphi_n \rangle = \langle f, \mathcal{P}_{\mathfrak{M}} g \rangle. \quad \square$$

Answer to Problem 4.19. (page 137):

$$\langle \mathcal{P}_{\mathfrak{M}} f, f \rangle = \langle \sum_{n \in M} \langle f, \varphi_n \rangle \varphi_n, f \rangle = \sum_{n \in M} \langle f, \varphi_n \rangle \langle \varphi_n, f \rangle = \sum_{n \in M} |\langle f, \varphi_n \rangle|^2 \ge 0. \quad \square$$

Answer to Problem 4.20. (page 139):

$$\left\| (\mathcal{P}_N f)(x) \right\| = \left\| \sum_{n=1}^{N} \langle f, \varphi_n \rangle \varphi_n \right\| \leq \left\| \sum_{n=1}^{\infty} \langle f, \varphi_n \rangle \varphi_n \right\| = \|f\|.$$

Therefore, $\|\mathcal{P}_N\|_{\text{op}} \leq 1$. Now choose $f(x) = \varphi_1(x)$ with $\|f\| = 1$, which results in $\|(\mathcal{P}_N f)(x)\| = 1$. Hence, $\|\mathcal{P}_N\|_{\text{op}} = 1$. Likewise,

$$\left\| ((\mathcal{I} - \mathcal{P}_N)f)(x) \right\| = \left\| \sum_{n=N+1}^{\infty} \langle f, \varphi_n \rangle \varphi_n \right\| \leq \left\| \sum_{n=1}^{\infty} \langle f, \varphi_n \rangle \varphi_n \right\| = \|f\|.$$

Therefore, $\|\mathcal{I} - \mathcal{P}_N\|_{\text{op}} \leq 1$. Now choose $f(x) = \varphi_{N+1}(x)$ with $\|f\| = 1$, which results in $\|(\mathcal{I} - \mathcal{P}_N)f\| = 1$. Hence, $\|\mathcal{I} - \mathcal{P}_N\|_{\text{op}} = 1$. $\qquad\square$

Answer to Problem 4.21. (page 139): From (a) to (b): we know that $\mathcal{P}_{\mathfrak{M}} f \in \mathfrak{M} \subset \mathfrak{N}$. As a result when we operate $\mathcal{P}_{\mathfrak{N}}$ on $\mathcal{P}_{\mathfrak{M}} f$, which already belongs to $\mathfrak{N}$, it will remain unchanged and we can assert that $\mathcal{P}_{\mathfrak{N}}(\mathcal{P}_{\mathfrak{M}} f) = \mathcal{P}_{\mathfrak{M}} f$ for all $f \in \mathcal{H}$. Therefore, $\mathcal{P}_{\mathfrak{N}} \circ \mathcal{P}_{\mathfrak{M}} = \mathcal{P}_{\mathfrak{M}}$.

From (b) to (c): since both $\mathcal{P}_{\mathfrak{M}}$ and $\mathcal{P}_{\mathfrak{N}}$ are self-adjoint we can write:

$$\mathcal{P}_{\mathfrak{M}} \circ \mathcal{P}_{\mathfrak{N}} = \mathcal{P}_{\mathfrak{M}}^{\star} \circ \mathcal{P}_{\mathfrak{N}}^{\star} = (\mathcal{P}_{\mathfrak{N}} \circ \mathcal{P}_{\mathfrak{M}}^{\star}) = \mathcal{P}_{\mathfrak{M}}^{\star} = \mathcal{P}_{\mathfrak{M}},$$

where we have used (b) for writing $(\mathcal{P}_{\mathfrak{N}} \circ \mathcal{P}_{\mathfrak{M}}^{\star}) = \mathcal{P}_{\mathfrak{M}}^{\star}$.

From (c) to (d): we show that $\mathcal{P}_{\mathfrak{N}} - \mathcal{P}_{\mathfrak{M}}$ is idempotent and self-adjoint, which can be easily established as

$$(\mathcal{P}_{\mathfrak{N}} - \mathcal{P}_{\mathfrak{M}})^{\star} = \mathcal{P}_{\mathfrak{N}}^{\star} - \mathcal{P}_{\mathfrak{M}}^{\star} = \mathcal{P}_{\mathfrak{N}} - \mathcal{P}_{\mathfrak{M}}$$

and

$$\begin{aligned}
(\mathcal{P}_{\mathfrak{N}} - \mathcal{P}_{\mathfrak{M}})^2 &= \mathcal{P}_{\mathfrak{N}}^2 + \mathcal{P}_{\mathfrak{M}}^2 - \mathcal{P}_{\mathfrak{N}} \circ \mathcal{P}_{\mathfrak{M}} - \mathcal{P}_{\mathfrak{M}} \circ \mathcal{P}_{\mathfrak{N}} \\
&= \mathcal{P}_{\mathfrak{N}} + \mathcal{P}_{\mathfrak{M}} - \mathcal{P}_{\mathfrak{N}}^{\star} \circ \mathcal{P}_{\mathfrak{M}}^{\star} - \mathcal{P}_{\mathfrak{M}} \\
&= \mathcal{P}_{\mathfrak{N}} + \mathcal{P}_{\mathfrak{M}} - (\mathcal{P}_{\mathfrak{M}} \circ \mathcal{P}_{\mathfrak{N}})^{\star} - \mathcal{P}_{\mathfrak{M}} = \mathcal{P}_{\mathfrak{N}} - \mathcal{P}_{\mathfrak{M}}.
\end{aligned}$$

From (d) to (e): since $\mathcal{P}_{\mathfrak{N}} - \mathcal{P}_{\mathfrak{M}}$ is a projection operator, any signal $f \in \mathcal{H}$ can be decomposed into $f = (\mathcal{P}_{\mathfrak{N}} - \mathcal{P}_{\mathfrak{M}})f + (\mathcal{P}_{\mathfrak{N}} - \mathcal{P}_{\mathfrak{M}})^{\perp} f$ and hence we can write

$$\langle (\mathcal{P}_{\mathfrak{N}} - \mathcal{P}_{\mathfrak{M}})f, f \rangle = \langle (\mathcal{P}_{\mathfrak{N}} - \mathcal{P}_{\mathfrak{M}})f, (\mathcal{P}_{\mathfrak{N}} - \mathcal{P}_{\mathfrak{M}})f \rangle = \left\| (\mathcal{P}_{\mathfrak{N}} - \mathcal{P}_{\mathfrak{M}})f \right\|^2 \geq 0.$$

From (e) to (f): since $\mathcal{P}_{\mathfrak{N}}$ and $\mathcal{P}_{\mathfrak{M}}$ are projection operators, any signal $f \in \mathcal{H}$ can be decomposed into $f = \mathcal{P}_{\mathfrak{N}} f + \mathcal{P}_{\mathfrak{N}}^{\perp} f$ and $f = \mathcal{P}_{\mathfrak{M}} f + \mathcal{P}_{\mathfrak{M}}^{\perp} f$ and hence we can write

$$\left\| \mathcal{P}_{\mathfrak{M}} f \right\|^2 - \left\| \mathcal{P}_{\mathfrak{N}} f \right\|^2 = \langle \mathcal{P}_{\mathfrak{M}} f, f \rangle - \langle \mathcal{P}_{\mathfrak{N}} f, f \rangle = \langle (\mathcal{P}_{\mathfrak{M}} - \mathcal{P}_{\mathfrak{N}})f, f \rangle \leq 0.$$

From (f) back to (a): choose a vector $f \in \mathfrak{M}$. Therefore, $\mathcal{P}_{\mathfrak{M}} f = f$ and $\|f\| = \|\mathcal{P}_{\mathfrak{M}} f\|$ and from (f) we conclude that $\|f\| = \|\mathcal{P}_{\mathfrak{M}} f\| \leq \|\mathcal{P}_{\mathfrak{N}} f\|$. Now any $f \in \mathcal{H}$ can be written as the sum of two orthogonal vectors $f = \mathcal{P}_{\mathfrak{N}} f + \mathcal{P}_{\mathfrak{N}}^{\perp} f$ with $\|\mathcal{P}_{\mathfrak{N}} f\| = \|f\| - \|\mathcal{P}_{\mathfrak{N}}^{\perp} f\| \leq \|f\|$, where the equality is achieved only when $f \in \mathfrak{N}$ and $\|\mathcal{P}_{\mathfrak{N}}^{\perp} f\| = 0$. In the end, from

$$\|f\| = \|\mathcal{P}_{\mathfrak{M}} f\| \leq \|\mathcal{P}_{\mathfrak{N}} f\| \leq \|f\|,$$

we conclude that $\|\mathcal{P}_{\mathfrak{N}}^{\perp} f\| = 0$ and hence f also belongs to $\mathfrak{N}$ and $\mathfrak{M} \subset \mathfrak{N}$ is proven. $\qquad\square$

Answer to Problem 4.22. (page 143):

$$\left\| (\mathcal{B} - \mathcal{B}_N)f \right\|^2 = \left\| \sum_{n=N+1}^{\infty} \frac{1}{n}\mathcal{P}_n f \right\|^2 = \sum_{n=N+1}^{\infty} \frac{1}{n^2}\|\mathcal{P}_n f\|^2 \leq \frac{1}{(N+1)^2}\|f\|^2,$$

from which we conclude that $\|(\mathcal{B} - \mathcal{B}_N)\|_{\mathrm{op}} \leq 1/(N+1)$. $\qquad\square$

Answers to problems in Chapter 5

Answer to Problem 5.1. (page 149): From (4.32), p. 128, we know that

$$k_{n,m}^{(\varphi)} = \langle K, \varphi_{n,m} \rangle.$$

Therefore,

$$\sum_{n=1}^{\infty}\sum_{m=1}^{\infty} |k_{n,m}^{(\varphi)}|^2 = \sum_{n=1}^{\infty}\sum_{m=1}^{\infty} \langle K, \varphi_{n,m}\rangle\langle\varphi_{n,m}, K\rangle = \|K\|^2 = \int_{\Omega\times\Omega} |K(x,y)|^2 dx\, dy,$$

where the norm of $K(x,y)$ is defined in $L^2(\Omega \times \Omega)$, which from (4.29), p. 127, we know is finite. Hence, the Hilbert-Schmidt operator $\mathcal{L}_K$ is compact. $\qquad\square$

Answers to problems in Chapter 6

Answer to Problem 6.1. (page 162): This just relies on the orthonormality of the $\{\varphi_n(x)\}$ on Ω:

$$\iint_{\Omega\times\Omega} \sum_{p=1}^{\infty}\sum_{q=1}^{\infty} \left(k_{q,p}^{(\varphi)}\varphi_q(x)\overline{\varphi_p(y)}\right)\varphi_m(y)\, dy\, \overline{\varphi_n(x)}\, dx$$

$$= \sum_{p=1}^{\infty}\sum_{q=1}^{\infty} \delta_{q,n}\delta_{p,m}k_{q,p}^{(\varphi)} = k_{n,m}^{(\varphi)}.$$

$\square$

Answer to Problem 6.2. (page 164): Just take the inner product of $\mathcal{L}_K\phi_m$ in (6.11), p. 164, with ϕ_n, which leads to $\lambda_n\langle\phi_m, \phi_n\rangle = \lambda_n\delta_{m,n}$ by orthonormality of the $\{\phi_n\}_{n\in\mathbb{N}}$. The expression for λ_n is the special case when $m = n$. $\qquad\square$

Answer to Problem 6.3. (page 167): This easily follows by comparing (6.15), p. 165, (after changing n index with m)

$$K(x,y) = \sum_{m=1}^{\infty} \left(\mathcal{L}_K\varphi_m\right)(x)\overline{\varphi_m(y)}$$

with (6.7), p. 161,

$$K(x,y) = \sum_{m=1}^{\infty}\sum_{n=1}^{\infty} k_{n,m}^{(\varphi)}\varphi_n(x)\overline{\varphi_m(y)},$$

from which we conclude that

$$(\mathcal{L}_K\varphi_m)(x) = \sum_{n=1}^{\infty} k_{n,m}^{(\varphi)}\varphi_n(x).$$

The interpretation is quite standard. Orthonormal function φ_m gets mapped or projected into different φ_n by $k_{n,m}^{(\varphi)} = \langle \mathcal{L}_K\varphi_m, \varphi_n \rangle$ and the overall action of the operator on φ_m is the superposition of these projections. □

Answer to Problem 6.4. (page 167): Noting that

$$f(x) = \sum_{m=1}^{\infty} \langle f, \varphi_m \rangle \varphi_m$$

and using the result of Problem 6.3, we can write

$$(\mathcal{L}_K f)(x) = \sum_{m=1}^{\infty} \langle f, \varphi_m \rangle (\mathcal{L}_K\varphi_m)(x) = \sum_{m=1}^{\infty} \langle f, \varphi_m \rangle \sum_{p=1}^{\infty} k_{p,m}^{(\varphi)}\varphi_p(x).$$

And therefore,

$$\langle \mathcal{L}_K f, \varphi_n \rangle = \sum_{m=1}^{\infty} \langle f, \varphi_m \rangle \sum_{p=1}^{\infty} k_{p,m}^{(\varphi)} \langle \varphi_p, \varphi_n \rangle = \sum_{m=1}^{\infty} \langle f, \varphi_m \rangle k_{n,m}^{(\varphi)}.$$

The interpretation is quite standard and similar to (4.13), p. 119, and steps described in page 117. □

Answer to Problem 6.5. (page 168): If the operator kernel is not Hilbert-Schmidt, then $\|K\|^2 = \infty$ and we cannot get any meaningful result about the norm of the operator and Corollary 6.4, p. 168, cannot be applied either. □

Answer to Problem 6.6. (page 168): This can be easily verified by using (6.7), p. 161,

$$\int_{\Omega} K(x,x)\,dx = \int_{\Omega} \sum_{m=1}^{\infty}\sum_{n=1}^{\infty} k_{n,m}^{(\varphi)}\varphi_n(x)\overline{\varphi_m(x)}\,dx = \sum_{m=1}^{\infty}\sum_{n=1}^{\infty} k_{n,m}^{(\varphi)}\delta_{m,n} = \sum_{n=1}^{\infty} k_{n,n}^{(\varphi)}.$$
□

Answer to Problem 6.7. (page 168): For the first part, using (6.17), p. 165, in (6.2), p. 160, we obtain

$$\iint_{\Omega\times\Omega} |K(x,y)|^2 dx\,dy = \sum_{n=1}^{\infty} \iint_{\Omega\times\Omega} |\varphi_n(x)\overline{\varphi_n(y)}|^2 dx\,dy = \sum_{n=1}^{\infty} 1 = \infty,$$

from which we conclude that $\delta(x-y)$ is not a Hilbert-Schmidt operator kernel.

For the second part, we substitute $K(x,y) = \delta(x-y) = \sum_{n=1}^{\infty}\varphi_n(x)\overline{\varphi_n(y)}$ into (6.1), p. 160, to obtain

$$(\mathcal{L}_K f)(x) = \sum_{n=1}^{\infty} \int_{\Omega} \varphi_n(x)\overline{\varphi_n(y)}f(y)\,dy = \sum_{n=1}^{\infty} \varphi_n(x)\langle f, \varphi_n \rangle = f(x),$$

which is the identity operator.

The operator is not compact because the identity operator is not compact. Refer to Section 5.4, p. 147, for more discussion.

□

Answers to problems in Part III

Answers to problems in Chapter 7

Answer to Problem 7.1. (page 177): Suppose $\mathbb{S}^2$ were a subspace of $\mathbb{R}^3$. Then if $\boldsymbol{u} \in \mathbb{S}^2$ then $\alpha \boldsymbol{u} \in \mathbb{S}^2$ for any scalar $\alpha \in \mathbb{R}$. But clearly $|\alpha \boldsymbol{u}| = |\alpha|$, which does not equal 1 if $\alpha \neq 1$. $\qquad\square$

Answer to Problem 7.2. (page 177): Given two vectors on 2-sphere such as

$$\widehat{\boldsymbol{u}} = (u_x, u_y, u_z)' = (\sin\theta\cos\phi, \sin\theta\sin\phi, \cos\theta)',$$
$$\widehat{\boldsymbol{v}} = (v_x, v_y, v_z)' = (\sin\vartheta\cos\varphi, \sin\vartheta\sin\varphi, \cos\vartheta)',$$

we write and simplify the angular distance as

$$
\begin{aligned}
\cos\Delta = \widehat{\boldsymbol{u}} \cdot \widehat{\boldsymbol{v}} &= u_x v_x + u_y v_y + u_z v_z \\
&= \sin\theta\cos\phi\sin\vartheta\cos\varphi + \sin\theta\sin\phi\sin\vartheta\sin\varphi + \cos\theta\cos\vartheta \\
&= \sin\theta\sin\vartheta(\cos\phi\cos\varphi + \sin\phi\sin\varphi) + \cos\theta\cos\vartheta \\
&= \sin\theta\sin\vartheta\cos(\phi - \varphi) + \cos\theta\cos\vartheta.
\end{aligned}
$$

$\qquad\square$

Answer to Problem 7.3. (page 188): Let us denote the rotation around the xyz-axes by (ϕ, θ, ρ), which is to be found and the known rotation around the zyz-axes as given in the problem by $(\varphi = 15°, \vartheta = 30°, \omega = 20°)$. Using (7.7), p. 182, (7.8), p. 182, and (7.9), p. 182, we can write the overall rotation matrix in the xyz-axes as

$$
\begin{aligned}
\boldsymbol{R} &= \begin{pmatrix} 1 & 0 & 0 \\ 0 & \cos\phi & -\sin\phi \\ 0 & \sin\phi & \cos\phi \end{pmatrix} \begin{pmatrix} \cos\theta & 0 & \sin\theta \\ 0 & 1 & 0 \\ -\sin\theta & 0 & \cos\theta \end{pmatrix} \begin{pmatrix} \cos\rho & -\sin\rho & 0 \\ \sin\rho & \cos\rho & 0 \\ 0 & 0 & 1 \end{pmatrix} \\
&= \begin{pmatrix} 1 & 0 & 0 \\ 0 & \cos\phi & -\sin\phi \\ 0 & \sin\phi & \cos\phi \end{pmatrix} \begin{pmatrix} \cos\theta\cos\rho & -\cos\theta\sin\rho & \sin\theta \\ \sin\rho & \cos\rho & 0 \\ -\sin\theta\cos\rho & \sin\theta\sin\rho & \cos\theta \end{pmatrix} \\
&= \begin{pmatrix} \cos\theta\cos\rho & -\cos\theta\sin\rho & \sin\theta \\ \cos\phi\sin\rho + \sin\phi\sin\theta\cos\rho & \cos\theta\cos\rho - \sin\phi\sin\theta\sin\rho & -\sin\phi\cos\theta \\ \sin\phi\sin\rho - \cos\phi\sin\theta\cos\rho & \sin\phi\cos\rho + \cos\phi\sin\theta\sin\rho & \cos\phi\cos\theta \end{pmatrix}.
\end{aligned}
$$

Now using (7.11), p. 186, the final matrix for rotation by ($\varphi = 15°$, $\vartheta = 30°$, $\omega = 20°$) around the zyz-axes is found to be

$$\boldsymbol{R}_{\text{ext}} = \begin{pmatrix} 0.6975 & -0.5293 & 0.4830 \\ 0.5410 & 0.8310 & 0.1294 \\ -0.4698 & 0.1710 & 0.8660 \end{pmatrix}.$$

By comparing the above two matrices, we find the rotation values around xyz-axes to be

$$\theta = \sin^{-1}(r_{1,3}) = 28.8791°,$$
$$\cos\theta = 0.8756.$$

Then

$$\sin\rho = -\frac{r_{1,2}}{\cos\theta} = 0.6045,$$

$$\cos\rho = \frac{r_{1,1}}{\cos\theta} = 0.7966,$$

from which we find $\rho = 37.1921°$. Similarly, ϕ can be found from

$$\sin\phi = -\frac{r_{2,3}}{\cos\theta} = -0.1478,$$

$$\cos\phi = \frac{r_{3,3}}{\cos\theta} = 0.9890,$$

to be $\phi = -8.4988° = 351.5012°$.   □

Answer to Problem 7.4. (page 194):

$$N_\ell^{-m}\frac{(\ell-m)!}{(\ell+m)!} = \sqrt{\frac{2\ell+1}{4\pi}\frac{(\ell+m)!}{(\ell-m)!}\frac{(\ell-m)!}{(\ell+m)!}} = \sqrt{\frac{2\ell+1}{4\pi}\frac{(\ell-m)!}{(\ell+m)!}} = N_\ell^m,$$

where we have substituted in (7.20), p. 189.   □

Answer to Problem 7.5. (page 194):

$$N_\ell^{-m}P_\ell^{-m}(\cos\theta) = N_\ell^{-m}(-1)^m\frac{(\ell-m)!}{(\ell+m)!}P_\ell^m(\cos\theta) = (-1)^m N_\ell^m P_\ell^m(\cos\theta),$$

where we used (7.32), p. 194, in Problem 7.4.   □

Answer to Problem 7.6. (page 194): From, (7.19), p. 189, and (7.20), p. 189,

$$Y_\ell^m(\theta,\phi) = N_\ell^m P_\ell^m(\cos\theta)e^{im\phi},$$

we have

$$Y_\ell^{-m}(\theta,\phi) = N_\ell^{-m}P_\ell^{-m}(\cos\theta)e^{-im\phi} = (-1)^m N_\ell^m P_\ell^m(\cos\theta)e^{-im\phi}$$
$$= (-1)^m \overline{N_\ell^m P_\ell^m(\cos\theta)e^{im\phi}} = (-1)^m \overline{Y_\ell^m(\theta,\phi)},$$

where we have used (7.24), p. 190, the identity from Problem 7.5 and noting that $N_\ell^m P_\ell^m(\cos\theta)$ is real.   □

Answer to Problem 7.7. (page 194): Just use the "general Leibniz rule" for

the n-times derivative of a product of two factors noting the combinatorial term

$$\binom{\ell+m}{s} = \frac{(\ell+m)!}{(s)!(\ell+m-s)!}.$$

□

Answer to Problem 7.8. (page 194): The proof in (Riley et al., 2006, pp. 588–589) amounts to showing

$$\left((-1)^{-m}P_\ell^{-m}\right)(x) = (-1)^m\frac{(\ell-m)!}{(\ell+m)!}\left((-1)^m P_\ell^m\right)(x),$$

which is equivalent to (7.24), p. 190. □

Answer to Problem 7.9. (page 194): That this holds for all m, that is, $m \in \{-\ell,\ldots,\ell\}$, is straightforward to check. □

Answer to Problem 7.10. (page 194): We can use the conjugation property (7.26), p. 192. Alternatively, (7.28), p. 192, is equivalent to

$$Y_\ell^{\mp m}(\theta,\phi) = N_\ell^m P_\ell^m(\cos\theta)e^{im\frac{(\pi-\phi)}{\phi}}, \quad m \in \{0,1,\ldots\}.$$

And therefore taking the ratio, we obtain

$$Y_\ell^{-m}(\theta,\phi) = e^{im(\pi-2\phi)}Y_\ell^m(\theta,\phi).$$

□

Answer to Problem 7.11. (page 196): This follows directly from

$$\langle f,g\rangle = \sum_n \langle f,Y_n\rangle\langle Y_n,g\rangle \equiv \sum_{\ell,m}\langle f,Y_\ell^m\rangle\langle Y_\ell^m,g\rangle = \sum_{\ell,m}(f)_\ell^m\overline{(g)_\ell^m}.$$

□

Answer to Problem 7.12. (page 197): Firstly, using the sifting property, we have

$$\int_{\mathbb{S}^2}\int_{\mathbb{S}^2}f(\widehat{\boldsymbol{u}})\delta(\widehat{\boldsymbol{u}},\widehat{\boldsymbol{v}})\,ds(\widehat{\boldsymbol{u}})\,\overline{g(\widehat{\boldsymbol{v}})}\,ds(\widehat{\boldsymbol{v}}) = \int_{\mathbb{S}^2}f(\widehat{\boldsymbol{v}})\overline{g(\widehat{\boldsymbol{v}})}\,ds(\widehat{\boldsymbol{v}}) = \langle f,g\rangle.$$

Secondly, using the delta function expansion, we have

$$\sum_{\ell,m}\int_{\mathbb{S}^2}f(\widehat{\boldsymbol{u}})Y_\ell^m(\widehat{\boldsymbol{u}})\,ds(\widehat{\boldsymbol{u}})\int_{\mathbb{S}^2}\overline{g(\widehat{\boldsymbol{v}})Y_\ell^m(\widehat{\boldsymbol{v}})}\,ds(\widehat{\boldsymbol{v}}) = \sum_{\ell,m}(f)_\ell^m\overline{(g)_\ell^m}.$$

Comparing the two gives the desired result. □

Answer to Problem 7.13. (page 207): This is a straightforward calculation using identities such as $\cos^2\theta = (1+\cos 2\theta)/2$. □

Answer to Problem 7.14. (page 207): Using the property (7.26), p. 192, and for all $m \in \{-\ell,\ldots,\ell\}$

$$\begin{aligned}Y_\ell^m(\theta,\phi) + (-1)^m Y_\ell^{-m}(\theta,\phi) &= Y_\ell^m(\theta,\phi) + \overline{Y_\ell^m(\theta,\phi)} \\ &= 2\,\mathfrak{Re}\left(Y_\ell^m(\theta,\phi)\right) \\ &= 2N_\ell^m P_\ell^m(\cos\theta)\cos m\phi, \quad m \in \{-\ell,\ldots,\ell\}.\end{aligned}$$

When $m \in \{-\ell,\ldots,-1\}$ we can use the first line of (7.58), p. 203, to demonstrate

$$Y_\ell^m(\theta,\phi) + (-1)^m Y_\ell^{-m}(\theta,\phi) = \sqrt{2}Y_{\ell,m}(\theta,\phi), \quad m \in \{-\ell,\ldots,-1\},$$

whereupon reorganizing, we have the desired first line of (7.68), p. 207,

$$Y_{\ell,m}(\theta,\phi) = \frac{1}{\sqrt{2}}\left(Y_\ell^m(\theta,\phi) + (-1)^m Y_\ell^{-m}(\theta,\phi)\right), \quad m \in \{-\ell,\ldots,-1\}.$$

Similarly, again using the property (7.26), p. 192, and for all $m \in \{-\ell, \ldots, \ell\}$

$$Y_\ell^m(\theta, \phi) - (-1)^m Y_\ell^{-m}(\theta, \phi) = Y_\ell^m(\theta, \phi) - \overline{Y_\ell^m(\theta, \phi)}$$
$$= 2i\,\mathfrak{Im}\big(Y_\ell^m(\theta, \phi)\big)$$
$$= 2i N_\ell^m P_\ell^m(\cos\theta)\sin m\phi, \quad m \in \{-\ell, \ldots, \ell\}.$$

When $m \in \{1, 2, \ldots, \ell\}$ we can use the third line of (7.58), p. 203, to demonstrate

$$Y_\ell^m(\theta, \phi) - (-1)^m Y_\ell^{-m}(\theta, \phi) = i\sqrt{2}\,Y_{\ell,m}(\theta, \phi), \quad m \in \{1, 2, \ldots, \ell\},$$

whereupon reorganizing, we have the desired third line of (7.68), p. 207,

$$Y_{\ell,m}(\theta, \phi) = \frac{1}{i\sqrt{2}}\big(Y_\ell^m(\theta, \phi) - (-1)^m Y_\ell^{-m}(\theta, \phi)\big), \quad m \in \{1, 2, \ldots, \ell\}.$$

Finally for $m = 0$, we have the equivalence between the second line of (7.58), p. 203, and the second line of (7.68), p. 207,

$$Y_{\ell 0}(\theta, \phi) = N_\ell^0 P_\ell^0(\cos\theta) = Y_\ell^0(\theta, \phi).$$

$\square$

Answer to Problem 7.15. (page 207): Equation (7.69c), p. 207, can be compared with (7.23), p. 190, to conclude

$$P_{\ell,m}(\mu) = (-1)^m P_\ell^m(x)\big|_{x=\mu}.$$

Then we see that (7.69b), p. 207, is given by

$$X_{\ell,m}(\theta) = N_\ell^m P_\ell^m(\cos\theta).$$

Therefore, (7.69a), p. 207, is identical to the real spherical harmonics in (7.58), p. 203.

$\square$

Answer to Problem 7.16. (page 207): For $m \in \{-\ell, \ldots, -1\}$, remembering that $Y_{\ell,m}(\theta, \phi)$ is real, we have

$$Y_\ell^m(\theta, \phi) = (-1)^m \overline{Y_\ell^{-m}(\theta, \phi)},$$
$$= (-1)^m \overline{\frac{(-1)^m}{\sqrt{2}} Y_{\ell,m}(\theta, \phi) + i\frac{1}{\sqrt{2}} Y_{\ell,-m}(\theta, \phi)},$$
$$= \frac{1}{\sqrt{2}} Y_{\ell,m}(\theta, \phi) - i\frac{(-1)^m}{\sqrt{2}} Y_{\ell,-m}(\theta, \phi),$$

which is (7.67), p. 206.

$\square$

Answer to Problem 7.17. (page 231): This is just a variant on the pre-rotation used to derive (7.125), p. 230, where the intermediate step uses the positive x-axis rather than the negative x-axis. Geometrically this means: the y-axis of the operand is pre-rotated first to the positive x-axis then to the positive z-axis. The ϑ rotation is applied about the z-axis. Then the operand is transferred back to the positive x-axis then to the y-axis, of course, being the inverse of the pre-rotation. All being equivalent to a y-axis rotation through ϑ, that is,

$$\mathcal{D}_y(\vartheta) = \mathcal{D}\Big(\frac{\pi}{2}, \frac{\pi}{2}, 0\Big) \circ \mathcal{D}_z(\varphi) \circ \mathcal{D}\Big(0, -\frac{\pi}{2}, -\frac{\pi}{2}\Big).$$

$\square$

Answer to Problem 7.18. (page 231): Just re-arrange (7.125), p. 230,

$$\mathcal{D}_z(\varphi) = \mathcal{D}\left(0, \frac{\pi}{2}, \frac{\pi}{2}\right) \circ \mathcal{D}_y(\vartheta) \circ \mathcal{D}\left(-\frac{\pi}{2}, -\frac{\pi}{2}, 0\right).$$

Geometrically this means: the z-axis of the operand is pre-rotated first to the negative x-axis then to the y-axis. The ϑ rotation is applied about the y-axis. Then the operand is transferred back to the negative x-axis then to the z-axis, of course, being the inverse of the pre-rotation. □

Answer to Problem 7.19. (page 233): Using (7.115), p. 227, and the property that the Wigner d-matrices are real-valued, the left-hand side of (7.127), p. 231 can be written

$$\int_0^{2\pi}\int_0^{2\pi} e^{-im\varphi} e^{-im'\omega} \overline{e^{-iq\varphi} e^{-iq'\omega}} \, d\omega \, d\varphi \int_0^{\pi} d_{m,m'}^{\ell}(\vartheta) d_{m,m'}^{p}(\vartheta) \sin\vartheta \, d\vartheta$$

$$= \underbrace{\int_0^{2\pi} e^{-i(m-q)\varphi} d\varphi}_{2\pi\delta_{m,q}} \underbrace{\int_0^{2\pi} e^{-i(m'-q')\omega} d\omega}_{2\pi\delta_{m',q'}} \frac{2}{2\ell+1}\delta_{\ell,p}$$

which is the desired result. □

Answer to Problem 7.20. (page 233): From the definition (7.115), p. 227, we have

$$D_{m,0}^{\ell}(\varphi, \vartheta, 0) = e^{-im\varphi} d_{m,0}^{\ell}(\vartheta),$$

which when compared to

$$\sqrt{\frac{4\pi}{2\ell+1}} \overline{Y_\ell^m(\vartheta, \varphi)} = \sqrt{\frac{4\pi}{2\ell+1}} N_\ell^m P_\ell^m(\cos\vartheta) e^{-im\varphi}$$

$$= e^{-im\varphi} \sqrt{\frac{(\ell-m)!}{(\ell+m)!}} P_\ell^m(\cos\vartheta)$$

gives the desired result. □

Answer to Problem 7.21. (page 237): If $f \in \mathcal{H}_\ell$ then $\langle f, Y_p^q \rangle = 0$ whenever $p \neq \ell$. So the generalized Parseval relation for spherical harmonics, (7.42), p. 196, becomes

$$\langle f, g \rangle = \sum_{p \neq \ell} \sum_{q=-p}^{p} \langle f, Y_p^q \rangle \langle Y_p^q, g \rangle + \sum_{q=-\ell}^{\ell} \langle f, Y_\ell^q \rangle \langle Y_\ell^q, g \rangle, \quad f, g \in \mathcal{H}_\ell,$$

where the first double summation is zero (all summands are zero). □

Answer to Problem 7.22. (page 237): For the symmetry relation we note that rotations are isometries so

$$\langle \mathcal{D}_y(\vartheta)f, g \rangle = \langle \mathcal{D}_y(-\vartheta) \circ \mathcal{D}_y(\vartheta)f, \mathcal{D}_y(-\vartheta)g \rangle = \langle f, \mathcal{D}_y(-\vartheta)g \rangle.$$

Next we have

$$d_{m,m'}^{\ell}(\vartheta) = \langle Y_\ell^{m'}, \mathcal{D}_y(-\vartheta)Y_\ell^m \rangle$$

$$= \overline{\langle \mathcal{D}_y(-\vartheta)Y_\ell^m, Y_\ell^{m'} \rangle} = \langle \mathcal{D}_y(-\vartheta)Y_\ell^m, Y_\ell^{m'} \rangle = d_{m',m}^{\ell}(-\vartheta),$$

where we have used that $d_{m,m'}^{\ell}(\vartheta)$ is real-valued. □

Answer to Problem 7.23. (page 238): The left-hand side of (7.143), p. 238, using (7.119), p. 228, can be written

$$\sum_{m'=-\ell}^{\ell} d^{\ell}_{m,m'}(\vartheta_1) d^{\ell}_{m',q}(\vartheta_2) = \sum_{m'=-\ell}^{\ell} \langle \mathcal{D}_y(\vartheta_1) Y^{m'}_{\ell}, Y^m_{\ell} \rangle \langle \mathcal{D}_y(\vartheta_2) Y^q_{\ell}, Y^{m'}_{\ell} \rangle$$

$$= \sum_{m'=-\ell}^{\ell} \langle \mathcal{D}_y(\vartheta_2) Y^q_{\ell}, Y^{m'}_{\ell} \rangle \langle Y^{m'}_{\ell}, \mathcal{D}_y(-\vartheta_1) Y^m_{\ell} \rangle$$

$$= \langle \mathcal{D}_y(\vartheta_2) Y^q_{\ell}, \mathcal{D}_y(-\vartheta_1) Y^m_{\ell} \rangle$$

$$= \langle \mathcal{D}_y(\vartheta_1 + \vartheta_2) Y^q_{\ell}, Y^m_{\ell} \rangle = d^{\ell}_{m,q}(\vartheta_1 + \vartheta_2)$$

where we have used (7.141), p. 237 (noting that $\mathcal{D}_y(\vartheta) Y^m_{\ell} \in \mathcal{H}_{\ell}$ since rotations do not mix degrees, $\mathcal{H}_{\ell}$ is the subspace in (7.140), p. 237) and (7.142), p. 237, twice.

In interpreting the results, the left-hand side is the $(2\ell+1) \times (2\ell+1)$ operator submatrix multiplication of two rotation operators about the y-axis through ϑ_2 followed by ϑ_1, which is clearly equivalent to a single y-axis rotation through $\vartheta_1 + \vartheta_2$. It is the ℓ-th block of the block-diagonal operator matrix version of $\mathcal{D}_y(\vartheta_1) \circ \mathcal{D}_y(\vartheta_2) = \mathcal{D}_y(\vartheta_1 + \vartheta_2)$. $\qquad \square$

Answer to Problem 7.24. (page 238): A proof can be found in (Edmonds, 1957, pp. 57–60). Paraphrasing the proof, from (7.139), p. 237, we have

$$d^{\ell}_{m,m'}(\pi) = \langle \mathcal{D}_y(\pi) Y^{m'}_{\ell}, Y^m_{\ell} \rangle = (-1)^{\ell-m} \delta_{m,-m'}$$

and this can be substituted into the degree-ℓ subspace generalized Parseval relation (7.141), p. 237. $\qquad \square$

Answer to Problem 7.25. (page 238): Applying operator $\mathcal{D}_y(\pi)$ corresponds to a rotation about the y-axis through an angle of π. The effect of such a rotation on spherical coordinates is to replace θ with $\pi - \theta$ and ϕ with $\pi - \phi$. Therefore,

$$\left(\mathcal{D}_y(\pi) Y^m_{\ell} \right)(\theta, \phi) = N^m_{\ell} P^m_{\ell} \left(\cos(\pi - \theta) \right) e^{jm(\pi - \phi)}$$

$$= N^m_{\ell} (-1)^{\ell+m} P^m_{\ell}(\cos\theta) (-1)^m e^{-jm\phi}$$

$$= (-1)^m N^{-m}_{\ell} (-1)^{\ell+m} P^{-m}_{\ell}(\cos\theta) (-1)^m e^{-jm\phi}$$

$$= (-1)^{\ell+m} Y^{-m}_{\ell}(\theta, \phi),$$

where we have used (7.33), p. 194, $N^{-m}_{\ell} P^{-m}_{\ell} = (-1)^m N^m_{\ell} P^m_{\ell}$, and $P^m_{\ell}(-x) = (-1)^{\ell+m} P^m_{\ell}(x)$. To prove the latter, let $x = \cos\theta$, then $\cos(\pi - \theta) = -x$, and substitute into the Rodrigues formula for the associated Legendre function, (7.23), p. 190. $\qquad \square$

Answer to Problem 7.26. (page 238): The two expressions can be shown as follows

$$d^{\ell}_{m,m'}(0) = \langle \mathcal{D}_y(0) Y^{m'}_{\ell}, Y^m_{\ell} \rangle = \langle Y^{m'}_{\ell}, Y^m_{\ell} \rangle = \delta_{m,m'}$$

and

$$d^{\ell}_{m,m'}(\pi) = \langle \mathcal{D}_y(\pi) Y^{m'}_{\ell}, Y^m_{\ell} \rangle = \langle (-1)^{\ell-m'} Y^{-m'}_{\ell}, Y^m_{\ell} \rangle = (-1)^{\ell-m'} \delta_{m,-m'},$$

where in the later we used (7.139), p. 237. $\qquad \square$

Answer to Problem 7.27. (page 239): For (7.144), p. 239, we can use (7.133), p. 235, to negate the ϑ angle in (7.138), p. 237. For (7.145), p. 239, we replace the symbol $-\vartheta$ with $+\vartheta$ in (7.144), p. 239. For (7.146), p. 239, we replace the symbol $-\vartheta$ with $+\vartheta$ in (7.138), p. 237. □

Answer to Problem 7.28. (page 244): We need to examine $d^\ell_{m,0}(0)$ in (7.116), p. 227, for $m' = 0$ and $\vartheta = 0$. For a given $m = -\ell + i$, $i \in \{0, 2, \ldots, 2\ell\}$, we first note that

$$\big(\sin(0)\big)^{2n+m}$$

is non-zero only when $n = -m/2 = (\ell - i)/2$ and since n is an integer, m has to be even. The denominator stays positive when all the following inequalities are satisfied:

$$n \leq \ell, \quad n \geq 0, \quad n \leq 2\ell - i, \quad \text{and} \quad n \geq \ell - i.$$

So on one hand $n \in \{\ell - i, \ldots, \ell\}$ and on the other hand n has to be equal to $(\ell - i)/2$. This is only possible when $\ell - i = 0$ or when $m = 0$, and, in this case, we have $d^\ell_{0,0}(0) = 1$. □

Answer to Problem 7.29. (page 244): Recall $Y^m_\ell(\theta, \phi) = N^m_\ell P^m_\ell(\cos\theta)e^{im\phi}$ from (7.19) and (7.20), p. 189. Operator $\mathcal{D}(0, 0, \omega)$ effects a rotation about the z-axis through ω so it is simple to infer that

$$\big(\mathcal{D}(0, 0, \omega)f\big)(\theta, \phi) = f(\theta, \phi - \omega) = \sum_{\ell,m}(f)^m_\ell Y^m_\ell(\theta, \phi - \omega)$$

$$= \sum_{\ell,m}(f)^m_\ell N^m_\ell P^m_\ell(\cos\theta)e^{im(\phi-\omega)}$$

Then

$$\frac{1}{2\pi}\int_0^{2\pi}\big(\mathcal{D}(0,0,\omega)f\big)(\theta,\phi)\,d\omega = \sum_{\ell,m}(f)^m_\ell N^m_\ell P^m_\ell(\cos\theta)e^{im\phi}\underbrace{\frac{1}{2\pi}\int_0^{2\pi}e^{-im\omega}d\omega}_{\delta_{m,0}}$$

$$= \sum_{\ell=0}^{\infty}(f)^0_\ell Y^0_\ell(\theta,\phi)$$

□

Answer to Problem 7.30. (page 248): For operator $\mathcal{B}_h$ we have

$$\langle \mathcal{B}_h f, f \rangle = \int_{\mathbb{S}^2} h(\widehat{\boldsymbol{u}})f(\widehat{\boldsymbol{u}})\overline{f(\widehat{\boldsymbol{u}})}\,ds(\widehat{\boldsymbol{u}}) = \int_{\mathbb{S}^2} h(\widehat{\boldsymbol{u}})\big|f(\widehat{\boldsymbol{u}})\big|^2 ds(\widehat{\boldsymbol{u}}),$$

using (7.98), p. 221.

For operator $\big(\mathcal{D}(\varphi, \vartheta, \omega)f\big)(\widehat{\boldsymbol{u}})$

$$\sum_{\ell,m}\sum_{q=-\ell}^{\ell}D^\ell_{m,q}(\varphi, \vartheta, \omega)(f)^q_\ell\overline{(f)^m_\ell}$$

follows directly from (7.166), p. 246, with substitution (7.117), p. 227.

For operator $\mathcal{B}^0$

$$\sum_{\ell=0}^{\infty}\big|(f)^0_\ell\big|^2$$

follows directly from (7.166), p. 246, with substitution (7.158), p. 242. □

Answer to Problem 7.31. (page 248): This is equivalent to the operator in Problem 7.32, p. 248, as is shown in (Kennedy et al., 2011). There is a transform pair

$$B(x) = \sum_{\ell=0}^{\infty} b_\ell \frac{(2\ell+1)}{4\pi} P_\ell(x),$$

and

$$b_\ell = 2\pi \int_{-1}^{+1} B(x) P_\ell(x)\, dx,$$

where one can identify $\widehat{\boldsymbol{u}} \cdot \widehat{\boldsymbol{v}}$ with $x \in [-1, +1]$; see (Kennedy et al., 2011). Furthermore, the operator matrix elements are $b_\ell \delta_{\ell,p}^{m,q}$, and the quadratic functional can be written

$$I_{\mathcal{B}_A}(f) = 2\pi \sum_{\ell=0}^{\infty} E_\ell \int_{-1}^{+1} B(x) P_\ell(x)\, dx,$$

where E_ℓ is the energy per degree, (7.46), p. 198.   □

Answer to Problem 7.32. (page 248): The output can be written

$$(\mathcal{B}_I f)(\widehat{\boldsymbol{u}}) = \sum_{\ell,m} b_\ell (f)_\ell^m Y_\ell^m(\widehat{\boldsymbol{u}}).$$

The kernel can be written

$$B_I(\widehat{\boldsymbol{u}}, \widehat{\boldsymbol{v}}) = \sum_{\ell=0}^{\infty} b_\ell \sum_{m=-\ell}^{\ell} Y_\ell^m(\widehat{\boldsymbol{u}}) \overline{Y_\ell^m(\widehat{\boldsymbol{v}})}$$

$$= \sum_{\ell=0}^{\infty} b_\ell \frac{(2\ell+1)}{4\pi} P_\ell(\widehat{\boldsymbol{u}} \cdot \widehat{\boldsymbol{v}}),$$

with the help of the addition theorem (7.30), p. 193.

The quadratic functional is a special case of (7.170), p. 247, as follows

$$I_{\mathcal{B}_I}(f) \triangleq \langle \mathcal{B}_I f, f \rangle = \sum_{\ell,m} b_\ell \left| (f)_\ell^m \right|^2 = \sum_{\ell=0}^{\infty} b_\ell E_\ell,$$

where E_ℓ is the energy per degree, (7.46), p. 198.   □

Answers to problems in Chapter 8

Answer to Problem 8.1. (page 276): Given $h(\widehat{\boldsymbol{u}})$ is real, then

$$\langle \mathcal{B}_h f, g \rangle = \int_{\mathbb{S}^2} h(\widehat{\boldsymbol{u}}) f(\widehat{\boldsymbol{u}}) \overline{g(\widehat{\boldsymbol{u}})}\, ds(\widehat{\boldsymbol{u}})$$

$$= \int_{\mathbb{S}^2} f(\widehat{\boldsymbol{u}}) \overline{h(\widehat{\boldsymbol{u}}) g(\widehat{\boldsymbol{u}})}\, ds(\widehat{\boldsymbol{u}}) = \langle f, \mathcal{B}_h g \rangle.$$

□

Answer to Problem 8.2. (page 276): From (7.109), p. 226,

$$(\mathcal{B}_h f)_\ell^m = \sum_{p,q} (f)_p^q \int_{\mathbb{S}^2} h(\widehat{\boldsymbol{u}}) Y_p^q(\widehat{\boldsymbol{u}}) \overline{Y_\ell^m(\widehat{\boldsymbol{u}})} \, ds(\widehat{\boldsymbol{u}})$$

$$= \sum_{p,q} h_{\ell,p}^{m,q} (f)_p^q$$

where $h_{\ell,p}^{m,q}$ is given in the problem statement. The other identity to be proved, (8.66), p. 276, is straightforward. □

Answer to Problem 8.3. (page 276): From Problem 8.1, p. 276, the operator $\mathcal{B}_h$ is self-adjoint and thereby the operator matrix satisfies $\mathbf{H} = \mathbf{H}^H$ and the result follows. □

Answer to Problem 8.4. (page 276): The case of spatial masking operator $\mathcal{B}_h$ $h(\widehat{\boldsymbol{u}})$ given in (8.57), p. 274, had already been studied in Chapter 7. The matrix elements, $s_{\ell,p}^{m,q}$, were given in (7.105), p. 222, and Table 7.2, p. 250. □

Answer to Problem 8.5. (page 279): Firstly choose the spatial mask $h(\widehat{\boldsymbol{u}})$ in the $\mathcal{B}_h$ operator, (8.56), p. 274, as the indicator function for $R \subset \mathbb{S}^2$, as follows:

$$h(\widehat{\boldsymbol{u}}) = 1_R(\widehat{\boldsymbol{u}}), \quad \widehat{\boldsymbol{u}} \in \mathbb{S}^2.$$

Then elements $h_{\ell,p}^{m,q}$ in (8.58), p. 275, are those of elements $D_{\ell,p}^{m,q}$ in (8.30), p. 262 at least for $\ell, p \leq L$. Let $\mathbf{D}$ be the first $(L+1)^2$ rows and $(L+1)^2$ columns of $\mathbf{H}$. The remainder of $\mathbf{H}$ is irrelevant since f is bandlimited to L. Furthermore, we write the $(L+1)^2$ finite-dimensional vector spectral representation of f as $\mathbf{f}$. Note that the functional $I_{\mathcal{B}_h}(f)$, (8.57), p. 274, simply gives the signal energy in R.

Secondly we need to determine the spectral operator matrix elements in the $\mathcal{B}_s$ operator, (8.61), p. 275. Again we need to only look at the $(L+1)^2$-dimensional portion since f is bandlimited to L. In this case we choose spectral operator matrix elements as $s_{\ell,p}^{m,q} = \delta_{\ell,p}^{m,q}$ for $\ell, p \leq L$. In the finite-dimensional setting we thus can replace $\mathbf{S}$ by the $(L+1)^2 \times (L+1)^2$ identity $\mathbf{I}$.

Bringing these results together, (8.76), p. 278, becomes the finite-dimensional version

$$-\mu_s \mathbf{f} - \mu_h \mathbf{D} \mathbf{f} = \mathbf{f},$$

whereupon if we define $\lambda \triangleq -(1+\mu_s)/\mu_h$ then we recover the eigenvector problem (8.29), p. 262. □

Answer to Problem 8.6. (page 279): Choose $h(\widehat{\boldsymbol{u}}) = 1_R(\widehat{\boldsymbol{u}})$, for $\widehat{\boldsymbol{u}} \in \mathbb{S}^2$, then given that f is spacelimited then it is unaffected by the $\mathcal{B}_h$ operator. Furthermore, if we limited attention to the spatial region $R \subset \mathbb{S}^2$ then $\mathcal{B}_h = \mathbf{I}$ can be substituted.

For the $\mathcal{B}_s$ operator we need to perform spectral truncation to degree L and this can be achieved by setting $b_{\ell,p}^{m,q} = \delta_{\ell,p}^{m,q}$ for $0 \leq \ell \leq L$ and zero otherwise. The corresponding finite-rank kernel $B_s(\widehat{\boldsymbol{u}}, \widehat{\boldsymbol{v}})$ was detailed in (7.93), p. 220, to be

$$B_s(\widehat{\boldsymbol{u}}, \widehat{\boldsymbol{v}}) = \sum_{\ell=0}^{L} \frac{(2\ell+1)}{4\pi} P_\ell(\widehat{\boldsymbol{u}} \cdot \widehat{\boldsymbol{v}}).$$

When restricted to the region R, we write this kernel as $D(\widehat{\boldsymbol{u}}, \widehat{\boldsymbol{v}})$, where $\widehat{\boldsymbol{u}}, \widehat{\boldsymbol{v}} \in R$, and it is identical to the kernel $D(\widehat{\boldsymbol{u}}, \widehat{\boldsymbol{v}})$ in (8.36), p. 264.

Bringing these results together, (8.73), p. 278, becomes the spatially limited version

$$-\mu_s \int_R D(\widehat{\boldsymbol{u}}, \widehat{\boldsymbol{v}}) f(\widehat{\boldsymbol{v}})\, ds(\widehat{\boldsymbol{v}}) - \mu_h f(\widehat{\boldsymbol{u}}) = f(\widehat{\boldsymbol{u}}), \quad \widehat{\boldsymbol{u}} \in R,$$

whereupon if we define $\lambda \triangleq -(1 + \mu_h)/\mu_s$ then we recover the eigenfunction problem (8.42), p. 265. $\qquad\square$

Answers to problems in Chapter 9

Answer to Problem 9.1. (page 296): Assuming the space of $L^2(\mathbb{R})$ with inner product defined as

$$\langle f, g \rangle = \int_{-\infty}^{\infty} f(t)\overline{g(t)}\, dt,$$

we elaborate as follows

$$\langle \mathcal{T}_s f, g \rangle = \int_{-\infty}^{\infty} (\mathcal{T}_s f)(t)\overline{g(t)}\, dt = \int_{-\infty}^{\infty} f(t - s)\overline{g(t)}\, dt$$

$$= \int_{-\infty}^{\infty} f(u)\overline{g(u + s)}\, du = \langle f, \mathcal{T}_{-s} g \rangle$$

to conclude that $\mathcal{T}_s^* = \mathcal{T}_{-s}$. $\qquad\square$

Answer to Problem 9.2. (page 296): With a change of variable $u = t + s$, we can write

$$b(s) = \int_{-\infty}^{\infty} h(s + t) f(t)\, dt = \int_{-\infty}^{\infty} h(u) f(u - s)\, du,$$

from which it becomes clear that the role of h and f in this definition cannot be switched. The same conclusion can be reached using the operator formulation

$$b(s) = \int_{-\infty}^{\infty} (\mathcal{T}_{-s} h)(t) f(t)\, dt = \int_{-\infty}^{\infty} (\mathcal{T}_s \circ \mathcal{T}_{-s} h)(t)(\mathcal{T}_s f)(t)\, dt$$

$$= \int_{-\infty}^{\infty} h(t)(\mathcal{T}_s f)(t)\, dt.$$

$\qquad\square$

Answer to Problem 9.3. (page 296): The steps are similar to problem Problem 9.2. $\qquad\square$

Answer to Problem 9.4. (page 301): The approach is similar to the proof of Theorem 9.1, p. 298. Essentially, we use the fact that all the operators in (9.13), p. 301, are linear and hence we can swap the order of integral over θ and some of the operators to write

$$(h * f)(\boldsymbol{u}) \triangleq \int_{\mathbb{R}^2} \mathcal{T}_{\boldsymbol{u}} \circ \mathcal{R}\Big(\frac{1}{2\pi} \int_0^{2\pi} (\mathcal{D}_\theta h)(\boldsymbol{v})\, d\theta\Big) f(\boldsymbol{o} + \boldsymbol{v})\, d\boldsymbol{v}, \quad \boldsymbol{u} \in \mathbb{R}^2.$$

And call the inner integral

$$h_0(v) \triangleq \frac{1}{2\pi} \int_0^{2\pi} (\mathcal{D}_\theta h)(v)\, d\theta$$

to arrive at

$$(h * f)(u) = \int_{\mathbb{R}^2} (\mathcal{T}_u \circ \mathcal{R} h_0)(v) f(o + v)\, dv = \int_{\mathbb{R}^2} h_0(u - v) f(o + v)\, dv$$

$$= (h_0 \star f)(u).$$

$h_0(v)$ is the averaged out or smoothed version of the original filter $h(v)$ over all possible rotation angles θ. □

Answer to Problem 9.5. (page 307): Consider

$$f(\widehat{u}) = \overline{g(\widehat{u})} = \sum_{\ell,m} \overline{(g)_\ell^m Y_\ell^m(\widehat{u})}$$

$$= \sum_{\ell,m} \overline{(g)_\ell^m} (-1)^m Y_\ell^{-m}(\widehat{u})$$

$$= \sum_{\ell,m} (-1)^m \overline{(g)_\ell^{-m}} Y_\ell^m(\widehat{u}),$$

where we used the conjugation property of the spherical harmonics (7.26), p. 192. From this we can infer

$$(f)_\ell^m = (-1)^m \overline{(g)_\ell^{-m}}.$$

Then the Parseval identity (9.29), p. 307, yields

$$\int_{\mathbb{S}^2} h(\widehat{u}) f(\widehat{u})\, ds(\widehat{u}) = \langle h, \overline{f} \rangle \equiv \langle h, g \rangle = \sum_{\ell,m} (h)_\ell^m \overline{(g)_\ell^m}$$

$$= \sum_{\ell,m} (-1)^m (h)_\ell^m (f)_\ell^{-m}.$$

□

Answer to Problem 9.6. (page 310):

$$\mathcal{D}(\varphi, \vartheta, \pi - \varphi) \circ \mathcal{D}(\varphi, \vartheta, \pi - \varphi) = \mathcal{D}_z(\varphi) \circ \mathcal{D}_y(\vartheta) \circ \mathcal{D}_z(\pi - \varphi)$$

$$\circ \mathcal{D}_z(\varphi) \circ \mathcal{D}_y(\vartheta) \circ \mathcal{D}_z(\pi - \varphi)$$

$$= \mathcal{D}_z(\varphi) \circ \mathcal{D}_y(\vartheta) \circ \mathcal{D}_z(\pi) \circ \mathcal{D}_y(\vartheta) \circ \mathcal{D}_z(\pi - \varphi)$$

$$= \mathcal{D}_z(\varphi) \circ \mathcal{D}_z(\pi) \circ \mathcal{D}_z(\pi - \varphi)$$

$$= \mathcal{D}_z(\varphi + \pi + \pi - \varphi) = \mathcal{D}_z(2\pi) = \mathcal{I},$$

where we have the property that in $\mathcal{D}_y(\vartheta) \circ \mathcal{D}_z(\pi) \circ \mathcal{D}_y(\vartheta)$, the two rotations about the y-axis by ϑ, with a π rotation about the z-axis in between, will cancel each other and only the π rotation about the z-axis remains. □

Answer to Problem 9.7. (page 310): We first focus on the product of middle two matrices

$$R_{-\varphi}^z R_{-\pi+\varphi}^z,$$

which using (7.9), p. 182, is expanded as

$$R_{-\varphi}^z R_{-\pi+\varphi}^z = \begin{pmatrix} \cos\varphi & \sin\varphi & 0 \\ -\sin\varphi & \cos\varphi & 0 \\ 0 & 0 & 1 \end{pmatrix} \begin{pmatrix} -\cos\varphi & \sin\varphi & 0 \\ -\sin\varphi & -\cos\varphi & 0 \\ 0 & 0 & 1 \end{pmatrix} = \begin{pmatrix} -1 & 0 & 0 \\ 0 & -1 & 0 \\ 0 & 0 & 1 \end{pmatrix}.$$

Now combining this result with the next two adjacent matrices and using (7.8), p. 182, we obtain

$$
\boldsymbol{R}^{y}_{-\vartheta}
\begin{pmatrix}
-1 & 0 & 0 \\
0 & -1 & 0 \\
0 & 0 & 1
\end{pmatrix}
\boldsymbol{R}^{y}_{-\vartheta} =
\begin{pmatrix}
\cos\vartheta & 0 & -\sin\vartheta \\
0 & 1 & 0 \\
\sin\vartheta & 0 & \cos\vartheta
\end{pmatrix}
\begin{pmatrix}
-\cos\vartheta & 0 & \sin\vartheta \\
0 & -1 & 0 \\
\sin\vartheta & 0 & \cos\vartheta
\end{pmatrix}
$$

$$
=
\begin{pmatrix}
-1 & 0 & 0 \\
0 & -1 & 0 \\
0 & 0 & 1
\end{pmatrix}.
$$

And finally combining this with the two outer matrices, we obtain

$$
\boldsymbol{R}^{z}_{-\pi+\varphi}
\begin{pmatrix}
-1 & 0 & 0 \\
0 & -1 & 0 \\
0 & 0 & 1
\end{pmatrix}
\boldsymbol{R}^{z}_{-\varphi} =
\begin{pmatrix}
-\cos\varphi & \sin\varphi & 0 \\
-\sin\varphi & -\cos\varphi & 0 \\
0 & 0 & 1
\end{pmatrix}
\begin{pmatrix}
-\cos\varphi & -\sin\varphi & 0 \\
\sin\varphi & -\cos\varphi & 0 \\
0 & 0 & 1
\end{pmatrix}
$$

$$
= \mathbf{I}_3.
$$

□

Answer to Problem 9.8. (page 310): By writing

$$
\mathcal{D}(\varphi, \vartheta, \pi - \varphi) = \mathcal{D}_z(\varphi) \circ \mathcal{D}_y(\vartheta) \circ \mathcal{D}_z(-\varphi) \circ \mathcal{D}_z(\pi),
$$

we can make a correspondence between the time reversal for the real-line convolution, turning $h(t)$ into $(\mathcal{R}h)(t) = h(-t)$, with direction reversal on the 2-sphere through the operator $\mathcal{D}_z(\pi)$, which reverses the direction of x-axis and y-axis. After this initial operation, rotation operators $\mathcal{D}_z(\varphi) \circ \mathcal{D}_y(\vartheta) \circ \mathcal{D}_z(-\varphi)$ on the 2-sphere are analogous of the translation for the real-line convolution. □

Answer to Problem 9.9. (page 320): Picking up from (9.49), p. 314, with $t = -m$ and t' we get

$$
d^{s}_{-m,t'}(\vartheta) = \sum_{n} (-1)^{n-t'-m} c_4(s, t', m, n)
$$
$$
\times \left(\cos\frac{\vartheta}{2}\right)^{2s-2n+t'+m} \left(\sin\frac{\vartheta}{2}\right)^{2n-t'-m},
$$

where

$$
c_4(s, t', m, n) \triangleq \frac{\sqrt{(s+t')!\,(s-t')!\,(s-m)!\,(s+m)!}}{(s+t'-n)!\,(n)!\,(s-n+m)!\,(s-t'-m)!}.
$$

Therefore, the integral in (9.48), p. 314, is expanded as

$$
\int_0^\pi d^{s}_{-m,t'}(\vartheta) d^{\ell}_{m,0}(\vartheta) \sin\vartheta\, d\vartheta = \sum_n \sum_{n'} (-1)^{(n+n'-t')} c_4(s, t', m, n)
$$
$$
\times c_2(\ell, m, n') \int_0^\pi \left(\cos\frac{\vartheta}{2}\right)^{2s+2\ell-2n-2n'+t'} \left(\sin\frac{\vartheta}{2}\right)^{2n+2n'-t'} \sin\vartheta\, d\vartheta.
$$

Upon a change of variable $\cos(\vartheta/2) = x$, the integral becomes

$$
\int_0^\pi \left(\cos\frac{\vartheta}{2}\right)^{2s+2\ell-2n-2n'+t'} \left(\sin\frac{\vartheta}{2}\right)^{2n+2n'-t'} \sin\vartheta\, d\vartheta
$$
$$
= 4 \int_0^1 x^{(2s+2\ell-2n-2n'+t'+1)} (1-x^2)^{(n+n'-t'/2)} dx
$$
$$
= 2 \frac{\Gamma(s+\ell-n-n'+t'/2+1)\Gamma(n+n'-t'/2+1)}{(s+\ell+1)!} \triangleq c_5(s, \ell, n, n', t'),
$$

where $\Gamma(\cdot)$ is the Gamma function. Substituting all the results back in (9.48), p. 314, we obtain

$$\langle D^s_{-m,t'}, Y^m_\ell \rangle = 2\pi(-1)^{t'} \sqrt{\frac{2\ell+1}{4\pi}} \sum_n \sum_{n'} (-1)^{(n+n'-t')} c_4(s,t',m,n)$$
$$\times c_2(\ell,m,n')\, c_5(s,\ell,n,n',t').$$

Finally, the spherical harmonic coefficient can be written as

$$\langle g_0, Y^m_\ell \rangle = \sqrt{\pi(2\ell+1)}(-1)^m \sum_{s,t'} (-1)^{t'} (f)^m_s (h)^{t'}_s \sum_n \sum_{n'} (-1)^{(n+n'-t')}$$
$$\times c_4(s,t',m,n)\, c_2(\ell,m,n')\, c_5(s,\ell,n,n',t').$$

$\square$

Answers to problems in Chapter 10

Answer to Problem 10.1. (page 324): The function space is separable and therefore has a countable dense subset. We only need to place a countable number of sticky-note labels on only those functions in that countable dense subset (complete orthonormal sequence). All other functions can be arbitrarily closely approximated in the norm by the labeled functions. If wanted, the sticky-note labels could be enumerated. $\square$

Answer to Problem 10.2. (page 326): We can show this by showing the parallelogram law fails for the sup-norm. For the two functions given in the Answer to Problem 2.10 on p. 366, the max-norm shown equals the sup-norm and this provides the counter-example. $\square$

Answer to Problem 10.3. (page 332): Because the positive real eigenvalues are upper bounded by B and $f \in \mathcal{H}_\lambda$ then

$$\frac{\|f\|^2}{B} = \frac{1}{B} \sum_{n\in\mathbb{N}} |(f)_n|^2 \le \sum_{n\in\mathbb{N}} \frac{|(f)_n|^2}{\lambda_n} < \infty.$$

by (10.12), p. 331. This implies $\|f\| < \infty$, that is, $f \in L^2(\Omega)$. $\square$

Answer to Problem 10.4. (page 332): From (10.6), p. 329, since λ_n is real $\forall n \in \mathbb{N}$,

$$\overline{\langle g, f \rangle}_{\mathcal{H}_\lambda} = \overline{\sum_{n\in\mathbb{N}} \frac{(g)_n \overline{(f)}_n}{\lambda_n}} = \sum_{n\in\mathbb{N}} \frac{(f)_n \overline{(g)}_n}{\lambda_n} = \langle f, g \rangle_{\mathcal{H}_\lambda},$$

which is (10.9a), p. 330. $\square$

Answer to Problem 10.5. (page 332):

$$\sum_{n\in\mathbb{N}} \frac{(\alpha_1 f_1 + \alpha_2 f_2)_n \overline{(g)}_n}{\lambda_n} = \sum_{n\in\mathbb{N}} \frac{(\alpha_1 (f_1)_n + \alpha_2 (f_2)_n)\overline{(g)}_n}{\lambda_n}$$

$$= \alpha_1 \sum_{n\in\mathbb{N}} \frac{(f_1)_n \overline{(g)}_n}{\lambda_n} + \alpha_2 \sum_{n\in\mathbb{N}} \frac{(f_2)_n \overline{(g)}_n}{\lambda_n} = \alpha_1 \langle f_1, g\rangle_{\mathcal{H}_\lambda} + \alpha_2 \langle f_2, g\rangle_{\mathcal{H}_\lambda},$$

which is (10.9b), p. 330. $\qquad\square$

Answer to Problem 10.6. (page 332): If $\lambda_n = 0$ for some n, then for the inner-product of f and g to be finite requires the $(f)_n = 0$ and $(g)_n = 0$ for all f and g. Then φ_n could be discarded or is unnecessary. Similarly if $\lambda_n = 0$ for all $n > N$ then we are basically working in a finite-dimensional space. Nothing new or interesting results from either case. $\qquad\square$

Answer to Problem 10.7. (page 334):

$$\langle \mathcal{L}_\lambda^{-1/2} f, f\rangle = \left\langle \sum_{n\in\mathbb{N}} \lambda_n^{-1/2}(f)_n \phi_n, \sum_{m\in\mathbb{N}} (f)_m \phi_m \right\rangle$$

$$= \sum_{n\in\mathbb{N}} \lambda_n^{-1/2} \sum_{m\in\mathbb{N}} \langle (f)_n \phi_n, (f)_m \phi_m\rangle = \sum_{n\in\mathbb{N}} \lambda_n^{-1/2} |(f)_n|^2 \geq 0,$$

since $\lambda_n^{-1/2} > 0$, for all $n \in \mathbb{N}$. $\qquad\square$

Answer to Problem 10.8. (page 334): Both $\lambda_n^{-1/2} > 0$ and $\lambda_n^{1/2} > 0$, $\forall n \in \mathbb{N}$. So the proof of $\mathcal{L}_\lambda^{1/2} \geq 0$ differs only trivially from the previous proof of $\mathcal{L}_\lambda^{-1/2} \geq 0$. If $f \neq o$, then at least one $(f)_n \neq 0$ and the summation must be greater than zero (since it consists of no negative terms and at least one positive term). So both $\mathcal{L}_\lambda^{-1/2} > 0$ and $\mathcal{L}_\lambda^{1/2} > 0$. $\qquad\square$

Answer to Problem 10.9. (page 335): Because $\mathcal{L}_\lambda^{1/2}$ and $\mathcal{L}_\lambda^{-1/2}$ are inverses then

$$f \in L^2(\Omega) \iff \mathcal{L}_\lambda^{1/2} f \in \mathcal{H}_\lambda$$

is the same as (10.19) and (10.20), p. 335: $\mathcal{L}_\lambda^{-1/2} f \in L^2(\Omega) \iff f \in \mathcal{H}_\lambda$. $\qquad\square$

Answer to Problem 10.10. (page 336): We have

$$\langle \mathcal{L}_\lambda f, g\rangle = \left\langle \sum_{n\in\mathbb{N}} \lambda_n (f)_n \phi_n, \sum_{m\in\mathbb{N}} (g)_m \phi_m \right\rangle$$

$$= \sum_{n\in\mathbb{N}} \lambda_n \sum_{m\in\mathbb{N}} \langle (f)_n \phi_n, (g)_m \phi_m\rangle = \sum_{n\in\mathbb{N}} \lambda_n \langle (f)_n \phi_n, (g)_n \phi_n\rangle$$

$$= \left\langle (f)_n \phi_n, \sum_{n\in\mathbb{N}} \overline{\lambda_n}(g)_n \phi_n \right\rangle = \left\langle (f)_n \phi_n, \sum_{n\in\mathbb{N}} \lambda_n (g)_n \phi_n\right\rangle, = \langle f, \mathcal{L}_\lambda g\rangle$$

noting $\overline{\lambda}_n = \lambda_n$. $\qquad\square$

Answer to Problem 10.11. (page 336): Building on the answer to the previous answer skipping steps (using $g = f$),

$$\langle \mathcal{L}_\lambda f, f\rangle = \sum_{n\in\mathbb{N}} \lambda_n \langle (f)_n \phi_n, (f)_n \phi_n\rangle = \sum_{n\in\mathbb{N}} \lambda_n \|(f)_n \phi_n\|^2 \geq 0,$$

since $\lambda_n > 0$, for all $n \in \mathbb{N}$. That is, given it is self-adjoint then $\mathcal{L}_\lambda \geq 0$. However, this can be only zero if $f = o$ so in fact $\mathcal{L}_\lambda > 0$. $\qquad\square$

Answer to Problem 10.12. (page 336): For a general input function expand it in terms of the eigenfunctions. The ϕ_n component gets scaled by $\lambda^{1/2}$ through the first $\mathcal{L}_\lambda^{1/2}$ operator and gets scaled again by $\lambda^{1/2}$ through the second $\mathcal{L}_\lambda^{1/2}$ operator. Overall it gets scaled by λ_n and this action is identical to $\mathcal{L}_\lambda$. $\qquad\square$

Answer to Problem 10.13. (page 338): We have

$$\left\| \{\lambda_n\}_{n \in \mathbb{N}} \right\|^2 = \sum_{n=1}^\infty \frac{1}{n^{1+2\epsilon}}$$

which satisfies the zeta function condition

$$1 + 2\epsilon > 1$$

when $\epsilon > 0$. So when ϵ is very small then the convergence is expected to be very slow. Of course, when $\epsilon = 0$ then we get the harmonic series, which is known to diverge. $\qquad\square$

Answer to Problem 10.14. (page 338): Firstly, from the previous question answer we note that $\{\lambda_n\}_{n \in \mathbb{N}} \in \ell^2$, but this fact is not crucial. Next

$$\frac{\left| (f)_n \right|^2}{\lambda_n} = \frac{n^{1/2+\epsilon}}{n^{2\nu}} = \frac{1}{n^{2\nu - 1/2 - \epsilon}},$$

therefore,

$$\left\| \left\{ \frac{(f)_n}{\sqrt{\lambda_n}} \right\}_{n \in \mathbb{N}} \right\|^2 = \sum_{n \in \mathbb{N}} \frac{\left| (f)_n \right|^2}{\lambda_n} = \zeta(2\nu - 1/2 - \epsilon)$$

where $\zeta(s)$ is the zeta function, and the argument is $s \triangleq 2\nu - 1/2 - \epsilon$, which is real. So we simply need $s > 1$, which means

$$\nu > (3/2 + \epsilon)/2 > 3/4.$$

As ϵ increases (and $\{\lambda_n\}_{n \in \mathbb{N}}$ converges faster) then the Fourier coefficients of f need to decay $\epsilon/2$ more quickly. $\qquad\square$

Answer to Problem 10.15. (page 351): We have by (10.35), p. 349, that $(K_y)_n = \lambda_n \overline{\phi_n(y)}$. Then

$$\|K_y\|_{\mathcal{H}_K}^2 = \sum_{n \in \mathbb{N}} \frac{\left| (K_y)_n \right|^2}{\lambda_n}$$

$$= \sum_{n \in \mathbb{N}} \frac{\left| \lambda_n \overline{\phi_n(y)} \right|^2}{\lambda_n},$$

$$= \sum_{n \in \mathbb{N}} \lambda_n \phi_n(y) \overline{\phi_n(y)} = K(y, y) < \infty,$$

since $K(\cdot, \cdot)$ is continuous on the bounded closed domain $\Omega \times \Omega$. $\qquad\square$

Answer to Problem 10.16. (page 351): We have

$$\langle f(\cdot), g(\cdot)\rangle_{\mathcal{H}_K} = \Big\langle \sum_{m=1}^{M} \alpha_m K_{y_m}(\cdot), \sum_{n=1}^{N} \beta_n K_{x_n}(\cdot)\Big\rangle_{\mathcal{H}_K}$$

$$= \sum_{m=1}^{M}\sum_{n=1}^{N} \alpha_m \overline{\beta_n} \langle K_{y_m}(\cdot), K_{x_n}(\cdot)\rangle_{\mathcal{H}_K}$$

$$= \sum_{m=1}^{M}\sum_{n=1}^{N} \alpha_m \overline{\beta_n} K_{y_m}(x_n), \quad \text{by (10.36), p. 350,}$$

$$= \sum_{m=1}^{M}\sum_{n=1}^{N} \alpha_m \overline{\beta_n} K(x_n, y_m),$$

which shows the use of kernel evaluations to compute inner products in the RKHS $\mathcal{H}_K$. □

Answer to Problem 10.17. (page 359): Identity (10.63), p. 359, enables us to construct a new kernel with the same asymptotic eigenvalue rate of decay as the previous Cui and Freden kernel:

$$\lambda_\ell = \begin{cases} \lambda_0 > 0 & \ell = 0, \\ \dfrac{1}{(4\ell^2 - 1)(2\ell + 3)} & \ell \in \{1, 2, \ldots\}. \end{cases}$$

Therefore

$$K(\widehat{\boldsymbol{x}}, \widehat{\boldsymbol{y}}) = \frac{(1 + 3\lambda_0)}{12\pi} - \frac{1}{8\pi}\sqrt{\frac{1 - (\widehat{\boldsymbol{x}} \cdot \widehat{\boldsymbol{y}})}{2}}$$

is a reproducing kernel for any $\lambda_0 > 0$. The choice $\lambda_0 = 1$ is the most natural one. □

Answer to Problem 10.18. (page 359): Comparing (10.64), p. 359, with (10.56), p. 356, we can identify the eigenvalues as

$$\lambda_\ell = \frac{4\pi \rho^\ell}{2\ell + 1}, \quad \ell \in \{0, 1, \ldots\}.$$

By taking $0 < \rho < 1$ then $\rho^\ell > 0$ and decays exponentially. Then

$$K(\widehat{\boldsymbol{x}}, \widehat{\boldsymbol{y}}) = \frac{1}{\sqrt{1 - 2(\widehat{\boldsymbol{x}} \cdot \widehat{\boldsymbol{y}})\rho + \rho^2}}$$

is a reproducing kernel for any $0 < \rho < 1$. □

Answer to Problem 10.19. (page 359): From (Seon, 2006, Equation (17)), we have

$$\lambda_\ell = \frac{1}{\sinh\kappa} \frac{d^\ell}{dx^\ell}\Big(\frac{\sinh(\kappa\sqrt{1 + 2x/\kappa})}{\sqrt{1 + 2x/\kappa}}\Big)\Big|_{x=0}, \quad \ell \in \{0, 1, \ldots\}.$$

Let $z = \kappa\sqrt{1 + 2x/\kappa}$ and consider the Taylor series of $\sinh(z)/z$ and its derivatives. From this one can show $\lambda_\ell > 0$ for all $\ell \in \mathbb{N}$ and all $\kappa \geq 0$. In fact $\lambda_0 = 1$, and $\lambda_1 = \coth\kappa - \kappa^{-1} > 0$ for all $\kappa \geq 0$. Then

$$K(\widehat{\boldsymbol{x}}, \widehat{\boldsymbol{y}}) = \frac{\kappa \exp\big(\kappa(\widehat{\boldsymbol{x}} \cdot \widehat{\boldsymbol{y}})\big)}{4\pi \sinh\kappa}$$

is a reproducing kernel for any $\kappa \geq 0$. □

Bibliography

ABHAYAPALA, T. D., POLLOCK, T. S., AND KENNEDY, R. A. (2003) "Characterization of 3D spatial wireless channels," In *Proc. IEEE Vehicular Technology Conference (Fall), VTC2003-Fall*, vol. 1, pp. 123–127. Orlando, FL.

ACZEL, A. D. (2001) *The Mystery of the Aleph: Mathematics, the Kabbalah, and the Search for Infinity.* Washington Square Press, New York, NY.

AIZERMAN, A., BRAVERMAN, E. M., AND ROZONER, L. I. (1964) "Theoretical foundations of the potential function method in pattern recognition learning," *Automation and Remote Control*, vol. 25, pp. 821–837.

ANTOINE, J.-P. AND VANDERGHEYNST, P. (1999) "Wavelets on the 2-sphere: A group-theoretical approach," *Appl. Comput. Harmon. Anal.*, vol. 7, no. 3, pp. 262–291.

ARMITAGE, C. AND WANDELT, B. D. (2004) "Deconvolution map-making for cosmic microwave background observations," *Phys. Rev. D*, vol. 70, no. 12, pp. 123007:1–123007:7.

ARONSZAJN, N. (1950) "Theory of reproducing kernels," *Trans. Amer. Math. Soc.*, vol. 68, no. 3, pp. 337–404.

AUDET, P. (2011) "Directional wavelet analysis on the sphere: Application to gravity and topography of the terrestrial planets," *J. Geophys. Res.*, vol. 116.

BELL, E. (1986) *Men of Mathematics.* Simon & Schuster, New York, NY.

BERNKOPF, M. (2008) "Schmidt, Erhard," *Complete Dictionary of Scientific Biography* http://www.encyclopedia.com/doc/1G2-2830903888.html.

BOUNIAKOWSKY, V. (1859) "Sur quelques inégalités concernant les intégrales ordinaires et les intégrales aux différences finies," *Mémoires de l'Acad. de St.-Pétersbourg (VIIe Série)*, vol. 1, no. 9.

BRECHBÜHLER, C. H., GERIG, G., AND KÜBLER, O. (1995) "Parametrization of closed surfaces for 3-D shape description," *Comput. Vision Image Understand.*, vol. 61, no. 2, pp. 154–170.

BÜLOW, T. (2004) "Spherical diffusion for 3D surface smoothing," *IEEE Trans. Pattern Anal. Mach. Intell.*, vol. 26, no. 12, pp. 1650–1654.

CAJORI, F. (1985) *A History of Mathematics.* Chelsea Publishing Company, London, 4th edn. (the first edition was published by The Macmillian Company, New York in 1893).

COHEN, L. (1989) "Time-frequency distributions–a review," *Proc. IEEE*, vol. 77, no. 7, pp. 941–981.

COHEN, L. (1995) *Time-Frequency Analysis: Theory and Applications.* Prentice-Hall, New York, NY.

COLTON, D. AND KRESS, R. (1998) *Inverse Acoustic and Electromagnetic Scattering Theory.* Springer-Verlag, Berlin, 2nd edn.

COURANT, R. AND HILBERT, D. (1966) *Methods of Mathematical Physics. Volume 1.* Interscience Publishers, New York, NY (the first German edition was published by Julius Springer, Berlin in 1924, and the first English edition by Interscience Publishers, New York in 1953).

CUCKER, F. AND SMALE, S. (2002) "On the mathematical foundations of learning," *Bull. Am. Math. Soc., New Ser.*, vol. 39, no. 1, pp. 1–49.

CUI, J. AND FREEDEN, W. (1997) "Equidistribution on the sphere," *SIAM J. Sci. Comput.*, vol. 18, no. 2, pp. 595–609.

DAHLEN, F. A. AND SIMONS, F. J. (2008) "Spectral estimation on a sphere in geophysics and cosmology," *Geophys. J. Int.*, vol. 174, no. 3, pp. 774–807.

DAUBEN, J. W. (1990) *Georg Cantor: His Mathematics and Philosophy of the Infinite.* Princeton University Press, Princeton, NJ.

DEBNATH, L. AND MIKUSIŃSKI, P. (1999) *Introduction to Hilbert Spaces with Applications.* Academic Press, San Diego, CA, 2nd edn.

DRISCOLL, J. R. AND HEALY, JR., D. M. (1994) "Computing Fourier transforms and convolutions on the 2-sphere," *Adv. Appl. Math.*, vol. 15, no. 2, pp. 202–250.

DURAK, L. AND ARIKAN, O. (2003) "Short-time Fourier transform: two fundamental properties and an optimal implementation," *IEEE Trans. Signal Process.*, vol. 51, no. 5, pp. 1231–1242.

EDMONDS, A. R. (1957) *Angular Momentum in Quantum Mechanics.* Princeton University Press, Princeton, NJ.

FARMELO, G. (2009) *The Strangest Man: The Hidden Life of Paul Dirac, Mystic of the Atom.* Basic Books, New York, NY.

FISCHER, E. (1907) "Sur la convergence en moyenne," *Comptes rendus de l'Académie des sciences, Paris*, vol. 144, pp. 1022–1024.

FRANKS, L. E. (1969) *Signal Theory*. Prentice-Hall, Englewood Cliffs, NJ.

FREEDEN, W. AND MICHEL, V. (1999) "Constructive approximation and numerical methods in geodetic research today. An attempt at a categorization based on an uncertainty principle," *J. Geodesy*, vol. 73, no. 9, pp. 452–465.

GÓRSKI, K. M., HIVON, E., BANDAY, A. J., WANDELT, B. D., HANSEN, F. K., REINECKE, M., AND BARTELMANN, M. (2005) "HEALPix: A Framework for high-resolution discretization and fast analysis of data distributed on the sphere," *Astrophys. J.*, vol. 622, no. 2, pp. 759–771.

HAN, C., SUN, B., RAMAMOORTHI, R., AND GRINSPUN, E. (2007) "Frequency domain normal map filtering," *ACM Trans. Graph.*, vol. 26, no. 3, pp. 28:1–28:12.

HARRIS, F. J. (1978) "On the use of windows for harmonic analysis with the discrete Fourier transform," *Proc. IEEE*, vol. 66, no. 1, pp. 51–83.

HELMBERG, G. (1969) *Introduction to Spectral Theory in Hilbert Space*. John Wiley & Sons, New York, NY (reprinted in 2008 by Dover Publications, Mineola, NY).

HUANG, W., KHALID, Z., AND KENNEDY, R. A. (2011) "Efficient computation of spherical harmonic transform using parallel architecture of CUDA," In *Proc. Int. Conf. Signal Process. Commun. Syst., ICSPCS'2011*, p. 6. Honolulu, HI.

KENNEDY, R. A., LAMAHEWA, T. A., AND WEI, L. (2011) "On azimuthally symmetric 2-sphere convolution," *Digit. Signal Process.*, vol. 5, no. 11, pp. 660–666.

KENNEDY, R. A., SADEGHI, P., ABHAYAPALA, T. D., AND JONES, H. M. (2007) "Intrinsic limits of dimensionality and richness in random multipath fields," *IEEE Trans. Signal Process.*, vol. 55, no. 6, pp. 2542–2556.

KHALID, Z., DURRANI, S., KENNEDY, R. A., AND SADEGHI, P. (2012a) "Spatio-spectral analysis on the sphere using spatially localized spherical harmonics transform," *IEEE Trans. Signal Process.*, vol. 60, no. 3, pp. 1487–1492.

KHALID, Z., KENNEDY, R. A., DURRANI, S., SADEGHI, P., WIAUX, Y., AND MCEWEN, J. D. (2013a) "Fast directional spatially localized spherical harmonic transform," *IEEE Trans. Signal Process.* (in print).

KHALID, Z., KENNEDY, R. A., AND SADEGHI, P. (2012b) "Efficient computation of commutative anisotropic convolution on the 2-sphere," In *Proc. Int. Conf. Signal Process. Commun. Syst., ICSPCS'2012*, p. 6. Gold Coast, Australia.

KHALID, Z., SADEGHI, P., KENNEDY, R. A., AND DURRANI, S. (2013b) "Spatially varying spectral filtering of signals on the unit sphere," *IEEE Trans. Signal Process.*, vol. 61, no. 3, pp. 530–544.

KOKS, D. (2006) "Using rotations to build aerospace coordinate systems," *Tech. Rep. DSTO–TN–0640*, Defence Science and Technology Organisation, Australia, Electronic Warfare and Radar Division, Systems Sciences Laboratory.

KREYSZIG, E. (1978) *Introductory Functional Analysis with Applications*. John Wiley & Sons, New York, NY.

LANDAU, H. J. AND POLLAK, H. O. (1962) "Prolate spheroidal wave functions, Fourier analysis and uncertainty–III: The dimension of the space of essentially time- and band-limited signals," *Bell Syst. Tech. J.*, vol. 41, no. 4, pp. 1295–1336.

LANDAU, H. J., SLEPIAN, D., AND POLLAK, H. O. (1961) "Prolate spheroidal wave functions, Fourier analysis and uncertainty–II," *Bell Syst. Tech. J.*, vol. 40, no. 1, pp. 65–84.

LEBEDEV, N. N. (1972) *Special Functions and Their Applications*. Dover Publications, New York, NY (revised edition, translated from the Russian and edited by Richard A. Silverman, unabridged and corrected republication).

MÁTÉ, L. (1989) *Hilbert Space Methods in Science and Engineering*. Adam Hilger, IOP Publishing, Bristol and New York.

MCEWEN, J. D., HOBSON, M. P., MORTLOCK, D. J., AND LASENBY, A. N. (2007) "Fast directional continuous spherical wavelet transform algorithms," *IEEE Trans. Signal Process.*, vol. 55, no. 2, pp. 520–529.

MCEWEN, J. D. AND WIAUX, Y. (2011) "A novel sampling theorem on the sphere," *IEEE Trans. Signal Process.*, vol. 59, no. 12, pp. 5876–5887.

MERCER, J. (1909) "Functions of positive and negative type, and their connection with the theory of integral equations," *Phil. Trans. R. Soc. Lond.*, vol. 209, no. 456, pp. 415–446.

MESSIAH, A. (1961) *Quantum Mechanics*. North-Holland, Amsterdam (translated from French by J. Potter).

NARCOWICH, F. J. AND WARD, J. D. (1996) "Non-stationary wavelets on the m-sphere for scattered data," *Appl. Comput. Harmonic Anal.*, vol. 3, no. 4, pp. 324–336.

NG, K.-W. (2005) "Full-sky correlation functions for CMB experiments with asymmetric window functions," *Phys. Rev. D*, vol. 71, no. 8, pp. 083009:1–083009:9.

OPPENHEIM, A. V., WILLSKY, A. S., AND NAWAB, S. H. (1996) *Signals and Systems*. Prentice-Hall, Upper Saddle River, NJ.

PAPOULIS, A. (1977) *Signal Analysis*. McGraw-Hill, New York, NY.

PIETSCH, A. (2010) "Erhard Schmidt and his contributions to functional analysis," *Math. Nachr.*, vol. 283, no. 1, pp. 6–20.

POLLOCK, T. S., ABHAYAPALA, T. D., AND KENNEDY, R. A. (2003) "Introducing space into MIMO capacity calculations," *J. Telecommun. Syst.*, vol. 24, no. 2, pp. 415–436.

RAFAELY, B. (2004) "Plane-wave decomposition of the sound field on a sphere by spherical convolution," *J. Acoust. Soc. Am.*, vol. 116, no. 4, pp. 2149–2157.

REID, C. (1986) *Hilbert-Courant.* Springer-Verlag, New York, NY (previously published as two separate volumes: *Hilbert* by Constance Reid, Springer-Verlag, 1970; *Courant in Göttingen and New York: The Story of an Improbable Mathematician* by Constance Reid, Springer-Verlag, 1976).

RIESZ, F. (1907) "Sur les systemes orthogonaux de fonctions," *Comptes rendus de l'Académie des sciences, Paris*, vol. 144, pp. 615–619.

RIESZ, F. AND SZ.-NAGY, B. (1990) *Functional Analysis.* Dover Publications, Mineola, NY, 2nd edn. (unabridged Dover republication of the edition published by Frederick Ungar Publishing Co., New York, 1955; translated from the original second French edition 1953).

RILEY, K. F., HOBSON, M. P., AND BENCE, S. J. (2006) *Mathematical Methods for Physics and Engineering.* Cambridge University Press, Cambridge, UK, 3rd edn.

RISBO, T. (1996) "Fourier transform summation of Legendre series and *D*-functions," *J. Geodesy*, vol. 70, no. 7, pp. 383–396.

SADEGHI, P., KENNEDY, R. A., AND KHALID, Z. (2012) "Commutative anisotropic convolution on the 2-sphere," *IEEE Trans. Signal Process.*, vol. 60, no. 12, pp. 6697–6703.

SAKURAI, J. J. (1994) *Modern Quantum Mechanics.* Addison-Wesley Publishing Company, Inc., Reading, MA.

SCHRÖDER, P. AND SWELDENS, W. (2000) "Spherical wavelets: Efficiently representing functions on a sphere," In R. Klees and R. Haagmans, eds., *Wavelets in the Geosciences*, vol. 90 of *Lecture Notes in Earth Sciences*, pp. 158–188. Springer, Berlin (reprinted from Computer Graphics Proceedings, 1995, 161-172, ACM Siggraph).

SEON, K.-I. (2006) "Smoothing of all-sky survey map with Fisher-von Mises function," *J. Korean Phys. Soc.*, vol. 48, no. 3, pp. 331–334.

SIMONS, F. J., DAHLEN, F. A., AND WIECZOREK, M. A. (2006) "Spatiospectral concentration on a sphere," *SIAM Rev.*, vol. 48, no. 3, pp. 504–536.

SIMONS, M., SOLOMON, S. C., AND HAGER, B. H. (1997) "Localization of gravity and topography: constraints on the tectonics and mantle dynamics of Venus," *Geophys. J. Int.*, vol. 131, no. 1, pp. 24–44.

SLEPIAN, D. (1978) "Prolate spheroidal wave functions, Fourier analysis and uncertainty–V: Discrete time case," *Bell Syst. Tech. J.*, vol. 57, no. 5, pp. 3009–3057.

SLEPIAN, D. AND POLLAK, H. O. (1961) "Prolate spheroidal wave functions, Fourier analysis and uncertainty–I," *Bell Syst. Tech. J.*, vol. 40, no. 1, pp. 43–63.

SLOANE, N. J. A. (2001) "The On-Line Encyclopedia of Integer Sequences," http://oeis.org/A059343 (table of Hermite polynomial coefficients).

SLOANE, N. J. A. (2002) "The On-Line Encyclopedia of Integer Sequences," http://oeis.org/A008316 (table of Legendre polynomial coefficients).

SPERGEL, D. N., BEAN, R., DORÉ, O., NOLTA, M. R., BENNETT, C. L., DUNKLEY, J., HINSHAW, G., JAROSIK, N., KOMATSU, E., PAGE, L., PEIRIS, H. V., VERDE, L., HALPERN, M., HILL, R. S., KOGUT, A., LIMON, M., MEYER, S. S., ODEGARD, N., TUCKER, G. S., WEILAND, J. L., WOLLACK, E., AND WRIGHT, E. L. (2007) "Three-year Wilkinson Microwave Anisotropy Probe (WMAP) observations: Implications for cosmology," *The Astrophysical Journal Supplement Series*, vol. 170, no. 2, pp. 377–408.

STARCK, J.-L., MOUDDEN, Y., ABRIAL, P., AND NGUYEN, M. (2006) "Wavelets, ridgelets and curvelets on the sphere," *Astronom. and Astrophys.*, vol. 446, no. 3, pp. 1191–1204.

STEWART, I. (1973) *Galois Theory*. Chapman and Hall, London.

STRANG, G. (2007) *Computational Science and Engineering*. Wellesley-Cambridge Press, Wellesley, MA.

TEGMARK, M., HARTMANN, D. H., BRIGGS, M. S., AND MEEGAN, C. A. (1996) "The angular power spectrum of BATSE 3B gamma-ray bursts," *Astrophys. J.*, vol. 468, pp. 214–224.

TRAPANI, S. AND NAVAZA, J. (2006) "Calculation of spherical harmonics and Wigner d functions by FFT. Applications to fast rotational matching in molecular replacement and implementation into AMoRe," *Acta Cryst. A*, vol. 62, no. 4, pp. 262–269.

WANDELT, B. D. AND GÓRSKI, K. M. (2001) "Fast convolution on the sphere," *Phys. Rev. D*, vol. 63, no. 12, p. 123002.

WATSON, G. N. (1995) *A Treatise on the Theory of Bessel Functions*. Cambridge University Press, Cambridge, UK.

WEI, L., KENNEDY, R. A., AND LAMAHEWA, T. A. (2011) "Quadratic variational framework for signal design on the 2-sphere," *IEEE Trans. Signal Process.*, vol. 59, no. 11, pp. 5243–5252.

WIAUX, Y., JACQUES, L., AND VANDERGHEYNST, P. (2005) "Correspondence principle between spherical and Euclidean wavelets," *Astrophys. J.*, vol. 632, no. 1, pp. 15–28.

WIAUX, Y., JACQUES, L., VIELVA, P., AND VANDERGHEYNST, P. (2006) "Fast directional correlation on the sphere with steerable filters," *Astrophys. J.*, vol. 652, no. 1, pp. 820–832.

WIAUX, Y., MCEWEN, J. D., VANDERGHEYNST, P., AND BLANC, O. (2008) "Exact reconstruction with directional wavelets on the sphere," *Mon. Not. R. Astron. Soc.*, vol. 388, no. 2, pp. 770–788.

WIECZOREK, M. (2007) "Gravity and topography of the terrestrial planets," *Treatise on Geophysics*, vol. 10, pp. 165–206.

WIECZOREK, M. A. AND SIMONS, F. J. (2005) "Localized spectral analysis on the sphere," *Geophys. J. Int.*, vol. 131, no. 1, pp. 655–675.

WIGNER, E. P. (1959) *Group Theory and its Application to the Quantum Mechanics of Atomic Spectra*. Academic Press, New York, NY (translated from the original German version first published in 1931; edited by J. J. Griffin, expanded and improved edition).

WIGNER, E. P. (1965) "On the matrices which reduce the Kronecker products of representations of S. R. groups," In L. C. Biedenharn and H. van Dam, eds., *Quantum Theory of Angular Momentum*, pp. 87–133. Academic Press, New York, NY (reprinted from a 1940 unpublished manuscript).

YEO, B. T. T., OU, W., AND GOLLAND, P. (2008) "On the construction of invertible filter banks on the 2-sphere," *IEEE Trans. Image Process.*, vol. 17, no. 3, pp. 283–300.

YU, P., GRANT, P. E., YUAN, Q., XIAO, H., SEGONNE, F., PIENAAR, R., BUSA, E., PACHECO, J., MAKRIS, N., BUCKNER, R. L., GOLLAND, P., AND FISCHL, B. (2007) "Cortical surface shape analysis based on spherical wavelets," *IEEE Trans. Med. Imag.*, vol. 26, no. 4, pp. 582–597.

ZUBER, M. T., SMITH, D. E., SOLOMON, S. C., MUHLEMAN, D. O., HEAD, J. W., GARVIN, J. B., ABSHIRE, J. B., AND BUFTON, J. L. (1992) "The Mars observer laser altimeter investigation," *J. Geophys. Res.*, vol. 97, no. E5, pp. 7781–7797.

Notation

Sets

$\mathbb{N}$	Set of natural numbers, $\mathbb{N} = \{1, 2, 3, \ldots\}$, 11
$\mathbb{Z}^\star$	Set of non-negative integers, $\mathbb{Z}^\star = \{0, 1, 2, 3, \ldots\}$, 11
$\mathbb{Z}$	Set of all integers, $\mathbb{Z} = \{\ldots, -3, -2, -1, 0, 1, 2, 3, \ldots\}$, 11
$\mathbb{Z}^2$	Set of two dimensional integer lattice points, 13
$\mathbb{Q}$	Set of rational numbers, 12
$\mathbb{Q}^\star$	Set of non-negative rational numbers (includes 0), 13
$\mathbb{R}$	Set of real numbers, 4
$\mathbb{R}^N$	Set of all N-vectors of real numbers, 4
$\mathbb{C}$	Set of complex numbers, 4
$\mathbb{C}^N$	Set of all N-vectors of complex numbers, 19
$\mathbb{S}^1$	1-sphere or unit circle, 85
$\mathbb{S}^2$	2-sphere or unit sphere, 87
$\cup$	Set union, 16
$\backslash$	Set difference, 177
$\subset$	Subset, 35
$|S|$	Cardinality of a set S, the number of elements in S, 12
$\aleph_0$	Cardinality of $\mathbb{N}$, the smallest transfinite number, $\aleph_0 = |\mathbb{N}|$, 9
$\mathfrak{c}$	Cardinality of continuum or the set of reals $\mathbb{R}$, $\mathfrak{c} = |\mathbb{R}|$, 9
$\mathcal{D}$	A dense set in $\mathcal{H}$, 31

Symbols

i	Unit imaginary number, $i = \sqrt{-1}$, 41
ϵ	An arbitrarily small positive real number, 20

Indices

$\ell, k, m, n, p,$ q, r, s, t, u	General-purpose indices belonging to $\mathbb{Z}$, $\mathbb{Z}^\star$, or $\mathbb{N}$, 13
p/q	A rational number belonging to $\mathbb{Q}$, or $\mathbb{Q}^\star$, 13
M, N, L	Dimension or size of a space, vector, etc., 4

Functions

e^x	Exponential function $e^x = \exp(x)$, 17
e^{ix}	Complex exponential function $e^{ix} = \cos x + i \sin x$, 59
$P_\ell(x)$	Legendre polynomial of degree ℓ, 53
$H_n(x)$	Hermite polynomial of degree n, 57
$\varrho_n(x)$	Hermite function of degree n, 59
$P_\ell^m(x)$	Associated Legendre function of degree ℓ and order m, 62, 190
$L_n(x)$	Laguerre polynomial of degree n, 96
$L_n^{(\alpha)}(x)$	Generalized Laguerre polynomial of degree n and parameter $\alpha \geq 0$, 96
$J_\nu(x)$	Bessel function of the first kind with parameter ν, 82
$Y_\ell^m(\theta, \phi)$	Spherical harmonic function of degree ℓ and order m defined on $\mathbb{S}^2$, 87, 189
$\Gamma(x)$	Gamma function, $\Gamma(x) = \int_0^\infty e^{-t} t^{x-1} dt$, 17
$n!$	Factorial function, $n! = n(n-1)(n-2)\cdots(2)(1) \equiv \Gamma(n+1)$, 17
$\mathbf{1}_X(x)$	Indicator or characteristic function, equals 1 if $x \in X$, 0 otherwise, 20
$\delta(\cdot)$	Dirac delta function, 93
$\delta(\widehat{\boldsymbol{u}}, \widehat{\boldsymbol{v}})$	Dirac delta function on 2-sphere, $\delta(\widehat{\boldsymbol{u}}, \widehat{\boldsymbol{v}}) = (\sin\theta)^{-1}\delta(\theta - \vartheta)\delta(\phi - \varphi)$, 95, 194
$\delta_{n,m}$	Kronecker delta function, $\delta_{n,m} = 1$ when $n = m$ and zero otherwise, 43

Miscellaneous

$'$	Transpose of a vector or a matrix, 4
$|\cdot|$	Absolute value of a scalar or a number, 16
$\overline{(\cdot)}$	Conjugate operation, 37
$(\cdot)^H$	Hermitian of a vector or a matrix, $(\cdot)^H = \overline{(\cdot)}'$, 132
$\mathrm{trace}(\cdot)$	Trace of a matrix, which is equal to the sum of its diagonal terms, 168
$\triangleq$	Definition or assignment, equality, 13
$\equiv$	Equivalence, 29
$\Longrightarrow$	Implies, 48
$\nRightarrow$	Does not imply, 48
$\Longleftrightarrow$	If and only if, is necessary and sufficient, 26
$\mapsto$	Maps to, 28
$\to$	Tends to, 22
$\to \infty$	Tends to infinity, 28
$\rightharpoonup$	Weakly converges to, 75, 115
$\to$	Strongly converges to, or converges in the mean, 28, 74, 115
$\alpha, \beta, \gamma, \delta$	Scalars, which unless otherwise stated belong to $\mathbb{C}$, 4
$a = (\alpha_1, \alpha_2, \ldots, \alpha_N)'$	A column vector in an N dimensional vector space $\mathbb{C}^N$ or $\mathbb{R}^N$, 4
$b = (\beta_1, \beta_2, \ldots, \beta_N)'$	A column vector in an N dimensional vector space $\mathbb{C}^N$ or $\mathbb{R}^N$, 4
$\mathfrak{Re}(\alpha)$	Real part of α, 38
$\mathfrak{Im}(\alpha)$	Imaginary part of α, 203
$O(\cdot)$	Big O – describing the limiting behavior of a function when the argument tends towards a particular value, 58
$\mathbf{I}_N$	Identity matrix of size N, 180
$\mathbf{I}$	Identity matrix of infinite size, 278

Spaces

$\mathcal{H}$	A linear space (vector, normed, inner product, Banach, Hilbert, etc.), 24
d, f, g, h	Vectors or elements of a space $\mathcal{H}$, if functions also written as $f(x)$, etc., 24
o	Element zero of a space $\mathcal{H}$, 25
$\|\cdot\|$	Norm of a vector in a normed space, 4, 26
$\{\mathcal{H}, \|\cdot\|\}$	Vector space $\mathcal{H}$ with norm $\|\cdot\|$, 27
$\mathcal{M}$	Metric space with metric $d(f, g)$ for $f, g \in \mathcal{M}$, 28
$f_1, f_2, f_3, \ldots$	A sequence of elements or vectors in $\mathcal{H}$, 28
$C[\alpha, \beta]$	Vector space of continuous functions on interval $[\alpha, \beta]$, 27
Ω	Domain of a function, refers to the closed interval on the real line $[\alpha, \beta]$. The limits α and β can be finite, but $\alpha = -\infty$ or $\beta = +\infty$ or both are also allowed, in which case the appropriate open interval is meant, 33
$C(\Omega)$	Vector space of continuous functions on interval Ω, 33
$L^2(\Omega)$	Space of square integrable functions on Ω, 33, 47
$L^2(\alpha, \beta)$	$L^2(\alpha, \beta)$ is only used for specific values of α and β, such as $L^2(0, 1)$ with $\Omega = [0, 1]$, 50
$L^2(\mathbb{R})$	Space of square integrable functions over the set of reals. That is $\Omega = (-\infty, \infty)$, 57
$L^2(\mathbb{S}^2)$	Space of square integrable functions on 2-sphere $\mathbb{S}^2$, 87, 189
$\langle \cdot, \cdot \rangle$	Inner product between two vectors in an inner product space, 5, 37
$\{\mathcal{H}, \langle \cdot, \cdot \rangle\}$	Vector space $\mathcal{H}$ with inner product $\langle \cdot, \cdot \rangle$, 38
$\theta(f, g)$	Angle between vectors f and g in $\mathcal{H}$, 38
$\langle \cdot, \cdot \rangle_\Omega$	Inner product over the space $L^2(\Omega)$, 128
$\langle \cdot, \cdot \rangle_{\Omega \times \Omega}$	Inner product over the space $L^2(\Omega \times \Omega)$, 160
$f \perp g$	Vector f is orthogonal to g, 43
$\mathcal{O}$	In Part I, an orthonormal set in space $\mathcal{H}$. In Part II, zero operator $\mathcal{O}f = o$, 43
$w(x)$	Weight function in $L^2(\Omega)$ in weighted inner product, 50
$\mathfrak{M}, \mathfrak{N}$	A subset, manifold, or subspace in $\mathcal{H}$, 64
$\bigvee \mathfrak{N}$	Subspace spanned by a subset $\mathfrak{N}$ of $\mathcal{H}$, 64
$\mathfrak{M}_1, \mathfrak{M}_2, \mathfrak{M}_3, \ldots$	A sequence of subsets, manifolds, or subspaces in $\mathcal{H}$, 65
$\mathfrak{M}^\perp$	Orthogonal subspace to $\mathfrak{M}$, 66
$\mathcal{P}_{\mathfrak{M}} f$	Projection of $f \in \mathcal{H}$ into subspace $\mathfrak{M}$, 66
$\mathcal{L}$	A bijective linear mapping from one Hilbert space to another, 99

Sequences

$a = \{\alpha_k\}_{k=1}^{\infty}$	An infinite dimensional sequence of scalars, 47
$b = \{\beta_k\}_{k=1}^{\infty}$	An infinite dimensional sequence of scalars, 47
$\mathbf{a} = $ $(\alpha_1, \alpha_2, \alpha_3, \ldots)'$	In Part II and Part III represents the vector form of sequence $a = \{\alpha_k\}_{k=1}^{\infty}$, 107
$(f)_n = \langle f, \varphi_n \rangle$	In Part II and Part III represents the n-th Fourier coefficient of f, 108
$\mathbf{f}$	In Part II and Part III represents the vector containing all Fourier coefficients of f, defined as $\big((f)_1, (f)_2, (f)_3, \ldots\big)'$, 108
ℓ^2	Space of square summable sequences, $a = \{\alpha_k\}_{k=1}^{\infty} \in \ell^2$ iff $\sum_{k=1}^{\infty} |\alpha_k|^2 < \infty$, 47
$a_1, a_2, a_3, \ldots$	A sequence of a's, 52
$e_k \triangleq \{\delta_{k,n}\}_{n=1}^{\infty}$	k-th element of the countably infinite orthonormal sequence within ℓ^2, 47
$\{\varphi_n\}_{n=1}^{\infty}$	An orthonormal sequence which may or may not be complete, 43, 70
$\big\{\xi_{n,m}^{(\varphi)}(x,y)\big\}_{n,m=1}^{\infty}$	The complete orthonormal sequence in $L^2(\Omega \times \Omega)$ induced by the complete orthonormal sequence $\{\varphi_n(x)\}_{n=1}^{\infty}$ in $L^2(\Omega)$, defined as $\{\varphi_n(x)\overline{\varphi_m(y)}\}_{n,m=1}^{\infty}$, 160, 128

Operators — general

$\mathcal{L}$	An operator, which is assumed to be linear unless otherwise stated, 108
$\mathcal{L}_1, \mathcal{L}_2, \mathcal{L}_3, \ldots$	A sequence of operators, 108
$\mathcal{L}_N \circ \cdots \circ \mathcal{L}_2 \circ \mathcal{L}_1 f$	Successive application of $\mathcal{L}_1$, then $\mathcal{L}_2, \cdots$, then $\mathcal{L}_N$ to f, 108
$\mathcal{L}^M f$	Equivalent to $\underbrace{\mathcal{L} \circ \cdots \circ \mathcal{L} \circ \mathcal{L}}_{M \text{ times}} f$, 139
$\|\mathcal{L}\|_{\text{op}}$	Norm of an operator $\mathcal{L}$, 112
$\mathcal{B}$	A bounded linear operator, 112
$\mathcal{B}^{-1}$	The inverse of the bounded linear operator $\mathcal{B}$ (if it exists), 113
$\mathcal{B}_1, \mathcal{B}_2, \mathcal{B}_3, \ldots$	A sequence of bounded linear operators, 115
$b_{n,m}^{(\varphi)} = \langle \mathcal{B}\varphi_m, \varphi_n \rangle$	Projection of $\mathcal{B}\varphi_m$ along φ_n, where $\{\varphi_n\}_{n=1}^{\infty}$ is a complete orthonormal sequence in $\mathcal{H}$, 117
$\mathbf{B}^{(\varphi)}$	Matrix representation of operator $\mathcal{B}$ with elements $b_{n,m}^{(\varphi)}$ at row n and column m, 118
$\mathcal{B}^{\star}$	Adjoint of the bounded linear operator $\mathcal{B}$, 123
λ_n	n-th eigenvalue of a linear operator $\mathcal{L}$, 142
ϕ_n	An eigenvector or eigenfunction of an operator corresponding to eigenvalue λ_n, 142
$\mathcal{C}$	A compact operator, 145
$\mathcal{C}_1, \mathcal{C}_2, \mathcal{C}_3, \ldots$	A sequence of compact linear operators, 150
$\mathcal{Q}$	Hilbert cube, 147

Specific operators

$\mathcal{I}$	Identity operator, $\mathcal{I}f = f$, $\forall f \in \mathcal{H}$, 108
$\mathcal{O}$	Zero operator, $\mathcal{O}f = o$, $\forall f \in \mathcal{H}$, 109
$\mathcal{S}_R$	Right shift operator in ℓ^2, 112
$\mathcal{S}_L$	Left shift operator in ℓ^2, 113
$\mathcal{T}_N$	Truncation operator in ℓ^2 to the first N elements, 113
$\mathcal{F}$	Fourier operator, which depends only on a complete orthonormal sequence, $\{\varphi_n\}_{n=1}^{\infty}$, and generates the Fourier coefficients of a given element $f \in \mathcal{H}$, 121
$K(x, y)$	Kernel for the Hilbert-Schmidt operator which belongs to $L^2(\Omega \times \Omega)$, 127, 160
κ	Norm of kernel $K(x, y)$ for the Hilbert-Schmidt operator, 127
$\mathcal{L}_K$	Hilbert-Schmidt integral operator on $L^2(\Omega)$ with kernel $K(x, y)$, 127, 160
$k_{n,m}^{(\varphi)}$	Projection of $\mathcal{L}_K \varphi_m$ along φ_n, defined as $\langle \mathcal{L}_K \varphi_m, \varphi_n \rangle = \langle K, \xi_{n,m}^{(\varphi)} \rangle_{\Omega \times \Omega}$, 128, 160
$\mathbf{K}^{(\varphi)}$	Matrix representation of operator $\mathcal{L}_K$ with elements $k_{n,m}^{(\varphi)}$ at row n and column m, 128
$\mathcal{P}_{\mathfrak{M}}, \mathcal{P}_{\mathfrak{N}}$	Projection operators into subspaces $\mathfrak{M}$ and $\mathfrak{N}$, respectively, 136, 139
$\mathcal{P}$	Shorthand for the projection operator, where subspace $\mathfrak{M}$ is understood, 136
$\mathcal{P}^{\perp}$	Shorthand for the projection operator into subspace $\mathfrak{M}^{\perp}$, 136
$\mathcal{P}_N$	Finite rank projection operator into an N dimensional subspace, 138
$\mathcal{L}_N$	A finite rank operator of rank N, 138

General notation on 2-sphere

x,y,z	Principal axes in 3D Euclidean space $\mathbb{R}^3$, 175
$\boldsymbol{u}$	A vector in $\mathbb{R}^3$ with coordinates $(u_x, u_y, u_z)'$ along the x, y, z axes, 175
$|\boldsymbol{u}|$	The Euclidean norm of $\boldsymbol{u}$, $|\boldsymbol{u}| \triangleq \sqrt{u_x^2 + u_y^2 + u_z^2}$, 175
$\mathbb{S}^2$	2-sphere: unit radius sphere embedded in $\mathbb{R}^3$: $\mathbb{S}^2 \triangleq \{\boldsymbol{u} \in \mathbb{R}^3 : |\boldsymbol{u}| = 1\}$, 175
$\widehat{\boldsymbol{u}}, \widehat{\boldsymbol{v}}$	Unit norm vectors on 2-sphere. For example, $\widehat{\boldsymbol{u}} = \boldsymbol{u}/|\boldsymbol{u}| \in \mathbb{S}^2$, 175
$\widehat{\boldsymbol{u}}_x$	Unit vector along direction x, $\widehat{\boldsymbol{u}}_x = (1,0,0)'$, 178
$\widehat{\boldsymbol{u}}_y$	Unit vector along direction y, $\widehat{\boldsymbol{u}}_y = (0,1,0)'$, 178
$\widehat{\boldsymbol{u}}_z$	Unit vector along direction z, $\widehat{\boldsymbol{u}}_z = (0,0,1)'$, 178
$\widehat{\boldsymbol{\eta}}$	North pole, $\widehat{\boldsymbol{\eta}} \triangleq \widehat{\boldsymbol{u}}_z = (0,0,1)'$, 176
$\theta \in [0, \pi]$	Co-latitude measured relative to positive z-axis ($\theta = 0$ corresponds to $\widehat{\boldsymbol{\eta}}$), 176
$\phi \in [0, 2\pi)$	Azimuth (longitude) measured relative to positive x-axis in x-y plane, , 176
$\widehat{\boldsymbol{u}}(\theta, \phi)$	$\widehat{\boldsymbol{u}}$ in spherical coordinates, $\widehat{\boldsymbol{u}}(\theta, \phi) = (\sin\theta\cos\phi, \sin\theta\sin\phi, \cos\theta)'$, 176
$\widehat{\boldsymbol{v}}(\vartheta, \varphi)$	$\widehat{\boldsymbol{v}}$ in terms of its spherical coordinates ϑ, φ, 176
Δ	Angular distance between the two points $\widehat{\boldsymbol{u}}$ and $\widehat{\boldsymbol{v}}$, 176
$ds(\widehat{\boldsymbol{u}})$	Area element on $\mathbb{S}^2$, $ds\widehat{\boldsymbol{u}} = \sin\theta d\phi d\theta$, 87, 177
R	A region on 2-sphere $\mathbb{S}^2$, 177
A	Area of region R, 177
$\bar{R}$	Complement of region R on 2-sphere, $\bar{R} \triangleq \mathbb{S}^2 \setminus R$, 177
R_{θ_0}	A symmetric polar cap region with a maximum co-latitude θ_0, 177
$SO(3)$	Special orthogonal group in 3D, 179
$\boldsymbol{R}$	A 3×3 matrix that describes relative change in orientation in $\mathbb{R}^3$, 179
$\boldsymbol{R}_\varphi^{(\widehat{\boldsymbol{w}})}$	Matrix representing rotation around unit vector $\widehat{\boldsymbol{w}}$ by angle φ, 180
$\boldsymbol{R}_\varphi^{(x)}$	Matrix representing rotation around x-axis ($\widehat{\boldsymbol{w}} = \widehat{\boldsymbol{u}}_x$) by φ, 182
$\boldsymbol{R}_\varphi^{(y)}$	Matrix representing rotation around y-axis ($\widehat{\boldsymbol{w}} = \widehat{\boldsymbol{u}}_y$) by φ, 182
$\boldsymbol{R}_\varphi^{(z)}$	Matrix representing rotation around z-axis ($\widehat{\boldsymbol{w}} = \widehat{\boldsymbol{u}}_z$) by φ, 182
X, Y, Z	Moving axes in intrinsic rotation, 183
$\boldsymbol{R}_{\varphi\vartheta\omega}^{(zyz)}$	Extrinsic rotation matrix by $(\varphi, \vartheta, \omega)$ around (fixed) zyz (right to left), 185

Signals on 2-sphere

$f(\theta,\phi) \equiv f(\widehat{\boldsymbol{u}})$	A signal defined on 2-sphere belonging to $L^2(\mathbb{S}^2)$, 189
$g(\theta,\phi) \equiv g(\widehat{\boldsymbol{u}})$	Another signal defined on 2-sphere belonging to $L^2(\mathbb{S}^2)$, 189
$h(\theta,\phi) \equiv h(\widehat{\boldsymbol{u}})$	Usually a filter defined on 2-sphere belonging to $L^2(\mathbb{S}^2)$ which modifies signal f or g, 221
$Y_\ell^m(\theta,\phi)$	Spherical harmonic of degree $\ell \in \{0,1,\dots\}$ and order $m \in \{-\ell,\dots,\ell\}$ on $\mathbb{S}^2$, $Y_\ell^m(\theta,\phi) \equiv Y_\ell^m(\widehat{\boldsymbol{u}})$, 87, 189
$P_\ell^m(x)$	Associated Legendre polynomial of degree ℓ and order m, $x = \cos\theta \in [-1,1]$, for $\theta \in [0,\pi]$, 62, 190
$N_\ell^m(x)$	Normalization corresponding to $Y_\ell^m(\theta,\phi)$, 189
$\breve{Y}_n^m(\theta,\phi)$	Spherical harmonic of degree n and order m as defined in (Colton and Kress, 1998), 198
$Y_{\ell,m}(\theta,\phi)$	Real spherical harmonic of degree ℓ and order m on $\mathbb{S}^2$, 203
$\widehat{Y}_{\ell,m}(\theta,\phi)$	Normalized real spherical harmonic of degree ℓ and order m such that its lies between $[-1,1]$, 208
$(f)_\ell^m$	Spherical harmonic coefficient of degree ℓ and order m for f, $(f)_\ell^m \triangleq \langle f, Y_\ell^m \rangle$, 195
$\displaystyle\sum_{\ell,m}$	Shorthand notation for $\displaystyle\sum_{\ell=0}^{\infty}\sum_{m=-\ell}^{\ell}$, 88, 195
$\displaystyle\sum_{\ell,m}^{L}$	Shorthand notation for $\displaystyle\sum_{\ell=0}^{L}\sum_{m=-\ell}^{\ell}$, 88, 195
n	Single index mapping of degree ℓ and order m, $n = \ell(\ell+1)+m$, $\ell = \lfloor\sqrt{n}\rfloor$ and $m = n - \lfloor\sqrt{n}\rfloor(\lfloor\sqrt{n}\rfloor+1)$, 195
r	Single index mapping of degree p and order q, $r = p(p+1)+q$
u	Single index mapping of degree s and order t, $u = s(s+1)+t$
$\mathbf{f}$	Vector spectral representation for f, defined as $((f)_0, (f)_1, \dots)' = ((f)_0^0, (f)_1^{-1}, \dots)'$, 198
$\delta_{\ell,p}^{m,q}$	Spherical harmonic Kronecker delta, $\delta_{\ell,p}^{m,q} \triangleq \delta_{\ell,p}\delta_{m,q}$ 191
$\delta(\widehat{\boldsymbol{u}},\widehat{\boldsymbol{v}})$	Dirac delta function on 2-sphere, $\delta(\widehat{\boldsymbol{u}},\widehat{\boldsymbol{v}}) = (\sin\theta)^{-1}\delta(\theta-\vartheta)\delta(\phi-\varphi)$, 95, 194
E_ℓ	Energy per degree of signal f, $E_\ell = \sum_{m=-\ell}^{\ell}|(f)_\ell^m|^2$, 198

Subspaces and operators on 2-sphere

$\mathcal{H}_L(\mathbb{S}^2)$	Bandlimited $(L+1)^2$ dimensional subspace of $L^2(\mathbb{S}^2)$ which is spanned by the spherical harmonics Y_ℓ^m with degree $\ell \in \{0, 1, \ldots, L\}$, 211
L_f	Maximum degree of a bandlimited signal $f \in \mathcal{H}_{L_f}$, 226
L_h	Maximum degree of a bandlimited signal $h \in \mathcal{H}_{L_h}$, 226
$\mathcal{H}_R(\mathbb{S}^2)$	Infinite-dimensional subspace of $L^2(\mathbb{S}^2)$ where the signals' spatial domain is limited to $R \subset \mathbb{S}^2$, 213
$\mathcal{H}^0(\mathbb{S}^2)$	Subspace of $L^2(\mathbb{S}^2)$ where signals are azimuthally symmetric, $f(\theta, \phi) = f(\theta)$, 213
$\mathcal{H}_{\theta_0}^0(\mathbb{S}^2)$	Subspace of $\mathcal{H}^0(\mathbb{S}^2)$ where signals are both spacelimited to a polar cap region R_{θ_0} and azimuthally symmetric, 214
$\mathcal{H}_L^0(\mathbb{S}^2)$	Subspace of $\mathcal{H}^0(\mathbb{S}^2)$ where signals are both bandlimited to a degree L and azimuthally symmetric, 214
$\mathcal{B}$	An operator on $L^2(\mathbb{S}^2)$ or its valid subspaces taking $f(\widehat{\boldsymbol{u}})$ to $g(\widehat{\boldsymbol{u}}) = (\mathcal{B}f)(\widehat{\boldsymbol{u}})$, 217
$\mathbf{B}$	Operator matrix corresponding to $\mathcal{B}$ where the complete orthonormal sequence is understood to be spherical harmonics, Y_ℓ^m, 217
$b_{\ell,p}^{m,q}$	Element of the operator matrix $\mathbf{B}$, defined as $\langle \mathcal{B}Y_p^q, Y_\ell^m \rangle$, 217
$B(\widehat{\boldsymbol{u}}, \widehat{\boldsymbol{v}})$	Operator kernel corresponding to $\mathcal{B}$, 218
$\mathcal{B}_L$	Spectral truncation operator of degree L, 219
$\mathcal{B}_R$	Spatial truncation operator to region R, 220
$\mathcal{B}_h$	Spatial masking operator with mask $h(\widehat{\boldsymbol{u}})$, 221
$\begin{pmatrix} j_1 & j_2 & j_3 \\ m_1 & m_2 & m_3 \end{pmatrix}$	Wigner $3j$ symbol, 223
$\mathcal{D}(\varphi, \vartheta, \omega)$	Rotation operator parametrized in the Euler angles in the zyz convention, defined as $\mathcal{D}_z(\varphi) \circ \mathcal{D}_y(\vartheta) \circ \mathcal{D}_z(\omega)$, 226
$D_{m,m'}^\ell(\varphi, \vartheta, \omega)$	Wigner D-matrix, defined as $\langle \mathcal{D}(\varphi, \vartheta, \omega)Y_\ell^{m'}, Y_\ell^m \rangle$, 227
$d_{m,m'}^\ell(\vartheta)$	Wigner d-matrix, defined as $\langle \mathcal{D}_y(\vartheta)Y_\ell^{m'}, Y_\ell^m \rangle$, 227
$D_{\varphi, \vartheta, \omega}(\widehat{\boldsymbol{u}}, \widehat{\boldsymbol{v}})$	Operator kernel corresponding to $\mathcal{D}(\varphi, \vartheta, \omega)$, 229
$\mathcal{B}^0$	Projection into the space of azimuthally symmetric signals $\mathcal{H}^0(\mathbb{S}^2)$, 242
$I_\mathcal{B}(f)$	Quadratic functional corresponding to operator $\mathcal{B}$ operating on f, 246

Signal concentration

ω	Angular frequency, 251
$F(\omega)$	Fourier transform of signal f in $L^2(\mathbb{R})$, 251
$\mathcal{H}_W$	Subspace of bandlimited functions with the maximum angular frequency $|\omega| \leq W$, 252
$\mathcal{B}_W$	Projection operator into the subspace $\mathcal{H}_W$, 252
$\mathcal{H}_T$	Subspace of timelimited functions within the time interval $|t| \leq T$, 253
$\mathcal{B}_T$	Projection operator into the subspace $\mathcal{H}_T$, 253
α	Time or spatial concentration ratio, 253, 261
β	Frequency or spectral concentration ratio, 253, 265
$D_{\ell,p}^{m,q}$	Elements of $\mathbf{D}$ matrix for the spatial concentration problem of bandlimited signals defined on $\mathbb{S}^2$, 261
$\mathbf{D}^{(m)}$	Submatrix for the spatial concentration problem of bandlimited signals in the polar cap for a fixed order m, 268
$N^{(m)}$	Number of significant eigenvalues for a given order m, 269
σ_S^2	Variance in spatial domain of an azimuthally symmetric function, 271
σ_L^2	Variance in spectral domain of an azimuthally symmetric function, 271
$g(\vartheta, \varphi; \ell, m)$	Spatially localized spherical harmonic transform (SLSHT) of degree ℓ and order m with the window centered at co-latitude ϑ and longitude φ, 280
$\mathbf{g}(\vartheta, \varphi)$	SLSHT distribution; a vector containing all $g(\vartheta, \varphi; \ell, m)$ for a fixed ϑ and φ, 282
E_ℓ'	Energy localization ratio per degree 290

Convolution

$\mathcal{T}_s$	Shift operator by s over the real line $\mathbb{R}$, 294
$\mathcal{R}$	Reversal operator over the real line $\mathbb{R}$, 294
$h \star f$	Conventional convolution between h and f over the real line $\mathbb{R}$ or $\mathbb{R}^N$, 293
$h \odot f$	Spherical convolution of type 1 between filter kernel h and signal f, 296
$h_0 \circledast f$	Spherical convolution of type 2 (isotropic convolution) between filter kernel h_0 and signal f, 301
$h \boxdot f$	Spherical convolution of type 3 with output in SO(3), 305
$g_\omega(\vartheta, \varphi)$	Anisotropic convolution with ω being either fixed or a function of ϑ and φ, 307
$h \odot f$	Commutative anisotropic convolution, 309

Reproducing kernel Hilbert space (RKHS)

$\mathcal{H}_\lambda$	Fourier weighted Hilbert space, 323
$\mathcal{H}_K$	Reproducing kernel Hilbert space (RKHS), 322
C^k	Class of functions where the function and its first $k \geq 0$ derivatives are continuous, 325
C^0	Class of continuous functions, 325
C^∞	Class of smooth functions where the function has derivatives of all orders, 325
$\langle f, g \rangle_{\mathcal{H}_\lambda}$	Inner product in the Fourier weighted Hilbert space, 329
$\|f\|_{\mathcal{H}_\lambda}$	Induced norm in the Fourier weighted Hilbert space, 331
$\langle f, g \rangle_{\mathcal{H}_K}$	Inner product in the RKHS, 347
$\|f\|_{\mathcal{H}_K}$	Induced norm in the RKHS, 347
$\mathcal{L} \geq 0$	Positive operator, 332
$\mathcal{L} > 0$	Strictly positive operator, 332
$\mathcal{L}_\lambda^{-1/2}$	The operator taking $f(x) \in \mathcal{H}_\lambda$ to $\left(\mathcal{L}_\lambda^{-1/2} f\right)(x) \in L^2(\Omega)$, 333
$\mathcal{L}_\lambda^{1/2}$	The operator taking $f(x) \in L^2(\Omega)$ to $\left(\mathcal{L}_\lambda^{1/2} f\right)(x) \in \mathcal{H}_\lambda$, 334
$\mathcal{L}_\lambda$	Fundamental positive operator given by $\mathcal{L}_\lambda^{1/2} \circ \mathcal{L}_\lambda^{1/2}$, 335
$\delta_K(x - y)$	Dirac delta function for RKHS, $\mathcal{H}_K$, 348

Index

Lightning Source UK Ltd.
Milton Keynes UK
UKOW07n0637091116
287191UK00005B/164/P